T0343251

PROCESS MODELLING AND MODEL ANALYSIS

This is Volume 4 of
PROCESS SYSTEMS ENGINEERING
A Series edited by George Stephanopoulos and John Perkins

PROCESS MODELLING AND MODEL ANALYSIS

K. M. Hangos
Systems and Control Laboratory
Computer and Automation Research Institute of the
Hungarian Academy of Sciences
Budapest, Hungary

I. T. Cameron
Cape Centre Department of Chemical Engineering
The University of Queensland
Brisbane, Queensland

San Diego San Francisco New York Boston London Sydney Tokyo

This book is printed on acid-free paper.

Academic Press
A Harcourt Science and Technology Company
Harcourt Place, 32 Jamestown Road, London NW1 7BY, UK
http://www.academicpress.com

Academic Press
A Harcourt Science and Technology Company
525 B Street, Suite 1900, San Diego, California 92101-4495, USA
http://www.academicpress.com

ISBN 0-12-156931-4

Library of Congress Catalog Number: 00-112073

A catalogue record of this book is available from the British Library
Typeset by Newgen Imaging Systems (P) Ltd., Chennai, India

01 02 03 04 05 06 BC 9 8 7 6 5 4 3 2 1

Transferred to digital printing in 2007.

Dedicated to

Misi, Ákos, Veronika

and

Lucille, James, Peter, Andrew

CONTENTS

2 A Systematic Approach to Model Building

3 Conservation Principles

4 Constitutive Relations

5 Dynamic Models—Lumped Parameter Systems

6 Solution Strategies for Lumped Parameter Models

7 Dynamic Models—Distributed Parameter Systems

8 Solution Strategies for Distributed Parameter Models

9 Process Model Hierarchies

II ADVANCED PROCESS MODELLING AND MODEL ANALYSIS

10 Basic Tools for Process Model Analysis

11 Data Acquisition and Analysis

12 Statistical Model Calibration and Validation

INTRODUCTION

Process modelling is one of the key activities in process systems engineering. Its importance is reflected in various ways. It is a significant activity in most major companies around the world, driven by such application areas as process optimization, design and control. It is a vital part of risk management, particularly consequence analysis of hazardous events such as loss of containment of process fluids. It is a permanent subject of conferences and symposia in fields related to process systems engineering. It is often the topic of various specialized courses offered at graduate, postgraduate and continuing professional education levels. There are various textbooks available for courses in process modelling and model solution amongst which are Himmelblau [1], Davis [2], Riggs [3] and Rice and Do [4]. These however are mainly devoted to *the solution techniques* related to process models and not to the problem on how to define, setup, analyse and test models. Several short monographs or mathematical notes with deeper insights on modelling are available, most notably by Aris [5] and Denn [6].

In most books on this subject there is a lack of a consistent modelling approach applicable to process systems engineering as well as a recognition that modelling is not just about producing a set of equations. There is far more to process modelling than writing equations. This is the reason why we decided to write the current book in order to give a more comprehensive treatment of process modelling useful to student, researcher and industrial practitioner.

There is another important aspect which limits the scope of the present material in the area of process modelling. It originates from the well-known fact that a particular process model depends not only on the process to be described but also on the modelling goal. It involves the intended use of the model and the user of that model. Moreover, the actual form of the model is also determined by the education, skills and taste of the modeller and that of the user. Due to the above reasons, the main emphasis has been on process models for dynamic simulation and process control purposes. These are principally lumped dynamic process models in the form of sets of differential—algebraic equations. Other approaches such as distributed parameter modelling and the description of discrete event and hybrid systems are also treated. Finally the use of empirical modelling is also covered, recognizing that our knowledge of many systems is extremely shallow and that input–output descriptions generated by analysing plant data are needed to complement a mechanistic approach.

Process modelling is an engineering activity with a relatively mature technology. The basic principles in model building are based on other disciplines in process engineering such as mathematics, chemistry and physics. Therefore, a good background in these areas is essential for a modeller. Thermodynamics, unit operations, reaction kinetics, catalysis, process flowsheeting and process control are the helpful prerequisites for a course in process modelling. A mathematical background is also helpful for the understanding and application of analysis and numerical methods in the area of linear algebra, algebraic and differential equations.

Structure of the book

The book consists of two parts. The first part is devoted to the building of process models whilst the second part is directed towards analysing models from the viewpoint of their intended use. The methodology is presented in a top–down systematic way following the steps of a modelling procedure. This often starts from the most general case. Emphasis is given in this book to identifying the key ingredients, developing conservation and constitutive equations then analysing and solving the resultant model. These concepts are introduced and discussed in separate chapters. Static and dynamic process models and their solution methods are treated in an integrated manner and then followed by a discussion on hierarchical process models which are related by scales of time or degree of detail. This is the field of multi-scale modelling.

The second part of the book is devoted to the problem of how to analyse process models for a given modelling goal. Three dominant application areas are discussed:

- control and diagnosis where mostly lumped dynamic process models are used,
- static flowsheeting with lumped static process models,
- dynamic flowsheeting where again mostly lumped dynamic process models are in use.

Special emphasis is given to the different but related process models and their properties which are important for the above application areas.

Various supplementary material is available in the appendices. This includes:

- Background material from mathematics covering linear algebra and mathematical statistics.
- Computer science concepts such as graphs and algorithms.

Each chapter has sections on review questions and application examples which help reinforce the content of each chapter. Many of the application exercises are suitable for group work by students.

The methods and procedures presented are illustrated by examples throughout the book augmented with MATLAB subroutines where appropriate. The examples are drawn from as wide a range of process engineering disciplines as possible. They include chemical processing, minerals process engineering, environmental engineering and food engineering in order to give a true process system's appeal. The model analysis methods in part two are applied to many of the same process systems used in part one. This method of presentation makes the book easy to use for both higher year undergraduate, postgraduate courses or for self study.

Making use of the book

This book is intended for a wide audience. The authors are convinced that it will be useful from the undergraduate to professional engineering level. The content of the book has been presented to groups at all levels with adaptation of the material for the particular audience. Because of the modular nature of the book, it is possible to concentrate on a number of chapters depending on the need of the reader.

For the undergraduate, we suggest that a sensible approach will be to consider Chapters 1–6, parts of 10–12 as a full 14-week semester course on basic process modelling. For advanced modelling at undergraduate level and also for postgraduate level, Chapters 7–9 can be considered. Chapter 17 on Modelling Applications should be viewed by all readers to see how the principles work out in practice.

Some industrial professionals with a particular interest in certain application areas could review modelling principles in the first half of the book before considering the specific application and analysis areas covered by Chapters 11–19. The options are illustrated in the following table:

Chapter	Undergraduate introduction	Postgraduate advanced	Professional introduction	Professional advanced
1. Role of Modelling in Process Systems Engineering	✓		✓	
2. A Modellling Methodology	✓	✓	✓	✓
3. Conservation Principles in Process Modelling	✓		✓	
4. Constitutive Relations for Modelling	✓		✓	
5. Lumped Parameter Model Development	✓		✓	
6. Solution of Lumped Parameter Models	✓		✓	
7. Distributed Parameter Model Development		✓		✓
8. Solution of Distributed Parameter Models		✓		✓
9. Incremental Modelling and Model Hierarchies		✓		✓
10. Basic Tools for Model Analysis	✓	✓	✓	✓
11. Data Acquisition and Analysis	✓	✓	✓	✓
12. Statistical Model Calibration and Validation	✓	✓	✓	✓
13. Analysis of Dynamic Process Models		✓		✓
14. Process Modelling for Control and Diagnosis		✓		✓
15. Modelling of Discrete Event Systems		✓		✓
16. Modelling of Hybrid Systems		✓		✓
17. Modelling Applications	✓	✓	✓	✓
18. Computer Aided Process Modelling		✓		✓
19. Empirical Model Building		✓	✓	

For instructors there is also access to PowerPoint presentations on all chapters through the website: `http://daisy.cheque.uq.edu.au/cape/modelling/index.html`.

Acknowledgements

The authors are indebted to many people who contributed to the present book in various ways. The lively atmosphere of the Department of Chemical Engineering of The University of Queensland and the possibility for both of us to teach the final year "Process Modelling and Solutions" course to several classes of chemical engineering students has made the writing of the book possible. Presenting sections to professional engineers within Australian industry and also to Ph.D. and academic staff in Europe through the Eurecha organization, has helped refine some of the content.

We would especially thank Christine Smith for the care and help in preparing different versions of the teaching materials and the manuscript. Also to Russell Williams and Steven McGahey for help in reviewing some of the chapters and to Gábor Szederkényi for his kind help with many of the LaTeX and figure issues.

We are conscious of the support of our colleagues working in the field of modelling in process systems engineering. These include Prof. John Perkins at Imperial College, Prof. George Stephanopoulos at MIT, Professors Rafique Gani and Sten Bay Jorgensen at the Danish Technical University, Prof. Heinz Preisig at Eindhoven and Prof. Wolfgang Marquardt at Aachen.

Special thanks to Dr. Bob Newell for many years of fruitful discussion and encouragement towards realism in modelling. We thank them all for the advice, discussions and sources of material they have provided.

We readily acknowledge the contribution of many whose ideas and published works are evident in this book. Any omissions and mistakes are solely the responsibility of the authors.

PART I
FUNDAMENTAL PRINCIPLES AND PROCESS MODEL DEVELOPMENT

1 THE ROLE OF MODELS IN PROCESS SYSTEMS ENGINEERING

Models?—really nothing more than an imitation of reality! We can relate to models of various types everywhere—some physical, others mathematical. They abound in all areas of human activity, be it economics, warfare, leisure, environment, cosmology or engineering. Why such an interest in this activity of model building and model use? It is clearly a means of gaining insight into the behaviour of systems, probing them, controlling them, optimizing them. One thing is certain: this is not a new activity, and some famous seventeenth-century prose makes it clear that some modelling was considered rather ambitious:

From man or angel the great Architect
Did wisely to conceal, and not divulge
His secrets to be scanned by them who ought
Rather admire; or if they list to try
Conjecture, he his fabric of the heav'ns
Hath left to their disputes, perhaps to move
His laughter at their quaint opinions wide
Hereafter, ***when they come to model heav'n***
And calculate the stars, how they will wield
The mighty frame, how build, unbuild, contrive
To save appearances, how gird the sphere
With centric and eccentric scribbled o'er,
Cycle and epicycle, orb in orb.

[John Milton (1608–74), English poet. *Paradise Lost*, Book. 8: 72–84]

Our task here is a little more modest, but nevertheless important for the field of process engineering. In this chapter, we want to explore the breadth of model use in the process industries—where models are used and how they are used. The list is almost unlimited and the attempted modelling is driven largely by the availability of high performance computing and the demands of an increasingly competitive marketplace.

However, we must admit that in many cases in process engineering much is concealed by our limited understanding of the systems we seek to design or manage. We can relate to the words of Milton that our efforts and opinions are sometimes quaint and laughable and our conjectures short of the mark, but the effort may well be worthwhile in terms of increased understanding and better management of the systems we deal with.

The emphasis in this book is on *mathematical modelling* rather than physical modelling, although the latter has an important place in process systems engineering (PSE) through small scale pilot plants to three dimensional (3D) construction models.

1.1. THE IDEA OF A MODEL

A model is an imitation of reality and a mathematical model is a particular form of representation. We should never forget this and get so distracted by the model that we forget the real application which is driving the modelling. In the process of model building we are translating our real world problem into an equivalent mathematical problem which we solve and then attempt to interpret. We do this to gain insight into the original real world situation or to use the model for control, optimization or possibly safety studies.

In discussing the idea of a model, Aris [5] considers the well-known concept of *change of scale* as being at the root of the word "model". Clearly, we can appreciate this idea from the wealth of scale models, be it of process plant, toys or miniature articles of real world items. However in the process engineering area the models we deal with are fundamentally mathematical in nature. They attempt to capture, in the form of equations, certain characteristics of a system for a specific use of that model. Hence, the concept of *purpose* is very much a key issue in model building.

The modelling enterprise links together a purpose $\mathcal{P}$ with a subject or physical system $\mathcal{S}$ and the system of equations $\mathcal{M}$ which represent the model. A series of experiments $\mathcal{E}$ can be applied to $\mathcal{M}$ in order to answer questions about the system $\mathcal{S}$. Clearly, in building a model, we require that certain characteristics of the actual system be represented by the model. Those characteristics could include:

- the correct response direction of the outputs as the inputs change;
- valid structure which correctly represents the connection between the inputs, outputs and internal variables;
- the correct short and/or long term behaviour of the model.

In some cases, certain characteristics which are unnecessary to the use of the model are also included. The resultant model has a specific region of applicability, depending on the experiments used to test the model behaviour against reality. These issues are more fully investigated in the following chapters.

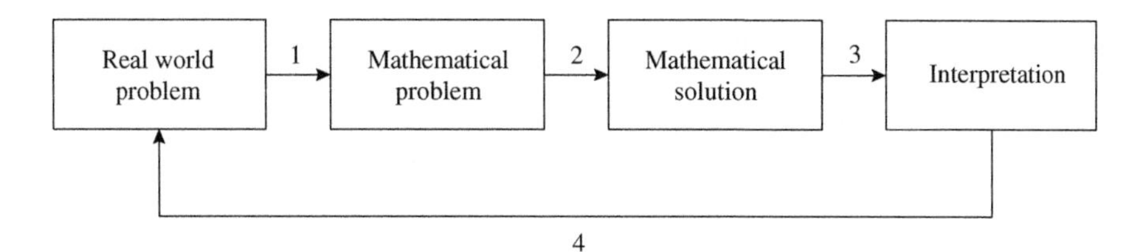

FIGURE 1.1 Real world modelling process.

The following definition derives from Minsky, who stated that:

DEFINITION 1.1.1. *A model ($\mathcal{M}$) for a system ($\mathcal{S}$) and an experiment ($\mathcal{E}$) is anything to which $\mathcal{E}$ can be applied in order to answer questions about $\mathcal{S}$.*

For those who wish to investigate the fuller implications of the etymology of the word "model", we refer to informative discussion given by Aris [5] and Minsky [7].

This overall process is represented schematically in Fig. 1.1 , which shows the four key steps in the overall modelling process.

Each of the steps in Fig. 1.1 has very important issues attached to them, which are covered in the subsequent chapters. At this point, it is worth mentioning a few of the issues to be raised at each of the steps seen in Fig. 1.1. These are:

1.1.1. Reality to Mathematics (Step 1)

Here we have to deal with the task of translating the real problem to one represented in mathematical terms. Some of the key issues which have to be dealt with here are:

- What do we understand about the real world problem?
- What is the intended use of the mathematical model?
- What governing phenomena or mechanisms are there in the system?
- What form of model is required?
- How should the model be structured and documented?
- How accurate does the model have to be?
- What data on the system are available and what is the quality and accuracy of the data?
- What are the system inputs, states, outputs and disturbances?

1.1.2. Mathematical Solution (Step 2)

Having generated some mathematical description of the real world system, it is then necessary to solve this for the unknown value of the variables representing that system. Key issues here are:

- What variables must be chosen in the model to satisfy the degrees of freedom?
- Is the model solvable?
- What numerical (or analytic) solution technique should be used?

- Can the structure of the problem be exploited to improve the solution speed or robustness?
- What form of representation should be used to display the results (2D graphs, 3D visualization)?
- How sensitive will the solution output be to variations in the system parameters or inputs?

1.1.3. Interpreting the Model Outputs (Step 3)

Here we need to have procedures and tests to check whether our model has been correctly implemented and then ask whether it imitates the real world to a sufficient accuracy to do the intended job. Key issues include:

- How is the model implementation to be verified?
- What type of model validation is appropriate and feasible for the problem?
- Is the resultant model identifiable?
- What needs to be changed, added or deleted in the model as a result of the validation?
- What level of simplification is justified?
- What data quality and quantity is necessary for validation and parameter estimates?
- What level of model validation is necessary? Should it be static or dynamic?
- What level of accuracy is appropriate?
- What system parameters, inputs or disturbances, need to be known accurately to ensure model predictive quality?

1.1.4. Using the Results in the Real World (Step 4)

Here we are faced with the implementation of the model or its results back into the real world problem we originally addressed. Some issues that arise are:

- For online applications where speed might be essential, do I need to reduce the model complexity?
- How can model updating be done and what data are needed to do it?
- Who will actually use the results and in what form should they appear?
- How is the model to be maintained?
- What level of documentation is necessary?

These issues are just some of the many which arise as models are conceptualized, developed, solved, tested and implemented. What is clear from the above discussion is the fact that *modelling is far more than simply the generation of a set of equations*. This book emphasizes the need for a much broader view of process modelling including the need for a model specification, a clear generation and statement of hypotheses and assumptions, equation generation, subsequent model calibration, validation and end use.

Many of the following chapters will deal directly with these issues. However, before turning to those chapters we should consider two more introductory aspects to help "set the scene". These are to do with model characterization and classification. The important point about stating these upfront is to be aware of these issues early

TABLE 1.1 Model Application Areas

Application area	Model use and aim
Process design	Feasibility analysis of novel designs Technical, economic, environmental assessment Effects of process parameter changes on performance Optimization using structural and parametric changes Analysing process interactions Waste minimization in design
Process control	Examining regulatory control strategies Analysing dynamics for setpoint changes or disturbances Optimal control strategies for batch operations Optimal control for multi-product operations Optimal startup and shutdown policies
Trouble-shooting	Identifying likely causes for quality problems Identifying likely causes for process deviations
Process safety	Detection of hazardous operating regimes Estimation of accidental release events Estimation of effects from release scenarios (fire etc.)
Operator training	Startup and shutdown for normal operations Emergency response training Routine operations training
Environmental impact	Quantifying emission rates for a specific design Dispersion predictions for air and water releases Characterizing social and economic impact Estimating acute accident effects (fire, explosion)

on and thus retain them as guiding concepts for what follows. Before we deal with these issues, we survey briefly where models are principally used in PSE and what is gained from their use.

1.2. MODEL APPLICATION AREAS IN PSE

It was mentioned that the list of applications in process engineering is almost endless. We can, however, categorize the use of models into several well-defined areas. These are outlined in Table 1.1, which sets out the typical application area and the aim of the modelling.

Clearly, the list can be extended in each application area and the individual categories can be extended too. It does show the wide range of applications and hence the importance of process modelling on the modern design, optimization and operation of process systems. What is sometimes not obvious is that each of the application areas may require quite different models to achieve the desired outcome. In the next section, we consider the issue of model classification and show the great diversity of model types which have been used.

However, we first give some examples of models and their application from several process systems areas.

EXAMPLE 1.2.1 (Fire radiation modelling). Safe operations are vital in the process industries and for land-use planning purposes. Liquefied petroleum gas can

FIGURE 1.2 BLEVE fireball caused by the rupture of an LPG tank (by permission of A.M. Birk, Queens University, Canada).

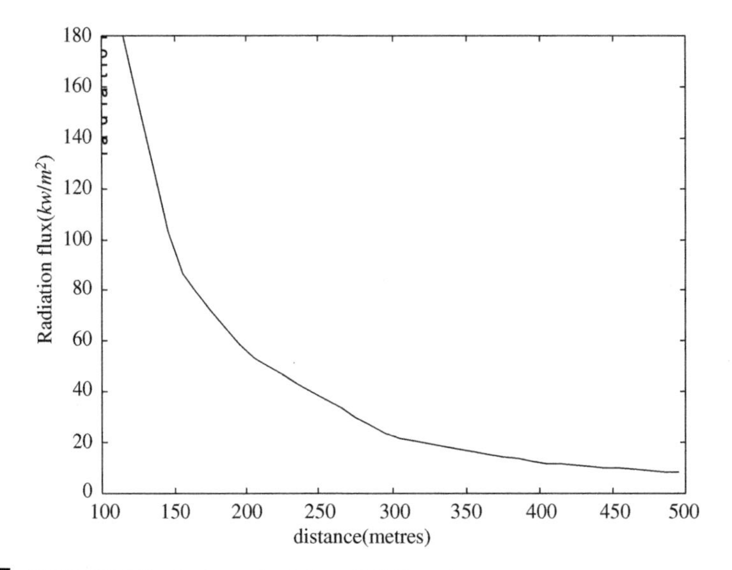

FIGURE 1.3 Radiation levels for a BLEVE incident.

be dangerous if released and ignited. One type of event which has occurred in several places around the world is the boiling liquid expanding vapour explosion (BLEVE).

Figure 1.2 shows the form of a BLEVE fireball caused by the rupture of an LPG tank. Of importance are radiation levels at key distances from the event as well as projectiles from the rupture of the vessel.

Predictive mathematical models can be used to estimate the level of radiation at nominated distances from the BLEVE, thus providing input to planning decisions. Figure 1.3 shows predicted radiation levels (kW/m^2) for a 50-tonne BLEVE out to a distance of 500 m. ■ ■ ■

EXAMPLE 1.2.2 (Compressor dynamics and surge control). Compressors are subject to surge conditions when inappropriately controlled. When surge occurs, it

FIGURE 1.4 Multi-stage centrifugal compressor (by permission of Mannesmann Demag Delaval).

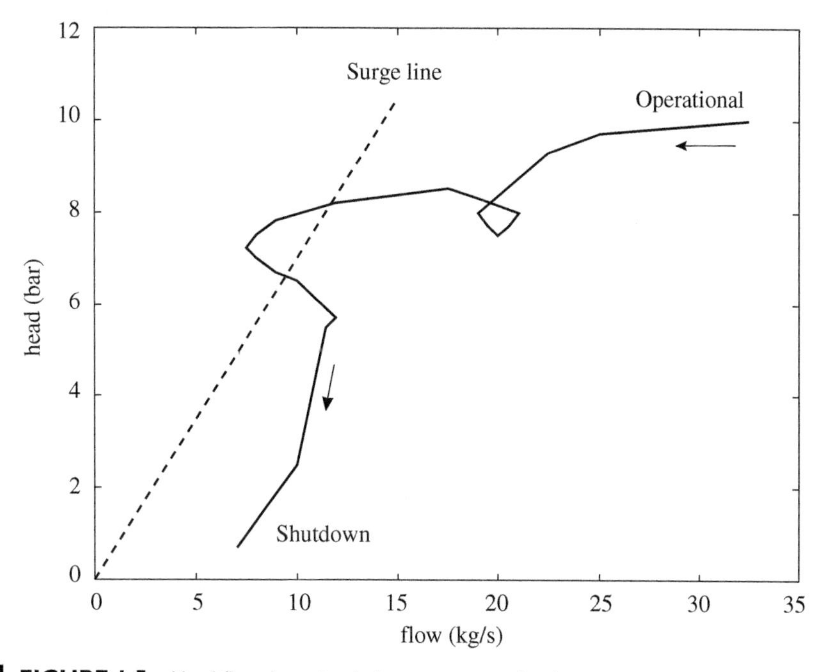

FIGURE 1.5 Head-flow dynamics during compressor shutdown.

can lead to serious damage or destruction of the equipment. Effective control systems are necessary to handle load changes. To test alternative control designs, accurate modelling and simulation are useful approaches. Figure 1.4 shows a large multi-stage compressor and Fig. 1.5 shows the predicted behaviour of the first stage head-flow dynamics under controlled shutdown.

1.3. MODEL CLASSIFICATION

We can devise several ways of classifying models. Each leads to a variety of model characteristics which have an impact on the solution techniques as well as the potential application areas where they can be used. Some model types are inappropriate in certain circumstances, such as a steady-state model for batch reactor start-up analysis. Table 1.2 gives an overview of model types, their basic characteristics and the final form of the models.

The next section explains some key aspects of these model types and later chapters are concerned with the development of several model types.

TABLE 1.2 Model Classification

Type of model	Criterion of classification
Mechanistic	Based on mechanisms/underlying phenomena
Empirical	Based on input–output data, trials or experiments
Stochastic	Contains model elements that are probabilistic in nature
Deterministic	Based on cause–effect analysis
Lumped parameter	Dependent variables not a function of spatial position
Distributed parameter	Dependent variables are a function of spatial position
Linear	Superposition principle applies
Nonlinear	Superposition principle does not apply
Continuous	Dependent variables defined over continuous space–time
Discrete	Only defined for discrete values of time and/or space
Hybrid	Containing continuous and discrete behaviour

1.3.1. Characteristic Nature of Process Models

Mechanistic models are also referred to as *phenomenological* models because of their basic derivation from system phenomena or mechanisms such as mass, heat and momentum transfer. Many commonplace models in process engineering applications are derived from a knowledge of the underlying mechanisms. However, most mechanistic models also contain empirical parts such as rate expressions or heat transfer relations. Mechanistic models often appear in design and optimization applications. They can be termed "white box" models since the mechanisms are evident in the model description.

Empirical models are the result of experiment and observation, usually not relying on the knowledge of the basic principles and mechanisms which are present in the system being studied. They employ essentially equation fitting where the parameters have little or no physical meaning. Empirical models are widely used where the actual underlying phenomena are not known or understood well. These models are often termed "black box" models, reflecting the fact that little is known about the real mechanisms of the process.

The most common form of model used in process engineering is a combination of mechanistic and empirical parts and hence is termed "grey box".

EXAMPLE 1.3.1 (Empirical BLEVE Model). In Example 1.2.1 the radiation levels from a BLEVE were illustrated. The size and duration of a BLEVE fireball have been estimated from the analysis of many incidents, most notably the major disaster in Mexico City during 1984. The empirical model is given by TNO [8] as

$$r = 3.24m^{0.325}, \quad t = 0.852m^{0.26},$$

where r is fireball radius (m), m the mass of fuel (kg) and t the duration of fireball (s).

Stochastic models arise when the description may contain elements which have natural random variations typically described by probability distributions. This characteristic is often associated with phenomena which are not describable in terms of cause and effect but rather by probabilities or likelihoods.

Deterministic models are the final type of models characterized by clear cause–effect relationships.

In most cases in process engineering the resultant model has elements from several of these model classes. Thus we can have a mechanistic model with some stochastic parts to it. A very common occurrence is a mechanistic model which includes empirical aspects such as reaction rate expressions or heat transfer relationships.

EXAMPLE 1.3.2 (Mechanistic compressor model). Example 1.2.2 showed the prediction of a compressor under rapidly controlled shutdown. The model used here was derived from fundamental mass, energy and momentum balances over the compressor plenum. By assuming 1D axial flow the model was reduced to a set of ordinary differential equations given by:

$$\text{Mass:} \quad \frac{dm}{dt} = m_1 - m_2;$$

$$\text{Energy:} \quad \frac{dE}{dt} = m_1 h_1 - m_2 h_2;$$

$$\text{Momentum:} \quad \frac{dM}{dt} = A_m(P_{t_1} - P_{t_2}) + F_{net}$$

where m_1, m_2 are inlet and outlet mass flows, h_1, h_2 the specific enthalpies, A_m is the inlet mean cross section, P_{t_i} are the inlet and outlet pressures and F_{net} is net force on lumped gas volume.

Table 1.2 also includes other classifications dependent on assumptions about spatial variations, the mathematical form and the nature of the underlying process being modelled.

1.3.2. Equation form of Process Models

We can also consider the types of equations which result from such models when we consider steady state and dynamic situations. These are shown in Table 1.3. The forms can involve linear algebraic equations (LAEs), nonlinear algebraic equations (NLAEs), ordinary differential equations (ODEs), elliptic partial differential equations (EPDEs) and parabolic partial differential equations (PPDEs).

Each of the equation forms requires special techniques for solution. This will be covered in subsequent chapters.

TABLE 1.3 Model Equation Forms

Type of model	Equation types	
	Steady-state problem	Dynamic problem
Deterministic	Nonlinear algebraic	ODEs/PDEs
Stochastic	Algebraic/difference equations	Stochastic ODEs or difference equations
Lumped parameter	Algebraic equations	ODEs
Distributed parameter	EPDEs	PPDEs
Linear	Linear algebraic equations	Linear ODEs
Nonlinear	Nonlinear algebraic equations	Nonlinear ODEs
Continuous	Algebraic equations	ODEs
Discrete	Difference equations	Difference equations

1.3.3. Characteristics of the System Volumes

When we develop models, it is necessary to define regions in the system where we apply conservation principles and basic physical and chemical laws in order to derive the mathematical description. These are the balance volumes. A basic classification relates to the nature of the material in those volumes. Where there are both temporal and spatial variations in the properties of interest, such as concentration or temperature, we call these systems "distributed". However, when there are no spatial variations and the material is homogeneous, we have a "lumped" system. The complexity of distributed parameter systems can be significant both in terms of the resulting model description and the required solution techniques. Lumped parameter models generally lead to simpler equation systems which are easier to solve.

1.3.4. Characteristics of the System Behaviour

When we consider system modelling, there are many situations where discrete events occur, such as turning on a pump or shutting a valve. These lead to discontinuous behaviour in the system either at a known time or at a particular level of one of the states such as temperature or concentration. These are called "time" or "state" events. A model which has both characteristics is termed a hybrid system. These are very common in process systems modelling.

Not only do we need to consider the classification of the models that are used in PSE applications but it is also helpful to look at some characteristics of those models.

1.4. MODEL CHARACTERISTICS

Here we consider some of the key characteristics which might affect our modelling and analysis.

- Models can be developed in hierarchies, where we can have several models for different tasks or models with varying complexity in terms of their structure and application area.

- Models exist with relative precision, which affect how and where we can use them.
- Models cause us to think about our system and force us to consider the key issues.
- Models can help direct further experiments and in-depth investigations.
- Models are developed at a cost in terms of money and effort. These need to be considered in any application.
- Models are always imperfect. It was once said by George E. Box, a well-known statistician, "All models are wrong, some are useful"!
- Models invariably require parameter estimation of constants within the model such as kinetic rate constants, heat transfer and mass transfer coefficients.
- Models can often be transferred from one discipline to another.
- Models should display the principle of parsimony, displaying the simplest form to achieve the desired modelling goal.
- Models should be identifiable in terms of their internal parameters.
- Models may often need simplification, or model order reduction to become useful tools.
- Models may be difficult or impossible to adequately validate.
- Models can become intractable in terms of their numerical solution.

We can keep some of these in mind when we come to develop models of our own for a particular application. It is clearly not a trivial issue in some cases. In other situations the model development can be straightforward.

1.5. A BRIEF HISTORICAL REVIEW OF MODELLING IN PSE

As a distinct discipline, PSE is a child of the broader field of systems engineering as applied to processing operations. As such, its appearance as a recognized discipline dates back to the middle of the twentieth century. In this section, we trace briefly the history of model building, analysis and model use in the field of PSE.

1.5.1. The Industrial Revolution

It was the industrial revolution which gave the impetus to systematic approaches for the analysis of processing and manufacturing operations. Those processes were no longer simple tasks but became increasingly complex in nature as the demand for commodity products increased. In particular, the early chemical developments of the late eighteenth century spurred on by the Franco-British wars led to industrial scale processes for the manufacture of gun powder, sulphuric acid, alkali as well as food products such as sugar from sugar beet. In these developments the French and the British competed in the development of new production processes, aided by the introduction of steam power in the early 1800s which greatly increased potential production capacity

In dealing with these new processes, it was necessary for the engineer to bring to bear on the problem techniques derived from many of the physical sciences and engineering disciplines. These analysis techniques quickly recognized the complex interacting behaviour of many activities. These ranged from manufacturing processes to communication systems. The complexities varied enormously but the approaches took on a "systems" view of the problem which gave due regard to the components in the process, the inputs, outputs of the system and the complex interactions which could occur due to the connected nature of the process.

Sporadic examples of the use of systems engineering as a sub-discipline of industrial engineering in the nineteenth and twentieth centuries found application in many of the industrial processes developed in both Europe and the United States. This also coincided with the emergence of chemical engineering as a distinct discipline at the end of the nineteenth century and the development of the unit operations concept which would dominate the view of chemical engineers for most of the twentieth century. There was a growing realization that significant benefits would be gained in the overall economics and performance of processes when a systems approach was adopted. This covered the design, control and operation of the process.

In order to achieve this goal, there was a growing trend to reduce complex behaviour to simple mathematical forms for easier process design—hence the use of mathematical models. The early handbooks of chemical engineering, e.g. Davis [9], were dominated by the equipment aspects with simple models for steam, fluid flow and mechanical behaviour of equipment. They were mainly descriptive in content, emphasizing the role of the chemical engineer, as expressed by Davis, as one who ensured:

> ... Completeness of reactions, fewness of repairs and economy of hand labour should be the creed of the Chemical Engineer.

Little existed in the area of process modelling aimed at reactor and separation systems. In the period from 1900 to the mid-1920s there was a fast growing body of literature on more detailed analysis of unit operations, which saw an increased reliance on mathematical modelling. Heat exchange, drying, evaporation, centrifugation, solids processing and separation technologies such as distillation were subject to the application of mass and energy balances for model development. Many papers appeared in such English journals as *Industrial & Engineering Chemistry*, *Chemistry and Industry*, and the *Society of the Chemical Industry*. Similar developments were taking place in foreign language journals, notably those in France and Germany. Textbooks such as those by Walker and co-workers [10] at the Massachusetts Institute of Technology, Olsen [11] and many monographs became increasingly analytic in their content, this also being reflected in the education system.

1.5.2. The Mid-twentieth Century

After the end of the Second World War there was a growing interest in the application of systems engineering approaches to industrial processes, especially in the chemical industry. The mid-1950s saw many developments in the application of mathematical modelling to process engineering unit operations, especially for the understanding and prediction of the behaviour of individual units. This was especially true in the area of chemical reactor analysis. Many prominent engineers, mathematicians and scientists were involved. It was a period of applying rigorous mathematical analysis to process systems which up until that time had not been analysed in such detail.

However, the efforts were mainly restricted to specific unit operations and failed to address the process as a "system".

This interest in mathematical analysis coincided with the early development and growing availability of computers. This has been a major driving force in modelling ever since. Some individuals, however, were more concerned with the overall process rather than the details of individual unit operations.

One of the earliest monographs on PSE appeared in 1961 as a result of work within the Monsanto Chemical Company in the USA. This was authored by T.J. Williams [12], who wrote:

> ... systems engineering has a significant contribution to make to the practice and development of chemical engineering. The crossing of barriers between chemical engineering and other engineering disciplines and the use of advanced mathematics to study fundamental process mechanisms cannot help but be fruitful.

He continued,

> ... the use of computers and the development of mathematical process simulation techniques may result in completely new methods and approaches which will justify themselves by economic and technological improvements.

It is interesting to note that Williams' application of systems engineering covered all activities from process development through plant design to control and operations. Much of the work at Monsanto centred on the use of advanced control techniques aided by the development of computers capable of performing online control. Computer-developed mathematical models were proposed as a basis for producing statistical models generated from their more rigorous counterparts. The statistical models, which were regression models, could then be used within a control scheme at relatively low computational cost. It is evident that significant dependence was placed on the development and use of mathematical models for the process units of the plant.

In concluding his remarks, Williams attempted to assess the future role and impact of systems engineering in the process industries. He saw the possibility of some 150 large-scale computers being used in the chemical process industries within the USA for repeated plant optimization studies, these computers being directly connected to the plant operation by the end of the 1960s. He wrote:

> ... the next 10 years, then, may see most of today's problems in these fields conquered.

He did, however, see some dangers not the least being

> ... the need for sympathetic persons in management and plant operations who know and appreciate the power of the methods and devices involved, and who will demand their use for study of their own particular plants.

1.5.3. The Modern Era

Clearly, the vision of T.J. Williams was not met within the 1960s but tremendous strides were made in the area of process modelling and simulation. The seminal work on transport phenomena by Bird *et al.* [13] in 1960 gave further impetus to the mathematical modelling of process systems through the use of fundamental principles of conservation of mass, energy and momentum. It has remained the pre-eminent book on this subject for over 40 years.

The same period saw the emergence of numerous digital computer simulators for both steady state and dynamic simulation. These were both industrially and academically developed tools. Many of the systems were forerunners of the current class of steady state and dynamic simulation packages. They were systems which incorporated packaged models for many unit operations with little ability for the user to model specific process operations not covered by the simulation system. The numerical routines were crude by today's standards but simply reflected the stage of development reached by numerical mathematics of the time. Efficiency and robustness of solution were often poor and diagnostics as to what had happened were virtually non-existent. Some things have not changed!

The development of mini-computers in the 1970s and the emergence of UNIX-based computers followed by the personal computer (PC) in the early 1980s gave a boost to the development of modelling and simulation tools. It became a reality that every engineer could have a simulation tool on the desk which could address a wide range of steady state and dynamic simulation problems. This development was also reflected in the process industries where equipment vendors were beginning to supply sophisticated distributed control systems (DCSs) based on mini- and microcomputers. These often incorporated simulation systems based on simple block representations of the process or in some cases incorporated real time higher level computing languages such as FORTRAN or BASIC. The systems were capable of incorporating large scale real time optimization and supervisory functions.

In this sense, the vision of T.J. Williams some 40 years ago is a reality in certain sectors of the process industries. Accompanying the development of the process simulators was an attempt to provide computer aided modelling frameworks for the generation of process models based on the application of fundamental conservation principles related to mass, energy and momentum. These have been almost exclusively in the academic domain with a slowly growing penetration into the industrial arena. Systems such as ASCEND, Model.la, gPROMS or Modelica are among these developments.

What continues to be of concern is the lack of comprehensive and reliable tools for process modelling and the almost exclusive slant towards the petrochemical industries of most commercial simulation systems. The effective and efficient development of mathematical models for new and non-traditional processes still remains the biggest hurdle to the exploitation of those models in PSE. The challenges voiced by T.J. Williams in 1961 are still with us. This is especially the case in the non-petrochemical sector such as minerals processing, food, agricultural products, pharmaceuticals, wastewater and the integrated process and manufacturing industries where large scale discrete–continuous operations are providing the current challenge.

Williams' final words in his 1961 volume are worth repeating,

> ... there are bright prospects ahead in the chemical process industries for systems engineering.

One could add, ... for the process industries in general and for modelling in particular!

1.6. SUMMARY

Mathematical models play a vital role in PSE. Nearly every area of application is undergirded by some form of mathematical representation of the system behaviour. The form and veracity of such models is a key issue in their use. Over the last 50 years, there has been a widespread use of models for predicting both steady state and dynamic behaviour of processes. This was principally in the chemical process industries but, more recently, these techniques have been applied into other areas such as minerals, pharmaceuticals and bio-products. In most cases these have been through the application of process simulators incorporating embedded models.

There is a maturity evident in traditional steady state simulators, less so in large scale dynamic simulators and little of real value for large scale discrete–continuous simulation. Behind each of these areas there is the need for effective model development and documentation of the basis for the models which are developed. Systematic approaches are essential if reliable model use is to be demanded.

The following chapters address a systematic approach to the mathematical development of process models and the analysis of those models. The idea that one model serves all purposes is fallacious. Models must be developed for a specific purpose, and that purpose will direct the modelling task. It can be in any of the areas of model application discussed in Section 1.2. This point is emphasized in the following chapter, which develops a systematic approach to model building. The rest of the book provides detailed information on underlying principles on which the models are built and analysed.

1.7. REVIEW QUESTIONS

Q1.1. What are the major steps in building a model of a process system? (Section 1.1)

Q1.2. What key issues might arise for each of the overall modelling steps? (Section 1.1)

Q1.3. What major areas of PSE often rely on the use of models? What are some of the outcomes in the use of those models? (Section 1.2)

Q1.4. How can models be classified into generic types? Are these categories mutually exclusive? If not, then explain why. (Section 1.3.1)

Q1.5. Explain the fundamental differences between stochastic, empirical and mechanistic models. What are some of the factors which make it easier or harder to develop such models? (Section 1.3.1)

Q1.6. What are some of the advantages and disadvantages in developing and using empirical versus mechanistic models for process applications? (Section 1.3.1)

1.8. APPLICATION EXERCISES

A1.1. Consider the model application areas mentioned in Section 1.2. Give some specific examples of models being used in those applications? What were the benefits derived from developing and using models? Was there any clear methodology used to develop the model for its intended purpose?

A1.2. What could be the impediments to effective model building in process systems applications? Discuss these and their significance and possible ways to overcome the impediments.

A1.3. Consider a particular industry sector such as food, minerals, chemicals or pharmaceuticals and review where and how models are used in those industry sectors. What forms of models are typically used? Is there any indication of the effort expended in developing these models against the potential benefits to be derived from their use?

A1.4. Consider the basic principles of mass, heat and momentum transfer and the types of models which arise in these areas. What are the key characteristics of such models describing, for example, heat radiation or heat conduction? How do you classify them in terms of the classes mentioned in Table 1.3.

2 A SYSTEMATIC APPROACH TO MODEL BUILDING

In the past, process modelling was often regarded as "more art than science" or "more art than engineering". However, it is increasingly an engineering activity with a growing maturity. A general systematic approach to modelling has emerged from the numerous models which have been set up and used in process engineering. This systematic approach endeavours to provide good engineering practice in process modelling. It can be regarded as providing some "golden rules" for the task.

This chapter contains the basic principles of model building, outlining the elements and procedures of a systematic approach. It addresses the following issues:

- The concept of a process system and the modelling goal, as well as the effect of the goals on the process model (Section 2.1).
- The general notion of a model, different types of models and mathematical models (Section 2.2).
- The description of the steps in a modelling procedure illustrated by simple examples and an explanation of the iterative nature of the model building process (Section 2.3).
- The necessary modelling ingredients, namely the necessary elements of a process model: the assumptions, the model equations and any initial and boundary conditions which are put into a systematic format (Section 2.4).

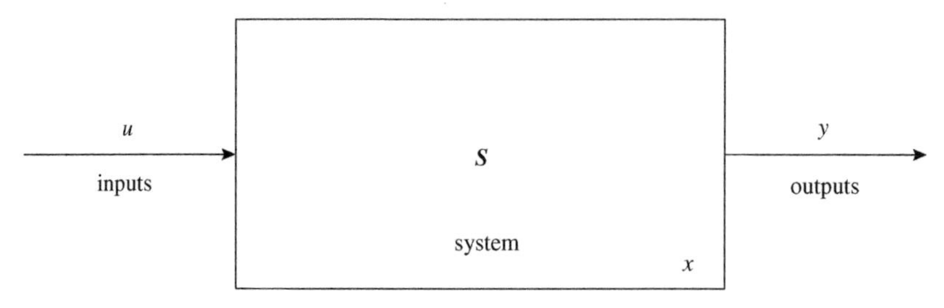

FIGURE 2.1 General system schematic.

2.1. THE PROCESS SYSTEM AND THE MODELLING GOAL

2.1.1. The Notion of a Process System

If we want to understand the notion of a *process system*, we can start from the general notion of a *system*. This can be defined in an abstract sense in system theory. A system is a part of the real world with well-defined physical boundaries. A system is influenced by its surroundings or environment via its *inputs* and generates influences on its surroundings by its *outputs* which occur through its boundary. This is seen in Fig. 2.1.

We are usually interested in the behaviour of the system *in time* $t \in \mathcal{T}$. A system is by nature a dynamic object. The system inputs u and the system outputs y can be single valued, giving a single input, single output (SISO) system. Alternately, the system can be a multiple input, multiple output (MIMO) system.

Both inputs and outputs are assumed to be time dependent possibly vector-valued functions which we call *signals*.

A system can be viewed as an operator in abstract mathematical sense transforming its inputs u to its outputs y. The states of the system are represented by the vector x and are usually associated with the mass, energy and momentum holdups in the system. Note that the states are also signals, that is time-dependent functions. We can express this mathematically as

$$u : \mathcal{T} \rightarrow \mathcal{U} \subseteq \mathcal{R}^r, \quad y : \mathcal{T} \rightarrow \mathcal{Y} \subseteq \mathcal{R}^v, \quad \mathbf{S} : \mathcal{U} \rightarrow \mathcal{Y}, \quad y = \mathbf{S}[u]. \tag{2.1}$$

Here, the vector of inputs u is a function of time and this vector-valued signal is taken from the set of all possible inputs $\mathcal{U}$ which are vector-valued signals of dimension r. The elements of the input vector at any given time t, $u_i(t)$ can be integer, real or symbolic values. Similarly, for the output vector y of dimension v.

The system S maps the inputs to the outputs as seen in Eq. (2.1). Internal to the system are the states x which allow a description of the behaviour at any point in time. Moreover $x(t)$ serves as a memory compressing all the past input–output history of the system up to a given time t.

A *process system* is then a system in which physical and chemical processes are taking place, these being the main interest to the modeller.

The system to be modelled could be seen as the whole process plant, its environment, part of the plant, an operating unit or an item of equipment. Hence, to define our system we need to *specify its boundaries, its inputs and outputs and the physico-chemical processes taking place within the system.* Process systems are conventionally

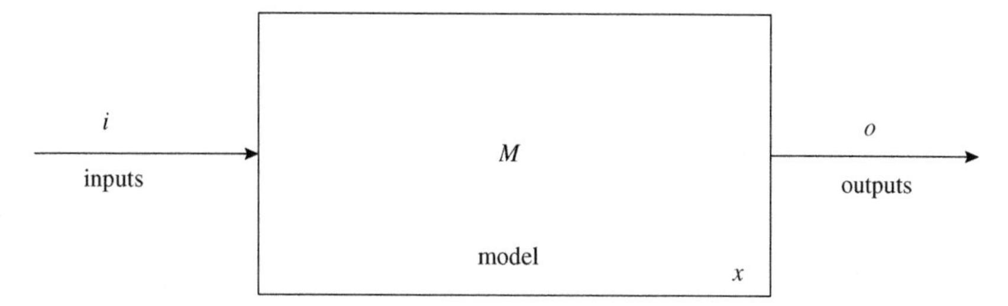

FIGURE 2.2 General model schematic.

specified in terms of a *flowsheet* which defines the boundaries together with inputs and outputs. Information is normally available about the internal structure of the system in terms of the operating units and their connections.

2.1.2. The Modelling Goal

The *modelling goal* specifies the intended use of the model. The modelling goal has a major impact on the level of detail and on the mathematical form of the model which will be built. Analogous to the idea of a process system shown in Fig. 2.1, a model acts in some way to mimic the behaviour of the real system it purports to represent. Thus, Fig. 2.2 shows the model with certain inputs and outputs. The use of the model can take various forms depending upon what is assumed to be known and what is to be computed.

Amongst the most important and widely used modelling goals in process engineering are the following:

Dynamic simulation
With the process model developed to represent changes in time, it is possible to predict the outputs o given all inputs i, the model structure M and parameters p.

Static or steady-state simulation
Here, the process system is assumed to be at steady state, representing an operating point of the system. Again the simulation problem computes the output values o given specific inputs i, a model structure M and its parameters p. This is sometimes known as a "rating" problem.

Design problem
Here, we are interested in calculating the values of certain parameters $\widehat{p}$ from the set of parameters p, given known inputs i and desired outputs o and a fixed structure M. This type of problem is normally solved using an optimization technique which finds the parameter values which generates the desired outputs. It is also called a "specification" problem.

Process control
The fundamental problems in process control are to consider a dynamic process model together with measured inputs i and/or outputs o in order to:

- design an input for which the system responds in a prescribed way, which gives a *regulation* or *state driving control problem*;

- find the structure of the model M with its parameters p using the input and output data, thus giving a *system identification problem*;
- find the internal states in M given a structure for the model, thus giving a *state estimation problem* which is typically solved using a form of least squares solution;
- find faulty modes and/or system parameters which correspond to measured input and output data, leading to *fault detection* and *diagnosis problems*.

In the second part of the book which deals with the analysis of process models, separate chapters are devoted to process models satisfying different modelling goals: Chapter 13 deals specifically with models for control, Chapter 15 with models for discrete event systems, while Chapter 16 deals with models for hybrid or discrete–continuous systems.

It follows from the above that a *problem definition* in process modelling should contain at least two sections:

- the specification of the process system to be modelled,
- the modelling goal.

The following example gives a simple illustration of a problem definition.

EXAMPLE 2.1.1 (Problem definition of a CSTR).

Process system
Consider a continuous stirred tank reactor (CSTR) with continuous flow in and out and with a single first-order chemical reaction taking place. The feed contains the reactant in an inert fluid.

Let us assume that the tank is adiabatic, such that its wall is perfectly insulated from the surroundings. The flowsheet is shown in Fig. 2.3.

Modelling goal
Describe the dynamic behaviour of the CSTR if the inlet concentration changes. The desired range of process variables will be between a lower value x^{L} and an upper value x^{U} with a desired accuracy of 10%.

2.2. MATHEMATICAL MODELS

In Chapter 1, we discussed briefly some key aspects of models and, in particular, mathematical models. We saw that models of various sorts are constructed and used in engineering especially where undertaking experiments would not be possible, feasible or desirable for economic, safety or environmental reasons. A model can be used to help design experiments which are costly or considered difficult or dangerous. Therefore, a model should be similar to the real process system in terms of its important properties for the intended use. In other words, the model should describe or reflect somehow the properties of the real system relevant to the modelling goal.

On the other hand, models are never identical to the real process system. They should be substantially less complex and hence much cheaper and easy to handle so that the analysis can be carried out in a convenient way. The reduced complexity of the model relative to the process system is usually achieved by simplification and

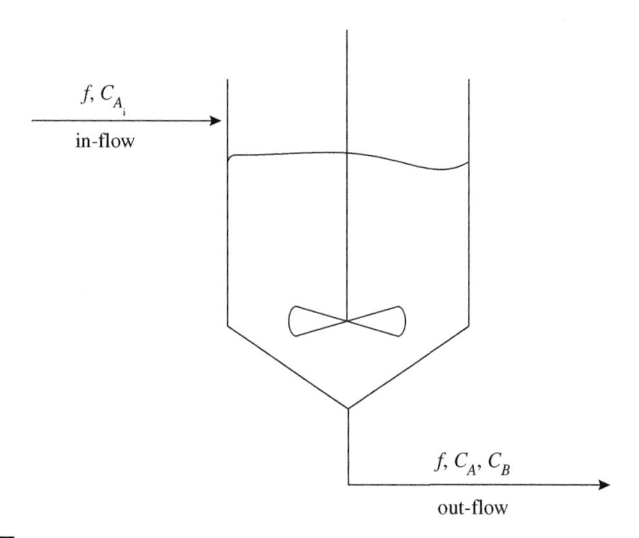

FIGURE 2.3 Flowsheet of a continuously stirred tank.

elimination of certain controlling mechanisms during the modelling process. It also relates to the number and type of inputs and outputs that are considered appropriate. We use *modelling assumptions* to describe the specific knowledge we have about the process system. This knowledge can help simplify and eliminate what we consider to be unnecessary aspects of the model's description. Because of the less complex nature of the model, the number of variables will be much smaller in dimension than the dimension of the real system. Hence, we can write that

$$\dim(y_M) = p_M \ll \dim(y_S) = p_S, \quad \dim(u_M) = r_M \ll \dim(u_S) = r_S,$$

where y_M, u_M denotes the input and output vectors of the model and y_S, u_S are those of the system being modelled.

2.2.1. White, Black and Grey-Box Models

In Chapter 1, we classified models using various criteria. In particular, models can be classified according to their physical nature. There are *analogue models* where the descriptive power of the model is based on physical, or physico-chemical analogues between the real process system and the model. Laboratory scale equipment or pilot plants use full physico-chemical analogy but there are also analogue models of process systems based on mechanical or electronic principles, using springs and damper pots or analogue computing components. In contrast, *mathematical models* are models consisting of mathematical objects, such as equations, graphs, or rules.

There are at least two fundamentally different ways of obtaining process models. First, we can use our process engineering knowledge to describe the physico–chemical processes taking place in the system on the level required by the modelling goal. This is a "first principles" engineering model. Such a model is termed a *white-box model* (mechanistic) to indicate that the model is totally transparent or understandable to a process engineer. In this case, we do not directly use any measured data about the process. However, we do make use of measured data indirectly, through the form and the value of the system parameters.

Second, the alternative means of obtaining process models is often dictated by a lack of accumulated engineering knowledge of the system. In this case, we can use measurement data of the inputs and outputs to build a process model. The need to stimulate or excite the system to obtain useful information for model identification is a major issue for any model builder. This includes the amount of data and its frequency content which enables the key responses to be captured in the resulting model. In this case, we can use structure and parameter estimation methods developed mainly in the field of process identification. This kind of model is called a *black-box model* (empirical) because knowledge of the process is only available through the measurement data and not from the underlying mechanisms. A good coverage of system identification methods is given by Ljung [14]

In process engineering practice, however, both purely white and purely black-box models are rare. In most cases, we use a suitable combination of our *a priori* process engineering knowledge to determine the structure and some of the parameter values of the model. We then use measured data to build the model. This involves the definition of kinetic and transport mechanisms, the estimation of key kinetic and transport parameters as well as validation of the model against the performance specifications. These combination models are termed *grey-box models*.

The terms used in this section are more recent inventions; however, the underlying principles and practice have existed for decades.

2.3. A SYSTEMATIC MODELLING PROCEDURE

Like other engineering tasks, good practice requires models to be constructed following a well defined sequence of steps. These steps are arranged in a "Seven (7) Step Modelling Procedure" which is introduced below and shown schematically in Fig. 2.4.

However, it should be noted that model development is inherently iterative in its nature. One must usually return to and repeat an earlier step in case of any problems, unusual or unwanted developments later in the process. *No one gets it right first time!* In fact, we never get a perfect model, just one that is usable.

Before starting to setup a process model, the problem definition should be clearly stated. This defines the process, the modelling goal and the validation criteria. This is part of the formal description in the SEVEN STEP MODELLING PROCEDURE, which is given in the form of an algorithmic problem. An algorithm being a systematic procedure for carrying out the modelling task, named after the Persian mathematician Abu Ja'far Mohammed ibn Mûsâ al Khowârizm of the ninth century [15].

In a formal description of an algorithmic problem, one should formally specify the following items:

- the inputs to the problem in the *Given* section,
- the desired output of the procedure in the *Find* or *Compute* section,
- the method description in the *Procedure* or *Solution* section.

Problem statements and their use in analysing computational or algorithmic complexity will be described later in Section 10.1 in Chapter 10.

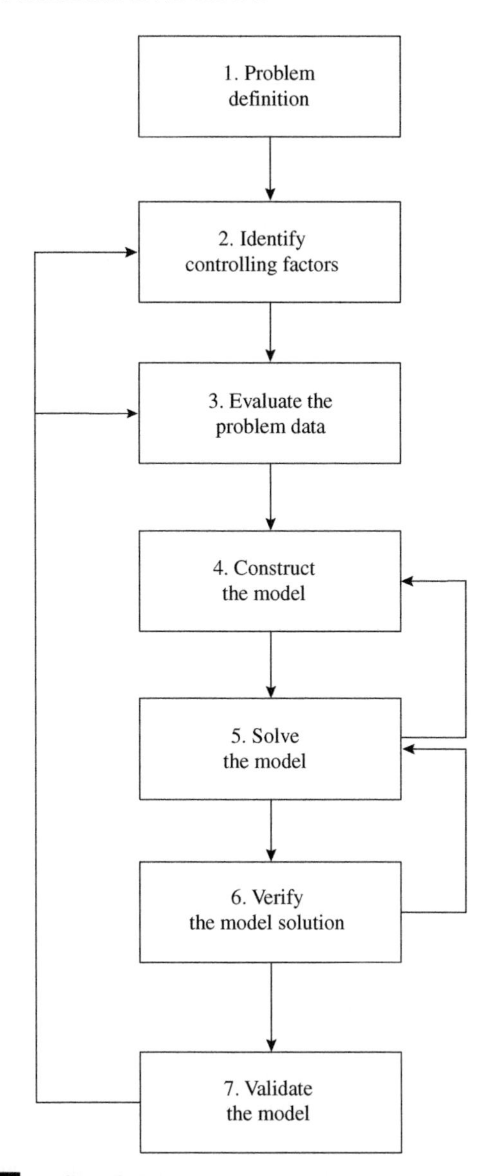

FIGURE 2.4 Systematic model building steps.

Applying the above principles to the general modelling procedure, the following algorithmic problem statement can be constructed.

SEVEN STEP MODELLING PROCEDURE

Given:

- a process system
- a modelling goal
- validation criteria

Find:

- a mathematical model

Procedure:
The model is built following a systematic procedure given below with seven steps. The steps of the modelling procedure are as follows:

1. *Define the problem*
This step refines the sections already present in the problem definition: the description of the process system with the modelling goal. Moreover, it fixes the degree of detail relevant to the modelling goal and specifies:
   - inputs and outputs
   - hierarchy level relevant to the model or hierarchy levels of the model in the case of hierarchical models (see Chapter 9 for hierarchy levels)
   - the type of spatial distribution (distributed or lumped model)
   - the necessary range and accuracy of the model and
   - the time characteristics (static versus dynamic) of the process model.

2. *Identify the controlling factors or mechanisms*
The next step is to investigate the physico-chemical processes and phenomena taking place in the system relevant to the modelling goal. These are termed *controlling factors or mechanisms*. The most important and common controlling factors include:
   - chemical reaction
   - diffusion of mass
   - conduction of heat
   - forced convection heat transfer
   - free convection heat transfer
   - radiation heat transfer
   - evaporation
   - turbulent mixing
   - heat or mass transfer through a boundary layer
   - fluid flow

Figure 2.5 shows schematically the situation which exists when attempting to model a process system. As we consider the system under study, we recognize the following issues:
   - There is a set of all process characteristics which are never fully identified.
   - In the set of all process characteristics for the system we often:
     - only identify and include in the model a subset of the essential characteristics. This means that some essential characteristics for the application can be missing from our model description;
     - include non-essential characteristics of the system in our models, which lead to unnecessary complexity and/or model order (or size).
   - We sometimes incorrectly identify process characteristics which are actually not part of the system and include them in our model.

The previous issues are often difficult to resolve and very dependent on our understanding of the system and our insight into what is important. Model validation and the principle of parsimony, which seeks the simplest representation for the task are the tools that need to be employed to address these issues.

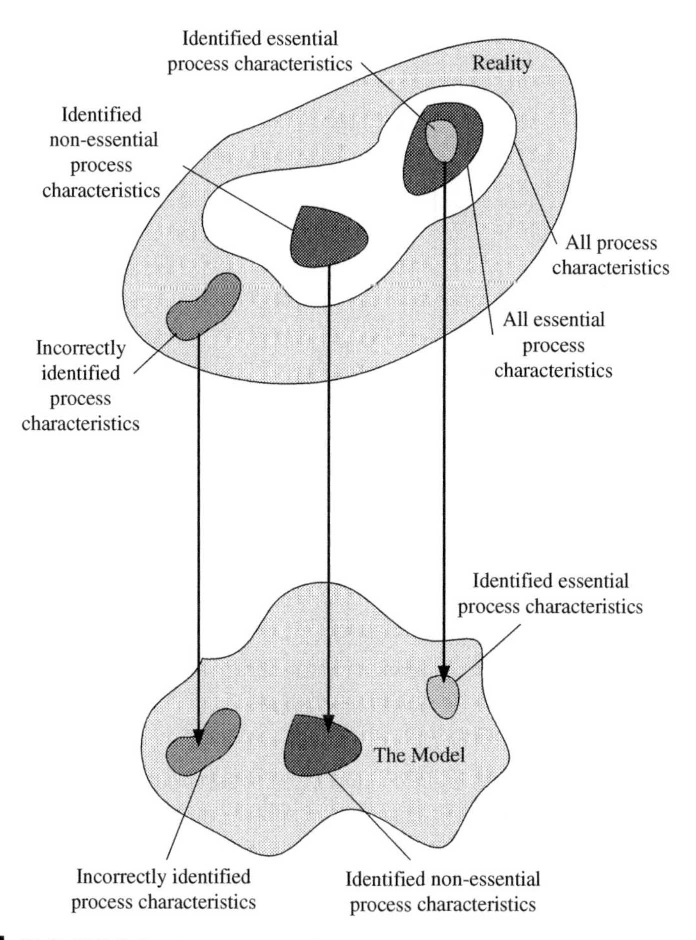

FIGURE 2.5 Characteristics of system and model.

It is very important to emphasize that one has to filter carefully the set of all possible controlling mechanisms taking into account the following key elements in the problem definition:

- the hierarchy level(s) relevant to the model,
- the type of spatial distribution,
- the necessary range and accuracy,
- the time characteristics.

The effect of the modelling goal on the selection of balance volumes are discussed separately in Section 3.3.

EXAMPLE 2.3.1 (Identifying key controlling mechanisms). As an example, consider the modelling of a jacketed tank which is well-stirred and heated using a hot oil feed as shown in Fig. 2.6. If we were to model this system to predict the dynamic behaviour of the liquid temperature, then some key controlling mechanisms could be:

- fluid flow of liquid into and out of the tank,
- fluid flow of hot oil in and out of the jacket,
- convective heat transfer between the jacket and tank contents,

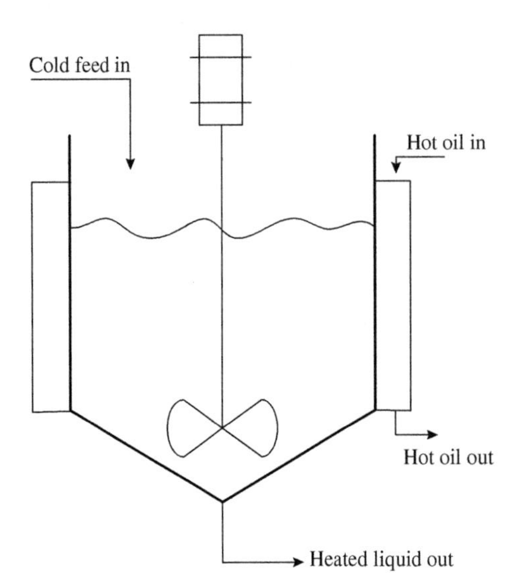

FIGURE 2.6 Heated tank.

- work input done through the agitator,
- convective heat losses from the jacket to the environment,
- convective heat losses from the surface of the heated liquid to the environment,
- evaporative losses of liquid from the tank contents to the environment.

We would need to assess the most important or essential characteristics required in the model for the stated modelling goal. This initially requires an order of magnitude assessment of mass and heat transfer rates to see what is of major importance.

The models for these controlling factors are developed in physical chemistry, reaction engineering, mass and heat transfer and unit operations. They form partial models within the overall process model. These partial models are generally black box in their nature. ■ ■ ■

3. *Evaluate the data for the problem*
As it has been already noted, models of real process systems are of the grey-box type, therefore, we almost always need to use either measured process data directly or estimated parameter values in our models.

It is very important to consider both measured process data and parameter values together with their uncertainties or precision. Some default precision values might be:

- industrial measured data is ± 10 to 30%,
- estimated parameters from laboratory or pilot–plant data is ± 5 to 20%,
- reaction kinetic data is ± 10 to 500%,

if nothing else is specified.

At this step, we may find out that there are neither suitable parameter values found in the literature nor measured data to estimate them. This situation may force us to reconsider our decisions in steps 1 and 2 and to return there to change them.

4. *Develop a set of model equations*
The model equations in a process model are either differential (both partial and ordinary differential equations may appear) or algebraic equations. The *differential*

equations originate from conservation balances; therefore they can be termed *balance equations*. The algebraic equations are usually of mixed origin: they will be called *constitutive equations*. Conservation balances are discussed in detail in Chapter 3, while constitutive equations are treated in Chapter 4.

A *model building sub-procedure* can be used to obtain a syntactically and semantically correct process model. The steps and properties of this sub-procedure are found in a separate Section 2.3.1 below.

5. *Find and implement a solution procedure*
Having set up a mathematical model, we must identify its mathematical form and find or implement a solution procedure. In all cases, we must ensure that the model is well posed such that the "excess" variables or degrees of freedom are satisfied. We also try and avoid certain numerical problems such as high index systems at this stage. Lack of solution techniques may prevent a modeller using a particular type of process model and can lead to additional simplifying assumptions to obtain a solvable model. This can be the case with distributed parameter process models.

6. *Verify the model solution*
Having a solution is just the start of the analysis. Verification is determining whether the model is behaving correctly. Is it coded correctly and giving you the answer you intended? This is not the same as model validation where we check the model against reality. You need to check carefully that the model is correctly implemented. Structured programming using "top-down" algorithm design can help here as well as the use of modular code which has been tested thoroughly. This is particularly important for large-scale models.

7. *Validate the model*
Once a mathematical model has been set up, one should try to validate it. This checks the quality of the resultant model against independent observations or assumptions. Usually, only partial validation is carried out in practical cases depending on the modelling goal.

There are various possibilities to validate a process model. The actual validation method strongly depends on the process system, on the modelling goal and on the possibilities of getting independent information for validation. The possibilities include but are not limited to:

- verify experimentally the simplifying assumptions,
- compare the model behaviour with the process behaviour,
- develop analytical models for simplified cases and compare the behaviour,
- compare with other models using a common problem,
- compare the model directly with process data.

The tools to help carry out this task include the use of sensitivity studies to identify the key controlling inputs or system parameters as well as the use of statistical validation tests. These can involve hypothesis testing and the use of various measures such as averages, variances, maxima, minima and correlations.

If the validation results show that the developed model is not suitable for the modelling goal then one has to return to step 2 and perform the sequence again. Usually, validation results indicate how to improve the model. We can often identify the inadequate areas in our model development and, therefore, not all modelling efforts are lost. Note that the final model is impossible to obtain in one pass through the modelling procedure. Some iterations should be expected.

A separate chapter in the second part of the book is devoted to model validation, see Chapter 12.

2.3.1. The Model Building Sub-procedure

Step 4 in the above SEVEN STEP MODELING PROCEDURE is a composite step in itself. It is broken down to sub-steps which are to be carried out sequentially but with inherent loops. The steps of this sub-procedure are as follows:

Model building sub-procedure in step 4.

4.0. *System and subsystem boundary and balance volume definitions*
The outcome of this step is the set of balance volumes for mass, energy and momentum leading to the conserved extensive quantities normally considered in process systems, such as total system mass, component masses or energy.

In order to define the balance volumes we need to identify within our system the regions where mass, energy or momentum are likely to accumulate. These are termed the system "holdups". The accumulation of conserved quantities such as energy or mass are often dictated by the physical equipment or phase behaviour within that equipment.

The following simple example of the heated tank illustrates the balance volume identification problem.

EXAMPLE 2.3.2 (Defining balance volumes). In Fig. 2.6, the example of a liquid heating system was given. What could be the relevant balance volumes to consider for this situation? One possibility is shown in Fig. 2.7, which gives two main balance volumes:

$\Sigma_1^{M,E}$: which is the balance volume (1) for mass (M) and energy (E) related to the liquid being heated;

$\Sigma_2^{M,E}$: which is the balance volume (2) for mass (M) and energy (E) related to the heating oil.

Other options are also possible depending on the characteristics of the system.

4.1. *Define the characterizing variables*
Here, we need to define the variables which will characterize the system being studied. These variables are associated with the inputs, outputs and internal states of the system as shown in Fig. 2.1. The variable we use will be strongly determined by our modelling goals. For the inputs and outputs we can consider component flows, total flow, mass or molar concentrations. Temperatures and pressures may also be important, depending on whether energy is considered important. For the internal states of the system, the characterizing variables are related to the variables representing the main mass, energy and momentum holdups.

4.2. *Establish the balance equations*
Here, we set up conservation balances for mass, energy and momentum for each balance volume.

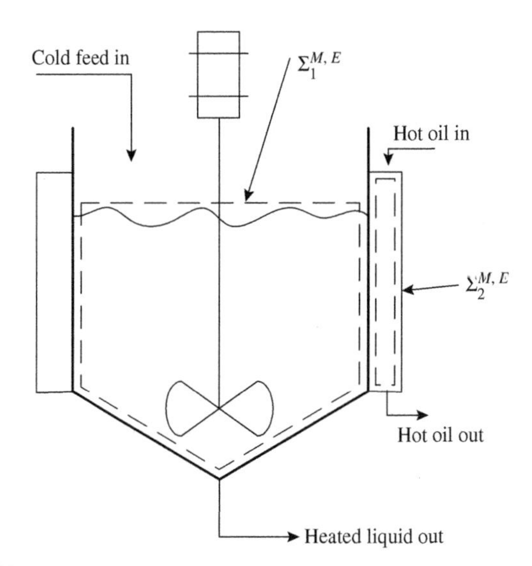

FIGURE 2.7 Balance volumes for heated tank.

4.3. *Transfer rate specifications*
The rate expressions for transfer of heat, mass and momentum between different balance volumes in the conservation balances are specified here, usually as functions of intensive quantities, such as concentrations and temperature.

4.4. *Property relation specifications*
These are mostly algebraic relationships expressing thermodynamic knowledge such as equations of state and the dependence of physico-chemical properties on thermodynamic state variables such as pressure, temperature and composition.

4.5. *Balance volume relation specifications*
An equipment with a fixed physical volume is often divided into several balance volumes if multiple phases are present. A balance volume relation describes the relationship between balance volumes and physical volumes.

4.6. *Equipment and control constraint specifications*
There is inevitably the need to define constraints on process systems. These are typically in the form of equipment operating constraints (for the pressures, temperatures etc.) and in terms of control constraints, which define relations between manipulated and controlled variables in the system.

4.7. *Modelling assumptions*
When developing a particular model we apply modelling assumptions to get our model equations. These assumptions form an integral part of the resultant model. Therefore, they are regarded as *ingredients* of a process model together with the model equations, their initial and boundary conditions. The ingredients of a process model will be discussed in more detail in the Section 2.4.

It is important to note that *modelling assumptions are usually built up incrementally in parallel and during sub-steps 4.0.–4.6.*

2.3.2. Incremental Model Building

As it has been emphasized already in Section 2.3.1, the steps of the *model building sub-procedure* should be carried out in a sequential–iterative manner. It means that the model equations are built up incrementally repeating sub-steps *4.2.–4.7.* in the following order of conserved extensive quantities:

1. *Overall mass sub-model*
   The conservation balances for the overall mass in each of the balance volumes and the variables therein appear in all other conservation balances. Therefore, this subset of model equation is built up first performing sub-steps *4.2.–4.7. of the model building sub-procedure.*
2. *Component mass sub-model*
   With the conservation balances for the overall mass in each of the balance volumes given, it is easy to set up the conservation balances for the component masses. This sub-set of model equation is added to the equations originated from the overall mass balances performing sub-steps *4.2.–4.7. of the model building sub-procedure.*
3. *Energy sub-model*
   Finally, the subset of model equations induced by the energy balance is added to the equations performing sub-steps *4.2.–4.7. of the model building sub-procedure.*

This way the kernel of the *the model building sub-procedure* is repeated at least three times for every balance volumes.

2.4. INGREDIENTS OF PROCESS MODELS

The SEVEN STEP MODELLING PROCEDURE described above shows that a process model resulting from this procedure is not simply a set of equations. It incorporates a lot more information. In order to encourage the clarity of presentation and the consistency of the process model, a *structural presentation* incorporating all key ingredients is now suggested.

1. *Assumptions*—which include but are not limited to

- time characteristics
- spatial characteristics
- flow conditions
- controlling mechanisms or factors
- neglected dependencies, such as
  - temperature dependence of physico-chemical properties
  - concentration dependence of physico-chemical properties
- required range of states and associated accuracy.

Here, we are careful to designate each assumption with a unique identifier $\mathcal{A}_i$, $i = 1, 2, \ldots, n$, where n is the total number of modelling assumptions.

2. *Model equations and characterizing variables*

Here, we have differential (balance) equations for

- overall mass, component masses for all components for all phases in all equipment at all hierarchical levels (where applicable),

- energy (or enthalpy),
- momentum;

constitutive equations for

- *transfer rates*: mass transfer, heat/energy transfer, and reaction rates,
- *property relations*: thermodynamical constraints and relations, such as the dependence of thermodynamical properties on the thermodynamical state variables (temperature, pressure and compositions), equilibrium relations and state equations,
- *balance volume relations*: relationships between the defined mass and energy balance regions,
- equipment and control constraints;

and the variables which characterize the system:

- flows, temperatures, pressures, concentrations, enthalpies,
- mass, energy and momentum holdups.

3. *Initial conditions* (where applicable)
Initial conditions are needed for the differential (balance) equations in dynamic process models. Initial conditions for all algebraic variables are also important since most solutions are numerical.

4. *Boundary conditions* (where applicable)
Boundary conditions must be specified for the differential (balance) equations in spatially distributed process models.

5. *Parameters*
As a result of step 3 in the SEVEN STEP MODELLING PROCEDURE the value and/or source of the model parameters are specified here with their units and precision. This also applies to all variables in the model.

We can return to our earlier Example 2.1.1 to illustrate how these ingredients are to be expressed.

EXAMPLE 2.4.1 (Modelling example: CSTR). Develop a process model of the continuously stirred tank specified in Example 2.1.1 for dynamic prediction and control purposes following the SEVEN STEP MODELLING PROCEDURE with all of its ingredients.

1. *Problem definition*
The *process system* to be modelled is a continuously stirred tank with continuous fluid flow in and out and with a single first order chemical reaction $A \rightarrow B$ taking place in an inert solvent. The tank is adiabatic with its wall perfectly insulated from the environment. The flowsheet schematic of this tank is shown in the Fig. 2.3.

(This description of the process system is exactly the same as the *Process system* section in the *Problem definition* of Example 2.1.1.)

The *modelling goal* is to predict the behaviour of the principal mass and energy states of the tank contents if the inlet concentration is changed over a stated range. The accuracy of the predictions should be $\pm 10\%$ of the real process. (It is again identical to the *Modelling goal* section of the *Problem definition*.)

2. *Controlling factors or mechanisms*

- chemical reaction
- perfect mixing

3. *Data for the problem*

No measured data is specified; therefore, we use the following parameter type data:

- reaction kinetic data, heat of reaction
- physico-chemical properties
- equipment parameters

from the literature or given by the process documentation.

4. *Process model*

Assumptions

$\mathcal{A}_1$: perfect mixing,
$\mathcal{A}_2$: constant physico-chemical properties
$\mathcal{A}_3$: equal inflow and outflow (implying constant liquid volume with $V = constant$),
$\mathcal{A}_4$: single first-order exothermic reaction, $\mathrm{A} \rightarrow \mathrm{P}$,
$\mathcal{A}_5$: adiabatic operation.

Model equations and characterizing variables

Differential (balance) equations in molar units

$$\frac{\mathrm{d}m_{\mathrm{A}}}{\mathrm{d}t} = f_{\mathrm{A_i}} - f_{\mathrm{A}} - rV, \tag{2.2}$$

$$V\rho c_p \frac{\mathrm{d}T}{\mathrm{d}t} = f c_{p_\mathrm{i}} \rho_\mathrm{i}(T_\mathrm{i} - T) + rV(-\Delta H_\mathrm{R}). \tag{2.3}$$

Constitutive equations

$$r = k_0\, \mathrm{e}^{-E/(RT)} C_A, \tag{2.4}$$

$$m_A = C_A V, \tag{2.5}$$

$$f_{\mathrm{A_i}} = f C_{A_\mathrm{i}}, \tag{2.6}$$

$$f_A = f C_A. \tag{2.7}$$

Variables

t	time [s]
C_A	concentration in the tank [mol/m^3]
V	liquid volume [m^3]
f	volumetric flowrate [m^3/s]
C_{A_i}	inlet concentration [mol/m^3]
r	reaction rate [mol/(s·m^3)]
c_p	specific heat of mixture [J/(mol K)]
c_{p_i}	specific heat of feed *i* [J/(mol K)]
T	temperature in the tank [K]
T_i	inlet temperature [K]
ΔH_R	heat of reaction [J/mol]
E	activation energy [J/mol]
k_0	pre-exponential factor [s^{-1}]
R	universal gas constant, 8.314 [J/(mol K)]
ρ	density of mixture [mol/m^3]

ρ_i density of feed i [mol/m^3]
m_A moles of A [mol]
f_{A_i} inlet flowrate of species A [mol/s]

Initial conditions

$$C_A(0) = C_{A_i}, \quad T(0) = T_i$$

Boundary conditions
None

Parameters
Values for the following parameters with 10% precision:

$$V,\ f,\ C_{A_i},\ T_i,\ c_p,\ c_{p_i},\ \rho,\ \rho_i$$

and for the reaction parameters with 30% precision:

$$k_0,\ E,\ \Delta H_R$$

5. *Solution procedure*
Solve using an ODE or differential-algebraic equation solver.

6. *Model verification*
Implement model equations using structured programming principles. Check code for correct execution. Check output trends against expected trends for reactor given step changes in feed variables. Also, check predicted steady-state values after a feed disturbance.

7. *Model validation*
Provide measured data from pilot plant or real process. Analyse plant data quality. Carry out validation of predicted outputs from step test of system using least squares estimation of error. Apply hypothesis testing to validate model based on least squares estimates. Refine model as required by performance criteria. ■ ■ ■

2.4.1. Incremental Evolution of the Ingredients of Process Models

As it is described in Section 2.3.2 above, the ingredients of a process model are set up incrementally in the following order:

- overall mass sub-model
- component mass sub-model
- energy sub-model

for every balance volume in the system.

This implies that the following ingredients are subject to incremental evolution:

- assumptions,
- model equations and characterizing variables including differential (balance) equations and constitutive equations,
- initial and boundary conditions,
- parameters.

The following example illustrates how this is done for the CSTR reactor example of Example 2.4.1.

EXAMPLE 2.4.2 (Incremental evolution of the ingredients of a CSTR model). *Show the incremental evolution of the process model of the continuously stirred tank modelled in Example 2.4.1.*

The following evaluation steps are carried out.

1. *List general modelling assumptions*

These are assumptions derived from the problem statement valid for the whole process system. In this case, these are assumptions $\mathcal{A}_1$ and $\mathcal{A}_2$.

2. *Overall mass balances*

There is an assumption on equal in- and outflow in the problem statement formalized as assumption $\mathcal{A}_3$ which implies constant volumetric and mass holdup. Therefore, the mass balance reduces to the equation $V =$ constant.

3. *Component mass balances*

The relevant assumptions derived from the problem statement are assumptions $\mathcal{A}_4$ and $\mathcal{A}_5$. With these assumptions, we set up the component mass balance equation (2.2) with its constitutive equations (2.4)–(2.7).

4. *Energy balances*

Finally, the energy balance Eq. (2.3) is added.

2.5. SUMMARY

In this chapter, we have discussed a systematic model building strategy. The strategy has been laid down in seven steps which are closely interrelated. As well, we noted that the process is cyclic requiring the modeller to retrace steps higher in the strategy when the modelling, solution or validation are unacceptable. Also, the importance of identifying key underlying mechanisms is emphasized in determining what is of importance in the modelling task. These often need to be revisited to confirm how significant their inclusion or exclusion might be.

It is vitally important that the model is both verified and validated. Various approaches were discussed. Later chapters will take up these issues in more detail.

In the following chapter, we introduce the basic underlying process and thermodynamic principles on which the model equations are based.

2.6. REVIEW QUESTIONS

Q2.1. Describe the general characteristics of a system. What is important to consider in engineering systems? (Section 2.1)

Q2.2. Describe typical modelling goals for a range of process engineering applications. (Section 2.1)

Q2.3. Discuss why modelling assumptions are important in the building of a model. (Section 2.2)

Q2.4. What is the difference between white, black- and grey-box models? (Section 2.2)

Q2.5. Outline the systematic approach to model building discussed in this chapter. What are the key steps and their characteristics? (Section 2.3)

Q2.6. What are some controlling factors or mechanisms which might be considered in a process modelling task? (Section 2.3)

Q2.7. In defining the modelling problem, what basic decisions need to be made before any mathematical modelling starts? (Section 2.3)

Q2.8. In developing the mathematical form of the model, what are the key equation sub-classes to be considered? (Section 2.3)

Q2.9. What techniques will help in model verification? (Section 2.3)

Q2.10. Discuss some general approaches to model validation. (Section 2.3)

Q2.11. List the five necessary ingredients that go to make up a full model of a process system! (Section 2.4)

2.7. APPLICATION EXERCISES

A2.1. Consider the following tank problem where liquid is pumped into an open tank and then flows out under gravity. The situation is shown in Fig. 2.8.

If this situation is to be modelled for analysis of disturbances in the inlet flow pressure from the pump, develop the problem specification by considering the ingredients needed for the model.

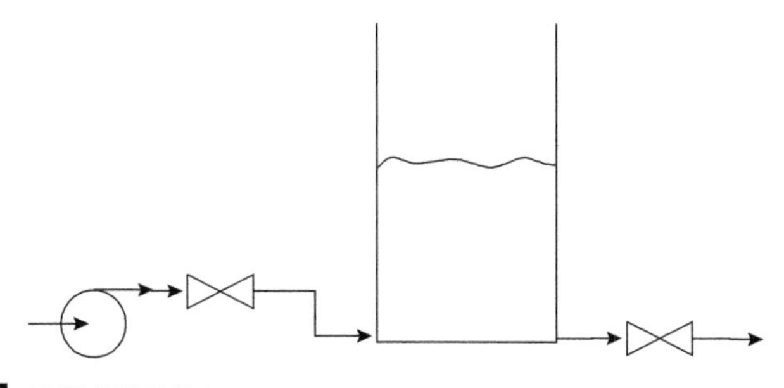

FIGURE 2.8 Simple tank and liquid flow system.

A2.2. Consider now a similar situation to that described in application A2.1, where now the tank is enclosed. This is shown in Fig. 2.9. Again develop a problem specification outlining all the ingredients for the modelling.

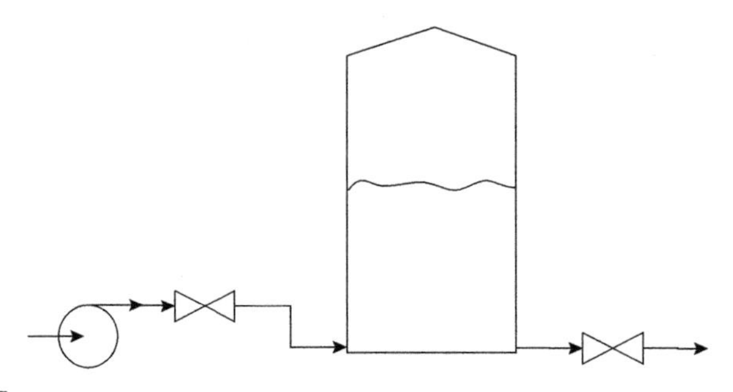

FIGURE 2.9 Enclosed tank.

A2.3. Consider a vessel (Figure 2.10), where a liquid feed is heated. Vapour and liquid are withdrawn. It is intended that a model of this process should be developed to investigate changes in the heat input Q from the steam coil as well as changes in feed conditions of temperature and composition. Develop a problem description for this situation discussing the necessary ingredients for the modelling.

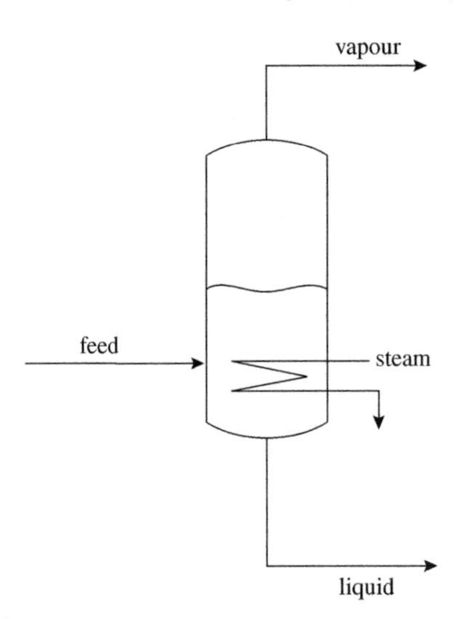

FIGURE 2.10 Liquid heating system.

A2.4. Consider Fig. 2.11, which shows a packed absorber for the removal of a solvent within an air stream. The vessel is packed with activated carbon particles held in place by screens. Develop a problem description, outlining the principal ingredients for modelling which considers the system response to changes in the incoming solvent concentration.

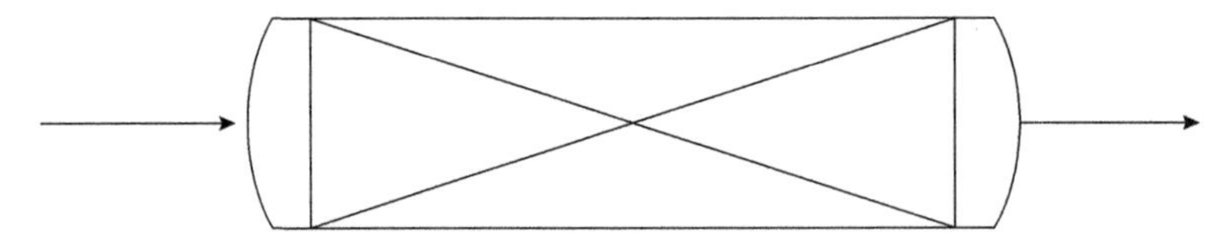

FIGURE 2.11 Packed bed absorber.

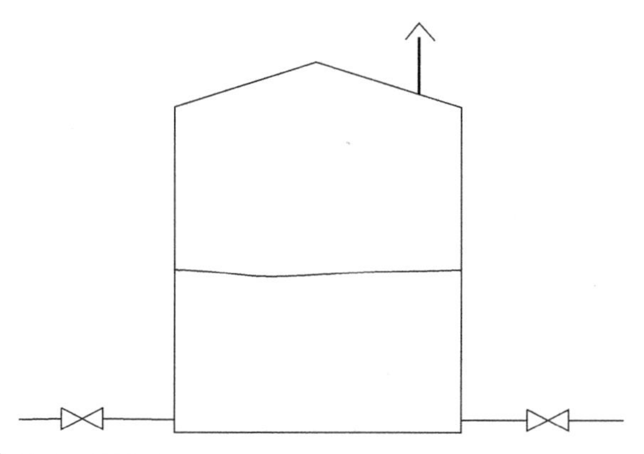

FIGURE 2.12 Vented tank.

A2.5. One of the potential sources of odours from storage facilities arises from the breathing of storage tanks as liquid is pumped in or out of the tank. Figure 2.12 shows a typical arrangement whereby vapours are discharged through a breather pipe with flame arrester attached to the end. The purpose of the modelling here is to predict the release rate of vapour as the liquid level changes in the tank. Develop a modelling description of this system.

A2.6. One of the key processes for making fertilizers and some pharmaceutical products involves particle size enlargement often carried out in a granulation drum. Figure 2.13 shows a typical inclined drum which rotates slowly in order for the granulation to occur. In many cases, a liquid binder is also added to control moisture and thus the rate of granulation. Develop a modelling description of this process including the relevant ingredients for the model.

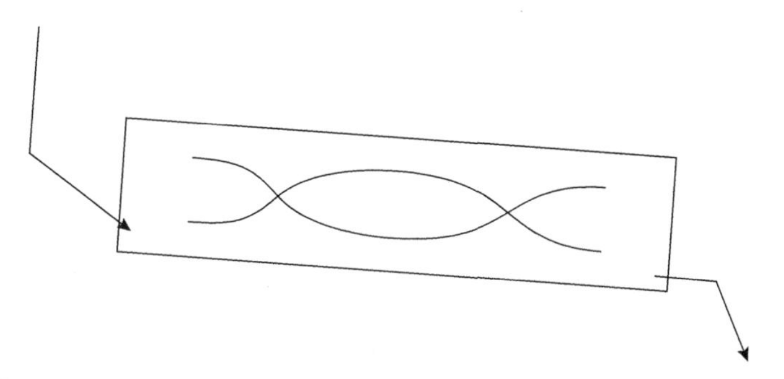

FIGURE 2.13 Rotating granulation drum.

A2.7. Figure 2.14 shows a conical mixing vessel for the blending of a liquid and solid stream. Develop a modelling description to include all relevant aspects to be able to predict the time varying behaviour of the system when changes occur in the inlet flowrates.

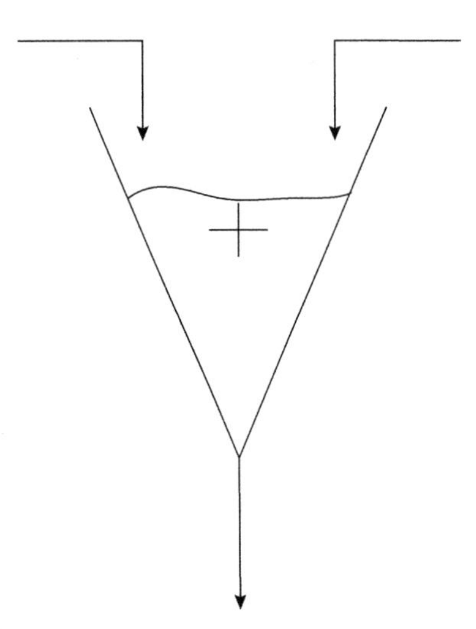

FIGURE 2.14 Conical blending tank.

A2.8. Heat transfer is a key unit operation in the process industries. Figure 2.15 shows a double-pipe heat exchanger for heat exchange between two liquid streams. Consider the modelling of such a system for design and optimization purposes. How would the model description change if stream temperatures change? Develop the model description.

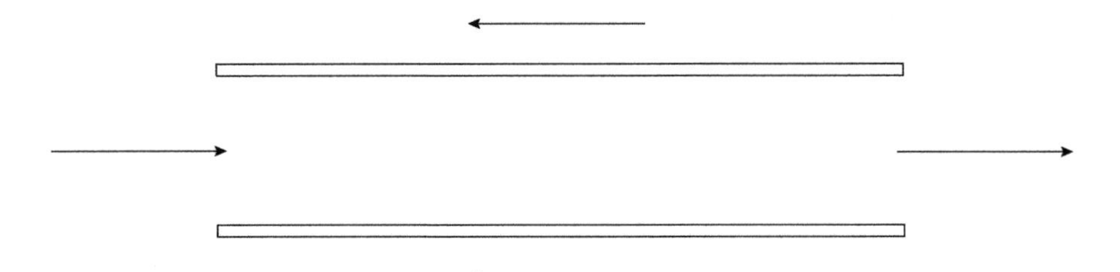

FIGURE 2.15 Double-pipe heat exchanger.

A2.9. Transport and storage of hazardous liquids is commonplace in many areas of modern society. Petroleum products and liquid chemicals such as solvents and pesticides can pose significant hazards if released into the environment through loss of containment from tanks. Consider the modelling of a pool of released solvent from the rupture of a storage tank in the open air. The pool is subsequently contained in a bunded enclosure around the tank. What are the key modelling ingredients which describe the model? Develop a model description. What other issues must be addressed if the spill is not bunded?

3 CONSERVATION PRINCIPLES

In the previous chapter, we introduced a systematic approach to mathematical modelling. Following the establishment of the purpose of the modelling and the underlying assumptions about the system and its behaviour, the next step was the development of the describing equations. One key aspect in this process is the application of *conservation principles* for conserved *extensive* quantities. Another important aspect is the development of the *constitutive relations* which provide a set of relations which are used to complete the model.

This chapter deals specifically with the underlying principles related to conservation principles. This is based on the fundamental principle of physics that mass, energy and momentum are neither destroyed nor created but simply change their form. The relationship between extensive and intensive quantities is discussed and the implications on their use in terms of model complexity and equation transformations are developed. The focus of this chapter is on macroscopic thermodynamic properties of process systems and how they are used for modelling purposes.

The application of conservation principles to typical gas–liquid–solid systems normally leads to relationships involving total system mass, component masses, total energy and system momentum. In particulate systems, we can also consider quantities such as particle number balances for the generation of population balance equations. This chapter describes the class of balance equations for total mass, component masses and energy. It covers the construction, general form and typical terms associated with these balances. This chapter also discusses the implications of conservation balances on the need for certain classes of constitutive equations for completing a model description.

It will become clear that in developing the conservation balances there are strong connections to other general process engineering topics such as transport phenomena, reaction kinetics and reaction engineering. The details of these are given in Chapter 4.

3.1. THERMODYNAMIC PRINCIPLES OF PROCESS SYSTEMS

The underlying principles on which all process modelling takes place are based on thermodynamic concepts [16,17]. Those thermodynamic concepts arise due to our current understanding about the nature of matter, confirmed by experiment and observation. In particular, the fundamental laws of physics provide us with a framework to describe the behaviour of systems through a consideration of the mass, energy and momentum of the system in 3D space. We have to consider the fourth dimension of time where we are also interested in the time varying behaviour of the system.

3.1.1. The Concept of a Space

For modelling purposes, thermodynamic properties are related to a defined region of 3D space which has an associated volume $\mathcal{V}$ and surface $\mathcal{F}$. This region is usually *a volume encapsulated by a closed surface which then defines the region of interest for the quantities of mass, energy and momentum.* It is sometimes termed the "balance volume" or "control volume". Associated with this volume is a volume surface, $\mathcal{F}$ which provides a connection to the surroundings or to other defined volumes of the process.

We can also classify the type of region by considering the character of its connections to its environment. We can identify three main regions of interest forming a system, namely,

- *Open systems*, where mass and energy can flow through the space of interest. These are the most common form of balance volume in PSE. Typical of these processes are continuous flow reactors, or liquid–gas phase separation systems.
- *Closed systems*, where there is no transfer of mass across the system boundaries but energy may transfer. We can readily identify many engineering systems where this is the situation. Typical of this system would be the batch reactor where there is no continuous addition of material during processing but temperature of the batch is controlled. Energy transfers to or from the contents.
- *Isolated systems*, where there is no transfer of energy or mass across the boundaries. Here the space under consideration is thermodynamically isolated from its surroundings. These are not common systems in the process industries.

Notice, however, that the volume $\mathcal{V}$ may change in time: think of a rising gas bubble in a liquid being the region of interest.

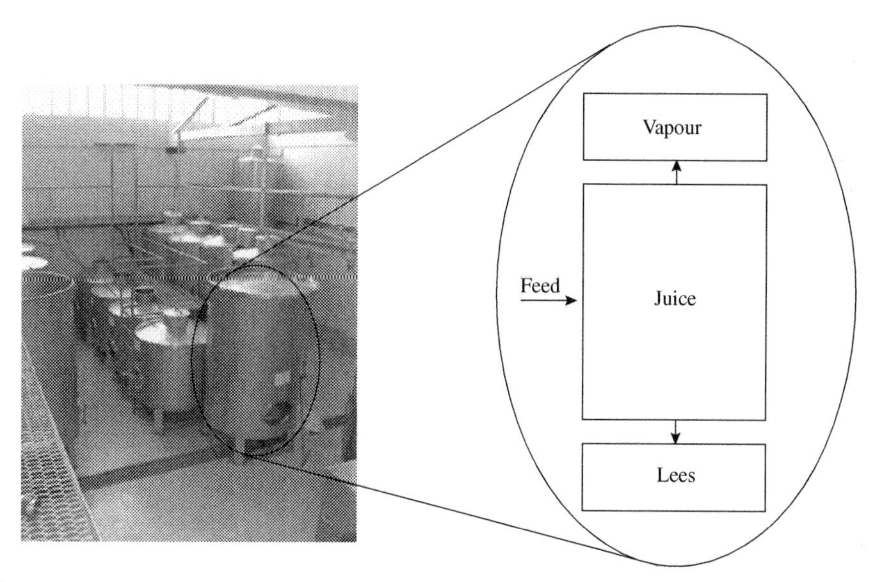

FIGURE 3.1 Fermentation tank and regions of interest (by permission of Stuart Range Estates).

Defining Regions of Interest

There are a number of ways in which we can define the regions of interest for our process system. These include:

- The physical equipment volumes, which define distinct regions in space which contain matter or energy. For example, the volume of a reactor in the case of mass or in the case of energy, the walls of the vessel.
- The separate phases or states of aggregation in which matter is identified. This includes gaseous, liquid and solid states of aggregation.

In addition to defining the region of interest, it will be necessary to have an associated local co-ordinate system with a defined origin, which allows us to locate a particular point in space and thereby the value of a state.

Note that in defining the spaces of interest we have defined a process system from the system theoretical viewpoint.

EXAMPLE 3.1.1 (Wine fermentation). In the case of grape juice fermentation, this is normally performed in stainless steel tanks similar to those shown in Fig. 3.1. Within the fermenter, we can identify three distinct phases of interest which may be used to develop a model based on conservation of mass and energy. These phases are vapour, juice and lees (solids). ■ ■ ■

Figure 3.2 shows that in some cases we can have a mix of lumped or distributed regions within the system of interest. We can also have levels of detail as seen in cases (a) and (b).

Another example of a somewhat more "flexible" region of interest is a bubble in a bulk liquid. Here, both the bubble volume and the surface associated to it may vary in time as the pressure and temperature in the bulk liquid around the bubble changes.

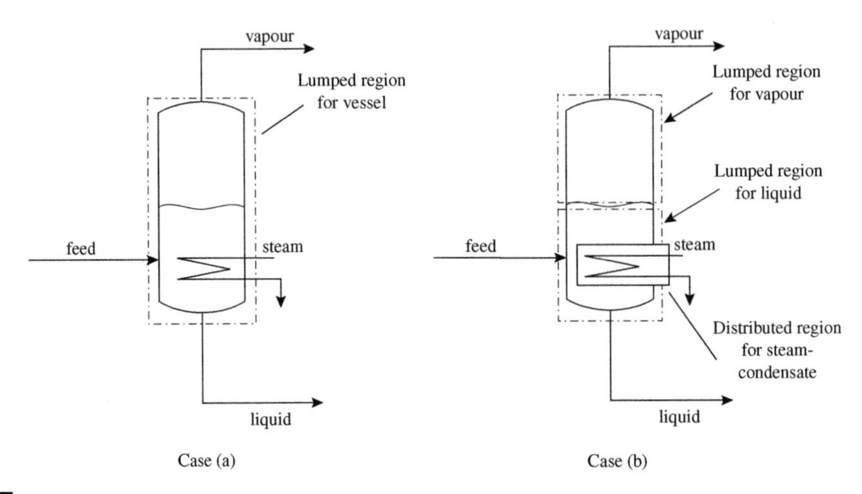

FIGURE 3.2 Possible balance volumes for a process evaporator.

Co-ordinate Systems in Space

A point in space can be located by a vector which originates from the origin of the co-ordinate system and terminates at the point. These co-ordinate systems can be defined in various ways. The most common systems in process engineering are given by:

- Cartesian or rectangular co-ordinates, where a point in space is given by the position along three co-ordinates: (x, y, z).
- Cylindrical co-ordinates, where a point is identified by a radial dimension, an angle of rotation and an axial distance: (r, θ, z).
- Spherical co-ordinates, where the point in space is given by a radial dimension, an angle of rotation and an angle of elevation: (r, θ, ϕ).

Figure 3.3 shows these basic co-ordinate systems and the definitions of the governing measures to locate a point in space.

The choice of co-ordinate system is largely dictated by the underlying geometry of the system being studied. For example, the volume represented by a tubular reactor can be easily considered in cylindrical co-ordinates, whereas the consideration of a slab of material is likely to be represented in Cartesian co-ordinates. Also, certain

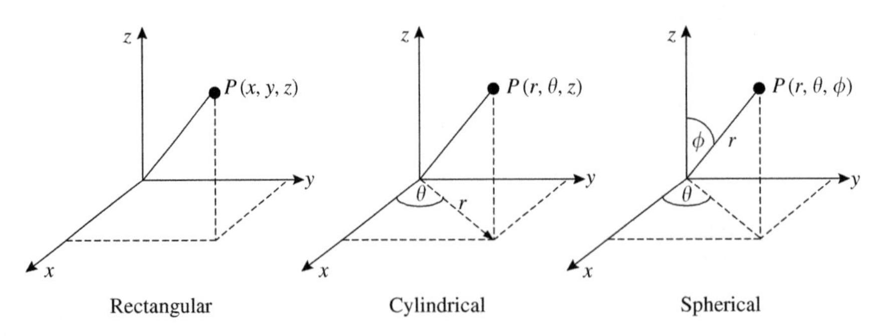

FIGURE 3.3 Principal co-ordinate systems.

properties of the system which are functions of several spatial coordinates in one system may only be a function of one co-ordinate in another system.

The two curvilinear co-ordinate systems of cylindrical and spherical geometry can be related to the popular rectangular co-ordinates given by x, y and z through the following relations:

Cylindrical co-ordinates:

$$x - r\cos(\theta), \quad y = r\sin(\theta), \quad z - z, \tag{3.1}$$

or

$$r = \sqrt{x^2 + y^2}, \quad \theta = \arctan\left(\frac{y}{x}\right), \quad z = z. \tag{3.2}$$

Spherical co-ordinates:

$$x = r\sin(\theta)\cos(\phi), \quad y = r\sin(\theta)\sin(\phi), \quad z = r\cos(\theta), \tag{3.3}$$

or

$$r = \sqrt{x^2 + y^2 + z^2}, \quad \theta = \arctan\left(\frac{\sqrt{x^2 + y^2}}{z}\right), \quad \phi = \arctan\left(\frac{y}{x}\right). \tag{3.4}$$

EXAMPLE 3.1.2 (Fixed bed reactor geometry). In Fig. 3.4, we see an industrial fixed bed catalytic reactor for the hydrogenation of an unsaturated hydrocarbon feed.

Reactants flow upwards through beds of nickel based catalysts. The reaction takes place throughout the length of the bed and the obvious co-ordinate system for application of mass and energy balances is cylindrical. It could be assumed that no variation of concentration and temperature occurs in the angular direction θ. Variations often occur in the radial direction, r.

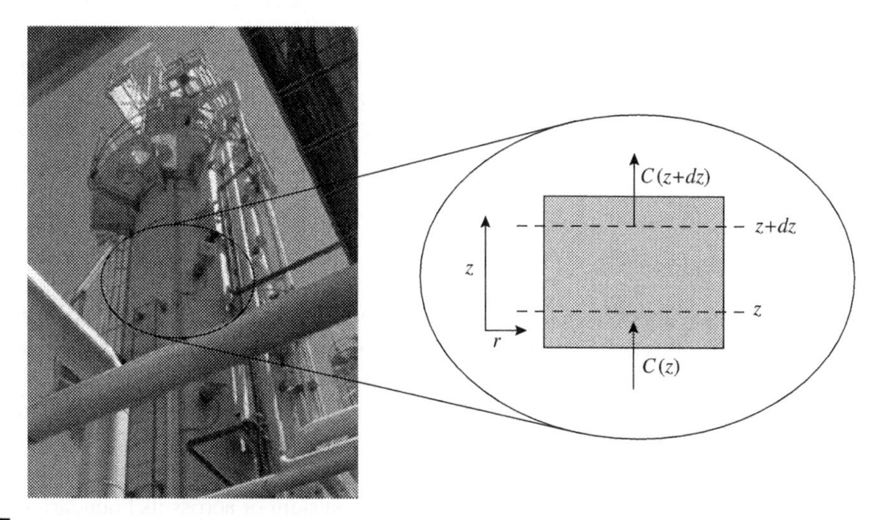

FIGURE 3.4 Fixed bed catalytic reactor (by permission of Suncor Energy).

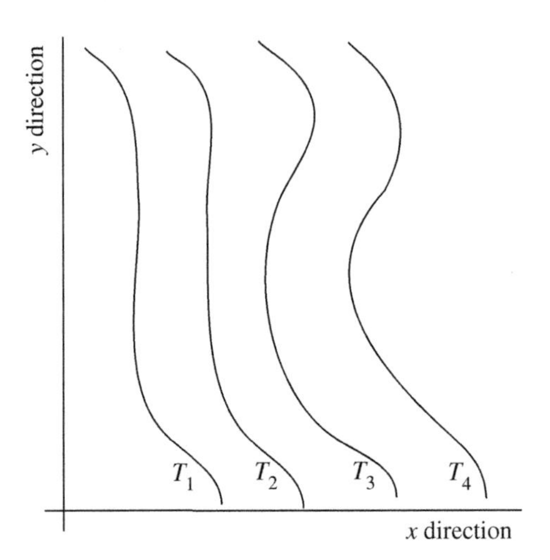

FIGURE 3.5 2D scalar temperature field.

3.1.2. Scalar Fields

Of great importance in thermodynamics is the idea of a scalar field. This is the value of a thermodynamic quantity at a given point within the region of interest. In this case, the value is a scalar with no directional property associated with the quantity.

For example, we can easily think about this in terms of the pressure within a specified volume, $\mathcal{V}$. In this case, we can define the pressure in terms of a Cartesian co-ordinate system as $P(x, y, z)$ or $P(r)$, where r is a vector, the position vector. Hence, for every point in the defined space we can associate a pressure value. We can also assume that this scalar quantity is time varying and hence write the scalar field as $P(x, y, z, t)$ or $P(r, t)$. The most common expression of such a time varying scalar field is seen in synoptic weather maps which show lines of constant atmospheric pressure or isobars. Clearly, each point in space can be assigned a specific scalar value of pressure.

Scalar fields are very common in process systems. Other examples of scalar fields include species concentration and system temperature. We can picture this in Fig. 3.5, where we see a scalar temperature field in a 2D space, such as the cross section of a vessel wall.

3.1.3. Vector Fields

Of equal importance is the concept of a vector field, where not only does the quantity in space possess a specific value but a direction is associated with that value. Hence, we can think of a vector field of fluid velocity $v(x, y, z, t)$ or $v(r, t)$ within our defined region in space. However, vector fields can be associated with the flow of mass and energy within the system. For example, heat flow can be regarded as a vector field and we can represent the variation in the flow of heat in a number of ways. Likewise, we can also regard mass transfer within a system or across its boundaries as a vector field.

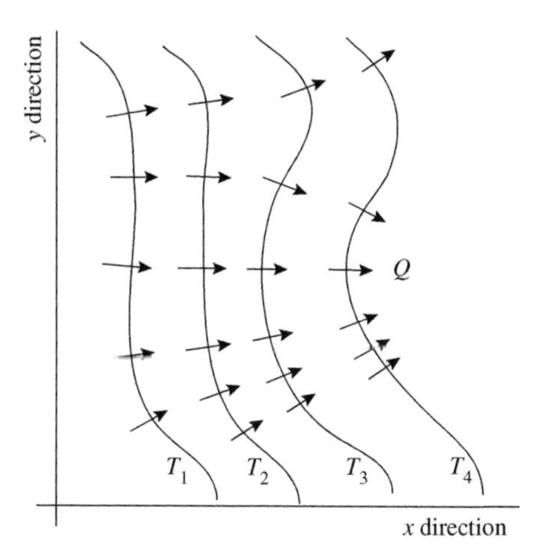

FIGURE 3.6 Vector field for heat flow.

The important point is that vector fields in process engineering are a result of differences or more precisely the gradients of an underlying scalar field. The underlying scalar field generates thermodynamical driving forces for the associated vector field.

Note, that the gradient of a scalar field $P(x, y, z)$ in terms of a Cartesian co-ordinate system is the following vector field:

$$\text{grad } P = \nabla P = \left(\frac{\partial P}{\partial x}, \frac{\partial P}{\partial y}, \frac{\partial P}{\partial z} \right). \tag{3.5}$$

How, we estimate the magnitude of that vector field will depend directly on the underlying scalar field which generates it. For example, we can relate heat flow to differences in the scalar temperature field. The vector field representing heat flux for the case shown previously in Fig. 3.5 is shown in Fig. 3.6.

3.1.4. State of Aggregation and Phases within Process Systems

The state of aggregation within a process system is related to the form of matter within the system. Typically, in process systems we identify three distinct states of aggregation, these being solid, liquid and gas. These states of aggregation show themselves as *phases* within a process. A phase can be regarded as a space where the local thermodynamical state is a continuous function of position. In a phase, there exists no discontinuities of the thermodynamical state within the space. This means that in the most general case the properties of a phase may vary as a function of both time and space. Where discontinuities occur, there is a change of phase and this represents a surface or boundary between the phases. In process systems, we can talk about a "dispersed phase", which represents material of a specific phase distributed within another phase. This could be liquid droplets dispersed in another liquid or in a gas as an aerosol.

EXAMPLE 3.1.3 Consider the case of two partially immiscible liquids and a vapour contained in a process system consisting of a vessel. For example, kerosene and water. Here, we could consider that there are three distinct phases within the space of interest for the case of modelling the system dynamics. We have two liquid phases (of different density) and a vapour phase. These phases can transfer both mass and energy between them depending on the thermodynamic driving forces. ■ ■ ■

In some instances, we can further refine our definition of a phase by imposing an equilibrium condition on the system. In this case, the phases consist of uniform states in both time and space.

3.1.5. Extensive and Intensive Thermodynamic Properties in Process Systems

One of the basic distinctions in process modelling is between extensive ($\mathcal{E}_i$) and intensive ($\mathcal{I}_i$) thermodynamic properties. *Extensive properties of a system are those properties which depend on the extent of the system.* As such, the value of $\mathcal{E}_i$ for the whole of the region of interest is the sum of the values for all the parts in the system. Such extensive properties include total mass and energy within the volume $\mathcal{V}$, since the mass or energy values for all the parts of the system must be added to get the whole.

Extensive properties are strictly additive, such that by composing a new system $\mathcal{S}$ by adding two systems $\mathcal{S}_1$ and $\mathcal{S}_2$, the extensive properties of $\mathcal{S}$ are the sum of the extensive properties of $\mathcal{S}_1$ and $\mathcal{S}_2$. Hence for property i we can write that

$$\mathcal{E}_i^{\mathcal{S}} = \mathcal{E}_i^{\mathcal{S}_1} + \mathcal{E}_i^{\mathcal{S}_2}. \tag{3.6}$$

There is also a *canonical set of extensive quantities* which is necessary and sufficient to characterize a process system from a thermodynamic viewpoint. The set can include total mass, volume, component masses, energy or enthalpy. The number of extensive quantities needed to define the thermodynamical state of a system and to determine all other macroscopic quantities depend on the individual system. The canonical set of extensive quantities can be viewed as thermodynamical state variables to the process system.

In contrast to extensive properties, *intensive properties do not depend on the extent of the system.* In this case, the value changes from point to point in the system. Typical intensive properties include temperature, pressure, densities and compositions. Intensive properties are quantities which are typically measured within a process system through the use of process sensors. As such they play an important role in process modelling.

Also, *intensive properties will equilibrate if we add two systems* $\mathcal{S}_1$ and $\mathcal{S}_2$. In this case, if we consider an intensive quantity $\mathcal{I}_i^{\mathcal{S}}$ for the system $\mathcal{S}$ and we have, for systems $\mathcal{S}_1$ and $\mathcal{S}_2$,

$$\mathcal{I}_i^{\mathcal{S}_1} < \mathcal{I}_i^{\mathcal{S}_2} \tag{3.7}$$

then we can state that

$$\mathcal{I}_i^{\mathcal{S}_1} < \mathcal{I}_i^{\mathcal{S}} < \mathcal{I}_i^{\mathcal{S}_2}. \tag{3.8}$$

In the case of a single component, single-phase system, an extensive property can be expressed in terms of the system mass (M) and two intensive properties as:

$$\mathcal{E}_i = Mf(\mathcal{I}_i, \mathcal{I}_j). \tag{3.9}$$

EXAMPLE 3.1.4 Consider the total internal energy (U) in a single component, single-phase system with mass M. Knowing the temperature and pressure of the system, the internal energy per unit mass (mass-specific internal energy) $\hat{U}$ can be calculated and hence the total internal energy for the system is given by

$$U = M\hat{U} = Mf(T, P),$$

where P is the pressure and T is the temperature of the system.

In the case of a multi-component system of n components, we can generalize the expression (3.9) of the extensive variable $\mathcal{E}_i$ by specifying in addition the $(n-1)$ independent composition values $x_1, x_2, \ldots, x_{n-1}$ to give

$$\mathcal{E}_i = Mf(x_1, x_2, \ldots, x_{n-1}, \mathcal{I}_i, \mathcal{I}_j). \tag{3.10}$$

We must also note that for a single component, single-phase system, specification of two intensive variables is sufficient to define any other intensive variable. Hence

$$\mathcal{I}_k = f(\mathcal{I}_i, \mathcal{I}_j), \tag{3.11}$$

or in the case of a multi-component system of n species, we need to specify in addition $(n-1)$ compositions to give

$$\mathcal{I}_k = f(x_1, x_2, \ldots, x_{n-1}, \mathcal{I}_i, \mathcal{I}_j). \tag{3.12}$$

In the majority of process system applications, *the principal intensive variables used for defining other intensive quantities are pressure P, temperature T and compositions* $(x_1, x_2, \ldots, x_{n-1})$. *This canonical set of intensive variables together with the total mass can be seen as a set of thermodynamical state variables.* Such is the basis for the many equations of state that can be defined.

EXAMPLE 3.1.5 Consider the typical definitions of key, pure component properties such as pressure P, mass-specific enthalpy $\hat{H}$ or mass-specific internal energy $\hat{U}$ with mass-specific volume $\hat{V} = 1/\rho$ with ρ being the density. We can write relations such as

$$P = f_1(T, \hat{V}), \quad \hat{H} = f_2(T, P), \quad \hat{U} = f_3(T, P).$$

Here, we can use two intensive variables to define another intensive variable.

There is clearly a direct link between intensive and extensive properties. An extensive property can be represented directly as an intensive property by dividing by the system mass:

$$\mathcal{I}_k = \frac{\mathcal{E}_k}{M}. \tag{3.13}$$

These are referred to as *mass-specific properties* and some have already been mentioned such as mass-specific volume, mass-specific enthalpy or mass-specific internal energy.

We may define other specific properties to an extensive property, for example volume-specific properties by dividing it by volume instead of mass.

As well as being the typical measured quantities, intensive properties are those used to express the degree of material and energy transfer between balance volumes or diffusive flow within balance volumes.

Material and energy flows or transfer rates are typically vector fields with an underlying scalar field of an intensive variable generating a thermodynamical driving force for them. *The underlying intensive variable of the flow or transfer rate of an extensive property is called the potential related to it.*

For example, heat flow is related to changes in the field value of temperature, whereas mass transfer is related to variations in chemical potential or concentrations. These are intensive properties. Therefore, we can say that *temperature is the potential to internal energy or enthalpy and chemical potential is the potential related to component mass.*

3.1.6. Phases and Independent Intensive Quantities

It has been mentioned that only two intensive quantities are needed to enable all other intensive quantities to be determined in a single phase, pure component system. These specifications are known as the thermodynamic *degrees of freedom* of the system. We can generalize this concept to any number of phases using the phase rule first formulated by Gibbs. For non-reacting systems where there are n components and π phases, the degrees of freedom (DOF) F (or independent intensive quantities) to be specified are

$$F = n + 2 - \pi. \tag{3.14}$$

In the case where there are ν independent reactions taking place, we have

$$F = n + 2 - \pi - \nu. \tag{3.15}$$

In some cases, this can be reduced further when we impose other constraints on the system.

EXAMPLE 3.1.6 Consider the case of a two component system, where there are two liquid phases and a vapour phase. What number of independent intensive variables can be specified for this system?

Butanol-water is such a system. Here,

$$F = n + 2 - \pi$$

with

$$n = 2 \quad \pi = 3,$$

so that

$$F = 2 + 2 - 3 = 1.$$

In this case, we can only specify one intensive quantity.

3.2. PRINCIPLE OF CONSERVATION

As a result of the first law of thermodynamics, energy is conserved within a system, although it may change its form. Also, both mass and momentum in the system will be conserved quantities in any space. In most process systems, we deal with open systems where mass, energy and momentum can flow across the boundary surface. As such, we can consider a space with volume $\mathcal{V}$ and boundary surface $\mathcal{F}$ as shown in Fig. 3.7

We can write a general conservation equation for an extensive system property Φ as

$$\left\{\begin{matrix}\text{net change}\\ \text{of quantity in time}\end{matrix}\right\} = \left\{\begin{matrix}\text{flow in}\\ \text{through boundary}\end{matrix}\right\} - \left\{\begin{matrix}\text{flow out}\\ \text{through boundary}\end{matrix}\right\} + \left\{\begin{matrix}\text{net}\\ \text{generation}\end{matrix}\right\} - \left\{\begin{matrix}\text{net}\\ \text{consumption}\end{matrix}\right\}.$$

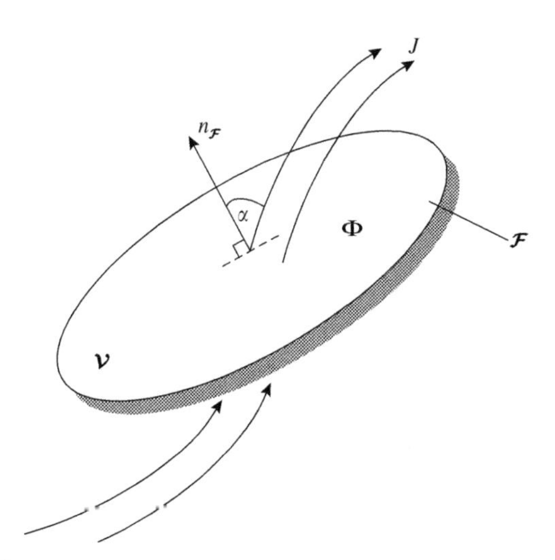

FIGURE 3.7 General balance volume $\mathcal{V}$.

3.2.1. Conservation Balances in Integral Form

In terms of a mathematical *balance in integral form*, using vector notation, we can write the above equation by considering the total extensive quantity Φ as the integral or sum of the volume-specific form of this property $\hat{\Phi}$ over the whole of the volume $\mathcal{V}$. This is written in time varying form as

$$\frac{\mathrm{d}}{\mathrm{d}t}\left\{\int_{\mathcal{V}} \widehat{\Phi(r,t)}\,\mathrm{d}v\right\} = -\oint_{\mathcal{F}} J(r,t)\cdot n_{\mathcal{F}}(r)\,\mathrm{d}f + \int_{\mathcal{V}} \widehat{q(r,t)}\,\mathrm{d}v. \tag{3.16}$$

The left-hand side represents the time variation of the extensive quantity Φ within the space $\mathcal{V}$. Note that the volume-specific form, $\hat{\Phi}$ of the extensive quantity Φ is both a function of position vector (r) and time (t).

The first term on the right-hand side accounts for the flow of the extensive quantity through the total closed boundary of the surface $\mathcal{F}$. It thus accounts for inflow and outflow from the volume under consideration. It consists of the value of the flow $J(r, t)$, which is a vector field varying from point to point on the surface $\mathcal{F}$, integrated over the whole closed surface. Here, the flow is regarded as passing through a unit area of the surface with an outward pointing unit (normal) vector $n_{\mathcal{F}}$ perpendicular to that surface. The normal vector is clearly a function of spatial position as we integrate over the complete surface. This term accounts for all possible modes of transfer through the closed surface. These could be convective or diffusive in nature.

The last term represents the net consumption or generation of the extensive quantity within the specified volume. This is typically related to energy or species sources or sinks associated with reactions in the system and is represented by the volume-specific source $\widehat{q(r, t)}$.

3.2.2. Conservation Balances in Differential Form

In the limit, as $\mathcal{V}$ becomes infinitesimally small, we can write the above equation *in differential form* after applying the Gauss divergence theorem to the surface integral to convert it to a volume integral. This is the usual form of the balance equation we see in most process modelling situations. We obtain

$$\frac{\partial \hat{\Phi}}{\partial t} = -\nabla \cdot J + \hat{q}, \tag{3.17}$$

where the divergence of J is given by

$$\nabla \cdot J = \text{div}(J) = \frac{\partial J_x}{\partial x} + \frac{\partial J_y}{\partial y} + \frac{\partial J_z}{\partial z}, \tag{3.18}$$

where J_x, J_y and J_z are the three components of the flow J in the direction of the principal co-ordinates. Note that the divergence of a vector field is a scalar field. This form of the balance equation is scalar in all terms.

Flow or Transport Terms

We have already considered in the general conservation equation that flows can enter and leave the volume $\mathcal{V}$. These flows can be of two general forms:

- convective flows, J_{C},
- diffusive or molecular flows, J_{D}.

Hence in the case of mass, the total flow J can be written as the sum of convective and molecular flows:

$$J = J_{\text{C}} + J_{\text{D}}, \tag{3.19}$$

where we can define the individual flow components as

$$J_{\text{C}} = \Phi(r, t)\, v(r, t), \tag{3.20}$$

$$J_{\text{D}} \cong -D \,\text{grad}\, \varphi(r, t) = -D\ \nabla\varphi(r, t). \tag{3.21}$$

Note that in the definition of the convective flow, the vector $v(r, t)$ is used to describe the convective flow pattern into or out of the system volume $\mathcal{V}$. This vector will typically be the velocity components of the flow in the appropriate co-ordinate system. The velocity components of the vector $v(r, t)$ can in principle be computed from the general transport equation for the system via a momentum conservation balance. In practice, however, simple flow patterns, such as plug flow or laminar flow are most often used which are stated as modelling assumptions.

Hence, for the overall mass balance in volume $\mathcal{V}$, we can write the differential conservation balance as

$$\frac{\partial \rho}{\partial t} = -\nabla \cdot (\rho v), \tag{3.22}$$

where ρ is the volume-specific mass and v is the velocity vector.

Note also, that the diffusive flow J_{D} is an approximate relation involving a diffusion coefficient D and the gradient of the potential $\varphi(r, t)$ which corresponds to the equivalent extensive property Φ. For example, when considering the diffusive molar flow of a species we can consider the driving force or potential to be the chemical potential of the species in the system. This is often reduced to the intensive quantity of molar concentration of the particular species. In the above relation for diffusive flow the cross terms (interaction terms) have been neglected. This can be vitally important in multi-component systems where the Stefan–Maxwell relations should be applied.

Hence, for the component i mass balance with no reaction we have

$$\frac{\partial \rho_i}{\partial t} = -\nabla \cdot (\rho_i v + j_i), \tag{3.23}$$

where ρ_i is the component i volume-specific mass and j_i is the diffusional flux vector.

Source Terms

The final term in the general conservation equation given by $\widehat{q(r, t)}$ is related to sources or sinks of mass, energy or momentum in the system. This term can arise in various ways including:

- Chemical reaction for component mass conservation, where species appear or are consumed due to reactions within the space of interest. This is an *internal* source term.
- Other *internal* source terms arise from energy dissipation or conversion plus compressibility or density changes.
- *External* energy sources include gravitational, electrical and magnetic fields as well as pressure fields.

It is clearly important to identify the key mechanisms in the space under consideration in order to capture the principal contributions to the source terms represented by $\widehat{q(r, t)}$. Note that the source term represented by $\widehat{q(r, t)}$ almost always depends on the potential $\varphi(r, t)$ which corresponds to the equivalent extensive properties characterizing our process system. Think of the dependence of the reaction rate on the temperature and compositions.

For the component i mass balance with chemical reaction we have

$$\frac{\partial \rho_i}{\partial t} = -\nabla \cdot (\rho_i v + j_i) + r_i, \tag{3.24}$$

where ρ_i is the component i volume-specific mass, j_i is the diffusional flux vector and r_i is the rate of generation or disappearance of species i, being equivalent to the source term $\widehat{q(r,t)}$.

It is possible to develop further conservation equations for total energy and momentum. In the case of conservation balances for vector quantities like momentum the balance equations can be quite complex . The interested reader is referred to the seminal work of Bird *et al.* [13] to obtain complete expressions in various co-ordinate systems.

Potential Variable Relations

As has been emphasized before, the potential φ which corresponds to the equivalent extensive property Φ is necessarily present in the conservation balance equations if we have either transfer or source terms or both. *Therefore, we need to add constitutive equations relating potentials to volume-specific extensive variables to make our process model complete.* These constitutive equations are called *potential variable relations* and are usually in the form of algebraic equations between intensive variables, like those shown in (3.12).

Particulate Systems

It is convenient in the context of conservation principles to also mention that discrete entities such as particle numbers are subject to conservation principles. This leads to the concept of "population balances". The concepts of conserved quantities in particulate systems has widespread application in minerals processing, food, pharmaceuticals and the chemical industries.

In the case of population based balances, the properties of the materials are characterized by population density functions (distribution functions) instead of a scalar quantity. To use this approach, the distribution function must contain information about the independent properties of the system. In this case, the particle co-ordinates are divided into external and internal co-ordinates where the external co-ordinates refer to the spatial distribution of the particles and the internal co-ordinates refer to the properties attached to the particles such as size, shape, age or chemical composition. The distribution function could be expressed as

$$n(t, x, y, z, \zeta_1, \zeta_2, \ldots, \zeta_m), \tag{3.25}$$

where m is the number of internal co-ordinates $\zeta_1, \zeta_2, \ldots, \zeta_m$ being used. In a manner similar to that for the integral form of the conservation balance we can write

$$\frac{\mathrm{d}}{\mathrm{d}t}\left\{\int_{\mathcal{V}} n(t,\overline{x})\,\mathrm{d}v\right\} = \int_{\mathcal{V}} (B(t,\overline{x}) - D(t,\overline{x}))\,\mathrm{d}v, \tag{3.26}$$

where B and D are birth and death terms for the system and $\overline{x} = [t, x, y, z, \zeta_1, \zeta_2, \ldots, \zeta_m]$.

Manipulation of the general balance equation leads to the microscopic population balance:

$$\frac{\partial n}{\partial t} = -\nabla \cdot (vn) + B - D, \tag{3.27}$$

where n is the number density function, being the number of particles per unit volume at time t. Other terms covering nucleation and breakage can also be included. Typically, the birth and death terms B, D are described by integral terms which include special kernel functions as seen in the following example.

EXAMPLE 3.2.1 (Granulation system modelling). In modelling the behaviour of granulation drums for the production of fertilizer, Adetayo *et al.* [18] used a population balance approach to predict the particle size distribution as a function of position and time. This involved the general population balance equation given by (3.27).

The general balance for the batch drum was given as

$$\frac{\partial n(v,t)}{\partial t} = B(v,t) - D(v,t), \tag{3.28}$$

where v is the particle volume, and the birth and death terms are given by

$$B(v,t) = \frac{1}{2N(t)} \int_0^{\infty} \beta(v-u,u)n(u,t)n(v-u,t)\,du, \tag{3.29}$$

$$D(v,t) = -\frac{1}{N(t)} \int_0^{v} \beta(v,u)n(u,t)n(v,t)\,du. \tag{3.30}$$

The coalescence kernel β can take various forms which reflect the understanding of the growth process. These include a continuous form suggested by Kapur [19]:

$$\beta(u,v) = \beta_0 \left[\frac{(u+v)^a}{uv^b}\right], \tag{3.31}$$

or a multi-stage, switched kernel given by Adetayo [20] as

$$\beta = \begin{cases} f(\text{moisture}) & t \le t_{s1} \\ f(u,v) & t > t_{s1} \end{cases} \tag{3.32}$$

switching at time t_{s1}.

The General Operator Form of the Conservation Equation in Differential Form

We can consider the general differential conservation expression given by Eq. (3.17) by expanding the terms as follows:

$$\frac{\partial \hat{\Phi}}{\partial t} + \nabla \cdot (J_C + J_D) = \hat{q}. \tag{3.33}$$

This can be further expanded by considering the form of the individual flow components in Eqs. (3.20) and (3.21) to obtain

$$\frac{\partial \hat{\Phi}}{\partial t} + \nabla \cdot \left(\widehat{\Phi(r,t)}\, v(r,t)\right) - \nabla \cdot (D\, \nabla \varphi(r,t)) = \hat{q}. \tag{3.34}$$

The above equation can be rearranged using that $D =$ constant to get

$$\frac{\partial \hat{\Phi}}{\partial t} = D\, \nabla \cdot (\nabla \varphi(r,t)) - \nabla \cdot \left(\hat{\Phi}(r,t)\, v(r,t)\right) + \hat{q}. \tag{3.35}$$

Further rearrangement applying the identities on the differential operators div and grad (both appear as ∇ in the equation) gives the final operator form of the conservation equation:

$$\frac{\partial \hat{\Phi}}{\partial t} = D\left(\nabla^2 \varphi(r,t)\right) - \nabla \cdot \left(\hat{\Phi}(r,t)\, v(r,t)\right) + \hat{q}. \tag{3.36}$$

Note that the above form is co-ordinate system independent. Its co-ordinate system dependent form is derived by expanding the differential operators in the equation in the co-ordinate system in question.

3.2.3. Conservation Equation in Rectangular (Cartesian) Co-ordinate System

We can consider the general differential conservation expression given by Eq. (3.36) in rectangular co-ordinates by expanding the differential operators in the corresponding terms as follows:

$$\frac{\partial \hat{\Phi}}{\partial t} = D\left(\frac{\partial^2 \varphi}{\partial x^2} + \frac{\partial^2 \varphi}{\partial y^2} + \frac{\partial^2 \varphi}{\partial z^2}\right) - \left(\frac{\partial \hat{\Phi} v_x}{\partial x} + \frac{\partial \hat{\Phi} v_y}{\partial y} + \frac{\partial \hat{\Phi} v_z}{\partial z}\right) + \hat{q}. \tag{3.37}$$

In Eq. (3.36), the term ∇^2 is termed the Laplacian operator, this being a scalar operator. The final conservation Eq. (3.37) is a scalar equation for the volume-specific form of the extensive quantity Φ, which includes the intensive potential counterpart φ arising from the diffusion term.

Properties of the Conservation Equations

This conservation equation possesses certain properties which are important in the modelling of process systems. First, Eq. (3.37) is a time-varying partial differential equation of parabolic form. It is generally nonlinear due to the form of the term $\hat{q}$ which is often a reaction rate term or heat transfer expression. It can, of course, be nonlinear even in the differential operator terms when the diffusion coefficient D is not a constant but a function of the coordinate system or the intensive variables.

As well, the conservation equations are normally coupled through the term $\hat{q}$ when both mass and energy conservation expressions are written. This is because the reaction term is normally a function of both concentration (mass intensive quantity) and temperature (energy intensive quantity) and these appear in both parabolic equations.

EXAMPLE 3.2.2 Consider the conservation equations for mass and energy for a reaction within a spherical catalyst pellet. Assume constant pressure and uniform distribution of the quantities along the angle coordinates φ and θ. The reaction rate expression is assumed to be in the form of

$$r = k_0 c \exp\left(\frac{E}{RT}\right) \tag{3.38}$$

with c being the molar concentration, T is the temperature and k_0, E, R are constants.

The component mass and energy conservation equations written for the volume-specific component mass (i.e. for the molar concentration) c and for the volume-specific internal energy and $\hat{U}$ are given by

$$\frac{\partial c}{\partial t} = \frac{1}{x^2}\frac{\partial}{\partial x}\left(x^2 \frac{\partial c}{\partial x}\right) - k_0 c \exp\left(\frac{E}{RT}\right), \quad (3.39)$$

$$\frac{\partial \hat{U}}{\partial t} = \frac{1}{x^2}\frac{\partial}{\partial x}\left(x^2 \frac{\partial \hat{U}}{\partial x}\right) - k_0 c \exp\left(\frac{E}{RT}\right) \Delta H_R, \quad (3.40)$$

where ΔH_R is the reaction enthalpy. Note that the convective term is missing because there is no convection within a pellet.

Observe the form of the diffusion term expressed in spherical co-ordinate system being the first term on the left-hand side of the equation.

Moreover, the potential relation for the temperature can be written as

$$\hat{U} = \mathrm{U}(T, c). \quad (3.41)$$

Clearly, the two conservation equations are closely coupled through the reaction term which appears in both equations. Both equations are parabolic type partial differential equations in nature.

Observe that we do not have any transfer term and the volume-specific source is identical to the reaction rate expression in this case. ■ ■ ■

3.2.4. Model Classes in Mathematical Terms

We can survey the types of mathematical models that arise from the application of conservation principles. Here, we can subdivide the problem into dynamic and steady-state models where either type can be distributed or lumped in nature. We can derive special forms of the general conservation equation (3.37) by applying mathematical transformations expressing our engineering assumptions of a system being dynamic or steady state, lumped or distributed parameter. In this manner, we can generate the various model equation forms which appear in Table 1.3.

EXAMPLE 3.2.3 (Distributed parameter static model form). Consider the general conservation equation (3.37) and assume, that there is no time variation in the system, i.e. the system is static. Then

$$\frac{\partial \hat{\Phi}}{\partial t} = 0;$$

We now get the following special form:

$$0 = D\left(\frac{\partial^2 \varphi}{\partial x^2} + \frac{\partial^2 \varphi}{\partial y^2} + \frac{\partial^2 \varphi}{\partial z^2}\right) - \left(\frac{\partial \hat{\Phi} v_x}{\partial x} + \frac{\partial \hat{\Phi} v_y}{\partial y} + \frac{\partial \hat{\Phi} v_z}{\partial z}\right) + \hat{q} \quad (3.42)$$

which is indeed an elliptic partial differential equation (EPDE) (cf. row 4 in Table 1.3). ■ ■ ■

More about the different process models in partial differential equation form will follow in Section 7.3.

The Induced Algebraic Equations

We have already mentioned before in Section 3.2.2 that the conservation calls for algebraic equations to make the formulated balance equation complete. These equations appear in the various classes of constitutive equations, such as

- *potential variable relations*—relating conserved extensive quantities and their associated potentials,
- *source terms*—relating the volume-specific source terms to the potential scalar fields of the system.

More about the constitutive equations will follow in Chapter 4.

3.3. BALANCE VOLUMES IN PROCESS SYSTEM APPLICATIONS

The role of balance volumes for mass, energy and momentum is crucial in process modelling. A balance volume is a basic element in process modelling as it determines the region in which the conserved quantity is contained. How are these regions chosen? Clearly, there are many possibilities. Figure 3.8 shows a simple evaporator vessel where liquid is being heated to produce a vapour. As can be seen, there are several options in the choice of the mass balance boundaries, given by the symbol $\sum_i^M$.

Balance Volumes and the Modelling Goal

Case (a) looks at developing the balances around the complete vessel, whereas case (b) considers each phase as a separate balance volume. In the second case, there will be mass and energy flows between the two balance volumes.

We could also consider other possibilities dependent on the degree of complexity demanded and on the characterizing variables called by the modelling goal. These could include mass balance volumes around bubbles within the liquid phase if mass transfer were an issue in multi-component systems.

In general, parts of a process system which are typical candidates for balance volume definition include those regions which:

- contain only one phase or pseudo-phase,
- can be assumed to be perfectly mixed, or
- can be assumed to have a uniform (homogeneous) flow pattern.

We use the term "pseudo-phase" for a region which we have coalesced for purposes of model simplification and the application of conservation balances. A pseudo-phase could constitute several distinct phases regarded as a single entity for modelling purposes.

Besides the thermodynamical and flow characteristics of the process system to be modelled the modelling goal can and will direct us to find the relevant set of balance volumes for the given modelling problem. The most common modelling goals were described and characterized in Section 2.1 in general terms. In a concrete modelling case *the modelling goal*, when properly stated, *implies*

- *The requested precision (accuracy) of the model output* thus determining the level of detail of the model. More about process models of the same system

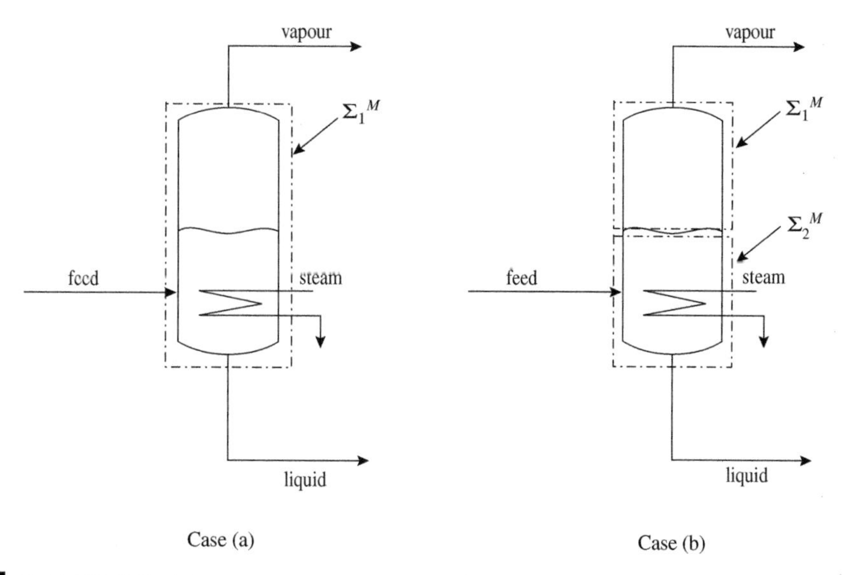

FIGURE 3.8 Mass and energy balance volumes for an evaporator.

with different level of detail is found in the chapter on model hierarchies (Chapter 9) in Section 9.1.

- The variables which should be present in the model equations to produce the model output to evaluate the modelling goal. These variables are called *characterizing variables*. Clearly, each variable is related to one or more balance volumes, therefore characterizing variables require some of the balance volumes to be present in the model.

The effect of the modelling goal on the selection of the balance volumes can be illustrated on our previous example of the simple evaporator vessel. If the modelling goal is to control the liquid level in the vessel then we should choose case (b), i.e. the case of the two balance volumes for the vapour and liquid phases separately.

Relation between the Balance Volumes

The *balance volumes* represent the elementary regions of a process system for which dynamic balances can be setup and assumptions can be made. In the case of systems with spatial variation in the conserved quantities, termed "distributed parameter" systems, we can define representative incremental volumes in which the conserved quantity is considered to be uniform for the purpose of developing the balance relations. It should also be noted that once balance volumes are defined for conservation of mass then the corresponding energy balance volumes are normally coincident with them. Mass balance volumes often dictate the maximal set of energy and momentum balance volumes. It is, however, feasible to have energy balance volumes which encompass several mass balance regions, or are additional to the mass balance volumes in the case of energy transfer through vessel walls. For example, in case (b) of Fig. 3.8, we can establish a single energy balance volume which encompasses the two mass balance volumes. There are significant implications in this type of definition. Let us formalize this concept.

A process system can be modelled based on a set of balance volumes, Σ. The full set consists of p subsets associated with the conserved quantities, such as overall and component masses, momentum and energy. Note that component mass balances are not independent of the overall mass balance, since the full set of component balances will sum to the total balance. This leads to problems of overspecification and subsequent numerical difficulties. All the mass balances, component and total balances should be setup over the same mass balance region. Therefore, we only consider a single balance volume set for all the mass balances.

We can consider three major conserved quantities (mass $(^M)$, energy $(^E)$, and momentum $(^{MM})$, i.e. $p = 3$). The full balance volume set Σ for a given modelling situation is then given by

$$\Sigma = \Sigma^M \cup \Sigma^E \cup \Sigma^{MM}, \tag{3.43}$$

where the individual boundary sets are

$$\Sigma^M = \left\{\Sigma_1^M, \ldots, \Sigma_M^M\right\} \quad \Sigma^E = \left\{\Sigma_1^E, \ldots, \Sigma_N^E\right\} \quad \Sigma^{MM} = \left\{\Sigma_1^{MM}, \ldots, \Sigma_P^{MM}\right\}. \tag{3.44}$$

As mentioned previously, it is important to note that energy balance volumes may include several full mass balance volumes, but none of the mass balance volumes contain more than one corresponding energy balance volume. Therefore, $N \leq M$ indicates that the mass balance volumes are primary balance volumes. There may, of course, be other independent energy balance volumes defined apart from those associated with mass holdups.

Furthermore, it is also important that the relation above puts restrictions on the corresponding energy balance volumes, if a new mass balance volume is formed by *coalescence* of several individual mass balance volumes. Coalescence in this context is the result of merging or fusing together distinct balance volumes into a single volume with homogeneous properties.

It is important to note that spatial and neighbouring relations can also be very important when formulating distributed parameter process models. These relations give relations between boundary conditions of the individual conservation equations over the balance volumes. More about this problem can be found in Chapters 7 and 9.

EXAMPLE 3.3.1 (Distillation column with trays). Figure 3.9 shows a typical industrial distillation column for hydrocarbon separations. The normal balance volume is shown as the plate p with various vapour and liquid flows arriving and leaving an individual stage. By aggregating all stages $1, \ldots, n$ in the column, a full model is developed.

In this case, the balance volume is a lumped region.

EXAMPLE 3.3.2 (Shell and tube heat exchanger). Figure 3.10 shows an industrial shell and tube heat exchanger and a typical balance volume for fluid in the tubes. This is a distributed parameter system with z representing the axial distance from the tube entry. Both the tube liquid T_L and coolant T_C are functions of time and position.

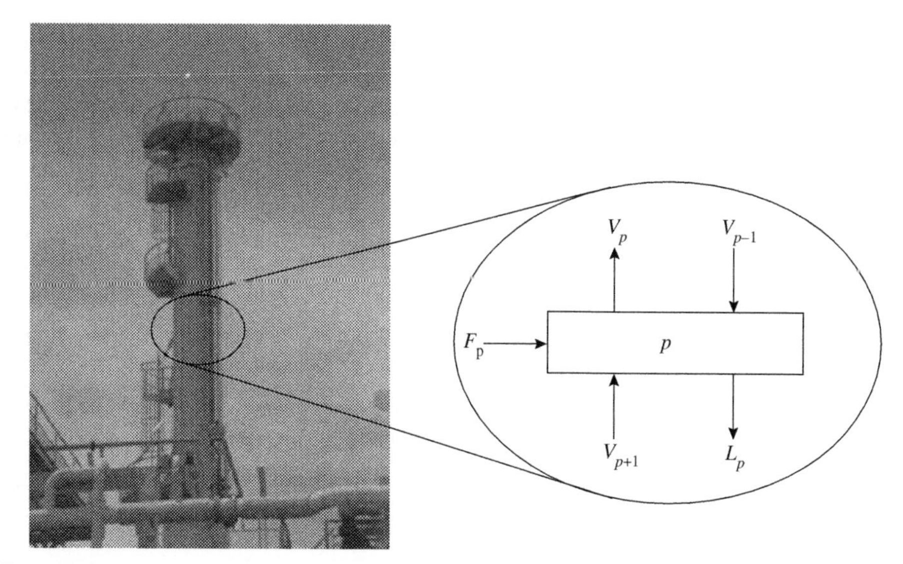

FIGURE 3.9 Industrial distillation column (by permission of Caltex).

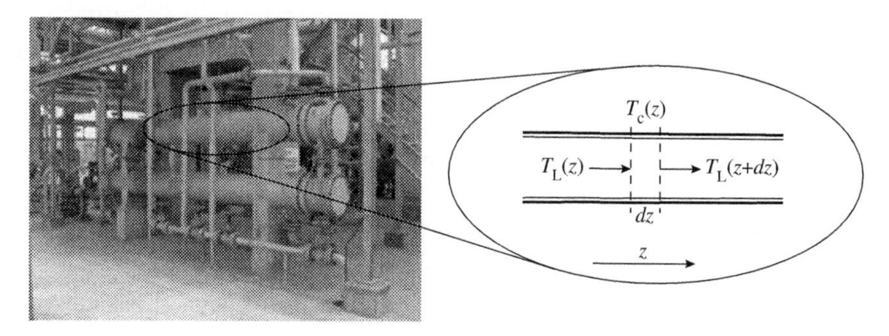

FIGURE 3.10 Industrial shell and tube heat exchanger (by permission of Caltex).

3.4. SUMMARY

In this chapter, we have dealt with the basic underlying principles of conservation applied to mass, energy and momentum. These principles form the basis for any modelling task. It was also shown that the quantities that are of interest can be classed as extensive and intensive. It is the extensive properties that constitute the role of conserved quantities whereas the intensive properties provide the basis for computing driving forces for material and energy transfer in the system.

The importance of the various co-ordinate systems in either rectangular, cylindrical or spherical geometry were introduced and their implications for modelling discussed. The final model form will be dependent on the assumed geometry of the system under study. This is a matter of judgement for the modeller but, generally, the physical or chemical phenomena will dictate the approach to be taken.

Finally, the definition of balance volumes in process engineering applications was considered, showing that there are indeed many ways of defining balances. Most, however, are dictated by the identification of system phases in which there is a condition of homogeneity. The next chapter introduces the concept of constitutive relations and definitions which are an important class of relations to help complete the full model description. Subsequent chapters then develop out the principles in particular application areas.

3.5. REVIEW QUESTIONS

Q3.1. Describe the difference between open, closed and isolated systems. What are process engineering examples of these systems? (Section 3.1)

Q3.2. What are the fundamental differences between scalar and vector fields. List examples of both types of fields? (Section 3.1)

Q3.3. Describe the role of intensive and extensive variables within a process system. What controls the independent specification of intensive variables for a particular system? (Section 3.1)

Q3.4. What role do balance volumes play in process modelling? Discuss ways in which they can be defined? (Section 3.3)

Q3.5. What role does the conservation principle play in modelling. In what two fundamental forms does the principle appear? Discuss the reasons that lead to these forms. (Section 3.2)

Q3.6. What are the principal terms that can appear within the conservation balance equation? What form do these terms take and how are they related to the physics and chemistry of the application? (Section 3.2)

Q3.7. Describe the role which conservation balances play in the mathematical modelling of process systems. (Section 3.3)

Q3.8. What are the key issues in deciding the location of balance volumes for a modelling application? (Section 3.3)

Q3.9. Why is it that energy balance volumes are normally coincident with mass balance volumes? Is there a natural hierarchy in defining balance volumes? (Section 3.3)

Q3.10. What are the implications of defining energy balance volumes which encompass several mass balance volumes? (Section 3.3)

3.6. APPLICATION EXERCISES

A3.1. For the case of the open tank given in Exercise A2.1 and illustrated in Fig. 2.8, discuss and define the likely balance volumes that could be used for the modelling problem. What are the principal characterizing variables given that we are interested in modelling the liquid level dynamics?

A3.2. Consider the balance volume definitions for the case of the closed tank problem dealt with in Exercise A2.2. How has this new process situation changed the potential balance volume definitions and what if anything is the relationship between balance volumes?

A3.3. In the case of the double-pipe heat exchanger discussed in Exercise A2.8, define and discuss how you would develop the key balance volumes for this piece of equipment. The equipment is shown in Fig. 2.15.

A3.4. The example of the absorption bed of Exercise A2.4 provides another interesting example of how we might define the balance volumes. Think about the possible balance volume definitions for this system and discuss the possible options that might be considered. The system is illustrated in Fig. 2.11.

A3.5. In Example A2.9 the spill of a hazardous solvent was considered. Discuss how you would define the balance volumes in such a scenario. Justify your development of the balance volumes.

A3.6. Consider the problem of the forced aeration of water in a wastewater pond. What are the possible balance volumes that could be defined for such a system? What simplifications could be applied to the balance volume definitions?

A3.7. A practical means of removing chemical species within retort water from the production of shale oil is to absorb the species onto spent shale from the retort (pyrolysis kiln). This can be done in large columns or using shale heaps. Consider how you would define the balance volumes for this application. Discuss the options available and the justification for them.

A3.8. Consider the possible balance volumes that could be defined in order to set up the conservation balances for a distillation column which uses conventional sieve trays as the separating devices. What could be the finest level of balance volume definition? What could be the most coarse? Discuss how the balance volume definitions could change when the column uses packing instead of conventional sieve trays. Discuss the characteristics of these balance volumes with regard to spatial variations.

A3.9. The production of wine is normally carried out in fermentation tanks over a period of days or weeks. Yeast reacts with the natural sugars in grape juice to produce alcohol, carbon dioxide and other compounds. Consider the initial definition of the balance volumes for such a modelling application. What simplifications could be made to the definitions and what justifications could be made for the decisions?

4 CONSTITUTIVE RELATIONS

Constitutive relations are usually algebraic equations of mixed origin. The underlying basic thermodynamic, kinetic, control and design knowledge is formulated and added to the conservation balances as constitutive equations. This is because when we write conservation balances for mass, energy and momentum, there will be terms in the equations which will require definition or calculation. These requirements give rise to a range of constitutive equations which are specific to the problem.

Constitutive equations describe five classes of relations in a model:

- *transfer rates*: mass transfer, heat/energy transfer;
- *reaction rate* expressions;
- *property relations*: thermodynamical constraints and relations, such as the dependence of thermodynamical properties on the thermodynamical state variables of temperature, pressure and composition. There are also equilibrium relations and equations of state;
- *balance volume relations*: relationships between the defined mass and energy balance regions;
- *equipment and control constraints*.

These constitutive equation classes are shown in Fig. 4.1.

4.1. TRANSFER RATE EQUATIONS

In many applications of the conservation principles for mass, energy and momentum there will be cases where mass, energy or momentum will be transferred into or out of

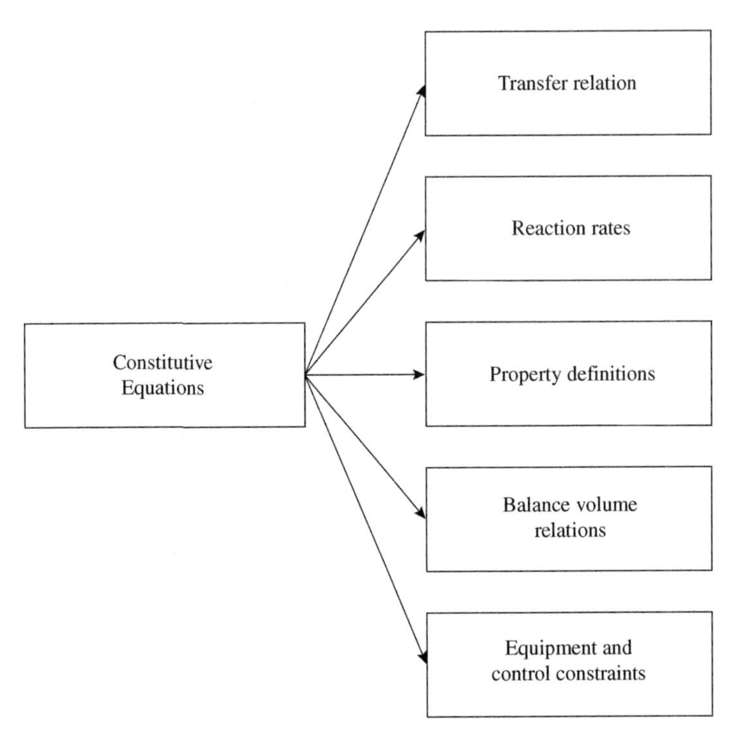

FIGURE 4.1 Classes of constitutive equations for process models.

our balance volumes. This is especially the case where we deal with interacting phases involving mass or energy. Here, there are driving forces at work which can transfer material and energy between the phases. These driving forces are generally in terms of the intensive properties of the system. They include concentration, temperature and pressure differentials but can be extended to electrical potentials and other driving forces such as chemical potential.

The generic form of the transfer equation is given by

$$\text{rate}_{\chi}^{(p,r)} = \psi^{(p,r)}(\zeta^{(p)} - \kappa^{(r)}), \tag{4.1}$$

where χ is related to the conserved quantity of either mass, energy or momentum which appears in the corresponding conservation equation for the balance volume (p). The mass, heat or momentum transfer can occur from another balance volume (r) or from within the same balance volume (p) if the scalar field of the intensive variable changes with spatial location. This is the situation for a distributed parameter system. An example would be the diffusive mass transfer which occurs for a pollutant in the atmosphere.

The variable $\zeta^{(p)}$ is an intensive property of the balance volume (p), whereas the variable $\kappa^{(r)}$ is either another intensive variable of the same type such as temperature or concentration or it is an intensive variable of another balance volume. It can also be a physico-chemical property related to the intensive variable $\zeta^{(p)}$ such as the saturated or equilibrium value $\zeta^{*(p)}$ of the quantity $\zeta^{(p)}$. This can be the case for mass transfer represented by the two-film model, where the driving force involves the bulk concentration and the concentration at the interface.

It is important to note that transfer can be in both directions. It can be to balance volume (p), or from balance volume (p) to another balance volume (r). This can be taken into account in the model. The driving force $(\zeta^{(p)} - \kappa^{(r)})$ is the same for both directions of transfer but the transfer coefficients, $\psi^{(p,r)}$ and $\psi^{(r,p)}$ may differ. Moreover, both transfer coefficients can be functions of the thermodynamical state variables in the two balance volumes, (p) and (r).

We now consider two specific instances of these types of interphase transfer.

4.1.1. Mass Transfer

Mass transfer involves a driving force in terms of chemical potential or equivalent intensive property such as concentration or partial pressure. The fundamental concept involves Fick's Law which describes the flux of a component (j) in terms of a diffusion coefficient D and a concentration gradient, $\mathrm{d}C/\mathrm{d}z$. It is given by

$$j = -D\frac{\mathrm{d}C}{\mathrm{d}z}, \tag{4.2}$$

where typical units are: j [kg · m^{-2} s^{-1}], D [m^2 s^{-1}], C [kg · m^{-3}], dz [m].

Typically, a two-film theory is proposed to describe the mass transfer process, where an interface (i) between the two phases is defined as well as bulk and interface concentrations for both phases. The films have theoretical thicknesses, d_{L} and d_{G}. This situation is seen in Fig. 4.2.

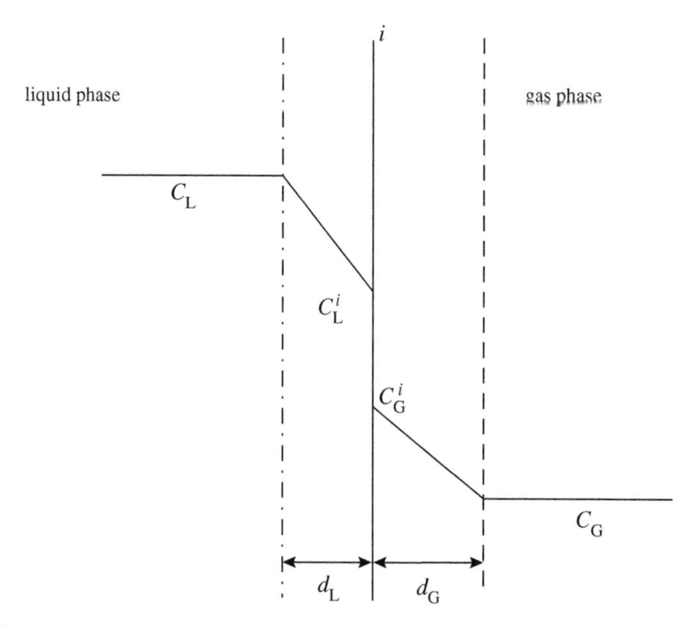

FIGURE 4.2 Two-film schematic for mass transfer.

The mass flux j can be written for each side of the film and equated assuming no accumulation at the interface to give

$$j = D_{\mathrm{L}} \frac{(C_{\mathrm{L}} - C_{\mathrm{L}}^{i})}{d_{\mathrm{L}}} = D_{\mathrm{G}} \frac{(C_{\mathrm{G}}^{i} - C_{\mathrm{G}})}{d_{\mathrm{G}}} \tag{4.3}$$

or in terms of film mass transfer coefficients:

$$j = k_{\mathrm{L}}(C_{\mathrm{L}} - C_{\mathrm{L}}^{i}) = k_{\mathrm{G}}(C_{\mathrm{G}}^{i} - C_{\mathrm{G}}), \tag{4.4}$$

where the mass transfer coefficients are defined as follows:

$$k_{\mathrm{L}} = \frac{D_{\mathrm{L}}}{d_{\mathrm{L}}}, \tag{4.5}$$

$$k_{\mathrm{G}} = \frac{D_{\mathrm{G}}}{d_{\mathrm{G}}}. \tag{4.6}$$

Sometimes the total mass flux, J is defined based on the available transfer area A [m^2], such that

$$J = jA, \tag{4.7}$$

or in terms of specific surface area a [m^2/m^3] such that

$$J = j \cdot (aV) \tag{4.8}$$

with V being the volume.

It is then possible to write Eq. (4.4) as a total flux equation:

$$J = k_{\mathrm{L}}a(C_{\mathrm{L}} - C_{\mathrm{L}}^{i})V = k_{\mathrm{G}}a(C_{\mathrm{G}}^{i} - C_{\mathrm{G}})V. \tag{4.9}$$

The combined terms of $k_{\mathrm{L}}a$ and $k_{\mathrm{G}}a$ are often those quoted in the mass transfer literature.

The problem with these equation forms is that the interface concentrations cannot be measured and the flux should be written in terms of measurable or easily computed quantities. To help achieve this, we can use the equilibrium relationship $C_{\mathrm{G}}^{*} = f(C_{\mathrm{L}})$ to write the mass transfer in terms of an overall mass transfer coefficient, K_{G}, giving for Eq. (4.4) the equivalent of:

$$j = k_{\mathrm{L}}(C_{\mathrm{L}} - C_{\mathrm{L}}^{i}) = k_{\mathrm{G}}(C_{\mathrm{G}}^{i} - C_{\mathrm{G}}) = K_{\mathrm{G}}(C_{\mathrm{G}}^{*} - C_{\mathrm{G}}) \tag{4.10}$$

with

$$\frac{1}{K_{\mathrm{G}}} = \frac{1}{k_{\mathrm{G}}} + \frac{m}{k_{\mathrm{L}}}. \tag{4.11}$$

Here, m is the slope of the equilibrium line and C_{G}^{*} is the gas phase concentration in equilibrium with the bulk liquid phase concentration C_{L}. Hence, most mass transfer can be written in terms of a concentration driving force and an overall mass transfer coefficient which is computed from correlations for the film mass transfer coefficients. Correlations for particular applications are found in [21] or standard mass transfer texts. Multi-component mass transfer is, in principle, more complex and correct approaches make use of the Stefan–Maxwell equation. There are several simplifications to the general approach which can make the modelling more tractable but at the expense of model accuracy. The reader is referred to the book by Taylor and Krishna [22] for a full discussion on this topic.

4.1.2. Heat Transfer

Heat transfer takes place through the three principal mechanisms: conduction, radiation and convection. Each mechanism can be represented by specific forms of constitutive equations. The following gives a brief outline of the principal heat transfer mechanisms and their mathematical representation.

Conduction

The rate of heat transfer by conduction is governed by the fundamental expression of Fourier given by

$$q_{\mathrm{CD}} = -kA\frac{\mathrm{d}T}{\mathrm{d}x} \tag{4.12}$$

which describes the energy transfer q_{CD} [$\mathrm{J \cdot s^{-1}}$ or W] in terms of a thermal conductivity, k [J/s · m · K], a heat transfer area, A [$\mathrm{m^2}$] and a temperature gradient $\mathrm{d}T/\mathrm{d}x$ [K/m]. This general heat transfer relation (4.12) can be integrated across the medium of conduction to give

$$\int_0^x q_{\mathrm{CD}}\,\mathrm{d}x = -kA\int_{T_1}^{T_2}\mathrm{d}T \tag{4.13}$$

or

$$q_{\mathrm{CD}} = \frac{kA}{x}(T_1 - T_2) = \frac{\Delta T}{x/(kA)}. \tag{4.14}$$

The term $x/(kA)$ in Eq. (4.14) is the thermal resistance to heat transfer and has units of K/W.

The form of the driving force will vary depending on the situation. For condensing steam with no subcooling of condensate the driving force is simply the difference between the steam temperature and the fluid temperature. For situations where the coolant or heating medium varies in temperature over the heat transfer area, an average or log mean temperature difference could be more useful. Standard texts on heat transfer such as Holman [23], Welty [24], and Incropera and DeWitt [25] give more details. References such as Coulson and Richardson [26] and McCabe *et al.* [27] discuss many issues related to heat transfer equipment.

Radiation

When heat is transferred from one body to another separated in space, then the key mechanism is radiative heat transfer. This is a common mechanism in high temperature process operations such as pyrometallurgical reactors or in steam boilers where flames radiate energy to heat transfer surfaces. Radiation occurs from all bodies and depends on the absolute temperature of the body and its surface properties. For a black body, the rate of emission of radiation is given by

$$q_{\mathrm{R_{BB}}} = \sigma A T^4, \tag{4.15}$$

where σ is the Stefan–Boltzman constant ($5.669 \times 10^{-8}\ \mathrm{W\,m^{-2}\,K^{-4}}$), A is the surface area [$\mathrm{m^2}$] and T is the black body absolute temperature [K]. Because real surfaces rarely act as true black bodies, the emissivity of the body (ε) is introduced to account for the reduced emission giving the grey body rate of radiation as

$$q_{R_{GB}} = \sigma A \varepsilon T^4. \tag{4.16}$$

For the general case of radiation between two black bodies we have

$$q_{R_{BB}} = \sigma A_1 F_{1-2}(T_1^4 - T_2^4), \tag{4.17}$$

where F_{1-2} is the geometric shape or "view" factor expressing the fraction of the total radiation leaving surface 1 which is intercepted by surface 2. In the case of two grey bodies, the situation is more complicated by the fact that absorptivity and emissivity are usually different for both bodies. This is captured by the use of a grey body shape factor $\mathcal{F}_{1-2}$ which incorporates both geometry and emissivity. This gives the two grey body radiation as

$$q_{R_{GB}} = \sigma A_1 \mathcal{F}_{1-2}(T_1^4 - T_2^4) \tag{4.18}$$

Calculation of shape factors for radiative heat transfer can be extremely complex, although for many common situations there are well established formulae and graphs in such references as Hottel [28] and Incropera and DeWitt [25].

Convection

Convective heat transfer occurs as a result of the transfer of energy between a moving gas or liquid phase and a solid phase. Typical examples include the energy transfer which occurs in process heat exchangers or in air cooled exchangers. Here, the mechanisms can be divided into natural convection and forced convection processes. In the first case, the fluid movement is induced by density differences created by temperature gradients. In the second case, the fluid movement is a result of mechanical motion induced by pumping action or driven by pressure differences. The rate of heat transfer is generally expressed as

$$q_{CV} = UA(\Delta T), \tag{4.19}$$

where an overall heat transfer coefficient U [W m^{-2} K] is used together with a temperature driving force ΔT [K] and a heat transfer area A [m^2]. This is a very common form of heat transfer expression for process applications. The computation of the overall heat transfer coefficient can be complicated depending on the complexity of the geometry and whether natural or forced convection occurs. Standard references such as Coulson and Richardson [29] give details of these calculations and also estimates of overall heat transfer coefficients.

4.2. REACTION KINETICS

Chemical reactions will give rise to the creation and consumption of chemical species within the mass balance volume. Where significant heat effects accompany chemical reaction such as the heats of reaction, these will have an impact on the related energy balance for the balance volume.

The reaction rate is a key concept in reaction engineering. It can be defined as the moles of component i appearing or being consumed per unit time and per unit volume. This can be expressed as

$$r_i = \frac{1}{V}\frac{dn_i}{dt}, \tag{4.20}$$

where n_i are the moles of species i.

Hence, the rate at which species are consumed or produced per time is simply

$$r_i V. \tag{4.21}$$

For a general reaction given by

$$\nu_X X + \nu_Y Y \rightarrow \nu_W W + \nu_Z Z \tag{4.22}$$

with stochiometric coefficients ν_i, the overall reaction rate $\hat{r}$ can be written in terms of the component reaction rates as

$$\hat{r} = \frac{1}{\nu_X V}\frac{dn_X}{dt} = \frac{1}{\nu_Y V}\frac{dn_Y}{dt} = \frac{1}{\nu_W V}\frac{dn_W}{dt} = \frac{1}{\nu_Z V}\frac{dn_Z}{dt} \tag{4.23}$$

remembering the convention that stochiometric coefficients of reactants are negative and those of products are positive. Hence, the reaction rates of any two individual species i, j are related by

$$r_j = r_i\left(\frac{\nu_i}{\nu_j}\right). \tag{4.24}$$

We should expect expressions such as Eq. (4.21) to appear in the conservation balances for component masses when modelling systems in which chemical reactions are taking place. They also appear indirectly in the energy balances as shown in Chapter 5.

In general, the reaction rate for species A will be a function of a reaction rate constant k_A and reactant species concentrations. Typical of this form would be:

$$r_A = k_A f(C_A^{\alpha}, C_B^{\beta}, \ldots), \tag{4.25}$$

where the reaction order is given by the sum of the coefficients $\alpha, \beta, \ldots$.

The reaction rate constant is generally temperature dependent and is written in the form of an Arrhenius expression:

$$k_A = k_0\,\mathrm{e}^{-E/(RT)}, \tag{4.26}$$

where E is the activation energy for the reaction, R is the gas constant and T is the system temperature.

For a first-order reaction where $A \rightarrow P$ we can write that

$$r_A = k_A C_A. \tag{4.27}$$

If the concentration is in $[\mathrm{mol/m^3}]$, then the units of k_A are typically $[1/\mathrm{s}]$.

4.3. THERMODYNAMICAL RELATIONS

Thermodynamic properties are absolutely essential to process systems modelling. In fact, they form the backbone of most successful simulations of process systems. Nearly all process based simulation systems have extensive physical properties packages attached to them. Typical of these are systems like ASPEN PLUS [30] or HYSIM/HYSYS [31]. As well, in stand-alone modelling and simulation applications we are often faced with the need to predict densities, viscosities, thermal conductivities or heat capacities for liquids, vapours and mixtures. In most cases, we will use simple correlations applicable to the range of validity of the model. In other cases, we might resort to the calculation of the property via a physical properties package like PPDS [32] or DIPPR [33].

4.3.1. Property Relations

Property relations can range from simple linear expressions to complex polynomials, in terms of the principal intensive thermodynamic properties of the system such as temperature, pressure and composition. The acceptability of the polynomial representations depends heavily on the range of use of these correlations. For control studies around a specific operating point, it may be possible to use a simple linear relationship in terms of temperature. Where the range of application is broad, then more complex polynomial forms might be justified. In general, we have physical properties for mixtures of the following kind:

Liquid density: $\rho_L = f(P, T, x_i);$ (4.28)

Vapour density: $\rho_V = f(P, T, y_i);$ (4.29)

Liquid enthalpy: $h = f(P, T, x_i);$ (4.30)

Vapour enthalpy: $H = f(P, T, y_i);$ (4.31)

where P, T are system pressure and temperature and x_i, y_i are mole or mass fractions in the liquid and vapour phase, respectively. The physical properties above are for mixtures, with the mixture properties being estimated from pure component data. For example, if we have a property Φ for a mixture of n components, then this can generally be estimated from the pure component properties $\phi_j (j = 1, \ldots, n)$ via a mixture rule. The most common is

$$\Phi = \sum_{j=1}^{n} x_j \phi_j. \tag{4.32}$$

In many cases, we can use simplified forms of the physical properties. In the case of enthalpy, we could use the following simplified forms for liquid and vapour phases where the reference state enthalpy $h(T_R)$ has been set to zero in the general expression:

$$h(T) = h(T_R) + \int_{T_R}^{T} c_p(T)\, dT. \tag{4.33}$$

This then leads to the following simplifications when the heat capacity c_p is constant.

Linear-in-temperature form:

$$h(T) = c_p T, \qquad H(T) = c_p T + \lambda_{\text{vap}}, \tag{4.34}$$

where c_p is the liquid heat capacity [J/kg K] and λ_{vap} is the heat of vapourization [J/kg]. Where the temperature change is significant, we can take this into account through use of a temperature dependent heat capacity relationship, giving

Nonlinear-in-temperature form:

$$c_p(T) = a_0 + a_1 T + a_2 T^2 + a_3 T^3 + \cdots, \tag{4.35}$$

and therefore

$$h(T) = h(T_0) + \int_{T_0}^{T} c_p(T)\,\mathrm{d}T, \tag{4.36}$$

$$h(T) = h(T_0) + \left[a_0 T + a_1 \frac{1}{2} T^2 + a_2 \frac{1}{3} T^3 + a_3 \frac{1}{4} T^4 + \cdots \right]_{T_0}^{T}. \tag{4.37}$$

In the case where there are significant variations from ideality due to high pressure operations or the nature of the components in the system, then the next option is to use a property prediction, which incorporates some form of correction term for the non-ideality. This generally relies on an equation of state. This is dealt with in Section 4.3.3.

4.3.2. Equilibrium Relations

Phase equilibrium is a common assumption used in process systems modelling. In its most fundamental form the condition relates to the equivalence of temperature, pressure and Gibbs free energy of a species in the phases considered to be in equilibrium. Thus,

$$\begin{aligned} T^{(1)} &= T^{(2)} = \cdots = T^{(p)}, \\ P^{(1)} &= P^{(2)} = \cdots = P^{(p)}, \\ \widetilde{G}_i^{(1)} &= \widetilde{G}_i^{(2)} = \cdots = \widetilde{G}_i^{(p)} \quad \text{for all species } i. \end{aligned} \tag{4.38}$$

These are general conditions, whether we are dealing with solid, liquid or gas phases. In the case of vapour–liquid systems which dominate many applications, the liquid and vapour states are often related through such estimations as bubble point and dew point calculations in order to compute the equilibrium concentrations, pressures or temperatures. Certain assumptions need to be made about the ideality of both vapour and liquid phase behaviour. It is quite common to assume ideal behaviour in both phases, which leads to the following models.

Raoult's law model:

Where the system pressure and vapour composition are given by

$$P = \sum_{j=1}^{n} x_j P_j^{\text{vap}}, \qquad y_j = \frac{x_j P_j^{\text{vap}}}{P}, \tag{4.39}$$

with the vapour pressure of component j given by a form of the Antoine equation:

$$\ln(P_j^{\text{vap}}) = A_j - \frac{B_j}{T + C_j}. \tag{4.40}$$

The coefficients of the Antoine equations are listed in standard books such as [34] or [29].

Another means of representing phase equilibrium is the use of relative volatility α_{ij}.

Relative volatility model:

$$\alpha_{ij} = \frac{y_i/x_i}{y_j/x_j} \tag{4.41}$$

and in the case of a binary system with constant α the relationship between vapour and liquid compositions can be expressed as:

$$y_i = \frac{\alpha_{ij}x_i}{1 + (\alpha_{ij} - 1)x_i}. \tag{4.42}$$

Using these expressions, which represent constitutive relations, phase equilibrium can be described.

One of the popular forms of phase equilibrium representation is the use of K values which are defined as follows.

K-value model:

$$K_j = \frac{y_j}{x_j}. \tag{4.43}$$

Hence, relative volatility can be expressed as $\alpha_{ij} = K_i/K_j$.

Where non-ideal liquid phase behaviour is evident, e.g. where there are mixtures of distinctly different chemical species (hydrocarbons and alcohols), non-ideal liquid phase equations need to be used when considering phase equilibrium. These involve the use of activity coefficients $\gamma_j,\ j = 1, \ldots, n$

Activity coefficient models:

$$P = \sum_{j=1}^{n} x_j P_j^{\text{vap}} \gamma_j \tag{4.44}$$

and

$$\gamma_j = f(T, P, x_i). \tag{4.45}$$

The individual circumstances will dictate the appropriate form by which phase equilibrium is represented in the form of a constitutive relation. Many types of activity coefficient models exist and are discussed in Walas [35] and Sandler [36].

4.3.3. Equations of State

Equations of state are used to relate pressure, volume and temperature for both systems of a pure component and also for multicomponent mixtures. There are many equations of state in the thermodynamic literature. The starting point is often the use of the simplest form, namely,

Ideal gas equation:

$$PV = nRT, \tag{4.46}$$

where P is system pressure [kPa], V the system volume [m^3], n is the number of moles [kg mol], T the temperature (K) and R is the gas constant with the value 8.314 [kPa m^3/kg mol K].

Many other equations of state can be used depending on the characteristics of the system under study. Amongst these are the Redlich–Kwong, Peng–Robinson and Soave–Redlich–Kwong equations. Most rely on a knowledge of the critical properties of the pure components and the acentric factor, which is related to the vapour pressure of the substance. Complex mixing rules can apply for multi-component mixtures (see in [35,36]).

4.4. BALANCE VOLUME RELATIONS

In some circumstances where mass balances are performed over phases and hence, balance volumes which when combined are restricted by a vessel or other constraints, then there are direct relations between the balance volumes. An instance of this is when a vapour and liquid phase are constrained to a process vessel of fixed volume. The two balance volumes must always sum to give the total, fixed vessel volume. This may ultimately determine certain intensive system properties such as the gas phase pressure.

4.5. EQUIPMENT AND CONTROL RELATIONS

In many applications, especially those in control studies, it is necessary to model the behaviour of the various elements in the control loops on the process. In fact the fidelity of the control system modelling can be crucial in determining the real dynamic behaviour of the system despite the best efforts in getting the underlying process model to be accurate. Typically, this will involve:

- ***Primary sensors*** *(thermocouples, resistance temperature detectors (RTD) or thermometers for temperature; orifice plates, turbine, magnetic flow meters for flow and differential pressure cells for pressure or level).*
- ***Transmitters*** *(for flow, temperature, pressure, etc.)*
- ***Controllers*** *(typically proportional (P), proportional-integral (PI) or proportional-integral-derivative (PID) forms or model based controllers)*
- ***Signal conditioning*** *(ensuring signals are normalized for processing (based on range and zero of instruments))*
- ***Final control elements*** *(normally control valves and the diaphragms or motors that drive the valve stem, plus the basic valve characteristic (proportional, equal percentage, quick opening)).*

We look at each of these in turn.

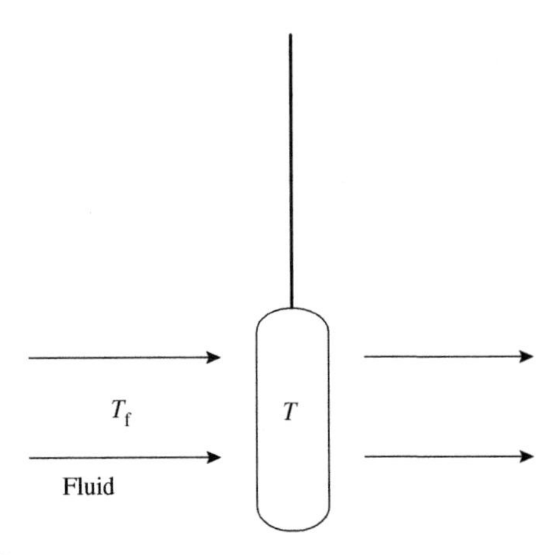

FIGURE 4.3 Sensor in a fluid field.

4.5.1. Sensors

Sensors are the key measurement elements in a process control loop. Typical of these are temperature elements where an element of mass M is usually immersed into the fluid requiring measurement. Figure 4.3 shows the situation of a sensor in a surrounding fluid.

The sensor temperature (T) dynamics are different from the fluid temperature T_f it is seeking to measure. To describe these dynamics we can perform an energy balance over the volume defined by the element to get

$$\frac{dU}{dt} = Q, \qquad \frac{dU}{dt} = \widetilde{U}A(T_f - T), \tag{4.47}$$

where U is the element internal energy [J], $\widetilde{U}$ is the overall heat transfer coefficient from the fluid to the sensor [W m^{-2} K], A is the element surface area [m^2].

We can convert the extensive form of the energy balance to the intensive form in terms of temperature by assuming that internal energy is equal to enthalpy ($H = U + PV$) and that the heat capacity c_p is constant. We then obtain the following relations:

$$\frac{d(Mc_pT)}{dt} = \widetilde{U}A(T_f - T), \qquad \frac{dT}{dt} = \frac{\widetilde{U}A}{Mc_p}(T_f - T), \tag{4.48}$$

where c_p is the sensor heat capacity [J/kgK].

We can write the sensor dynamics in terms of a time constant (τ) for the sensor as

$$\frac{dT}{dt} = \frac{(T_f - T)}{\tau}, \tag{4.49}$$

where the time constant τ is

$$\frac{Mc_p}{\widetilde{U}A}. \tag{4.50}$$

This typically has a value expressed in seconds and it is relatively straightforward to estimate the time constant from a simple step test on the sensor.

It can be easily seen that the Laplace transform of Eq. (4.49) leads to

$$sT = \frac{T_\mathrm{f} - T}{\tau} \tag{4.51}$$

giving the sensor transfer function:

$$\frac{T(s)}{T_\mathrm{f}(s)} = \frac{1}{\tau s + 1} \tag{4.52}$$

or a typical first-order response.

4.5.2. Transmitters

Transmitters convert the sensor signal into a control signal for use by recorders, indicators or controllers. Typically, the sensor signal is in millivolts, a pressure differential value or other appropriate unit. This is converted into an air pressure signal (20–100 kPa) or current (4–20 mA). Hence, in modelling these elements we need an algebraic relation which describes the output signal as a function of the sensor signal. This is dependent on the transmitter zero and its measurement span. In many cases, this is simply a linear relationship:

$$O_p = O_{p_\mathrm{min}} + (I_p - z_0)G, \tag{4.53}$$

where O_p is the output signal (e.g. mA), O_{p_min} the minimum transmitter output signal, I_p the input value (e.g. 200°C), z_0 the minimum input value (e.g. 100°C), G the transmitter gain ($(O_{p_\mathrm{max}} - O_{p_\mathrm{min}})/span$) and *span* is ($I_{p_\mathrm{max}} - z_0$).

In some cases, it is possible to allow a time lag by imposing a first-order response as seen in the previous section.

4.5.3. Actuators

Control valve actuators respond to the controller output signal (I) and cause the valve stem to be positioned to achieve a particular flowrate. These elements typically behave as second-order systems of the form:

$$\tau^2 \frac{\mathrm{d}^2 \mathcal{S}}{\mathrm{d}t^2} + 2\xi\tau \frac{\mathrm{d}\mathcal{S}}{\mathrm{d}t} + \mathcal{S} = G_a \cdot I \tag{4.54}$$

or as the second-order differential equation:

$$\frac{\mathrm{d}^2 \mathcal{S}}{\mathrm{d}t^2} = \frac{(G_\mathrm{a} \cdot I - \mathcal{S})}{\tau^2} - 2\frac{\xi}{\tau}\frac{\mathrm{d}\mathcal{S}}{\mathrm{d}t}, \tag{4.55}$$

where $\mathcal{S}$ is the stem movement (usually from 0 to 1), G_a the actuator gain and τ and ξ are the time constant and damping factor for the actuator, respectively.

To solve this second-order ODE, we need to convert it to two first-order equations. This is simply done by letting

$$\frac{\mathrm{d}\mathcal{S}}{\mathrm{d}t} = w; \tag{4.56}$$

then after substituting into the original equation (4.55), we obtain the other ODE:

$$\frac{\mathrm{d}w}{\mathrm{d}t} = \frac{(G_a \cdot I - \mathcal{S})}{\tau^2} - 2\frac{\xi}{\tau}w. \tag{4.57}$$

These last two equations can be used to model the behaviour of an actuator for a control valve.

4.5.4. Control Valves

Control valves are the final control element in any control loop. A simple form of relationship describing their flowrate behaviour is

$$F = C_{\mathrm{V}} c(\mathcal{S})(\Delta P)^{0.5}, \tag{4.58}$$

where the characteristic of the valve $c(\mathcal{S})$ is a function of the stem position $\mathcal{S}$. This is typically from 0 to 1. The valve coefficient C_{V} will be dependent on the particular valve and manufacturer. The pressure drop is the other term of importance. The following valve characteristics are the most commonly used:

$$\begin{aligned} c(\mathcal{S}) &= \mathcal{S} \quad \text{(linear)}, \\ c(\mathcal{S}) &= a^{\mathcal{S}-1} \quad \text{(equal percentage)}, \\ c(\mathcal{S}) &= \sqrt{\mathcal{S}} \quad \text{(square root)}. \end{aligned}$$

In a static valve a similar relationship can be used, typically of the form:

$$F = C_{\mathrm{V}}(\Delta P)^{0.5}. \tag{4.59}$$

This is normally adequate for most applications.

4.5.5. Controllers

The controllers that can be used in a process modelling exercise can be very simple P, PI or PID controllers up to complex model based control strategies such as Generic Model Control which uses a nonlinear model of the process inside the controller. Here, we review the simplest conventional controllers.

Proportional (P) Control

The proportional controller has the form

$$O_{\mathrm{C}} = B + K_{\mathrm{C}}(S_p - O_p) = B + K_{\mathrm{C}}\varepsilon, \tag{4.60}$$

where B is the bias signal of the controller, K_{C} the controller gain (can be negative), S_p the controller setpoint, O_p the signal from the transmitter and ε the error between setpoint and input signal.

Proportional-Integral (PI) Control

This controller adds an integral action term and has the form:

$$O_C = B + K_C\varepsilon + \frac{K_C}{\tau_I}\int \varepsilon \, dt, \tag{4.61}$$

where τ_I is the integral time (usually set in minutes).

Proportional-Integral-Derivative (PID) Control

Here, we add another term involving the derivative of the error to get:

$$O_C = B + K_C\varepsilon + \frac{K_C}{\tau_I}\int \varepsilon \, dt + K_C\tau_D\frac{d\varepsilon}{dt}, \tag{4.62}$$

where τ_D is the derivative time (minutes).

These expressions for the controllers can be used to compute the output signal to the final control element. In the case of the PI and PID controllers, it will be necessary to convert the integral term into a differential equation to allow it to be easily computed. Hence, if we define:

$$\mathcal{I} = \int \varepsilon \, dt, \tag{4.63}$$

then,

$$\frac{d\mathcal{I}}{dt} = \varepsilon, \tag{4.64}$$

and the controller consists of the differential equation (4.64) plus the original controller equation with the appropriate substituted term for the integral of the error.

4.6. SUMMARY

This chapter has introduced the concept of constitutive equations in a model. These are typically nonlinear algebraic relations which help complete the final model. In this chapter, they have been categorized into five classes for convenience. The constitutive equations naturally arise to help define unknown terms on the right-hand side of the conservation balances. In many cases, one constitutive equation will spawn other equations until a full model description is obtained. When this occurs, it is then possible to carry our model analysis to check the consistency of the model equations as well as the issue of degrees of freedom. These matters are considered in Chapters 5 and 6 which sets out in more detail the development, analysis and solution of lumped parameter process models.

4.7. REVIEW QUESTIONS

Q4.1. Describe the need for constitutive equations in process modelling. (Section 4.0)

Q4.2. How is the need for constitutive equations recognized at the early stages of model development? (Section 4.0)

Q4.3. What are the principal classes of constitutive equations in the mathematical modelling of process systems? (Section 4.0)

Q4.4. Describe the typical form of mass and heat transfer relations and identify difficulties in applying them. (Section 4.1)

Q4.5. Give the typical form of a reaction rate expression describing the terms and stating the common units of measurement used. (Section 4.2)

Q4.6. How are physico-chemical properties often represented in process modelling? What alternatives are there for calculation of these properties? What difficulties can arise in using the various approaches? (Section 4.3)

Q4.7. Describe the elements in a standard control loop. What are the characteristics of each? Which elements are dynamic or steady state relations? (Section 4.5)

Q4.8. Describe the dynamics of a sensor element. What are the important dynamic features that require representation or measurement? (Section 4.5)

Q4.9. Outline the constitutive equations that describe the behaviour of the main controllers designated P, PI and PID. (Section 4.5)

Q4.10. How would you represent the behaviour of a control valve in a control loop? What about a static valve? (Section 4.5)

4.8. APPLICATION EXERCISES

A4.1. Consider the dynamics of an enclosed tank which has a varying liquid level as flow in and out change. Assume that the outlet pressure of the pump changes, thus affecting the inlet flowrate to the tank. This is shown in Fig. 4.4.

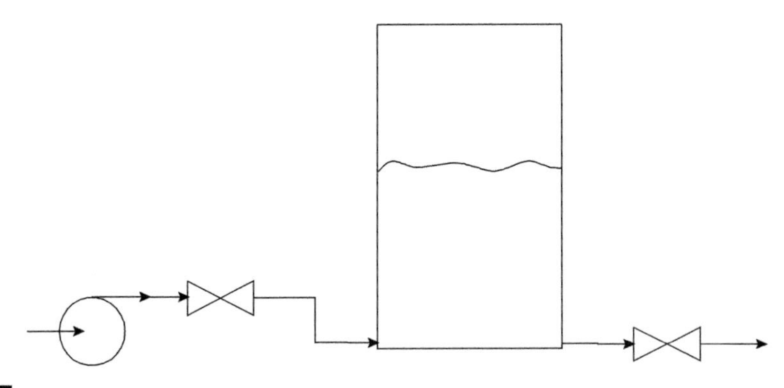

FIGURE 4.4 Enclosed tank.

(a) Develop a mathematical model of the system assuming that there are no dynamics in the vapour phase.
(b) Consider how the model changes if it is assumed that temperature in the vapour space is an important system variable.

A4.2. Consider the multi-component reactor shown in Fig. 4.5.

The CSTR carries out a complex multi-component reaction sequence given by

$$A + B \rightarrow X, \quad B + X \rightarrow Y, \quad B + Y \rightarrow Z. \tag{4.65}$$

The reactions have rate constants of k_1, k_2 and k_3.

(a) Given a fresh feed consisting of A and B, with inlet concentrations of C_{A_0}, C_{B_0}, develop a mathematical model which describes the process dynamics assuming that the reactor operates isothermally.

(b) If the reactions are exothermic with heat of reaction ΔH_R, develop the model for the non-isothermal case.

(c) Develop a further enhancement of the model where a cooling water coil of surface area A is inserted into the reactor to control the reactor temperature to a nominated setpoint.

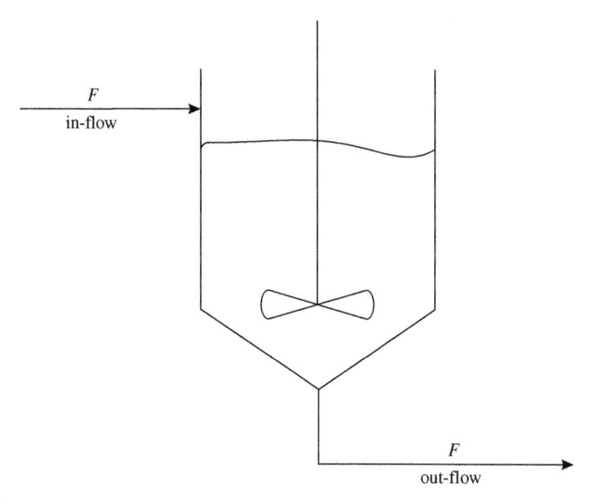

FIGURE 4.5 Continuous stirred tank reactor.

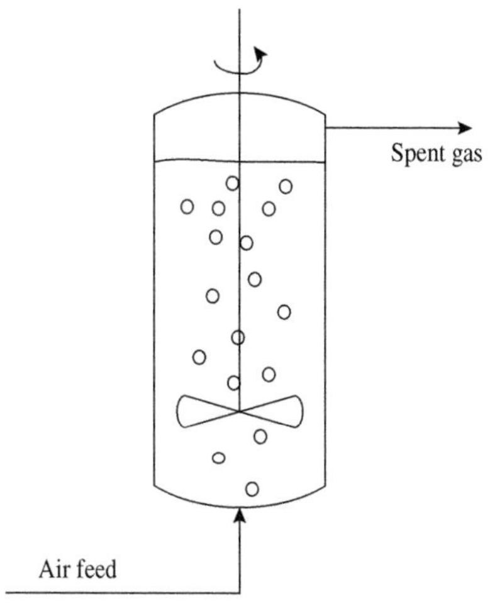

FIGURE 4.6 Oxidation reactor.

A4.3. Consider the oxidation reactor in Fig. 4.6, where air is sparged through a hydrocarbon mixture and the following reaction takes place [37]:

$$2A + O_2 \rightarrow P. \tag{4.66}$$

The reaction rate is given by

$$r_A = -kC_AC_O. \tag{4.67}$$

(a) Develop a mathematical model of the reactor dynamics assuming the reactor is isothermal.
(b) Add a cooling control system to the reactor system and develop the model if the reaction is exothermic.
(c) Add a dissolved oxygen concentration controller which adjusts air rate to the reactor.

A4.5. Discuss the types of constitutive equations which are needed in the description of a model which considers the dynamics of an evaporating pool of hazardous solvent. What particular problems do you foresee with availability or access to data?

5 DYNAMIC MODELS—LUMPED PARAMETER SYSTEMS

This chapter develops the underlying modelling principles for lumped parameter systems (LPSs) and the subsequent analysis of those models. Here, we are concerned about the time varying behaviour of systems which are considered to have states which are homogeneous within the balance volume $\mathcal{V}$. Hence, the concept of "lumping" the scalar field into a representative single state value. Sometimes the term "well mixed" is applied to such systems where the spatial variation in the scalar field, which describes the state of interest, is uniform. In some cases, we may also consider the stationary behaviour of such systems which leads to steady state model descriptions.

In this section, the most general case is treated. Special emphasis is put on the various forms of initial and boundary conditions that arise with their engineering meaning as well as the effect of these conditions on the solution properties of the process model.

Lumped parameter dynamic models, or *compartmental models* are widely used for control and diagnostic purposes, these applications being considered in more detail in Chapters 13 and 14. As well, they are frequently used as the basis for engineering design, startup and shutdown studies as well as assessing safe operating procedures.

5.1. CHARACTERIZING MODELS AND MODEL EQUATION SETS

As has been mentioned in Chapters 1 and 2, the modelling assumptions that are made will lead to various dynamic and steady-state model forms.

Dynamic Models

In terms of dynamic models we have two clearly identifiable classes:

- distributed parameter dynamic models,
- lumped parameter dynamic models.

In the above classes, we identify the distributed parameter dynamic models with various forms of PDEs, principally parabolic partial differential equations (PPDEs). The derivation of these models are dealt with in detail in Chapter 7 and techniques for their numerical solution are given in Chapter 8. For these models, it is vital to carefully specify *both* initial conditions at $t = 0$ for the states as well as boundary conditions over the balance volume of the system.

The lumped parameter dynamic models result in systems of ODEs often coupled with many nonlinear and linear algebraic constraints. The total system is referred to as a differential-algebraic equation (DAE) set. The equations need to have a specified set of consistent initial conditions for all states. This can be a challenging problem due to the effects of the nonlinear constraints which can impose extra conditions on the choice of initial values. The analysis and the solution of such systems using numerical techniques are discussed in the following chapter.

Steady-State Models

In many applications, the dynamic behaviour of the system is not necessarily the key issue to be examined. The steady-state solution may be of more importance. In the case of lumped parameter models which are subject to a steady-state assumption, the ordinary differential equations are transformed to a set of nonlinear algebraic equations (NLAEs). The steady-state assumption has the effect of eliminating all derivative terms dx_i/dt by setting them to zero. The initial conditions of the original equation set then take on the role of initial estimates for the algebraic solution technique.

In the case where steady-state assumptions are applied to distributed parameter dynamic models in one dimension, these result in equations which are classed as ODE boundary value problems (BVP). In this case it is necessary to solve a set of ODEs plus algebraic constraints which are subject to the system's boundary conditions.

Figure 5.1 shows the classes of problems normally encountered in process systems engineering.

5.2. LUMPED PARAMETER MODELS—INITIAL VALUE PROBLEMS (IVPs)

These models originate from *Lumped Mass/Component Mass/Energy Balances* and lead to ODE models.

5.2.1. Properties of the Modelling Problem

The ingredients for the model and the outcomes include:

Assumptions such as

1. lumped dynamic model implying no spatial variation in states
2. all balances can be taken into account

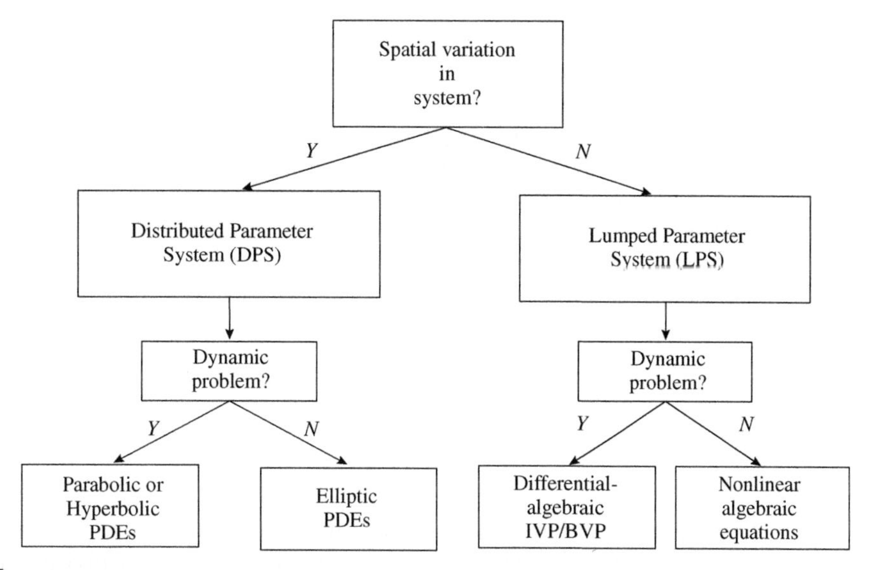

FIGURE 5.1 Characterization of models and equation sets.

3. flow pattern through connections
4. principal mechanisms stated

Model equations developed for

- every operating unit
- every phase
- every component
- overall mass and energy quantities

Initial conditions and design variable selection given for

- all state variables
- other variables to satisfy the degrees of freedom (DOF)

Solution and use including

- model verification
- model validation and parameter estimation
- application to the identified problem

5.2.2. Lumped Conservation Balances

Lumped parameter models do *not* incorporate the spatial variation of states within the balance volume; therefore, the scalar field of the intensive quantities (most often concentrations or temperature) is a function of time only. What this means is that the application of the general conservation equations leads to models which are represented by ODEs in time. Moreover, the closed boundary surface encapsulating the balance volume is also homogeneous in space.

Chapter 3 introduced the general conservation balance over a 3D volume $\mathcal{V}$ with a closed boundary surface $\mathcal{F}$ and its integral (3.16) and differential (3.17) form. *For lumped parameter models we apply the integral form of the conservation balances and use the condition of homogeneity of all variables in space.* This enables us to take the variables for the volume-specific extensive quantity ($\hat{\Phi}$) and its source ($\hat{q}$) out from the volumetric integral terms in Eq. (3.16) as well as considering the flow J as an overall flow through the whole surface $\mathcal{F}$ with no diffusive but only convective and transfer flows present. In this case, $J = J_C$. Therefore, the general form of the conservation balances for a conserved extensive quantity Φ in the case of LPSs is as follows:

$$\frac{d\Phi}{dt} = J_C + q, \tag{5.1}$$

with

$$J_C = J_C^{(i)} - J_C^{(o)}, \tag{5.2}$$

where $J_C^{(i)}$ is the inflow and $J_C^{(o)}$ is the outflow of the conserved extensive quantity Φ.

The following sections outline the basic conservation principles and constitutive equation setup underlying lumped parameter model development.

5.3. CONSERVATION BALANCES FOR MASS

Let us develop the general expressions for both *total and component mass balances* in a process system. Figure 5.2 shows a general balance volume for mass, given as a defined space with boundary Σ^M. Here, the component flows into the system are given as $f_{i,j}$ for component i in stream j, whilst the total stream flow is F_j $(\equiv \sum_i f_{i,j})$. Mass flows out of the system are shown as $f_{i,k}$ with total stream flows as F_k.

Within the balance volume $\mathcal{V}$ we have the total component mass m_i $(i = 1, \ldots, n)$ and total mass M $(\equiv \sum_i m_i)$.

From the figure, we can identify that there are several ways mass can enter and leave the balance volume through the boundary surface Σ^M. In most cases, the transfers are convective flows where material is entering and leaving the system via stream flows, typically through piped connections. The flows might also be associated with diffusional mechanisms where mass moves under the influence of a potential or

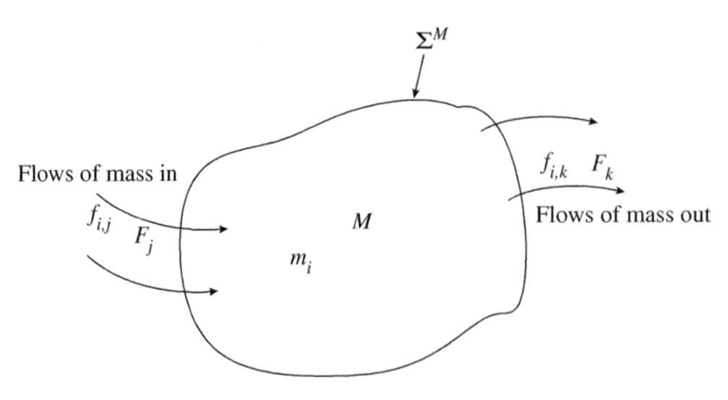

FIGURE 5.2 General mass balance volume.

driving force across the boundary surface. This could be a vapour–liquid, liquid–liquid or liquid–solid interface.

We can write either total or component mass balances as required by the modelling purpose. Where there are multi-component streams it will usually be necessary to use component mass balances, especially where reactions or mass transfer processes are involved.

5.3.1. Total Mass Balance

The general expression for a *total mass balance* can be written in word form as

$$\left\{\begin{matrix}\text{rate of accumulation}\\ \text{of mass}\end{matrix}\right\} = \left\{\begin{matrix}\text{mass flow}\\ \text{in}\end{matrix}\right\} - \left\{\begin{matrix}\text{mass flow}\\ \text{out}\end{matrix}\right\},$$

or in the case of a lumped parameter system, the equation form for p input streams and q output streams is

$$\frac{\mathrm{d}M}{\mathrm{d}t} = \sum_{j=1}^{p} F_j - \sum_{k=1}^{q} F_k. \tag{5.3}$$

Note that the above equation is a special case of the general lumped conservation balance equations (5.1)–(5.2) for the total mass.

5.3.2. Component Mass Balances

The general expression for the *component mass balances* in word form is

$$\left\{\begin{matrix}\text{rate of mass accumulation}\\ \text{of component } i\end{matrix}\right\} = \left\{\begin{matrix}\text{mass flow in of}\\ \text{component } i\end{matrix}\right\} - \left\{\begin{matrix}\text{mass flow out of}\\ \text{component } i\end{matrix}\right\} + \left\{\begin{matrix}\text{rate of formation or consumption}\\ \text{of component } i\end{matrix}\right\},$$

where the last term accounts for the creation or disappearance of component i via chemical reaction.

In the case of an LPS, we have the general equation in the form

$$\frac{\mathrm{d}m_i}{\mathrm{d}t} = \sum_{j=1}^{p} f_{i,j} - \sum_{k=1}^{q} f_{i,k} + g_i, \quad i = 1, \ldots, n, \tag{5.4}$$

where m_i is the mass holdup of component i within the balance volume $\mathcal{V}$ and g_i is the mass rate of generation or consumption of species i in the balance volume due to reaction. Clearly, the magnitude of the term g_i depends on the reaction rate within the system volume.

We can also write the general mass balance in molar terms n_i by introducing the molar flowrates instead of the mass flowrates of species i in stream j. The general molar balance is written as

$$\frac{\mathrm{d}n_i}{\mathrm{d}t} = \sum_{j=1}^{p} \tilde{f}_{i,j} - \sum_{k=1}^{q} \tilde{f}_{i,k} + \tilde{g}_i, \quad i = 1, \ldots, n. \tag{5.5}$$

The generation term can be expressed in various ways:

- In Chapter 4, the reaction rate concept was introduced and from this we can write

$$\tilde{g}_i = \left(\frac{\mathrm{d}n_i}{\mathrm{d}t}\right)_{\text{reaction}} = r_i V = \nu_i r V. \tag{5.6}$$

  This is a very common means of expressing species generation in reacting systems.

- Alternately, we can define the generation rate in terms of the *molar extent of reaction* ξ as the moles produced scaled by the stoichiometric coefficient. In this case, $n_i^{(o)}$ represents the initial moles of i present,

$$\xi = \frac{n_i - n_i^{(o)}}{\nu_i} \tag{5.7}$$

  giving

$$\tilde{g}_i = \nu_i \frac{\mathrm{d}\xi}{\mathrm{d}t}. \tag{5.8}$$

  If we define the extent of reaction per unit volume as $\hat{\xi} = \xi/V$, then we can write

$$\tilde{g}_i = \nu_i V \frac{\mathrm{d}\hat{\xi}}{\mathrm{d}t} \tag{5.9}$$

  Note that the term $\mathrm{d}\hat{\xi}/\mathrm{d}t$ is the same as the overall or intrinsic reaction rate, since

$$\frac{\mathrm{d}\hat{\xi}}{\mathrm{d}t} = \frac{1}{V}\frac{\mathrm{d}\xi}{\mathrm{d}t} = \frac{1}{V\nu_i}\frac{\mathrm{d}n_i}{\mathrm{d}t} = r.$$

  For multiple reactions involving species i, we can sum all possible contributions from the various reactions to calculate the generation rate.

It should also be remembered that the flows into and out of the balance volume can be of two types:

- convective mass flows (e.g. stream flows in physical connections into or out of the volume);
- diffusive or molecular mass flows (e.g. flows driven by chemical potentials, concentrations, or partial pressure gradients across the boundary).

Note that the above component mass balance equation is a special case of the general lumped conservation balance equations (5.1)–(5.2).

Mass and Molar Quantities in Conservation Balances

Mass flows and molar flows are very different quantities when considering mass conservation principles. Care must be exercised when using molar quantities for conservation balances. Only total mass is a true extensive property in a general system and as such there can be no generation nor consumption for this quantity.

In the case of the total mass balance being applied in molar terms, the conservation equation (5.3) is only valid for *non-reacting systems*.

For both mass and molar *component mass balances*, care must be exercised to ensure that all generation and consumption terms are included otherwise the conservation balances are not valid.

5.4. CONSERVATION BALANCES FOR ENERGY

In a manner similar to that in Section 5.3, we can develop a general energy balance based on conservation of energy principles. Again we can define a system boundary $\sum^E$ and the balance volume $\mathcal{V}$. Figure 5.3 illustrates this for the general energy balance volume.

In this case, the total system energy is given by E, the mass flows into and out of the balance volume are F_j and F_k and the total energy per unit mass of the inlet and outlet streams are $\hat{E}_j$ and $\hat{E}_k$.

We need to ask how this volume is defined. In most cases, the energy balance volume Σ^E will be coincident with the corresponding mass balance volume Σ^M. There are circumstances where the energy balance volume Σ^E can also encompass two or more mass balance volumes. There are also situations in which the mass balance volumes will not be of importance because mass is invariant. However, the coincident energy balance volume will be of significance. This is particularly the case with heat transfer situations where the wall mass is invariant and no mass balance is performed but the conservation of energy is crucial due to changes in the energy holdup of the wall. The choice of balance volumes is discussed in detail in Section 3.3.

5.4.1. Total Energy Balance

The general conservation balance for total energy over the balance volume $\mathcal{V}$ with surface $\sum^E$ is given by

$$\begin{Bmatrix}\text{rate of change}\\ \text{of total energy}\end{Bmatrix} = \begin{Bmatrix}\text{flow of energy}\\ \text{into the system}\end{Bmatrix} - \begin{Bmatrix}\text{flow of energy}\\ \text{out of system}\end{Bmatrix}$$

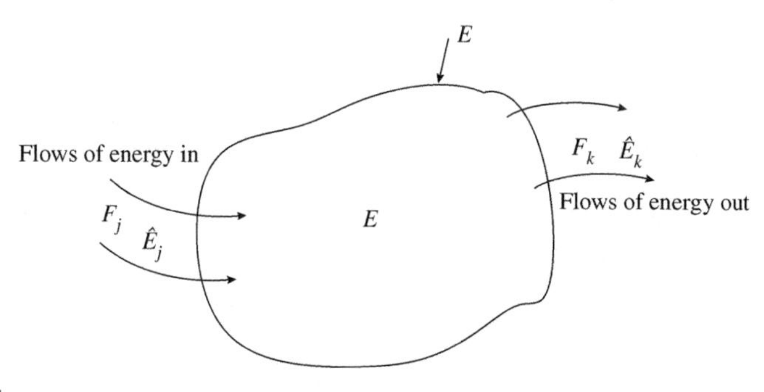

FIGURE 5.3 General energy balance volume.

The total energy E [J] of the system comprises three principal components in process systems:

- internal energy U
- kinetic energy K_E
- potential energy P_E

Hence, we can write

$$E = U + K_E + P_E. \tag{5.10}$$

It is necessary to define in what ways energy flows into and out of the system. We consider the principal components of these energy flows as

- convective energy flows
- conductive and radiative heat flows
- work terms

Convective Energy Flows

These are associated with the energy carried by each mass flow F_j and F_k into and out of the balance volume. We associate with each mass flow an energy flow which is made up of internal, kinetic and potential energy. Thus the convective energy flows can be written for inlet and outlet flows as

$$\sum_{j=1}^{p} F_j(\hat{U} + \hat{K}_E + \hat{P}_E)_j \tag{5.11}$$

and

$$\sum_{k=1}^{q} F_k(\hat{U} + \hat{K}_E + \hat{P}_E)_k, \tag{5.12}$$

where $\hat{U}$, $\hat{K}_E$, $\hat{P}_E$ are specific energy quantities. Typically, they are defined as energy per unit mass [J/kg]. They can also be written in molar units [J/mol] and designated as $\widetilde{U}$, $\widetilde{K}_E$ and $\widetilde{P}_E$.

Conductive and Radiative Heat Flows

These are given as follows:

- Conductive energy which flows either in or out of the system through conductive heat transfer mechanisms associated with the interfaces of the balance volume. We can designate these as Q_C.
- Radiative energy flows either in or out of the system through radiation heat transfer mechanisms, given by Q_R.

The two previous categories of energy flows are conventionally combined together and written as

$$Q = Q_C + Q_R. \tag{5.13}$$

Work Terms

The energy balance may contain terms describing the work done by the system on the surroundings which consists of several terms. We note that the accepted convention is that work done on the system is regarded as positive work. The work terms of interest are now discussed:

- Shaft work (W_S) which accounts for mechanical energy flow to or from the system. Shaft work done on the system by the surroundings raises the energy content of the system and is thus positive work.
- Expansion work (W_E) which accounts for work done in expansion or contraction of the balance volume. This term is often not present especially when the balance volume is fixed such as in the case of a piece of process equipment. However, in many circumstances where balance volumes share an interface and the interface is movable then this term can be important.

  Where the system *pressure remains constant*, the expansion work is related to the force applied on the boundary and the rate of boundary movement giving:

$$W_E = -P\frac{dV}{dt}, \tag{5.14}$$

where *pressure is not constant* over the boundary, it is necessary to carry out a volumetric integral over the volume $\mathcal{V}$ to compute the expansion work given by

$$W_E = -\int_V P(r,t)\,dv. \tag{5.15}$$

- Flow work (W_F) which relates to work done on the fluid as it moves into and out of the balance volume. This depends on the inlet and outlet pressures, the specific volumes of the fluids at these conditions and the mass flows of the streams. This flow work contribution on the system can be expressed as

$$W_F = \sum_{j=1}^{p} F_j(P\hat{V})_j - \sum_{k=1}^{q} F_k(P\hat{V})_k \tag{5.16}$$

The Total Energy Balance

The conservation balance for energy over the balance volume $\mathcal{V}$ can now be written as

$$\frac{dE}{dt} = \sum_{j=1}^{p} F_j\hat{E}_j - \sum_{k=1}^{q} F_k\hat{E}_k + Q + W. \tag{5.17}$$

We can expand the work quantity to obtain

$$\frac{dE}{dt} = \sum_{j=1}^{p} F_j(\hat{U} + \hat{K}_E + \hat{P}_E)_j - \sum_{k=1}^{q} F_k(\hat{U} + \hat{K}_E + \hat{P}_E)_k + Q \tag{5.18}$$

$$+ \left\{ \sum_{j=1}^{p} F_j(P\hat{V})_j - \sum_{k=1}^{q} F_k(P\hat{V})_k + W_E + W_S \right\}, \tag{5.19}$$

and after rearrangement we get

$$\frac{\mathrm{d}E}{\mathrm{d}t} = \sum_{j=1}^{p} F_j(\hat{U} + P\hat{V} + \hat{K}_E + \hat{P}_E)_j - \sum_{k=1}^{q} F_k(\hat{U} + P\hat{V} + \hat{K}_E + \hat{P}_E)_k + Q + \hat{W}, \quad (5.20)$$

where $\hat{W} = W_{\mathrm{E}} + W_{\mathrm{S}}$.

Using the thermodynamic relationship for enthalpy H given by $H = U + PV$, it is possible to write the general energy balance using mass specific enthalpy $\hat{H}$ [J/kg] as

$$\frac{\mathrm{d}E}{\mathrm{d}t} = \sum_{j=1}^{p} F_j(\hat{H} + \hat{K}_E + \hat{P}_E)_j - \sum_{k=1}^{q} F_k(\hat{H} + \hat{K}_E + \hat{P}_E)_k + Q + \hat{W}. \quad (5.21)$$

The general energy balance equation (5.21) can also be written in equivalent molar terms, using $\widetilde{H}$ to represent stream enthalpy per unit mole. Other terms in the equation are similarly defined in molar units.

It should also be noted that Eq. (5.21) contains implicitly any energy contributions from internal processes, specifically those associated with chemical reactions. The next section will develop a number of key simplifications of the general energy balance which are applicable for a range of common process engineering applications.

5.4.2. Simplifications and Modifications of the General Energy Balance

In many process systems applications, we can make a number of simplifications to the terms shown in Eq. (5.21). We now review some of these:

Assumption 1 In many cases, the kinetic and potential energy components can be neglected:

- $K_E \simeq 0$, since many flows are low velocity and the energy contribution of the term $(F_j v^2/2)$ is small.
- $P_E \simeq 0$, since elevation differences for many process systems are small and hence the energy contribution ($F_j z$, z = elevation) is small.

These simplifications lead to a final simplified form of the energy balance where the right-hand side terms are now written in terms of specific enthalpies. The left-hand side represents the total internal energy of the system. This is a common representation in chemical process systems where internal energy content often dominates the total energy content of the system.

$$\frac{\mathrm{d}U}{\mathrm{d}t} = \sum_{j=1}^{p} F_j\hat{H}_j - \sum_{k=1}^{q} F_k\hat{H}_k + Q + \hat{W}. \quad (5.22)$$

Computation of the specific enthalpies for all inlet and outlet streams is usually done using thermodynamic prediction packages which can take into account fluid phase non-idealities. Other simplifications of the enthalpy representation can also be made as seen in Section 4.3.

Assumption 2 We normally do not deal directly with the internal energy U, in the general energy balance such as Eq. (5.22) but prefer to use alternate properties. Using the definition of enthalpy we can write Eq. (5.22) as

$$\frac{\mathrm{d}U}{\mathrm{d}t} = \frac{\mathrm{d}(H - PV)}{\mathrm{d}t} = \sum_{j=1}^{p} F_j \hat{H}_j - \sum_{k=1}^{q} F_k \hat{H}_k + Q + \hat{W}, \tag{5.23}$$

$$\frac{\mathrm{d}H}{\mathrm{d}t} - \frac{d(PV)}{\mathrm{d}t} = \sum_{j=1}^{p} F_j \hat{H}_j - \sum_{k=1}^{q} F_k \hat{H}_k + Q + \hat{W}. \tag{5.24}$$

If P and V are constant, then we can write,

$$\frac{\mathrm{d}H}{\mathrm{d}t} = \sum_{j=1}^{p} F_j \hat{H}_j - \sum_{k=1}^{q} F_k \hat{H}_k + Q + \hat{W}. \tag{5.25}$$

This is often not the case in gas phase systems.

Assumption 3 In Eq. (5.25), the specific enthalpy of the balance volume and the specific enthalpy at the outlet are not necessarily equal. In the case where pressure variations within the balance volume are small, such as in liquid systems or where enthalpy variations due to pressure are small, we can assume that the specific enthalpies are equal and hence:

$$\hat{H} = \hat{H}_k \quad k = 1, \ldots, q. \tag{5.26}$$

We can then write the energy balance as

$$\frac{\mathrm{d}H}{\mathrm{d}t} = \sum_{j=1}^{p} F_j \hat{H}_j - \sum_{k=1}^{q} F_k \hat{H} + Q + \hat{W}. \tag{5.27}$$

This is the most common form of the energy balance for a liquid system.

Assumption 4 We know that in Eq. (5.27), the enthalpies are evaluated at the temperature conditions of the feeds (T_j) and also at the system temperature (T). By making certain assumptions about the enthalpy representation, we can make further simplifications. In particular, we note that the enthalpy of the feed $\hat{H}_j$ can be written in terms of the system temperature T.

$$\hat{H}_j(T_j) = \hat{H}_j(T) + \int_{T}^{T_j} c_{p_j}(T)\,\mathrm{d}T. \tag{5.28}$$

If we assume that c_p is a constant, then

$$\hat{H}_j(T_j) = \hat{H}_j(T) + c_{p_j}(T_j - T). \tag{5.29}$$

Hence, we can write our modified energy balance as

$$\frac{\mathrm{d}H}{\mathrm{d}t} = \sum_{j=1}^{p} F_j \left[\hat{H}_j(T) + c_{p_j}(T_j - T)\right] - \sum_{k=1}^{q} F_k \hat{H}(T) + Q + \hat{W}. \tag{5.30}$$

In this case, the first term is the energy contribution needed to adjust the feed enthalpy at its inlet conditions to the conditions of the balance volume.

Assumption 5 It has already been mentioned that when considering reacting systems no explicit appearance of the heat of reaction is seen in the general energy balance. This is because the energy gain or loss is seen in the value of the outlet enthalpy evaluated at the system temperature T. We can now develop the energy balance in a way which makes the reaction term explicit in the energy balance.

We first note that enthalpy is a function of temperature, pressure and the moles of species, n_i, $i = 1, \ldots, nc$ giving $H = f(T, P, n_i)$ and hence the time derivative of this relationship is given by

$$\frac{\mathrm{d}H}{\mathrm{d}t} = \frac{\partial H}{\partial T}\frac{\mathrm{d}T}{\mathrm{d}t} + \frac{\partial H}{\partial P}\frac{\mathrm{d}P}{\mathrm{d}t} + \sum_i \frac{\partial H}{\partial n_i}\frac{\mathrm{d}n_i}{\mathrm{d}t}. \tag{5.31}$$

Here, the term $\partial H/\partial n_i$ is the partial molar enthalpy of species i, which we can write as $\bar{H}_i$.

Now, in the case of liquid systems, the term $\partial H/\partial P$ is close to zero and is identically zero for ideal gases. We can also note that the definition of specific heat per unit mass at constant pressure is given by

$$c_p = \left(\frac{\partial H}{\partial T}\right)_P. \tag{5.32}$$

This then allows us to write the general equation (5.31) as

$$\frac{\mathrm{d}H}{\mathrm{d}t} = V\rho c_p \frac{\mathrm{d}T}{\mathrm{d}t} + \sum_i \bar{H}_i \frac{\mathrm{d}n_i}{\mathrm{d}t} \tag{5.33}$$

The last term in Eq. (5.33) can be expanded using the mass balances for a reacting system where the species n_i appear or are consumed. For a general reaction given by

$$\nu_X X + \nu_Y Y \rightarrow \nu_W W + \nu_Z Z, \tag{5.34}$$

we can write the species mole balances given by Eq. (5.5) as

$$\frac{\mathrm{d}n_i}{\mathrm{d}t} = \sum_{j=1}^{p} \tilde{f}_{i,j} - \sum_{k=1}^{q} \tilde{f}_{i,k} + \nu_i V r, \quad i = 1, \ldots, nc. \tag{5.35}$$

The last term in Eq. (5.33) then becomes

$$\sum_{i=1}^{nc} \bar{H}_i \frac{\mathrm{d}n_i}{\mathrm{d}t} = \sum_{i=1}^{nc} \bar{H}_i \sum_{j=1}^{p} \tilde{f}_{i,j} - \sum_{i=1}^{nc} \bar{H}_i \sum_{k=1}^{q} \tilde{f}_{i,k} + rV \sum_{i=1}^{nc} \bar{H}_i \nu_i. \tag{5.36}$$

The term $\sum_{i=1}^{nc} \bar{H}_i \nu_i = (\nu_W \bar{H}_W + \nu_Z \bar{H}_Z - \nu_X \bar{H}_X - \nu_Y \bar{H}_Y)$ is known as the change in molar enthalpy for the reaction and is denoted as ΔH_R, or heat of reaction. Hence, the general equation (5.33) can now be written as

$$\frac{dH}{dt} = V\rho c_p \frac{dT}{dt} + \sum_{i=1}^{nc} \bar{H}_i \sum_{j=1}^{p} \widetilde{f}_{i,j} - \sum_{i=1}^{nc} \bar{H}_i \sum_{k=1}^{q} \widetilde{f}_{i,k} + rV\Delta H_R. \tag{5.37}$$

If we now equate Eqs (5.30) and (5.37) and rearrange terms we obtain

$$V\rho c_p \frac{dT}{dt} = \sum_{j=1}^{p} F_j \left[\hat{H}_j(T) + c_{p_j}(T_j - T)\right] - \sum_{k=1}^{q} F_k \hat{H}(T) \tag{5.38}$$

$$- \sum_{i=1}^{nc} \bar{H}_i \sum_{j=1}^{p} \widetilde{f}_{i,j} + \sum_{i=1}^{nc} \bar{H}_i \sum_{k=1}^{q} \widetilde{f}_{i,k} - rV\Delta H_R + Q + \hat{W}. \tag{5.39}$$

Noting that

$$\sum_{i=1}^{nc} \bar{H}_i \sum_{k=1}^{q} \widetilde{f}_{i,k} = \sum_{k=1}^{q} F_k \hat{H} \quad \text{and} \quad \sum_{i=1}^{nc} \bar{H}_i \sum_{j=1}^{p} \widetilde{f}_{i,j} = \sum_{k=1}^{p} F_j \hat{H},$$

Eq. (5.38) simplifies to

$$V\rho c_p \frac{dT}{dt} = \sum_{j=1}^{p} F_j c_{p_j}(T_j - T) + \sum_{j=1}^{p} F_j(\hat{H}_j(T) - \hat{H}) + rV(-\Delta H_R) + Q + \hat{W}. \tag{5.40}$$

The second term represents the difference in partial molar enthalpies of the feed and balance volume contents at balance volume conditions. This is zero for ideal mixtures and small in relation to the reaction heat term for non-ideal mixtures. As such we can normally neglect it to give the final expression for the energy balance as

$$V\rho c_p \frac{dT}{dt} = \sum_{j=1}^{p} F_j c_{p_j}(T_j - T) + rV(-\Delta H_R) + Q + \hat{W}. \tag{5.41}$$

This is the most common form of energy balance for reacting systems.

The first term on the right-hand side represents the energy needed to adjust all feeds to the reactor conditions. The second represents the energy generation or consumption at the reactor temperature. The last two terms are the relevant heat and work terms.

5.5. CONSERVATION BALANCES FOR MOMENTUM

In many systems it is also important to consider the conservation of momentum. This is particularly the case in mechanical systems and in flow systems where various forces act. These can include pressure forces, viscous forces, shear forces and gravitational forces. Momentum is the product of mass and velocity. We can thus write the general form of the balance applied to a similar balance volume as shown in Fig. 5.3 as

$$\begin{Bmatrix}\text{rate of change}\\ \text{of momentum}\end{Bmatrix} = \begin{Bmatrix}\text{rate of momentum}\\ \text{into system}\end{Bmatrix} - \begin{Bmatrix}\text{rate of momentum}\\ \text{out of system}\end{Bmatrix} + \begin{Bmatrix}\text{rate of momentum}\\ \text{generation}\end{Bmatrix}.$$

The last term is the summation of all the forces acting on the system. In considering momentum, it is important to consider all components of the forces acting on the system under study. This means that the problem is basically a 3D problem. In reality we often simplify this to a 1D problem. This alternative expression of the momentum balance is given by

$$\begin{Bmatrix}\text{rate of change}\\ \text{of momentum}\end{Bmatrix} = \begin{Bmatrix}\text{rate of momentum}\\ \text{into system}\end{Bmatrix} - \begin{Bmatrix}\text{rate of momentum}\\ \text{out of system}\end{Bmatrix} + \begin{Bmatrix}\text{sum of all forces}\\ \text{on the system}\end{Bmatrix}.$$

We can write the general momentum balance equation as

$$\frac{\mathrm{d}\mathcal{M}}{\mathrm{d}t} = \mathcal{M}^{(\mathrm{i})} - \mathcal{M}^{(\mathrm{o})} + \sum_{k=1}^{p} \mathcal{F}_k, \tag{5.42}$$

where $\mathcal{M}^{(\mathrm{i})}$, $\mathcal{M}^{(\mathrm{o})}$ are momentum terms in and out of the system and $\mathcal{F}_k$, $k = 1, \ldots, p$ are the forces acting on the system.

Momentum balances in models for lumped parameter systems appear most often in equations relating convective flows to forces generated by pressure, viscous and gravity gradients. These are typically expressed by some form of the general Bernoulli equation which incorporates various simplifications.

EXAMPLE 5.5.1 (Simple momentum balances). Consider the application of a general momentum balance for an incompressible ideal fluid volume V. The momentum balance gives:

$$\frac{\mathrm{D}(\rho V v)}{\mathrm{D}t} = \frac{\mathrm{D}\mathcal{M}}{\mathrm{D}t} = -V\nabla P + \rho V g, \tag{5.43}$$

where $\mathrm{D}\mathcal{M}/\mathrm{D}t$ is the substantial derivative of the momentum, for a point flowing with the fluid and ∇P is the pressure gradient vector. This is covered in length in Bird *et al.* [13]. The equation can be written in terms of a unit volume with the body force g being replaced by $-g\nabla h$. After taking the scalar product of the momentum balance with the velocity vector v and assuming constant density we obtain:

$$\rho\frac{\mathrm{D}(v^2/2)}{\mathrm{D}t} = -v \cdot \nabla(P + \rho g h). \tag{5.44}$$

For a steady-state flow the right-hand side is set to zero giving the condition that:

$$\rho\frac{v^2}{2} + P + \rho g h \tag{5.45}$$

should be constant. This is the well-known static Bernoulli momentum balance without the presence of viscous forces producing friction losses in the flow.

Equation (5.45) is often used for modelling flows in piping systems and through valves. A particular form of the equation leads to a simple steady-state constitutive equation for the flow through a valve as

$$F = C_V\sqrt{\Delta P}, \tag{5.46}$$

where C_V is the valve coefficient.

Momentum balances also have an important role in modelling of mechanical system where various forces are at work. One of the simplest examples of this is the dynamics of a control valve actuator and stem, as shown in the following example.

EXAMPLE 5.5.2 (The dynamics of a control valve). Consider the actuator and stem of a control valve shown in Fig. 5.4. The diaphragm moves the stem position x at a velocity v, under the force exerted by changing the air pressure P on the diaphragm. This is resisted by a spring and the friction associated with the stem moving through the packing. A momentum balance can be carried out of the stem assembly of mass M noting that there are three principal forces at work. These are the force on the diaphragm due to air pressure (F_d), the compression force exerted by the spring (F_s) and frictional forces due to the stem packing (F_p).

The momentum balance gives

$$\frac{d\mathcal{M}}{dt} = \frac{d(Mv)}{dt} = F_d - F_s - F_p. \tag{5.47}$$

Substituting for the forces, we have

$$M\frac{dv}{dt} = PA - kx - cv, \tag{5.48}$$

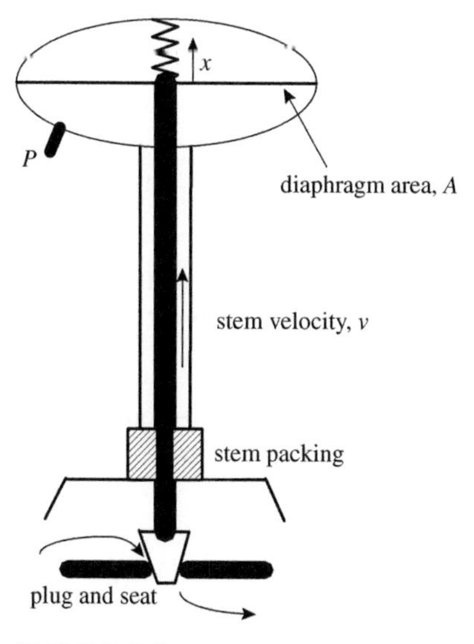

FIGURE 5.4 Control valve actuator and stem.

where k is the spring constant and c is a coefficient of friction for the packing. Final rearrangement of the equation leads to

$$\left(\frac{M}{k}\right)\frac{\mathrm{d}^2x}{\mathrm{d}t^2}+\left(\frac{c}{k}\right)\frac{\mathrm{d}x}{\mathrm{d}t}+x=\left(\frac{PA}{k}\right), \tag{5.49}$$

which is a form of a second-order system response given by

$$\tau^2\frac{\mathrm{d}^2x}{\mathrm{d}t^2}+2\xi\tau\frac{\mathrm{d}x}{\mathrm{d}t}+x=K \tag{5.50}$$

■ ■ ■ with $\tau=\sqrt{M/k}$ and $\xi=1/2\sqrt{(c^2/(kM))}$.

5.6. THE SET OF CONSERVATION BALANCES FOR LUMPED SYSTEMS

The process model of an LPS consists of a set of conservation balance equations that are ODEs equipped with suitable constitutive equations. The balance equations are usually coupled through reaction rate and transfer terms. This is illustrated on the following example of a CSTR which can be seen as a modelling unit of all lumped parameter system models.

EXAMPLE 5.6.1 (Lumped parameter modelling of a CSTR).

A CSTR is shown in Fig. 5.5 with reactant volume V, component mass holdup m_A for component A, feed flowrate $F^{(\mathrm{i})}$ [m^3/s] at temperature $T^{(\mathrm{i})}$. Feed concentration of component A is $c_A^{(\mathrm{i})}$. Outlet flowrate $F^{(\mathrm{o})}$ is in units [m^3/s].

Assumptions

$\mathcal{A}1$. perfect mixing implying no spatial variations,
$\mathcal{A}2$. incompressible fluid phase,
$\mathcal{A}3$. constant physical properties,
$\mathcal{A}4$. all flows and properties given in mole units,
$\mathcal{A}5$. equal molar densities,
$\mathcal{A}6$. reactions and reaction rates given by

$$A\rightarrow P,\quad r_A=k_0\,\mathrm{e}^{-E/(RT)}c_A. \tag{5.51}$$

Model equations
First, we set up the model equations in their *extensive form* written in molar quantities. The following state the *overall mass balance, the component balance for A and the total energy balance.*

$$\begin{aligned}
\text{accumulation-rate} &= \text{in-flow} \quad - \text{out-flow} + \text{source},\\
\frac{\mathrm{d}(\rho V)}{\mathrm{d}t} &= F^{(\mathrm{i})}\rho^{(\mathrm{i})} - F^{(\mathrm{o})}\rho,\\
\frac{\mathrm{d}(m_A)}{\mathrm{d}t} &= f_A^{(\mathrm{i})} - f_A^{(\mathrm{o})} - Vr_A,\\
V\rho c_P\frac{\mathrm{d}(T)}{\mathrm{d}t} &= F^{(\mathrm{i})}\bar{H}^{(\mathrm{i})}\rho - F^{(\mathrm{i})}\bar{H}\rho - Vr_A\Delta H_{\mathrm{R}}.
\end{aligned} \tag{5.52}$$

A set of *constitutive equations* accompanies the conservation balances. These include:

$$f_A^{(i)} = F^{(i)}c_A^{(i)}, \quad f_A^{(o)} = F^{(o)}c_A, \quad M = V\rho, \quad m_A = c_A V,$$

$$r_A = k_0\,\mathrm{e}^{-E/(RT)}c_A, \quad H = Mc_P T, \quad \bar{H} = \frac{H}{M}, \quad \bar{H}^{(i)} = c_P^{(i)}T^{(i)}.$$

The model equations above have to be solved for the following set of state variables:

$$\underline{x} = [V \;\; m_A \;\; T]^{\mathrm{T}}. \tag{5.53}$$

Note that a set of *initial conditions* is needed for the solution.

Initial conditions:

$$V(0) = V_0; \quad m_A(0) = m_{A_0}; \quad T(0) = T_0. \tag{5.54}$$

■ ■ ■

As shown in the example above, the *general form of the lumped parameter model equations* is an initial value problem for a set of ODEs with algebraic constraints and initial conditions $x(0)$. This is called a *DAE-IVP* problem:

$$\frac{\mathrm{d}x}{\mathrm{d}t} = f(x, y, t), \tag{5.55}$$

$$0 = g(x, y, t), \tag{5.56}$$

$$x(0) = x_0. \tag{5.57}$$

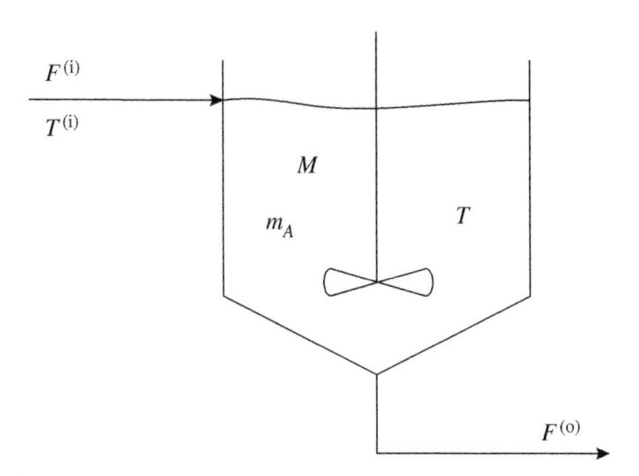

FIGURE 5.5 Continuous stirred tank reactor.

5.7. CONSERVATION BALANCES IN INTENSIVE VARIABLE FORM

In Section 3.1.5, the difference between extensive and intensive variables was discussed and the role they play in the modelling of process engineering systems. We formulate all conservation balances in extensive form using total mass, component mass and energy. The intensive variables normally enter the equation set through the constitutive equations which define reaction rates, thermo-physical properties and other relations. This is recommended modelling practice.

However, it is sometimes necessary or expedient to transform the extensive form into an equivalent intensive form. This is normally done by algebraic transformations on the original equations. It has already been mentioned that the extensive variables are usually related in a nonlinear manner to the intensive variable. For example, the intensive molar concentration c_i is related to the total mass holdup M or the volume V and density ρ via the relation

$$c_i = \frac{m_i}{M} = \frac{m_i}{\rho V} \tag{5.58}$$

or for a component i in a stream j,

$$c_{i,j} = \frac{f_{i,j}}{F_j}. \tag{5.59}$$

These algebraic transformations can be made on the original balance equations (5.4) and (5.22), as the next sections demonstrate. We also discuss the modelling and solution implications of such transformations.

5.7.1. Intensive Form of the Component Mass Balance

Here, we deal with the transformation of Eq. (5.4) using Eq. (5.58). Applying the transformation gives the following component mass balance equations:

$$\frac{\mathrm{d}(c_i M)}{\mathrm{d}t} = c_i^{(\mathrm{i})} F^{(\mathrm{i})} - c_i F^{(\mathrm{o})} + g_i \quad i = 1, \ldots, n, \tag{5.60}$$

and we note that both c_i and M terms on the left-hand side can be a function of time. If we were to expand the left-hand side, we obtain

$$c_i \frac{\mathrm{d}M}{\mathrm{d}t} + M \frac{\mathrm{d}c_i}{\mathrm{d}t} = c_i^{(\mathrm{i})} F^{(\mathrm{i})} - c_i F^{(\mathrm{o})} + g_i \quad i = 1, \ldots, n. \tag{5.61}$$

and only if $\mathrm{d}M/\mathrm{d}t = 0$ can we simplify the problem as

$$M \frac{\mathrm{d}c_i}{\mathrm{d}t} = c_i^{(\mathrm{i})} F^{(\mathrm{i})} - c_i F^{(\mathrm{o})} + g_i \quad i = 1, \ldots, n. \tag{5.62}$$

This, however, is a unlikely case and we are normally left to deal with the more complex relation given by Eq. (5.61). We can solve these problems in several ways:

1. Find an expression for $\mathrm{d}M/\mathrm{d}t$ and solve that along with the component balance equations to give the value of M.
2. Rearranging Eq. (5.61) to eliminate the unwanted derivative term and leave only $\mathrm{d}c_i/\mathrm{d}t$ by using the total mass balance

$$\frac{\mathrm{d}M}{\mathrm{d}t} = F^{(\mathrm{i})} - F^{(\mathrm{o})}$$

to get

$$M\frac{\mathrm{d}c_i}{\mathrm{d}t} = c_i^{(\mathrm{i})}F^{(\mathrm{i})} - c_iF^{(\mathrm{i})} + g_i, \quad i = 1, \ldots, n. \tag{5.63}$$

Observe the difference between Eqs (5.62) and (5.63). The latter has lost its conservation balance form because the term $c_iF^{(\mathrm{i})}$ is *not* a component mass flow being a product of a mass flow (the inlet) and a concentration at another flow (outlet).

3. Solve the original Eq. (5.60) and compute M by summing up all the component molar values (c_iM).

What is clear is that having balances in intensive variable form can be more complex both from the development point of view as well as from the solution perspective.

EXAMPLE 5.7.1 (CSTR model in intensive form). Consider the previous CSTR Example 5.6.1 and put into intensive form.

By using the definitions of the intensive variables and taking into account that ρ and c_P are constants, we can transform the original set of model equations into the intensive state form:

$$\begin{aligned}
\frac{\mathrm{d}(V)}{\mathrm{d}t} &= F^{(\mathrm{i})} - F^{(\mathrm{o})} \\
\frac{\mathrm{d}c_A}{\mathrm{d}t} &= \frac{c_A^{(\mathrm{i})}F^{(\mathrm{i})}}{V} - \frac{c_AF^{(\mathrm{i})}}{V} - r_A, \\
\frac{\mathrm{d}T}{\mathrm{d}t} &= \frac{F^{(\mathrm{i})}T^{(\mathrm{i})}}{V} - \frac{F^{(\mathrm{i})}T}{V} - \frac{r_A\Delta H_R}{\rho c_P}, \\
r_A &= \mathcal{R}(c_A, T).
\end{aligned} \tag{5.64}$$

■ ■ ■

5.8. DIMENSIONLESS VARIABLES

Variables in a balance equation or in a process model usually correspond to different physico-chemical quantities or parameters measured in different engineering units. This fact may lead to misunderstandings in comparing the magnitude of different terms in a model equation and wrong assumptions in neglecting terms if these terms are given in different unit systems, for example, in SI or traditional British units. Therefore, good engineering practice means that we should *normalize variables.* That is, we relate them to standard variable ranges relevant to the modelling problem in question.

In order to illustrate the procedure used to normalize variables let us assume that a model variable x is measured in units [unit]. For example, M denoting overall mass is measured in [kg]. The following problem statement describes how to normalize the variable x.

NORMALIZATION OF VARIABLES

Given:

A model variable x measured in some units [unit] and minimum (x_{min}) and maximum (x_{max}) possible values for x such that

$$x_{\mathrm{min}} \le x \le x_{\mathrm{max}}.$$

Compute:
The normalized variable $\bar{x}$, where

$$\bar{x} = \frac{x - x_{\min}}{x_{\max} - x_{\min}}.$$

It is easy to see, that the normalized variable $\bar{x}$ has the following advantageous properties:

- its range is normalized, i.e.

$$0 \le \bar{x} \le 1;$$

- it is unitless (dimensionless).

The procedure can also be applied to the time variable so that time is also normalized between 0 and 1.

Unitless or dimensionless model equations are those where all variables and parameters are normalized. In doing this all variables are seen to be between 0 and 1. Certain variables can also be grouped together leading to dimensionless numbers. Many are well known such as Reynolds and Froude numbers. Others arise from the application of the normalizing principle to different application areas such as reaction engineering where Biot and Damköhler numbers are important. Dimensionless numbers also play a major part in the correlation of experimental data.

5.9. NORMALIZATION OF BALANCE EQUATIONS

The normalization of balance equations uses the techniques outlined in the previous section and applies the procedure to all conserved quantities. In order to illustrate the technique and some of the important outcomes, we consider Example 5.7.1, which gives both mass and energy balance equations for a reaction in a continuous stirred tank.

The mass balance equation can be written in intensive form as

$$V\frac{\mathrm{d}c_A}{\mathrm{d}t} = F^{(\mathrm{i})}(c_A^{(\mathrm{i})} - c_A) - rV, \tag{5.65}$$

where V was the volume of reactants [m^3], $F^{(\mathrm{i})}$ was the volumetric flowrate [$\mathrm{m}^3\, s^{-1}$], r was the reaction rate [$\mathrm{mol} \cdot \mathrm{s}^{-1} \cdot \mathrm{m}^3$], $c_A^{(\mathrm{i})}$ was the feed concentration [$\mathrm{mol} \cdot \mathrm{m}^{-3}$], and c_A was the concentration in the reactor [$\mathrm{mol} \cdot \mathrm{m}^{-3}$].

We can easily rearrange this equation by dividing through by the volume V, to give

$$\frac{\mathrm{d}c_A}{\mathrm{d}t} = \frac{F^{(\mathrm{i})}(c_A^{(\mathrm{i})} - c_A)}{V} - r \tag{5.66}$$

or

$$\frac{\mathrm{d}c_A}{\mathrm{d}t} = \frac{(c_A^{(\mathrm{i})} - c_A)}{\tau} - r, \tag{5.67}$$

where τ has the units of [s], being a characteristic time constant for the reactor.

We know from reaction engineering that the inlet reactant concentration c_{A_i} will be the maximum value of reactant A. Hence, we can use this value as our reference and define a new dimensionless concentration:

$$\hat{c}_A = \frac{c_A}{c_A^{(i)}}. \tag{5.68}$$

For time, t we can use the characteristic time constant τ as a normalizing quantity to obtain:

$$\hat{t} = \frac{t}{\tau}. \tag{5.69}$$

This value will, however, not be confined to the range (0, 1), but this is not critical. Both $\hat{t}$ and $\hat{c}_A$ are dimensionless. Using these normalizing definitions, we can now substitute for the variables in the original equation (5.67) to give the following expression:

$$\frac{c_A^{(i)}}{\tau}\frac{d\hat{c}_A}{d\hat{t}} = \frac{(c_A^{(i)} - c_A^{(i)}\hat{c}_A)}{\tau} - k\hat{c}_A c_A^{(i)}, \tag{5.70}$$

where k is the reaction rate constant ($k_0\, e^{E/(RT)}$) which on rearranging gives:

$$\frac{d\hat{c}_A}{d\hat{t}} = 1 - \hat{c}_A - (k\tau)\hat{c}_A. \tag{5.71}$$

This is now a dimensionless equation with the major parameter being $k\tau$, which itself is dimensionless. This is called the Damköhler number. In this case, the most important variable is the Damköhler number as it controls the dynamics of the reactor. Using experimental data and plotting the log of $\hat{c}_A$ versus τ will lead to an estimate of the Damköhler number as the slope and hence an estimate of the reaction rate constant k.

5.10. STEADY-STATE LUMPED PARAMETER SYSTEMS

In some circumstances we might be interested only in the steady state of the process. The general mass and energy balances can be modified to give the steady-state balances by simply setting all derivative (time varying) terms to zero. Hence, we arrive at the equivalent steady-state mass, component mass and total energy balances corresponding to Eqs (5.3), (5.4) and (5.21). These are:

Steady-state total mass balance

$$0 = \sum_{j=1}^{p} F_j - \sum_{k=1}^{q} F_k; \tag{5.72}$$

Steady-state component mass balance

$$0 = \sum_{j=1}^{p} f_{i,j} - \sum_{k=1}^{q} f_{i,k} + g_i, \quad i = 1, \ldots, n; \tag{5.73}$$

Steady-state energy balance

$$0 = \sum_{j=1}^{p} F_j(\hat{U} + P_1\hat{V}_1 + \hat{K}_E + \hat{P}_E)_j - \sum_{k=1}^{q} F_k(\hat{U} + P_2\hat{V}_2 + \hat{K}_E + \hat{P}_E)_k + Q + \hat{W}. \tag{5.74}$$

These equations are typically solved using some form of iterative numerical solver such as Newton's method or a variant. Steady-state balances form the basis for the substantial number of process flowsheeting programs which are routinely used in the process industries. These include ASPEN PLUS [30], HYSIM [31] and PRO II.

5.11. ANALYSIS OF LUMPED PARAMETER MODELS

5.11.1. Degrees of Freedom Analysis

In the same way that algebraic equations require a degree of freedom analysis to ensure they are properly posed and solvable, dynamic models also require a similar analysis.

The basis concept of DOF analysis is to determine the difference between the number of variables (unknowns) in a given problem, and the number of equations that describe a mathematical representation of the problem.

Thus,

$$N_{\mathrm{DF}} = N_{\mathrm{u}} - N_{\mathrm{e}},$$

where N_{DF} is the number of DOF, N_{u} the number of independent variables (unknowns) and N_{e} the number of independent equations.

There are three possible values for N_{DF} to take:

(a) $N_{\mathrm{DF}} = 0$. This implies that the number of independent unknowns and independent equations is the same.

A *unique* solution may exist.

(b) $N_{\mathrm{DF}} > 0$. This implies that the number of independent variables is greater than the number of independent equations.

The problem is *underspecified* and a solution is possible only if some of the independent variables are "fixed" (i.e. held constant) by some external considerations in order that N_{DF} be reduced to zero. Some thought must be given to which system variables are chosen as fixed. In the case of optimization these DOF will be adjusted to give a "best" solution to the problem.

(c) $N_{\mathrm{DF}} < 0$. This implies that the number of variables is less than the number of equations.

The problem is *overspecified*, meaning that there are less variables than equations. If this occurs it is necessary to check and make sure that you have included all relevant equations. If so, then the solution to such a problem is one which best "fits" all the equations. This is often called a "least squares" solution.

When calculating the DOF for a process model, there are several schemes which can be used to count the variables and associated equations. The key is to be consistent with your approach.

Variables in the Model

Typical variables in the model have already been discussed in the previous sections. As well as the variables there are parameters and constants which also appear. These parameters are normally fixed by the designer but they can also be varied in order to optimize the process system. They could be included in the DOF analysis for completeness. In the section that follows we can develop our analysis using both approaches. We should get the same result! The constants in the model equations are not normally alerted and as such should not be included in the DOF analysis.

Equations in the Model

All equations must be *independent*. For example, one common mistake in writing mass balances is to write the individual component balances *as well as* the overall mass balance. These are dependent since the sum of the individual component balances gives the overall mass balance.

Initial Conditions for Dynamic Models

It is important that the initial condition of *all* the differential variables is set by the user before any attempt is made to solve the problem. The initial conditions should not be included as part of the DOF analysis but should be assumed as essential to the correct establishment of the model.

We will see that there are circumstances which arise where it is not possible to *independently* select the initial condition for each state.

Selecting Variables to Satisfy the Degrees of Freedom

In many situations you will not be free to select just *any variable* you please as a specified or design variable. In larger systems which are composed of a number of components or processing units, it is not valid to overspecify one unit and leave another underspecified so that the global DOF is satisfied. Hence, it is important to check that in large problems, the subsystems are not overspecified.

Lumped parameter models generally take the form (semi-explicit):

$$\frac{\mathrm{d}y}{\mathrm{d}t} = f(y, z, t), \tag{5.75}$$

$$0 = g(y, z, t), \tag{5.76}$$

where y is the vector of differential or state variables, $[y_1, y_2, \ldots, y_n]^{\mathrm{T}}$ and z the vector of algebraic variables (auxiliary variables), $[z_1, z_2, \ldots, z_q]^{\mathrm{T}}$. f and g are general nonlinear functions.

Here, we have n differential (state) equations and m algebraic equations, where $n = \dim f$ and $m = \dim g$. All up we have $n + m$ independent equations.

The number of independent equations is then:

$$N_e = \dim f + \dim g, \tag{5.77}$$

where $\dim f$ is the number of independent differential equations and $\dim g$ the number of independent algebraic equations. It is normally the case that we have n states in our system with q auxiliary or algebraic variables ($q \geq n$).

Therefore, the number of DOF in this case is

$$N_{DF} = n + q - \dim f - \dim g \tag{5.78}$$

Consequently, if $N_{DF} > 0$, one has to specify

$$n + q - \dim f - \dim g \tag{5.79}$$

variables before any attempt can be made to solve the problem. The variables that are selected to be specified before solving the dynamic model are often called *design* variables or simply *specified* variables.

EXAMPLE 5.11.1 (*A simple tank model*). Consider a simple tank model shown in Fig. 5.6. Here, we want to check how the DOF requirement is satisfied with different specifications.

The process can be described by the following four equations:

$$\frac{dh}{dt} = \frac{1}{A}(F_1 - F_2), \tag{f_1}$$

$$F_1 - C_V\sqrt{P_1 - P_2} = 0, \tag{g_1}$$

$$F_2 - C_V\sqrt{P_2 - P_3} = 0, \tag{g_2}$$

$$P_2 - P_0 - \rho g h = 0. \tag{g_3}$$

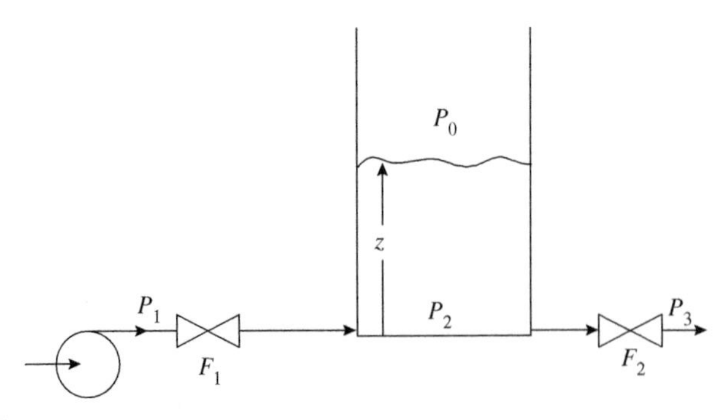

FIGURE 5.6 Simple tank problem.

Analysis shows that:

- The state variable is h.
- Auxiliary variables: $P_0, P_1, P_2, P_3, F_1, F_2$ with h being the only *differential* variable, the others being *algebraic*.
- The model parameters are: A, C_V and ρ.

Hence, the number of variables is $N_u = 7$, the number of model equations is $N_e = 4$, and thus the DOF are $N_{DF} = N_u - N_e = 3$. ■ ■ ■

Note that we could have also included the parameters in the list of variables giving a total of ten (10) system variables and parameters. In doing this, we then have to recognize that we have three extra equations (or assignments) since each parameter has to be given a specific value, e.g. $A = 12.5\,\text{m}^2$. The final degrees of freedom however remain unchanged (i.e. $N_{DF} = N_u - N_e = 10 - 7 = 3$). What is important is a consistent approach. It is recommended that both variables and parameters be included in any DOF analysis.

In the above example, the most common choice of the three design variables would be P_0, P_1 and P_3 (remembering that h at time zero is specified). The system equations can then be solved for the four (4) unknowns h, F_1, F_2 and P_2. Other choices are possible but some combinations may lead to difficulties when an attempt is made to solve the resulting equation set with the commercially available numerical methods. To illustrate these difficulties, let us compare the specification above with a slightly different one:

$$\text{Specification 1: } S_1 = [P_0,\ P_1,\ P_3]^T, \tag{5.80}$$

$$\text{Specification 2: } S_2 = [P_0,\ P_1,\ P_2]^T. \tag{5.81}$$

Specification 1

In case of Specification 1, the remaining variables in the equation set are h, F_1, F_2, P_2. Most commercial packages solve the algebraic set g for the algebraic variables z (pretending the differential variables have already been initialized), and then use the value of the algebraic variables in the differential equation set f to calculate the derivative $\dot{x}$. This is then used in the integration step to calculate the new value of the differential variable x at time $(t_o + \Delta t)$.

To start this procedure, the algebraic equation set has to be solvable in the algebraic variables z. In other words, the Jacobian $g_z = (\partial g/\partial z)$ must not be singular. That is, there are no rows or columns which are fully zero or no row or column is simply a multiple or a linear combination of another row/column(s).

In the case of Specification 1, the algebraic set has the following structure. Here, we mark an entry if the particular variable occurs in a given equation. This is commonly known as "incidence" matrix or "occurrence" matrix.

$$\begin{array}{cccc} & F_1 & F_2 & P_2 \\ g_1 & \times & & \times \\ g_2 & & \times & \times \\ g_3 & & & \times \end{array} \tag{5.82}$$

The above occurrence matrix shows that the Jacobian is non-singular. Obviously, (g_3) can be solved for P_2, (g_2) for F_2 and (g_1) for F_1.

This specification is consistent and leads to "smooth" solution procedures. The initial value of h can be given arbitrarily.

Specification 2

Specification 2, however, leads to some difficulties if we try to apply the same procedure to solve the resulting equation set.

The first problem is encountered when we try to solve the algebraic equation set for the algebraic variables. Although we specified the value of independent variables and the degrees of freedom requirement was satisfied, g is not solvable in the remaining algebraic variables! We can see this from the occurrence matrix given by

$$\begin{array}{cccc} & F_1 & F_2 & P_3 \\ g_1 & \times & & \\ g_2 & & \times & \times \\ g_3 & & & \end{array} \tag{5.83}$$

The Jacobian of the algebraic equation set is clearly singular, since the occurrence matrix contains no entry in the row of g_3.

However, equation

$$P_2 - P_0 - \rho g h = 0$$

contains the differential variable h. Since P_2 and P_0 are specified, and an initial value is assigned to h, this equation in itself is also over-specified. In other words, if P_2 and P_0 are both specified, the initial estimate for h must not be an arbitrary value, but should be consistent with g_3.

This example gives rise to the question whether such problems can be identified and tackled with any general procedure. To answer this question we have to introduce a new term, the *index* of a DAE set.

5.11.2. High-Index Differential-Algebraic Equations

DEFINITION 5.11.1. *The index is the minimum number of differentiations with respect to time that the algebraic system of equations has to undergo to convert the system into a set of ODEs.*

The index of a pure ODE system is zero by definition. If the index of a DAE is one (1), then the initial values of the differential variables can be selected arbitrarily, and easily solved by conventional methods such as Runge–Kutta and Backward Differentiation methods.

If, however, the index is higher than 1, special care should be taken in assigning the initial values of the variables, since some "hidden" constraints lie behind the problem specifications.

The requirement of index-1 for a DAE set is equivalent to the requirement that the algebraic equation set should have Jacobian of full rank with respect to the algebraic variables. That is,

$$g_z = \left(\frac{\partial g}{\partial z}\right) \tag{5.84}$$

must be non-singular.

This statement can be easily verified. For the general DAE system we have

$$\frac{dy}{dt} = f(y, z, t), \tag{5.85}$$

$$0 = g(y, z, t). \tag{5.86}$$

Differentiating the second equation with respect to time once leads to

$$0 = g_y \cdot \frac{dy}{dt} + g_z \cdot \frac{dz}{dt}, \tag{5.87}$$

$$0 = g_y \cdot f + g_z \cdot \frac{dz}{dt}. \tag{5.88}$$

Hence,

$$\frac{dz}{dt} = -g_z^{-1} g_y f, \tag{5.89}$$

and so we have produced a set of pure ODEs given by

$$\frac{dy}{dt} = f(y, z, t), \tag{5.90}$$

$$\frac{dz}{dt} = -g_z^{-1} g_y f(y, z, t), \tag{5.91}$$

if and only if g_z is of full rank. In this case, g_z is invertible and one differentiation led to an ordinary differential equation set.

If g_z is singular, then the index is greater than one, i.e. INDEX > 1. This is often referred to as a "higher-index" problem.

A typical higher-index problem arises if some algebraic variables are missing from some of the algebraic equations.

Let us take the extreme case when all algebraic variables are missing from all algebraic equations:

$$\frac{dy}{dt} = f(y, z, t), \tag{5.92}$$

$$0 = g(y). \tag{5.93}$$

Differentiating the algebraic constraints once leads to

$$0 = g_y \frac{dy}{dt} = g_y \cdot f(y, z). \tag{5.94}$$

Since dz/dt has not appeared yet, we have to differentiate the above equation once again:

$$0 = g_{yy} \frac{dy}{dt} \cdot f(y, z) + g_y \cdot f_z(y, z) \cdot \frac{dz}{dt} + g_y f_y \frac{dy}{dt}. \tag{5.95}$$

If dz/dt is to exist, then the coefficient $g_y f_z$ must be non-zero and hence invertible. If this is the case, the DAE system is of index 2.

EXAMPLE 5.11.2 (A linear DAE system). We consider a simple linear DAE system given by

$$\dot{x}_1 = x_1 + x_2 + z_1, \qquad (f_1)$$

$$\dot{x}_2 = x_1 - x_2 - z_1, \qquad (f_2)$$

$$0 = x_1 + 2x_2 - z_1; \qquad (g_1)$$

let us investigate the index of this system. To do so, we differentiate the algebraic constraint g_1 to get

$$0 = \dot{x}_1 + 2\dot{x}_2 - \dot{z}_1$$

or

$$\dot{z}_1 = \dot{x}_1 + 2\dot{x}_2.$$

Substitute for $\dot{x}_1$ and $\dot{x}_2$ from (f_1) and (f_2) to get

$$\dot{z}_1 = 3x_1 - x_2 - z_1.$$

Hence, this algebraic constraint has been converted to an ODE after 1 differentiation. This system is INDEX = 1.

As an alternative, consider a change in the algebraic constraint (g_1) to

$$0 = x_1 + 2x_2.$$

Differentiate this for the first time to get

$$0 = \dot{x}_1 + 2\dot{x}_2$$

and substitute from (f_1) and (f_2) to get

$$0 = 3x_1 - x_2 - z_1.$$

Clearly, this first differentiation has not produced a differential equation in z_1, hence we differentiate once more to get

$$0 = 3\dot{x}_1 - \dot{x}_2 - \dot{z}_1,$$

$$\dot{z}_1 = 3\dot{x}_1 - \dot{x}_2$$

and

$$\dot{z}_1 = 2x_1 + 4x_2 + 4z_1.$$

■ ■ ■ This result shows that the DAE set is INDEX = 2.

Why the Problem?

It might be asked why the index is of importance in DAE systems. It is *not* an issue for pure ODE systems but when the INDEX > 1 the numerical techniques which are used to solve such problems fail to control the solution error and can fail completely.

5.11.3. Factors Leading to High-Index Problems

It has already been seen that inappropriate specifications lead to problems with high-index. There are at least three main reasons why high-index problems arise. These include:

(i) Choice of specified (design) variables.
(ii) The use of forcing functions on the system.
(iii) Modelling issues.

It must be said that in all the above situations there may be inappropriate cases which lead to a high-index problem. Other situations are valid and truly lead to high-index problems.

However, numerical routines are generally incapable of handling these high-index situations. We prefer to model in such a way that we obtain an index-one (1) problem.

Choice of Design Variables

In case (i), we have seen that the choice of design variables can lead to high-index problems. This was shown in the example with the simple tank problem. We can follow this through the following example.

EXAMPLE 5.11.3 (Specifications on the simple tank model). Consider again the simple tank example of Example 5.11.1. Here, we chose as Specification 2 P_0, P_1 and P_2 as design variables leaving the unknowns as F_1, F_2 and F_3, plus the state variable h. If we differentiate the algebraic constraints we obtain

$$\frac{dF_1}{dt} = 0,$$

$$\frac{dF_2}{dt} = -\frac{C_v}{2}(P_2 - P_3)^{-1/2}\frac{dP_3}{dt},$$

$$0 = 0 - \rho g \frac{dh}{dt}.$$

The last equation, after substitution gives

$$F_1 - F_2 = 0.$$

This does not yet constitute a set of first-order differential equations, so we can differentiate again to give

$$\frac{dF_1}{dt} = \frac{dF_2}{dt},$$

and hence

$$\frac{dF_1}{dt} = \frac{dF_2}{dt} = 0$$

and

$$\frac{dP_3}{dt} = 0.$$

This is an index-2 problem. A cursory observation on the system will easily show that the suggested specification leads to a steady-state situation which is reflected in the above analysis.

Use of Forcing Functions and Modelling Issues

These reasons can also lead to high-index problems. In the case of forcing functions we can add extra equations to the DAE set which are expressed only in terms of the existing variables.

Inconsistent equations due to poor modelling practice can also lead to high-index problems.

More details on these issues can be found in Gani and Cameron [38] and Hangos and Cameron [39].

5.11.4. Consistent Initial Conditions

Accompanying the issue of high-index is that of consistent initial conditions. With a pure ODE system we are required to set the initial conditions of the states (differential variables) x, in order to solve the problem. That is,

$$\dot{x} = f(x, t), \tag{5.96}$$

we need $x(0) = x_0$.

We can, in this case, set $x(0)$ to arbitrary values.

However, it can be the case with DAE systems that the additional algebraic equations impose added constraints such that it is not possible to arbitrarily set the values of $x(0)$. This can be seen from the previous examples.

EXAMPLE 5.11.4 (Consistent initial conditions of a linear DAE system). The previous Example 5.11.2 with the algebraic constraint

$$0 = x_1 + 2x_2 \tag{5.97}$$

led to the relation

$$0 = \dot{x}_1 + 2\dot{x}_2. \tag{5.98}$$

So, the initial conditions of the states are related and

$$x_1(0) = -2x_2(0), \tag{5.99}$$

$$\dot{x}_1(0) = -2\dot{x}_2(0). \tag{5.100}$$

This means that only one state can be specified in an arbitrary way.

EXAMPLE 5.11.5 (Consistent initial conditions for a tank problem). In the simple tank model (see in Example 5.11.1) we gave a specification (Specification 2) $[P_{\text{atm}}, P_1, P_2]^T$ which led to an index-2 problem.

The implication of the analysis is that there is no possibility of selecting $h(0)$ to be an arbitrary value, since

$$\dot{h}(0) = 0. \tag{5.101}$$

One other major issue is to do with the initialization of the *algebraic* variables z. It is often the case that the numerical routine will fail at the starting point if the values $x(0)$ are not sufficiently close to the steady-state values corresponding to $x(0)$.

Thus, we have a consistent set $x(0)$ and often need to ensure that the value of the algebraic residuals given by

$$g(x(0), z, t) \tag{5.102}$$

is close to zero by choosing the initial values $z(0)$ to make

$$\| g(x(0), z(0), t) \| < \epsilon.$$

■ ■ ■ In some cases, it can mean solving an algebraic set to get an initial steady state.

The problem of specifying consistent initial conditions for DAEs was first addressed by Pantelides [40] and has been the subject of various approaches since then [41]–[43]. It can often be a frustrating issue and is very dependent on the problem structure.

5.11.5. Model Stability and Stiffness (Ill-Conditioning)

One of the major problems which arises in process modelling is the issue of model stiffness or ill-conditioning. This characteristic of the model in turn puts significant demands on the numerical techniques which can be used to solve such problems.

The characteristic of ill-conditioning can be seen in a number of ways:

(i) the "time constants" of the process,
(ii) the eigenvalues of the model equations.

Many physical systems display behaviour which is characterized by a wide range of time constants in the system. The "time constant" reflects in some way just how slowly or quickly a particular component or phenomenon reacts to a disturbance in the system or to the action attributed to it. For example, in a reaction system where we have a CSTR, there may be reactions which occur very quickly, whilst others are very slow. The *fast reactions* are associated with *small time constants* whilst the *slow reactions* are related to the *large time constants* of the system. These time constants are related to the eigenvalues of the system equations as

$$\tau_i \propto \frac{1}{\lambda_i}, \tag{5.103}$$

where τ_i is the time constant and λ_i is the real part of the system eigenvalue.

Hence, components (small time constants) correspond to large eigenvalues and vice versa. Where we have a system with a mixture of widely varying time constants we expect to see ill-conditioned behaviour.

To appreciate the reason for the difficulties encountered in some problems we turn to investigate the *underlying stability* of the mathematical model. It tells us something about how errors are propagated as we solve the problem.

5.12. STABILITY OF THE MATHEMATICAL PROBLEM

Key Concept

The propagation of errors is not only dependent on the type of method used but is influenced dramatically by the behaviour of the problem, notably by the integral curves, which represent the family of solutions to the problem. This is clearly dependent on the individual problem.

Before looking at analysing the problem to appreciate various behaviours, we consider a general linear problem which gives some insight into solving general problems.

EXAMPLE 5.12.1 (Stability of a simple linear ordinary differential system). Before addressing the general nonlinear ODE system, let us look at a simple 2 variable problem to illustrate some basic characteristics.

Consider the problem

$$\frac{dy}{dt} = y' = \mathbf{A}y + \phi, \tag{5.104}$$

$$\begin{pmatrix} y_1' \\ y_2' \end{pmatrix} = \begin{pmatrix} -2000 & 999.75 \\ 1 & -1 \end{pmatrix} \begin{pmatrix} y_1 \\ y_2 \end{pmatrix} + \begin{pmatrix} 1000.25 \\ 0 \end{pmatrix}, \tag{5.105}$$

where $y_1(0) = 0;\ y_2(0) = -2$

The *exact* solution is given by

$$y_1(t) = -1.499e^{-0.5t} + 0.499e^{-20005t} + 1, \tag{5.106}$$

$$y_2(t) = -2.999e^{-0.5t} - 0.0025e^{-20005t} + 1. \tag{5.107}$$

Note that the first terms in each equation represent the slow transients in the solution whilst the second terms are the fast transients (see the exponential terms). Finally, the constant terms represent the steady state values. The slow transients determine just how long it takes to reach steady state. We can note that the fast transient is over at $t \cong 0.002$ whilst the slow transient is over at $t \cong 10$.

If we solved this with a classical Runge–Kutta method, we would need about 7000 steps to reach steady state. Even though the fast transient has died out quickly, the eigenvalue associated with this component still controls the steplength of the method when the method has a finite limit on the solution errors.

We need to analyse the general problem to appreciate the implications.

5.12.1. The Mathematical Problem

Now let us consider the general nonlinear set of ODEs given by

$$y' = f(t, y) \quad y(a) = y_0, \qquad t \in (a, b). \tag{5.108}$$

The behaviour of the solution to the problem near a particular solution $g(t)$ can be qualitatively assessed by the linearized variational equations given by:

$$y' = \mathbf{J}(t, g(t))[y - g(t)] + f(t, g(t)), \tag{5.109}$$

where $\mathbf{J} = \partial f_i/\partial y_j$ is evaluated at $(t, g(t))$.

Since the *local* behaviour is being considered, the Jacobian could be replaced by a constant matrix $\mathbf{A}$ provided the variation of $\mathbf{J}$ in an interval of t is small. This represents a "snap shot" of the problem.

Assuming that the matrix $\mathbf{A}$ has distinct eigenvalues λ_i, $i = 1, 2, \ldots, n$ and that the eigenvectors are v_i, $i = 1, 2, \ldots, n$ the general solution of the variational equation has the form

$$y(t) = \sum_{i=1}^{n} c_i e^{\lambda_i t} v_i + g(t). \quad (5.110)$$

It is the value of $e^{\lambda_i t}$ which characterizes the local response to perturbations about the particular solution $g(t)$.

There are *three important cases* related to the eigenvalues, which illustrate the three major classes of problems to be encountered.

Unstable Case

Here, some λ_i are positive and large, hence the solution curves fan out. A very difficult problem for any ODE method. This is *inherent instability* in the mathematical problem.

This phenomenon can be seen in the following example.

EXAMPLE 5.12.2 (Unstable ordinary differential equation). Consider the solution of the following ordinary differential equation (ODE-IVP):

$$y' = y - t, \quad y(0) = 1. \quad (5.111)$$

This has the general solution

$$y = Ae^{t} + t + 1, \quad (5.112)$$

so that the dominant term Ae^t is eliminated by applying the initial conditions to give $A = 0$. Hence, the dominant term is *theoretically absent* but when small errors are introduced in the *numerical* solution this term can grow faster than the true solution.

This is an inherently unstable problem with a positive eigenvalue of $\lambda = 1$. It is clear that as this problem is integrated numerically, the solution will continue to grow without bound as time heads for infinity.

Some process engineering problems have this type of characteristic. Some catalytic reactor problems can exhibit thermal runaway which leads to an unstable situation when a critical temperature in the reactor is reached. Certain processes which have a control system installed can also exhibit instability due to unsatisfactory controller tuning or design.

Stable Case

Here the λ_i have negative real parts and are small in magnitude and hence the solution curves are roughly parallel to $g(t)$. These are reasonably easy problems to solve, using conventional explicit techniques like Euler or Runge–Kutta methods. Stable problems are also common in process engineering.

Ultra Stable Case

Here, some λ_i are large and negative (there are others that are small and negative) and the solution curves quickly converge to $g(t)$. This behaviour is good for propagation of error in the ODE but not for a numerical method. This class of problems is called "*stiff*". When inappropriate numerical methods such as Euler's method is applied to an ultra stable problem then there is bound to be difficulties.

These problems occur frequently in process dynamics, reaction kinetics, control system responses, heat transfer, etc.

Other terms for "stiffness" are "ill-conditioning", or "widely separated time constants" or "system with a large Lipschitz constant".

It should be noted that stiffness is a property of the mathematical problem not the numerical method.

The phenomenon of stiffness is illustrated by the following example.

EXAMPLE 5.12.3 (An example of a stiff ordinary differential equation). Consider the problem given by

$$\frac{dy}{dt} = \mathbf{A}y, \quad y(0) = [2, 1]^{\mathrm{T}}, \quad t \in [0, 10], \tag{5.113}$$

where the matrix $\mathbf{A}$ is given by

$$\mathbf{A} = \begin{pmatrix} -500.5 & 499.5 \\ 499.5 & -500.5 \end{pmatrix}. \tag{5.114}$$

The exact solution to this linear ODE is

$$y_1 = 1.5e^{-t} + 0.5e^{-1000t}, \tag{5.115}$$

$$y_2 = 1.5e^{-t} - 0.5e^{-1000t}. \tag{5.116}$$

Eigenvalues of the Jacobian of $\mathbf{A}$ are $\lambda_1 = -1000$; $\lambda_2 = -1$. Hence, the problem has the initial transient followed by the integration of the slower transient as seen in the analytic solutions. However, in the "stiff" region, which is close to when the steady state is being approached, the fast component that has already decayed to a very small value still controls the steplength since

$$h_{\max} = \frac{\text{stability bound for method}}{\text{Re}(\lambda_{\max})}. \tag{5.117}$$

In the case of Euler's method, the stability bound is located at -2.

5.12.2. Comments

1. It can be seen from the above example that the solution contains a very fast component e^{-1000t} and a slow component e^{-t}. Obviously, the fast component quickly decays to zero and during this time, the numerical method will be taking small steps to preserve accuracy while integrating this component. Once this component has decayed, it could be thought that the steplength of the numerical method could be increased solely on the basis of integrating the slow component until steady state is reached. However, in many cases this

is not so, since the fast component can quickly increase without bound, thus destroying the solution. It is important to minimize the steplength restriction of the method by using an appropriate numerical scheme *so that only solution accuracy determines the steplength.* In general, we can state that:

- the largest negative eigenvalue controls integration steplength, while
- the smallest eigenvalue controls the integration time to the steady state.

2. It must also be emphasized that the "stiff-region" is where the particular solution is dominating. That is, *after* the initial transient has decayed.
3. It is quite common in the literature to see reference made to the "stiffness ratio" of a problem, this being an attempt to quantify the degree of stiffness. It is generally defined as

$$S(t) = \frac{\max_i \mid \lambda_i \mid}{\min_i \mid \lambda_i \mid}. \tag{5.118}$$

This is *unsatisfactory* for several reasons, and should *not* be used to quantify the difficulty in solving a problem, since

- often some $\lambda_i = 0$ due to conservation laws;
- it is not generally true that $h^{-1} \cong \min_i(\mathrm{Re}(-\lambda_i))$ as the definition implies.

The fact that is of *greatest importance* is max $(-\lambda_i h)$, where h is the stepsize, that accuracy would allow if stability was not a limitation. Hence, as an alternative we could use the definition that

$$S(t) = \max_i(\mathrm{Re}(-\lambda_i h)) \tag{5.119}$$

noting that *the definition is dependent on the particular method being used.*

4. The example used to illustrate the problem of stiffness is linear. Most problems are in fact *nonlinear* and this means that the "stiff" region can change with the independent variable. A problem may be initially "stiff" and then become "non-stiff" or vice versa.
5. For the ultra-stable case application of an inappropriate method for stiff equations (i.e. Euler's method) can produce a numerical solution which becomes unstable.

5.13. SUMMARY

This chapter has introduced the idea of lumped parameter balance volumes in order to establish conservation equations for mass, energy or momentum. The particular balances written will depend on the purpose of the modelling as well as the physics and chemistry of the process system.

The idea of conservation balances was developed showing the general form of each balance as well as the importance of individual terms within the balance equations. In all cases, the conservation balances were written using *extensive* variables. It was shown that in some circumstances it is necessary to have the balance equations in *intensive* variable form. However, the transformation to intensive variable form can lead to more complex relationships as well as complications in the numerical solution of such systems. These issues will be dealt with in Chapter 6.

5.14. REVIEW QUESTIONS

Q5.1. What are the general expressions for total mass and component mass balances? Describe the significance of each term. (Section 5.3)

Q5.2. What is the general expression for the energy balance over a defined region? Describe the various terms in the balance equation and the significance of each. (Section 5.4)

Q5.3. When is it likely that a momentum balance would be used in modelling a process system? (Section 5.5)

Q5.4. What are dimensionless variables? Why is the concept important? (Section 5.8)

Q5.5. What is the aim of normalizing conservation balance equations? (Section 5.8)

Q5.6. Describe why conservation balances might need to be written using intensive rather than extensive variables. (Section 5.7)

Q5.7. What are the advantages and disadvantages of writing conservation balances in intensive variable form? (Section 5.7)

Q5.8. Describe the importance of DOF in modelling practice. What are the key steps in carrying out the analysis? (Section 5.11.1)

Q5.9. Why is the index of a DAE system important and how can it be assessed? What factors give rise to high-index problems? (Section 5.11.2)

Q5.10. What are consistent initial conditions as applied to DAE systems? Why are they important? (Section 5.11.4)

Q5.11. Describe what is meant by model stability and how is it assessed? What does this tell you about the underlying behaviour of the model of the system? (Section 5.12)

5.15. APPLICATION EXERCISES

A5.1. For the Exercise A2.1 in Chapter 2, develop the mass and energy balances for the system after defining the balance volumes for mass and energy.

A5.2. Consider the energy balance of Example 5.6.1. Normalize this equation by a suitable choice of normalizing variables.

A5.3. For Exercise A2.2 in Chapter 2, define balance boundaries and then develop the conservation balances for both mass and energy for those balance volumes.

A5.4. Take the problem in Exercise A3.1 of Chapter 3, and consider the case where the tank is now enclosed. Develop both mass and energy balances for this new system. What has changed? Is it significant in terms of the dynamics?

A5.5. Develop a mathematical model for the domestic water heater shown in Fig. 5.7. Water flowrate is variable but the temperature needs to be controlled to a desired setpoint. To do this, an electric immersion heating coil is adjusted by a temperature control loop. Develop the model in a structured way following the model development framework:

(a) define the problem,
(b) identify all relevant controlling factors,

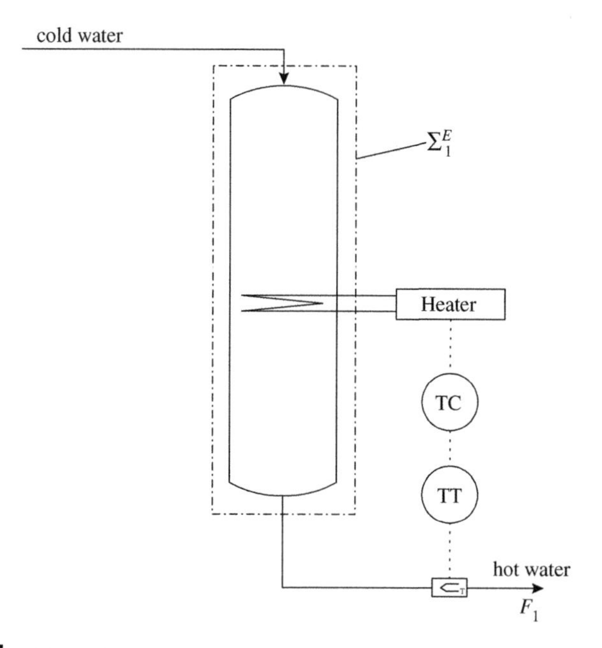

FIGURE 5.7 Domestic hot-water heater.

(c) state the data you require and where you might access it,
(d) state all your modelling assumptions and justify them,
(e) clearly indicate your balance boundaries,
(f) develop the mathematical model and clearly state the development steps,
(g) analyse the equations for degrees of freedom.

A5.6. Consider the process of blending two streams of liquid in a tank where all streams can vary in flowrate. The main stream (1) contains two components (A and B) whilst the second stream (2) contains only component B. Both streams are at different temperatures. The goal is to add stream 2 in order to maintain a fixed concentration of B in the outlet from the blending tank. The outlet from the tank is pumped under level control. Figure 5.8 shows the process.

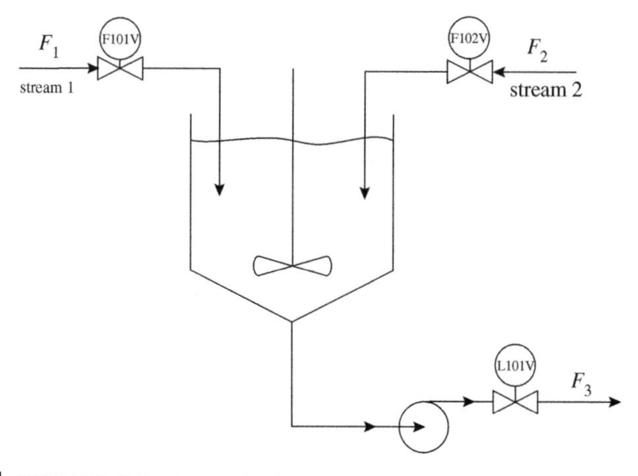

FIGURE 5.8 Liquid blending system.

Develop the model in a structured way following the model development framework and the key steps given in Exercise A5.5

A5.7. An ice cube is placed into a cylindrical plastic container of water which is initially at room temperature and begins to dissolve. Consider the case where the tumbler contents are stirred to help dissolve the ice cube. Develop a model, using the modelling framework, to predict the dynamics of the system over the short term and long term.

One of the key parameters of the system will be the overall heat transfer coefficient of the melting ice cube. Make an estimate or predict its value.

Use typical values for the system of 25°C ambient temperature for the water, container liquid volume of 250 ml and a rectangular shaped ice cube of approximately 25 cm^3. Typical aspect ratio of the ice-cube is 1:1.25:2.5. It will be necessary to state clearly the value and accuracy of all parameters used in the model.

A5.8. A CSTR is a common piece of processing equipment. In this application, consider the modelling of a CSTR for startup and control purposes. In this case, there is a cascade of three jacketted CSTRs in series with cooling provided in either a cocurrent or countercurrent direction. The cascade is shown in Fig. 5.9 and indicates a countercurrent flow situation.

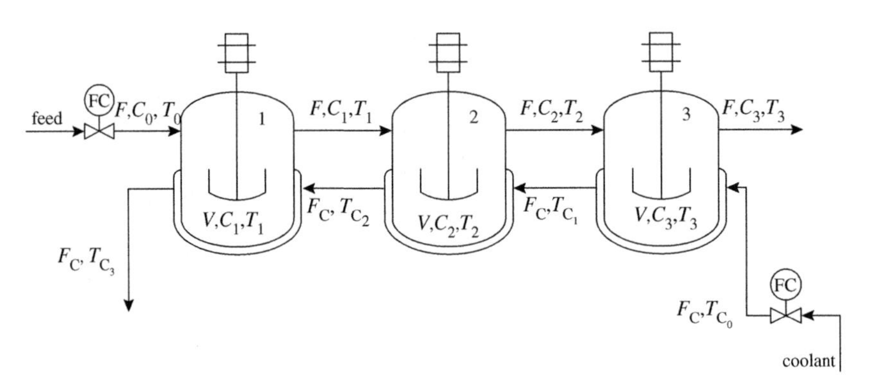

FIGURE 5.9 Cascade of continuous stirred tank reactors.

Because of the physico-chemical nature of this system, multiple steady states can occur due to the balance between cooling and heating in each reactor. This directly affects the performance of the reactors. In one case, low temperatures in both reactor and cooling systems leads to low conversions of around 20%. In contrast, the high-temperature scenario leads to conversions around 99%. Other intermediate states are also possible in this system.

Data on the problem are as follows:

- The reaction is first-order exothermic, $A \rightarrow B$, with $r = kC_A$, and $k = k_0 e^{-E/(RT)}$, $k_0 = 2000\,s^{-1}$.
- Heat transfer area per reactor, $A_H = 0.09\,m^2$.
- Reactor volume, $V_i = 0.8\,m^3$, cooling jacket volume, $V_{C_i} = 0.1\,m^3$.
- Heat capacity of mixture, $C_P = 0.565\,kJ/kgK$, heat capacity of coolant, $C_{PC} = 4.1868\,kJ/kgK$.
- Density of mixture, $\rho = 850\,kg/m^3$, density of coolant, $\rho_C = 1000\,kg/m^3$

- Feed flowrate, $F = 0.0025\,\text{m}^3/\text{s}$, coolant flowrate, $F_C = 0.002\,\text{m}^3/\text{s}$.
- Overall heat transfer coefficient, $U = 167.5\,\text{W/m}^2\,\text{K}$.
- Initial feed concentration, $C_0 = 0.6\,\text{kg mol/m}^3$.
- Heat of reaction, $\Delta H_R = 146{,}538\,\text{kJ/kg mol}$.
- Activation energy, $E = 42{,}287\,\text{kJ/kg mol}$.
- Gas constant, $R = 8.314\,\text{kJ/kg mol} \cdot \text{K}$.

A5.9. Consider the following set of equations (see [44]) derived from the modelling of a forced circulation evaporator shown in Fig. 5.10.

The model equations for the forced circulation evaporator are given by

$$\rho A_s \frac{dL_2}{dt} = F_1 - F_4 - F_2, \quad M\frac{dX_2}{dt} = F_1X_1 - F_2X_2,$$

$$C\frac{dP_2}{dt} = F_4 - F_5, \quad T_2 = a_2P_2 + b_2X_2 + c_2,$$

$$T_3 = a_3P_2 + c_3, \quad F_4 = \frac{[Q_{100} + F_1C_{p1}(T_1 - T_2)]}{\lambda},$$

$$T_{100} = a_TP_{100} + c_T, \quad Q_{100} = a_q(F_1 + F_3)(T_{100} - T_2),$$

$$F_{100} = \frac{Q_{100}}{\lambda_S}, \quad Q_{200} = \frac{UA_2(T_3 - T_{200})}{[1 + (UA_2/2C_PF_{200})]},$$

$$T_{201} = T_{200} + \frac{Q_{200}}{C_PF_{200}}, \quad F_5 = \frac{Q_{200}}{\lambda}.$$

The seven (7) parameters are specified: $\rho A_S, M, C, C_P, \lambda, \lambda_S, UA_2$.

All constants a_i, b_j, c_k in the thermo-physical correlations are known. Other variables are given on the schematic.

(a) Carry out a DOF analysis for the equation system and choose a set of design variables which satisfies the DOF.

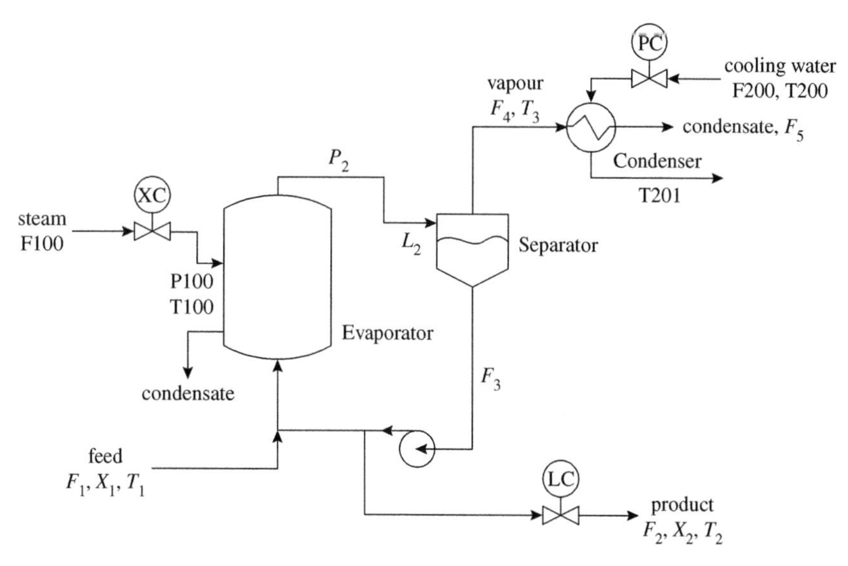

FIGURE 5.10 Forced circulation evaporator system.

(b) Check that the choice of design variables leads to an index-1 problem.
(c) Are there choices which lead to index-2 or higher systems? Explain the significance of the specifications which lead to these situations.

A5.10. Develop and analyse a dynamic model for the behaviour of a pool of diethylamine (DEA) $(CH_3CH_2)_2NH$, which forms after a spill to ground. Assume that the bounded area is 25 m^2 and that the initial spill gives a pool depth of 100 mm. A wind of 3 m/s is blowing and the ambient temperature is 26°C. The pool is formed over a concrete pad whose temperature is constant at 40°C. The incoming solar radiation level is 950 $W\,m^{-2}$.

6 SOLUTION STRATEGIES FOR LUMPED PARAMETER MODELS

This chapter considers the numerical solution of lumped parameter models resulting from the application of either steady-state or dynamic conservation balances on a process system. In particular, we consider the numerical solution of both nonlinear algebraic equations (NLAEs) and ordinary differential equations-initial value problems (ODE-IVPs). In the context of process engineering, dynamic modelling often leads to mixed systems of ODEs and NLAEs. These are termed differential-algebraic equations or DAEs. Approaches to the numerical solution of such systems are also discussed. In some cases, lumped parameter models lead to ordinary differential equations-boundary value problems (ODE-BVPs). The solution of these types of problems is discussed in Chapter 8.

This chapter commences with some typical examples of the types of problems often requiring numerical solution. It is followed by an introduction to some basic definitions of important terms frequently used in the context of numerical solution of ODEs. Consideration is then given to stability analysis of the methods and how this concept interacts with the stability of the actual problem. Several practical implications of these issues are discussed.

In presenting the numerical solution approaches, two major classes of numerical methods are introduced which have widespread application and availability. These are classified as linear multi-step methods (LMMs) and single-step methods. Several examples of both methods are introduced and their characteristics are discussed.

To conclude, we discuss the numerical solution of DAE systems using a variety of approaches and discuss the advantages and disadvantages of the techniques.

6.1. PROCESS ENGINEERING EXAMPLE PROBLEMS

The following two problems are typical of those which arise from the modelling of process engineering systems. In the first case, we have a set of ODEs whilst the second example deals with a set of DAEs.

EXAMPLE 6.1.1 (A chemical kinetics problem (Robertson [45])). The following three ODEs were derived from the analysis of a chemical reaction system involving three species (y_1, y_2, y_3) and their kinetic rate constants k_i. The initial conditions $y_i(0)$ are also stated.

We have the following system of normalized equations with their initial conditions:

$$y_1' = \frac{dy_1}{dt} = k_1 y_1 + k_2 y_2 y_3; \quad y_1(0) = 1, \tag{6.1}$$

$$y_2' = \frac{dy_2}{dt} = k_1 y_1 - k_2 y_2 y_3 - k_3 y_2^2; \quad y_2(0) = 0, \tag{6.2}$$

$$y_3' = \frac{dy_3}{dt} = k_3 y_2^2; \quad y_3(0) = 0. \tag{6.3}$$

We would like to know the time varying behaviour of this reaction system over the period of time $t = 0$ to 1000. This will show us the transient behaviour as well as the final new steady-state condition.

Typical rate constants are $k_1 = 0.04$, $k_2 = 10^4$ and $k_3 = 3 \times 10^7$. It is important to note that the rate constants range over nine (9) orders of magnitude. This puts special emphasis on the types of numerical methods we can use to solve this problem. In particular, this problem has the characteristic of being "stiff". It is also a strongly coupled problem requiring all equations to be solved simultaneously. It is typical of many process engineering applications.

EXAMPLE 6.1.2 (Tank dynamics). If we consider the dynamics of the simple tank system shown in Fig. 6.1, we have a DAE set comprising the following equations:

$$\begin{aligned}
\frac{dz}{dt} &= \frac{(F_1 - F_2)}{A}, \quad z(0) = z_0, \\
F_1 &= C_{v_1}\sqrt{P_1 - P_2}, \\
F_2 &= C_{v_2}\sqrt{P_2 - P_3}, \\
P_2 &= P_0 + \rho g z,
\end{aligned}$$

where z is the liquid height, P_i the system pressures, F_i the volumetric flowrates into and out of tank and C_{v_i} are the valve coefficients.

The challenge here is the procedures needed to solve such a system numerically. The key issue is how to handle the NLAEs along with the ODEs. Several approaches can be used which exploit the structure of the equations to break the problem down in smaller sequential calculational tasks. Other methods tackle the equations as a coupled set by solving both the ODEs and the NLAEs simultaneously. These are discussed later in the chapter.

We need to develop reliable techniques which will solve such problems easily.

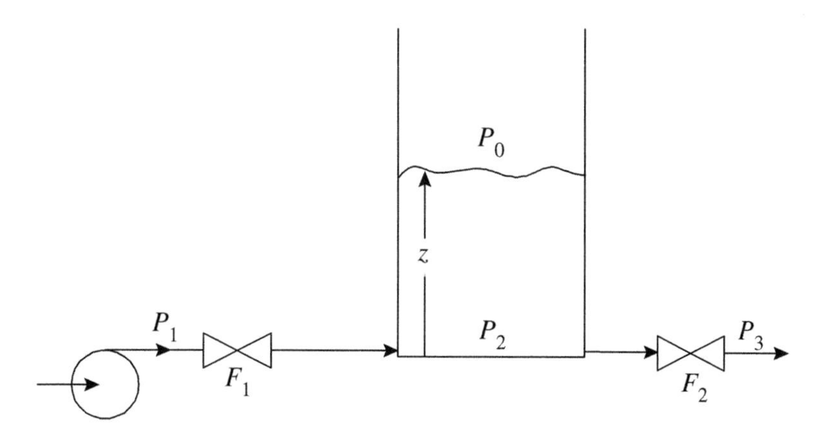

FIGURE 6.1 Simple tank dynamics.

6.2. ORDINARY DIFFERENTIAL EQUATIONS

The basic problem we wish to solve is a set of ODEs given in general terms by

$$y' = f(y, t), \quad y(a) = y_0, \tag{6.4}$$

where $t \in (a, b)$ and $y \in R^n$ (y is a n-dimensional vector of real numbers). This represents a system of first-order ODEs with initial values.

This is a vector of equations in the variables y. Note that it is necessary to state the initial values of all the y variables at $t = a$. The solution is required over the time interval (a, b).

6.2.1. Higher Order Equations

Sometimes, after modelling a process system, we have equations containing higher order derivatives. For the case of higher order systems, we can reduce these to first-order equations, which are then easily solved by standard techniques.

If we consider the following higher order ODE

$$y^{\{n\}} + f(y^{\{n-1\}}, y^{\{n-2\}}, \ldots, y^{\{2\}}, y^{\{1\}}, y) = 0, \tag{6.5}$$

where $y^{\{n\}}$ represents the nth derivative of y.

By making a series of substitutions for the derivative terms of the form:

$$y_i = y^{\{i-1\}} = \frac{\mathrm{d}^{\{i-1\}}y}{\mathrm{d}t^{\{i-1\}}}, \tag{6.6}$$

we can transform the higher order equation into a set of n first-order ODEs.

Normally the initial conditions for the higher order equation are given as

$$G_i\left(y^{\{n-1\}}(0), y^{\{n-2\}}(0), \ldots, y^{\{1\}}(0), y(0)\right) = 0, \tag{6.7}$$

and using the same substitutions we can reduce these conditions to

$$y_i(0) = y^{\{i-1\}}(0). \tag{6.8}$$

EXAMPLE 6.2.1 (A second-order system). Consider the higher (2nd) order equation

$$\frac{d^2y}{dt^2} + f(y) = 0 \quad \text{with} \ \frac{dy}{dt}(0) = c_1; \ \ y(0) = c_2. \tag{6.9}$$

To transform to a set of first-order ODEs, we let

$$y_1 = y, \tag{6.10}$$

$$y_2 = y' = \frac{dy_1}{dt}, \tag{6.11}$$

hence, we get two ODEs with their respective initial conditions:

$$\frac{dy_2}{dt} + f(\bar{y}) = 0; \quad y_2(0) = c_1, \tag{6.12}$$

$$\frac{dy_1}{dt} = y_2; \quad y_1(0) = c_2. \tag{6.13}$$

■■■

6.2.2. Conversion to Autonomous Form

This removes the explicit presence of the independent variable t from the right-hand side functions. This we call the "autonomous" form of the problem. We define a new variable y_{n+1} so that we have

$$y' = \frac{dy}{dt} = f(y), \tag{6.14}$$

$$y'_{n+1} = 1 \quad \text{and} \quad y_{n+1}(0) = 0. \tag{6.15}$$

In most modern numerical methods, the value of the independent variable can be retained in the right-hand side functions.

6.3. BASIC CONCEPTS IN NUMERICAL METHODS

Key concept

A discretization or numerical solution of an initial value problem is a sequence of discrete values $\{y_n\}_{n \in N}$*, where N is the set of integers, computed via a method* F_n *which solves at each step the algebraic equations given by*

$$F_n y_n = 0, \quad n \in N. \tag{6.16}$$

Note that F_n represents some general form of algebraic equation which is solved for y_n at each step.

EXAMPLE 6.3.1 (Euler's single-step method). Consider Euler's method where the numerical method is described by the formula:

$$F_n y_n = y_n - y_{n-1} - h_{n-1} f(y_{n-1}) = 0.$$

Here, h_{n-1} is the steplength given by $(t_n - t_{n-1})$, and $y_{n-1}, f(y_{n-1})$ are known values of the solution and the gradient from the current solution point (t_{n-1}). Euler's method simply projects the old solution y_{n-1} ahead by h_{n-1} with slope $f(y_{n-1})$. All we need to advance the step is simply to evaluate the derivative at t_{n-1} and then compute the new solution value y_n.

6.3.1. Categories of Methods

When considering the vast number of numerical methods available in the literature, it is useful to categorize the techniques into a number of distinct classes, based on criteria related to the way the solution is progressed from one step to another as well as the form of difference equations to be solved at each step. Figure 6.2 shows a simple classification of most methods although some combine aspects of both classes.

It is obvious that the algebraic equations to be solved in generating $\{y_n\}_{n \in N}$ can be of two forms, which divides numerical methods into:—*explicit* and *implicit*. We now consider these two methods plus a further criterion for classification.

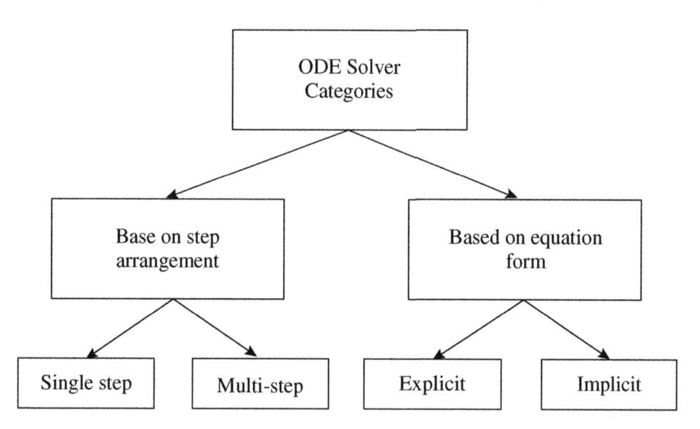

FIGURE 6.2 Categories of ODE solvers.

Explicit Methods

A numerical method F_n is *explicit* if the algebraic equations $F_n y_n$ are *explicit* in y_n.

EXAMPLE 6.3.2 (A simple single-step explicit method). Euler's method was given in the previous example. It is an example of an explicit method. In this case, we can simply substitute known values (y_{n-1}) to calculate y_n. Here, there is no need to iterate to get the next solution y_n.

Implicit Methods

A numerical method F_n is *implicit* if the algebraic equations $F_n y_n$ are *implicit* in y_n. In this case, we normally need to iterate the equations to get the solution for y_n.

EXAMPLE 6.3.3 (A simple single-step implicit method). Consider the Backward Euler method, which is an implicit version of the standard Euler method:

$$F_n y_n = y_n - y_{n-1} - h_{n-1} f(y_n) = 0.$$

Note that in this case the gradient $f(y_n)$ is being evaluated *at the end of the current step*, at time t_n. Since we do not know y_n which is the solution at the end of the current step, we have to guess it and iterate until we converge to the solution. Clearly, there is significantly more work per step in this implicit method than in the equivalent explicit method. In the case of multiple equations, we need to solve a set of NLAEs for each output point. It will be evident that there are characteristics of this method that justify its use even though it is expensive computationally.

The discretization methods besides being either "explicit" or "implicit" can also be classified as either "single step" or "multi-step" methods, depending on the amount of information used in computing the next step. Thus we have the following two methods.

Single-Step Methods

A numerical method is a *single-step* method, if it *only uses information from the last solution point* or *within the current step* to compute the new output value. Single-step methods have been the most well known of all numerical integration routines and date back many centuries.

EXAMPLE 6.3.4 (Runge–Kutta or Euler one-step methods). Euler's method or the Runge-Kutta methods are regarded as single step methods because they only use information at t_{n-1} to get to t_n. They use information from the last computed solution point y_{n-1} or within the current step (t_{n-1} to t_n) to generate the new solution value at t_n. They do not use past points of the solution from $t_{n-2}, t_{n-3}, \ldots, t_{n-k}$.

We can consider a Runge–Kutta method to see this characteristic, where the solution for y_n is given by

$$y_n = y_{n-1} + h \sum_{i=1}^{s} b_i f\left(t_{n-1} + c_i h, y_{n-1} + \sum_{j=1}^{s} a_{ij} k_j\right). \tag{6.17}$$

In this method, there is no occurrence of past solution points, $t_{n-2}, \ldots$.

Multi-step Methods

A numerical method is a multi-step method if it uses information from several past solution points to calculate the new output point.

EXAMPLE 6.3.5 (Linear multi-step methods). LMMs (of k steps) are typical examples of these methods. They are expressed as

$$\sum_{j=0}^{k} \alpha_j y_{n+j} = h \sum_{j=0}^{k} \beta_j f_{n+j},$$

where α_j and β_j are constants (either α_0 or β_0 are non-zero. Generally, $\alpha_k = 1$).

Thus, for $k - 2$ we have

$$y_{n+2} = -(\alpha_0 y_n + \alpha_1 y_{n+1}) + h(\beta_0 f_n + \beta_1 f_{n+1} + \beta_2 f_{n+2}),$$

which is a two-step method.

Remember that all the numerical methods involve computational errors since they are methods which attempt to *approximate* the solution over the range of interest. Hence, we must appreciate what type of errors are involved in our numerical procedure as well as the way these errors propagate as we attempt to solve the problem.

The key issue which we have to face is choosing an appropriate numerical method to solve our problem. The method we choose is strongly dictated by the characteristics of the problem we are solving. Appreciating that aspect of the solution procedure is vital. Hence, we consider the issues of truncation error and stability.

6.4. LOCAL TRUNCATION ERROR AND STABILITY

Here, we want to develop the important relationships between stability of the solution and the factors which affect it. Stability refers to how the errors committed by the numerical approximation method propagate as the solution progresses. We want to avoid solutions which "blow up" because of inappropriate methods or steplengths. In order to see what is important in determining this, we consider a simple mathematical problem of known solution and compare that with the solution generated by a numerical method.

First, we establish some nomenclature by letting

$$y(t_n) = \text{the exact solution at } t_n, \tag{6.18}$$

$$y_n = \text{the numerical solution at } t_n, \tag{6.19}$$

as a convention which will be used throughout this chapter.

We will consider a simple example problem by looking at its *true solution* as well as its *numerical solution*. Comparison of the two shows us how errors propagate.

6.4.1. Analytical Solution

Consider solving the linear system

$$y' = -\mathbf{A}y, \quad 0 \leq t \leq t_f \tag{6.20}$$

with initial conditions

$$y(0) = y_0. \tag{6.21}$$

The true solution from point t_{n-1} to t_n is easily calculated as

$$y(t_n) = e^{-\mathbf{A}h} y(t_{n-1}), \tag{6.22}$$

where $t_n = nh$ and $h > 0$. The exponential matrix is readily calculated.

Let us introduce the operator $S(z) = e^{-z}$, we can rewrite Eq. (6.22) so that

$$y(t_n) = S(\mathbf{A}h) y(t_{n-1}). \tag{6.23}$$

Hence, the analytical solution of the linear ODE from t_n to t_{n-1} is obtained by multiplying the last solution by the value $S(\mathbf{A}h)$, which is the matrix exponential for the set of equations. Now let us turn to the numerical solution.

6.4.2. Numerical Solution

For the numerical solution, let us use Euler's method as an example to see how the solution is progressed from step to step.

The Euler method applied to the original problem in Eq. (6.20) gives

$$y_n = y_{n-1} + hf_{n-1} = y_{n-1} - \mathbf{A}hy_{n-1} \tag{6.24}$$

with the initial conditions

$$y(0) = y_0. \tag{6.25}$$

Let the operator $K(z) = (1 - z)$ be the "amplification" factor, then we can write the numerical solution as

$$y_n = K(\mathbf{A}h) y_{n-1}. \tag{6.26}$$

Hence, the numerical solution progresses from step to step using the value $K(\mathbf{A}h)$ which is dependent on the form of the method as well as the problem.

6.4.3. Global Error

We can now define the global error as the difference between the true solution and the numerical solution. It is this value which we want to control during the solution procedure:

$$\varepsilon_n = y(t_n) - y_n. \tag{6.27}$$

Hence,

$$\varepsilon_n = S(\mathbf{A}h) y(t_{n-1}) - K(\mathbf{A}h) y_{n-1}, \tag{6.28}$$

$$\varepsilon_n = S(\mathbf{A}h) y(t_{n-1}) - K(\mathbf{A}h) y(t_{n-1}) - K(\mathbf{A}h) y_{n-1} + K(\mathbf{A}h) y(t_{n-1}), \tag{6.29}$$

$$\varepsilon_n = K\varepsilon_{n-1} + (S - K) y(t_{n-1}), \tag{6.30}$$

$$\varepsilon_n = K\varepsilon_{n-1} + Ty(t_{n-1}). \tag{6.31}$$

Here, we let $T =$ truncation operator.

Notice that the global error at t_n consists of two parts. The first is the term $K\varepsilon_{n-1}$ which represents that fraction of the previous global error which is carried forward into the present step. Obviously, this should not be magnified and hence we require

$\|K\| < 1$. The second part of the global error is the truncation error $Ty(t_{n-1})$, which represents the error committed over the current step because we are using a numerical approximation of a particular finite accuracy.

We can represent the global error over all n steps as

$$\varepsilon_n = \sum_{j=1}^{n} K^{j-1} Ty\left(t_{n-j}\right). \tag{6.32}$$

That is,

$$\varepsilon_1 = Ty(t_n), \tag{6.33}$$

$$\varepsilon_2 = KT(y(t_{n-1})), \tag{6.34}$$

and so on.

A *bound* for this error can be written as

$$\|\varepsilon_n\| \leq n \left\{ \max_{0 \leq j \leq n-1} \|K\|^j \cdot \max_{0 \leq j \leq n-1} \|Ty(t_j)\| \right\}. \tag{6.35}$$

EXAMPLE 6.4.1 (Error propagation in a numerical solution). To illustrate the effect of K on the error propagation, consider solving the ODE:

$$y' = -2y, \quad y(0) = 1, \tag{6.36}$$

using Euler's method, where

$$y_n = y_{n-1} + hf(y_{n-1}), \tag{6.37}$$

$$y_n = (1 - 2h)y_{n-1}. \tag{6.38}$$

Using steplengths of $h = 0.1, 0.25, 1.0, 1.5$, we get the following when we integrate to $t_f = 3.0$.

t	$y(t)$	$y_{(h=0.1)}$	$y_{(h=0.25)}$	$y_{(h=1)}$	$y_{(h=1.5)}$
0	1.0000	1.0000	1.0000	1.0000	1.0000
0.5	0.3678	0.3277	0.2500	—	—
1.0	0.1353	0.1074	0.0625	−1.000	—
1.5	0.0498	0.0352	0.0156	—	−2.0000
2.0	0.0183	0.0115	0.0039	+1.000	—
2.5	0.0067	0.0038	0.0010	—	—
3.0	0.0025	0.0012	0.0002	−1.0000	+4.0000

Clearly, the results show that for $h = 0.1$ and 0.25, we obtain a stable solution, with the smaller steplength giving improved accuracy. When $h = 1$, the solution oscillates and in fact the method is just on the stability bound where $|K| < 1$. That is,

$$h = 1, \quad |K| = |1 - 2h| = |-1| = 1. \tag{6.39}$$

When h is further increased to $h = 1.5$, the solution becomes completely unstable and at this steplength $|K| = 2$.

It should also be noted that as h increases the truncation error is also increasing. However the propagation of the error is ultimately controlled by the factor K. The key point is that in solving any process problem we do not want the stability boundary (characterized by the K value) to control the steplength, rather we want the desired accuracy to be the controlling factor.

From the above discussion we obtain the following important concepts.

6.4.4. Stability

A numerical method is called *stable* if

$$\|K\| \leq 1, \tag{6.40}$$

so that the global error is not amplified at each step.

6.4.5. Consistency

A numerical method is *consistent* (or accurate) of order p if

$$\|Ty\| = 0(h^{p+1})\|y\| \tag{6.41}$$

meaning that as we reduce the steplength h, we expect that the truncation error tends to zero at a rate proportional to the order.

Hence, the bound developed for the global error suggests that

$$\|\varepsilon_n\| = 0(h^p) \tag{6.42}$$

provided both stability and consistency conditions are true. Thus, we can define the local truncation error of a numerical method.

6.4.6. Truncation Error

The local truncation error ℓ_n, of a numerical method F_n is given by

$$\ell_n = F_n \Delta_n y(t) \quad (\equiv Ty(t_{n-1})), \tag{6.43}$$

where y is the true solution of the ODE and Δ_n is an operator which makes y a discrete valued function. Thus, the local truncation error is the amount by which the true solution fails to satisfy the numerical formula.

EXAMPLE 6.4.2 (Truncation error of Euler's method). Taking Euler's method we can substitute the true solution into the numerical formula to get

$$\ell_n = y(t_n) - y(t_{n-1}) - h_{n-1}y'(t_{n-1}) \tag{6.44}$$

$$= \left[y(t_{n-1}) + h_{n-1}y'(t_{n-1}) + \frac{h_{n-1}^2}{2!}y''(t_{n-1}) + \cdots \right] \tag{6.45}$$

$$- y(t_{n-1}) - h_{n-1}y'(t_{n-1}),$$

$$\ell_n = \frac{h_{n-1}^2}{2!}y''(t_{n-1}) \ldots . \tag{6.46}$$

Euler's method is, thus, first-order accurate (consistent) with a second order truncation error. ■ ■ ■

6.4.7. Comments and Observations

From the above concepts, it is worth making some comments on the importance and significance of these ideas:

(a) The control of global error depends on the value of the amplification factor $K(\mathbf{A}h)$ and the closeness of $K(\mathbf{A}h)$ to the solution operator $S(\mathbf{A}h)$.
(b) For stable error propagation, we require the absolute value or norm of the amplification factor for a numerical method to be less than 1 and that the truncation error be bounded.
(c) The closeness of K to S is a measure of the accuracy of the numerical method or truncation error.
(d) Global errors propagate in a complex manner.
(e) If the bound for the *local* error is $O(h^{p+1})$, then the bound for the global error is $O(h^p)$.

Finally, we consider the fact that very few ODE numerical codes actually control global error. Rather the codes control *local error* and we thus need to define this term.

6.4.8. Local Error

The *local error* of a numerical method is given by

$$u(t_n) - y_n, \tag{6.47}$$

where the *local* solution $u(t)$ is the solution to the ODE initial value problem:

$$u' = f(t, u); \quad u(t_{n-1}) = y_{n-1}. \tag{6.48}$$

Thus, the local error takes the solution through the last computed point (y_{n-1}, t_{n-1}). The numerical method then attempts to approximate the local solution over the step and then controls the error by demanding that

$$\|u(t_n) - y_n\| \leq \tau_n, \tag{6.49}$$

where τ_n is a user set tolerance.

Graphically, we have the situation shown in Fig. 6.3 which shows what happens from step to step. At each step in the numerical method the solution is passing to a neighbouring solution curve. The character of these "integral curves" largely determines the behaviour of the local error. It is important to analyse the behaviour of the integral curves because their behaviour has a significant effect on the choice of numerical methods to be used.

6.5. STABILITY OF THE NUMERICAL METHOD

Having analysed the problem and discovered a number of basic problem classes in Chapter 5, which range from stable to ultra-stable, we turn to analyse the numerical

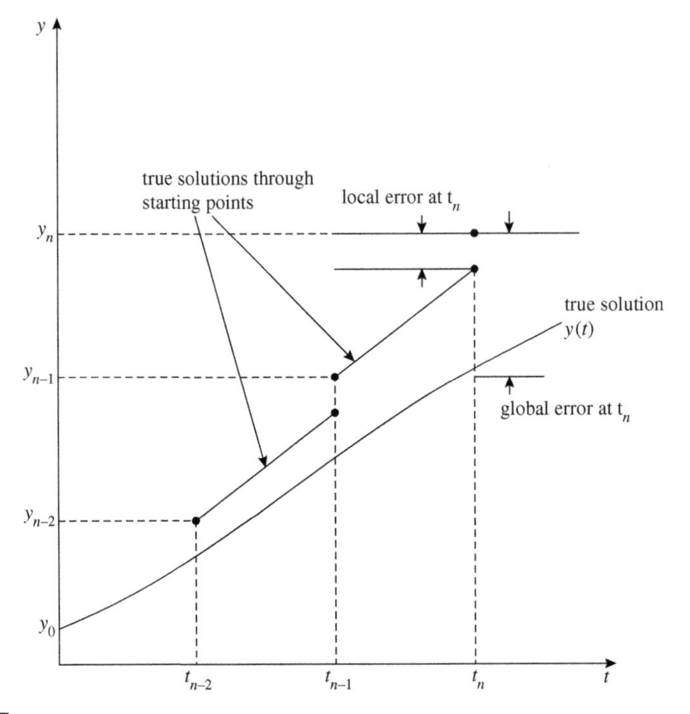

FIGURE 6.3 Local and global error concepts.

techniques which can be applied to those problems. This is important as the application of inappropriate methods to a particular problem can lead to gross inefficiencies in the solution or even to spurious results.

We noted that numerical stability is related to the propagation of perturbations throughout the solution trajectory. These perturbations include local truncation errors, errors in initial values and round-off errors in the computed values.

To ensure *convergence* to the true solution, the method must be *consistent* and *zero-stable*. Consistency ensures that the local truncation is bounded as $h \to 0$ and zero-stability ensures stable error propagation in the limit as $h \to 0$.

However, in practical methods, we want *finite steplength stability* so we desire that the method be absolutely stable for a given steplength h.

6.5.1. Absolute Stability

A numerical method is said to be *absolutely stable* for a given steplength and for a *given initial value problem*, if the global error ε_n, remains bounded as $n \to \infty$.

This is problem dependent. Hence we need some simple, standard ODE to use as a test equation.

Appropriate Test Equations

Returning to the variational equations of the general IVP, we can write that

$$y' = \mathbf{A}[y - g(t)] + f(t, g(t)). \tag{6.50}$$

Two simplifications of this equation can be made. The most usual test equation is the linear, autonomous test equation

$$y' = \lambda y, \tag{6.51}$$

although other variants have been used. Here, λ is a particular eigenvalue of the Jacobian $\mathbf{J}(t)$ of the original equation evaluated at t_n.

6.5.2. Absolute Stability for Numerical Methods

A region R_A of the complex $h\lambda$ plane is said to be a *region of absolute stability* of the method, if the method is absolutely stable for all $h\lambda \in R_A$.

Assessing the Stability Region of a One-Step Method

We must be able to assess the size and shape of the stability region so that, we can choose intelligently the right method for the problem.

For one-step methods the region of absolute stability is the region where the stability function $R(h\lambda)$ is less than one. We derive the stability function by applying the numerical method to the simple linear test problem.

This is illustrated on the example below.

EXAMPLE 6.5.1 (Stability region of Euler's method). Consider Euler's method applied to $y' = \lambda y$. Here we obtain

$$y_n = y_{n-1} + h\lambda y_{n-1} = R(h\lambda)y_{n-1}, \tag{6.52}$$

where the stability function is given by

$$R(h\lambda) = 1 + h\lambda. \tag{6.53}$$

Plotting the function $|R\,(h\lambda)\,|$ in the complex $h\lambda$ plane, we can observe the stability region in Fig. 6.4. We consider complex eigenvalues because in oscillatory problems we expect complex conjugate eigenvalues in the problem.

Here the maximum value for $h\lambda = -2$.

For a one-step method the stability function $R\,(h\lambda)$ is interpreted as a rational approximation to the exponential function. We note that the true solution for the simple test problem can be written as

$$y(t_n) = y_0\, e^{\lambda t_n} = y_0(e^{h\lambda})^n, \tag{6.54}$$

whilst the numerical solution is given by

$$y_n = y_0 R(h\lambda)^n. \tag{6.55}$$

By direct comparison of the two equations, it is clear that the rational function $R(h\lambda)$ is an approximation to $e^{(h\lambda)}$.

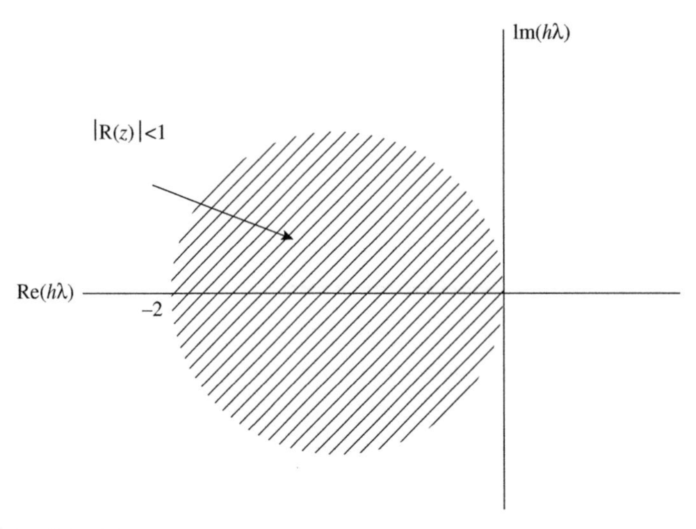

FIGURE 6.4 Absolute stability region for Euler's method.

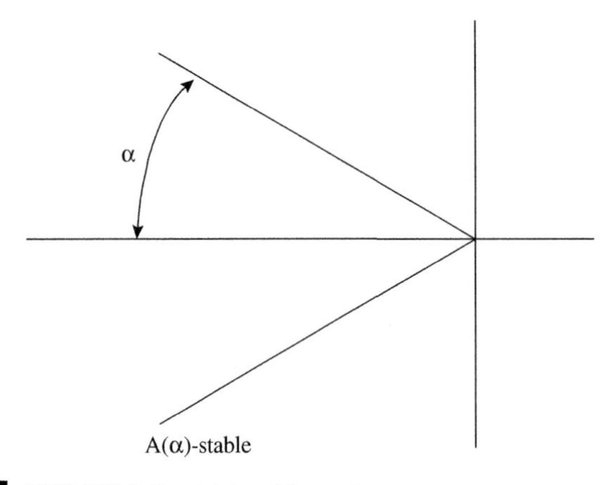

FIGURE 6.5 A (α)-stability region.

6.5.3. Stability Definitions

There are several definitions of stability which depend on the shape of the stability regions in the $h\lambda$-plane. These are given by

- Dahlquist introduced the term *A-stability* to denote the whole left-hand plane.
- Widlund called a method $A(\alpha)$*-stable* if its region of stability contained an infinite wedge.
- Cryer introduced $A(0)$*-stability* for very small α.
- Ehle and Chipman introduced *L-stability* if it is A-stable and $R(h\lambda) \to 0$ as $\text{Re}(h\lambda) \to -\infty$. This is also called "stiffly A-stable".
- Gear introduced the term "stiffly stable", if the method is stable in a region where $\text{Re}(h\lambda) \leq D$ and accurate in a region R_B, where $D \leq \text{Re}(h\lambda)\alpha$, $\text{Im}(h\lambda)\theta$.

These various regions are illustrated in Figs. 6.5 and 6.6.

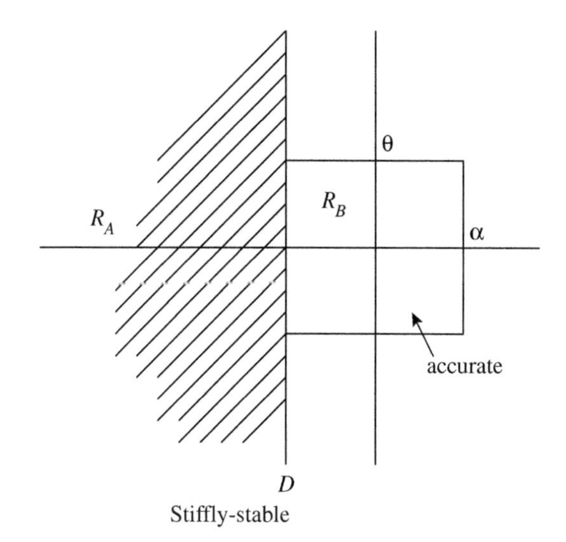

FIGURE 6.6 Stiffly stable region for Gear's method.

6.5.4. Implications of Stability Analysis

(a) The current stability definitions are based on a linear equation and hence do not necessarily apply to nonlinear problems.
(b) The point where $R(h\lambda) = 1$ gives the value of $h\lambda$ for stable error propagation.
(c) The rational approximation $R(h\lambda)$ compared to $e^{h\lambda}$ gives a measure of the accuracy of the method.
(d) If $R(h\lambda) < 0$, then the error at each step can oscillate, since there is a change of sign at each step.
(e) The stability bound for a particular problem can be assessed by considering the largest eigenvalue.
(f) For nonlinear problems, the region of stiffness can change with the independent variable (normally time).

6.6. KEY NUMERICAL METHODS

In this section, we investigate the characteristics of the most popular classes of single-step and multi-step methods. In both classes there are explicit and implicit formulae. As expected, when implicit methods are considered, we are generally forced to solve NLAEs at each step whereas for explicit methods, simple substitution allows us to solve for the next output point.

We consider in the next section the Runge–Kutta methods and in the following section we look at LMMs.

6.6.1. Runge–Kutta Methods

Of the single-step methods the most widely known and used are the Runge–Kutta formulae.

Key concept
The Runge–Kutta methods are constructed by evaluating the gradient of the differential equations at various points, called "stage" calculations and then taking a weighted average to get the new solution. In this way, the methods attempt to match an equivalent Taylor Series up to a specified order of accuracy.

We can write the general s-stage Runge–Kutta method calculating from t_n to t_{n+1} as

$$y_{n+1} = y_n + \sum_{i=1}^{s} b_i f(y_{n,i}, t_{n,i}), \tag{6.56}$$

$$y_{n,i} = y_n + h \sum_{j=1}^{s} a_{ij} f(y_{n,j}, t_{n,j}), \quad i = 1, \ldots, s, \tag{6.57}$$

where

$$t_{n,j} = t_n + c_j h \tag{6.58}$$

represent the points on the time axis where the derivative evaluations take place. These are generally located between t_n and t_{n+1}.

The above general formulation can also be written in alternative form as

$$y_{n+1} = y_n + \sum_{i=1}^{s} b_i k_i, \tag{6.59}$$

$$k_i = hf\left(t_n + c_i h, y_n + \sum_{j=1}^{s} a_{ij} k_j\right). \tag{6.60}$$

In many cases, it is convenient to represent the general Runge–Kutta methods in the form of a Butcher block, which displays the coefficients a_{ij}, b_i and c_i of the methods. This Butcher block is given by

$$\begin{array}{c|cccccc} c_1 & a_{11} & a_{12} & a_{13} & \cdots & a_{1s} \\ c_2 & a_{21} & a_{22} & a_{23} & \cdots & a_{2s} \\ \vdots & \vdots & \vdots & \vdots & \vdots & \vdots \\ c_s & a_{s1} & a_{s2} & a_{s3} & \cdots & a_{ss} \\ \hline & b_1 & b_2 & b_3 & \cdots & b_s \end{array}$$

The b_i values are the weights of the Runge–Kutta method and the values of c_i are the row sums.

By placing restrictions on the values of the parameters a_{ij} we can develop a variety of Runge–Kutta methods with different characteristics.

The principal methods are:

(a) *Explicit methods*: $a_{ij} = 0, \ j \geq i$
Here the upper triangular part of the A matrix is zero. These methods are very common and are easily implemented. They however only possess limited stability regions and are thus only suitable for non-stiff problems. Typical of these methods are the codes used in MATLAB such as ODE 23 and ODE 45.

(b) *Semi-implicit methods*: $a_{ij} = 0,\ j > i$
Here the diagonal of the matrix A is non-zero, with the upper triangular elements zero. These methods have greatly improved stability regions, which generally include the whole left hand $h\lambda$-plane. They however have the disadvantage that it is necessary to solve s sets of n nonlinear equations to advance the solution. There are certain techniques available to reduce the computational load per step.

In the case, where all the diagonal elements are equal such that $a_{ii} = \gamma$, the methods are called Diagonally Implicit Runge–Kutta methods or DIRK methods.

(c) *Fully implicit methods*: No restrictions on a_{ij}
The implicit methods are attractive from the point of view of accuracy since an s-stage method can possess $2s$ order or accuracy. They can also possess very strong stability characteristics. However, the disadvantage is that to progress the step, it is necessary to solve ns equations per step. This is a significant amount of work per step. Again, there are sophisticated ways of reducing the computational effort per step.

EXAMPLE 6.6.1 (A two-stage explicit Runge–Kutta method). Consider developing a two-stage ($s = 2$) explicit Runge–Kutta method. From the general formula we can write that

$$y_{n+1} = y_n + h(b_i f(y_{n,1}) + b_2 f(y_{n,2})), \tag{6.61}$$

$$y_{n,1} = y_n, \tag{6.62}$$

$$y_{n,2} = y_n + h a_{21} f(y_{n,1}). \tag{6.63}$$

Here, we can see that the first-stage calculation is at the start of the step (t_n) and then this value is used inside the evaluation of the second-stage calculation. These stage values are then combined to give the new solution. The advantage is that each stage calculation can be done in turn and is computationally cheap. However, the explicit Runge–Kutta methods have very restricted stability regions in the left hand $h\lambda$-stability plane.

Evaluating the Runge–Kutta Parameters

The Runge–Kutta methods are characterized by their parameters. Once the type of method as well as the number of stages have been selected, it is then necessary to evaluate the parameters a_{ij}, b_i and c_i to determine the accuracy of the method.

To do this, we expand the Runge–Kutta method and then compare terms in powers of h with an equivalent Taylor series expansion. In doing this comparison, we establish a set of algebraic relations which give the values of the parameters.

We can see this in the following example.

EXAMPLE 6.6.2 (Two-stage explicit Runge–Kutta method development). Consider the two-stage explicit Runge–Kutta method in the previous example. Let us develop a second-order method from this formula.

For the method, we can use the alternate form of the general Runge–Kutta expression and expand it as follows:

$$y_{n+1} = y_n + b_1 k_1 + b_2 k_2, \tag{6.64}$$

$$k_1 = hf\,(y_n) = hf_n, \tag{6.65}$$

$$k_2 = hf\,(y_n + a_{21}k_1) = hf_n + h^2 a_{21}\frac{\partial f}{\partial y}\frac{\mathrm{d}y}{\mathrm{d}t} = hf_n + h^2 a_{21}\left(f_y\, f\right)_n. \tag{6.66}$$

So that on substituting into the main equation (6.64) we obtain the expansion:

$$y_{n+1} = y_n + (b_1 + b_2)hf_n + b_2 a_{21} h^2 (f_y f)_n. \tag{6.67}$$

We can now perform an expansion of the general Taylor series (exact solution) for the problem $y' = f\,(y)$ to get

$$y_{n+1} = y_n + h\left.\frac{\mathrm{d}y}{\mathrm{d}t}\right|_n + \left.\frac{h^2}{2!}\frac{\mathrm{d}^2 y}{\mathrm{d}t^2}\right|_n + \cdots, \tag{6.68}$$

where the derivatives need to be expanded in the following manner, because they are functions of y:

$$\left.\frac{\mathrm{d}y}{\mathrm{d}t}\right|_n = f\,(y)|_n = f_n, \tag{6.69}$$

$$\left.\frac{\mathrm{d}^2 y}{\mathrm{d}t^2}\right|_n = \left.\frac{\partial f}{\partial y}\frac{\mathrm{d}y}{\mathrm{d}t}\right|_n = (f_y f)\big|_n, \quad \left[f_y = \frac{\partial f}{\partial y}\right]. \tag{6.70}$$

Substituting these derivatives back into the Taylor series expansion gives

$$y_{n+1} = y_n + hf_n + \frac{h^2}{2!}(f_y f) + \cdots. \tag{6.71}$$

We can now evaluate the parameters $b_1,\ b_2$ and a_{21} by matching the terms in h and h^2 from the two expansions given by Eqs. (6.67) and (6.71). This matching process gives what are termed the "order conditions" for the Runge–Kutta parameters. In the above case, we have

$$h\colon b_1 + b_2 = 1, \qquad h^2\colon b_2 a_{21} = \tfrac{1}{2}.$$

■ ■ ■

In the case of the Example 6.6.2 above, we have two equations and three unknowns, so we have one degree of freedom for the selection of one of the parameters. There are two cases we could consider; however, it is clear that there are an infinite number of *2 stage, order 2 explicit Runge–Kutta methods*.

1. Choose $a_{21} = c_2 = 0.5$, giving $b_1 = 0$ and $b_2 = 1$. Hence,

$$y_{n+1} = y_n + hf(y_n + 0.5hf_n, t_n + 0.5h) \tag{6.72}$$

This is known as "mid-point rule" and dates from the late 1700s!

2. Choose $c_2 = 1$, giving $b_1 = b_2 = 0.5$. Hence,

$$y_{n+1} = y_n + \frac{h}{2} f_n + \frac{h}{2} f(y_n + h f_n, t_n + h). \tag{6.73}$$

This is known as the "Improved Euler method". Both these methods have order two accuracy and hence the truncation error is $O(h^3)$.

The classical *4th order Runge–Kutta method* is well known and has a Butcher Block of the form

0				
0.5	0.5			
0.5	0	0.5		
1	0	0	1	
	1/6	1/3	1/3	1/6

High Order Methods

Higher order Runge–Kutta methods are given by extending the matching procedure to more and more terms. However, the higher order derivatives contain many more terms and short cut techniques are needed. Generally, the best way to get the order conditions is via the use of arborescences or tree theory.

The scale of the problem can be appreciated by realizing that for order 3 there are another two conditions to be satisfied. For order 4 there are another 4 and for order 5 there are 9 further equations giving a total of 17 NLAEs.

It must also be appreciated that is not always possible to construct explicit methods where the order of accuracy is the same as the number of stage calculations. This is because the number of order conditions increases more rapidly than the number of available Runge–Kutta parameters. The relationship between p (order) and s (stage number) is given in the following table:

Explicit Runge–Kutta methods

$p =$	1	2	3	4	5	6	7	8
$s =$	1	2	3	4	6	7	9	11

6.6.2. Implicit Runge–Kutta Methods

As previously mentioned, there are numerous varieties of Runge–Kutta methods, depending on the choice of the values of the parameters in the **A** matrix of the Butcher Block. We will look at another implicit class of methods which have improved stability properties.

Diagonally Implicit Runge–Kutta Methods

These are a subset of the semi-implicit Runge–Kutta methods already discussed. In the case of the Diagonally Implicit Runge–Kutta (DIRK) methods, all the diagonal elements in the **A** matrix are identical, giving $a_{ii} = \gamma$. The DIRK methods can be written as

$$y_{n+1} = y_n + h\sum_{i=1}^{s} b_i f(y_{n,i}, t_{n,i}), \tag{6.74}$$

$$y_{n,i} = y_n + h\sum_{j=1}^{i=s} a_{ij} f(y_{n,j}, t_{n,j}) + h\gamma f(y_{n,i}, t_{n,i}), \quad i = 1, \ldots, s, \tag{6.75}$$

where

$$t_{n,j} = t_n + c_j h, \quad j = 1, \ldots, s.$$

The Butcher Block for these methods looks like

$$\begin{array}{c|cccccc} c_1 & \gamma & & & & \\ c_2 & a_{21} & \gamma & & & \\ \cdots & \cdots & \cdots & \gamma & & \\ \cdots & \cdots & \cdots & \cdots & \ddots & \\ c_s & a_{s1} & a_{s2} & a_{s3} & \cdots & \gamma \\ \hline & b_1 & b_2 & b_3 & \cdots & b_s \end{array}$$

The DIRK methods can possess A-stability as well as stronger stability properties such as S-stability which is based on a nonlinear test function. For A-stable methods of an order p method, the value of γ needs to lie within certain bounds as given in the following table:

p	Range of γ
1	$[0.5, \infty]$
2	$[0.25, \infty]$
3	$[1/3, 1.06858]$
4	$[0.394339, 1.28057]$

The DIRK methods can also be derived so that their accuracy is $p = s + 1$. However, it means that they cannot possess stronger stability properties than A-stability. The following table shows the relationship between the stability and order for a given number of stages:

Stages in DIRK	A-stable ($p = s + 1$)	L-stable ($p = s$)
1	Yes	Yes
2	Yes	Yes
3	Yes	Yes
4	No	No

EXAMPLE 6.6.3 (Development of a single stage Diagonally Implicit Runge–Kutta method). Consider developing a one stage DIRK method. From the general formula we can write that

$$y_{n+1} = y_n + hb_1 f(y_{n,1}), \tag{6.76}$$

$$y_{n,1} = y_n + h\gamma f(y_{n,1}). \tag{6.77}$$

Here we know that $b_1 = 1$. However, we can choose γ to be either 0.5 or 1.0. If γ is 1.0 then we have an order one method, which is in fact the Backward Euler method! If γ is 0.5 then the order can be greater than one. ■ ■ ■

Solving the Implicit Formulae

Because the DIRK methods are implicit, it involves solving a set of n equations for each stage calculation using an iterative method. The most common way of doing this is via a Newton iteration. We can see that an implicit stage calculation for a DIRK method looks like

$$\bar{y} = h\gamma f(\bar{y}) + \bar{\psi}. \tag{6.78}$$

A Newton iteration applied to this equation gives the following iteration:

$$(\mathbf{I} - h\gamma\mathbf{J})\,\Delta y_{n,i}^{k+1} = h\gamma f\left(y_{n,i}^{k}\right) + \psi - y_{n,i}^{k}. \tag{6.79}$$

Here, $\mathbf{J}$ is the Jacobian of the right-hand side of the system equations. Since γ is the same from one stage calculation to the next, it is wise to use the factored iteration matrix $(\mathbf{I} - h\gamma\mathbf{J})$ for as long as possible. This linear system is usually solved using some form of Gaussian factorization technique.

6.6.3. Estimating Local Errors

A major problem with Runge–Kutta methods is the stepsize control and error estimation in the solution. Many codes use fixed steplength implementation but this is inefficient and can produce unacceptable results. There are two basic approaches for estimating error per step:

(a) *Extrapolation*: where the step is repeated by using two half steps and the resultant solution compared with that of the single step.
(b) *Embedding*: where two methods of differing order are used over the step and the difference used as an error estimator. This is the technique used in many codes such as MATLAB's ODE23 (2nd + 3rd order methods) or ODE45 (4th + 5th order methods).

Extrapolation

This is based on the Richardson extrapolation technique. Here, the integration from t_{n-1} to t_n is done twice. A full step is taken to give a solution $y_{n,h}$ and then two steps of $h/2$ are taken to give another solution at the end of the step designated by $y_{n,h/2}$.

The numerical solution of order p with step h can be written as

$$y_{n,h} = y(t_n) + h^{p+1}\psi(t_n) + O(h^{p+2}), \tag{6.80}$$

where $\psi(t_n)$ is the principal error term of the pth order formula.

For two steps at $h/2$ we obtain

$$y_{n,h/2} = y(t_n) + 2\left(\frac{h}{2}\right)^{p+1}\psi(t_n) + O(h^{p+2}). \tag{6.81}$$

If we want to estimate the principal error in the last equation, i.e. the more accurate solution $y_{n,h/2}$, we can subtract the two expressions and rearrange to get an estimate of the error (E_n):

$$E_n = \frac{(y_{n,h} - y_{n,h/2})}{2^p - 1}. \tag{6.82}$$

This error can now be added to the more accurate solution to get an "extrapolated" or more accurate value:

$$\hat{y} = y_{n,h/2} + E_n. \tag{6.83}$$

EXAMPLE 6.6.4 (Error estimator for second-order Runge–Kutta method). In applying the extrapolation technique to a second-order Runge–Kutta method the error estimator would be

$$E_n = \frac{(y_{,n,h} - y_{n,h/2})}{3}, \tag{6.84}$$

since $p = 2$.

Embedding

In this case, each step is integrated using a pth order method as well as a $(p+1)$th order method. The error in the lower order can then be estimated as

$$E_n = \|y_n^{p+1} - y_n^p\|.$$

It is then possible to proceed with either solution although it is normally advisable to use the higher order method which will be more accurate.

The advantage of embedding is that Runge–Kutta methods can be constructed so that the two solutions can be obtained by the same Butcher block. Hence, it can be a very cheap way of getting an error estimate compared with the extrapolation technique.

6.6.4. Controlling Step Size Automatically

Using either extrapolation or embedding, we can obtain an error estimator E_n. If we have a user specified error tolerance per step (τ), we can compare the two and then decide how to change the steplength to control error automatically. It is clear that if the error estimate is smaller than our desired error, then we can increase the steplength. The reverse is true. However, by how much do we change the steplength? This is clearly controlled by the accuracy (order) of the method being used. We can analyse this situation and develop a strategy to change steplength as follows.

The error in a Runge–Kutta method of order p at t_n is given by

$$E_n = h_n^{p+1}\psi(t_n) + O\left(h_n^{p+2}\right). \tag{6.85}$$

Consider that the new steplength is related to the previous step by

$$h_{n+1} = qh_n. \tag{6.86}$$

If we assume that the value of $\psi(t)$ does not change significantly over the step, then the principal error term for the next step becomes:

$$(qh_n)^{p+1}\psi(t_n) + O\left(h_n^{p+2}\right). \tag{6.87}$$

We know that we want the error on the next step to be less than or equal to the tolerance τ, so we can demand that

$$\tau = (qh_n)^{p+1}\psi(t_n) + O\left(h_n^{p+2}\right), \tag{6.88}$$

and thus,

$$q^{p+1} = \frac{\tau}{E_n}, \tag{6.89}$$

$$h_{n+1} = h_n\left(\frac{\tau}{E_n}\right)^{1/(p+1)}. \tag{6.90}$$

In practice, this new value h_{n+1} is multiplied by a safety margin of 0.8–0.9 to address the basic assumption of constant $\psi(t_n)$ in developing the technique.

6.6.5. Linear Multi-step Methods and their Solution

The Adams–Bashforth and Adams–Moulton methods are LMMs which require iterative solution techniques to solve the implicit difference equations. As previously noted in Section 6.3.1 we can write the methods as

$$\sum_{j=0}^{k}\alpha_j y_{n+j} = h\sum_{j=0}^{k}\beta_j f_{n+j}, \tag{6.91}$$

if $\beta_k = 0$ then the methods are *explicit*,

if $\beta_k \neq 0$ then the methods are *implicit*.

In the above equation we need to solve for y_{n+k}. We can rewrite this in generic form as

$$y = h\gamma f(y) + \psi, \tag{6.92}$$

where ψ contains all the past (known) information.

There are now two basic ways of solving this nonlinear system.

Functional Iteration

We have

$$y^{n+1} = h\gamma f(y^n) + \psi, \tag{6.93}$$

and we iterate to convergence so that

$$\|y^{n+1} - y^n\| \leq \varepsilon. \tag{6.94}$$

However, we know that this iterative technique will only converge when

$$|h\gamma L| < 1, \tag{6.95}$$

where L is the Lipschitz constant of f, which is given by $\|\partial f/\partial y\|$. Depending on the value of L, the steplength h will be bounded by $1/\gamma L$. In the case of a "stiff" problem, L is large and hence this iterative technique limits h.

Newton Iteration

A more robust method for large h is the Newton-like iteration which when applied to the method yields

$$(\mathbf{I} - h\gamma\mathbf{J})\Delta y^{n+1} = \psi + h\gamma f(y^n) - y^n, \tag{6.96}$$

where

$$\mathbf{J} = \frac{\partial f_i}{\partial y_j}; \quad \mathbf{I} = \text{unit matrix}. \tag{6.97}$$

Normally, this is used in production codes for Adam's methods although an option to use functional interaction is also included for solving "stable" problems, since this avoids the use of a Jacobian matrix.

Error Estimates and Step-Changing in LMMs

The derivation of LMMs can be done via Taylor series matching. The constant C_{p+1} can be identified as the principal error constant and the local truncation error can be written in the form

$$y(t_{n+k}) - y_{n+k} = C_{p+1}h^{p+1}y^{(p+1)}(t_n) + 0(h^{p+2}), \tag{6.98}$$

where $y^{(p)}$ is the pth derivative of y. Hence, for an explicit and implicit method we can write

$$L[y(t); h] = C_{p+1}h^{p+1}y^{(p+1)}(t) + 0(h^{p+2}), \tag{6.99}$$

$$L^*[y(t); h] = \hat{C}_{p^*+1}h^{p^*+1}y^{(p^{*}+1)}(t) + 0(h^{p+2}). \tag{6.100}$$

If we assume that the order of the methods is the same ($p = p^*$), then we can write that

$$C_{p+1}h^{p+1}y^{(p+1)}(t_n) = y(t_{n+k}) - y_{n+k}^{(0)} + 0(h^{p+2}) \tag{6.101}$$

and

$$\hat{C}_{p+1}h^{p+1}y^{(p+1)}(t_n) = y(t_{n+k}) - y_{n+k}^{(m)} + 0(h^{p+2}), \tag{6.102}$$

where $y_{n+k}^{(0)}$ is the predicted value of the solution by the explicit method and $y_{n+k}^{(m)}$ is the final value computed by its implicit counterpart. On subtracting and rearranging we can obtain an estimate of the principal truncation error for the implicit method.

$$\hat{C}_{p+1}h^{p+1}y^{(p+1)}(t_n) = \frac{\hat{C}_{p+1}}{C_{p+1} - \hat{C}_{p+1}}\left(y_{n+k}^{(m)} - y_{n+k}^{(0)}\right). \tag{6.103}$$

This is known as *Milne's device* for estimation of error.

We can also obtain the principal error term for the explicit method as

$$C_{p+1}h^{p+1}y^{(p+1)}(t_n) = \frac{C_{p+1}}{C_{p+1} - \hat{C}_{p+1}}\left(y_{n+k}^{(m)} - y_{n+k}^{(0)}\right). \tag{6.104}$$

We can thus use previous error estimator as a means of controlling error in the corrector (implicit) equation by using the difference between the predicted value of the solution $y^{(0)}$, and the final value $y^{(m)}$ from the corrector. Hence, the error estimate E_n is given by

$$E_n = \frac{\hat{C}_{p+1}}{C_{p+1} - \hat{C}_{p+1}}(y^{(m)} - y^{(0)}). \tag{6.105}$$

Normally, we want to control the error per step so that $E_n < \tau$. We would like to adjust the steplength h to achieve this error. Hence, we can write that

$$\hat{C}_{p+1}h_{\text{old}}^{p+1}y^{(p+1)}(t_n) \simeq E_n \tag{6.106}$$

and on the next step we want to have the error $E_{\text{new}} \simeq E_n$ so that

$$\hat{C}_{p+1}h_{\text{new}}^{p+1}y^{(p+1)}(t_n) = \tau \tag{6.107}$$

$$E_n = \tfrac{1}{2}(y^{(m)} - y^{(0)}). \tag{6.108}$$

Using the previous expressions we obtain

$$h_{\text{new}} = h_{\text{old}}\left(\frac{\tau}{E_n}\right)^{1/(p+1)} \tag{6.109}$$

in a similar manner to the Runge–Kutta methods.

EXAMPLE 6.6.5 (Error estimator for a linear multi-step pair). Consider the use of an Euler method as a predictor and the Backward Euler method as a corrector as a means of controlling error automatically.

We know that for the Euler method that $C_{p+1} = \frac{1}{2}$ and for the Backward Euler method $\hat{C}_{p+1} = -\frac{1}{2}$ so that the principal error estimate for the backward Euler method is

$$E_n = \tfrac{1}{2}(y^{(m)} - y^{(0)}) \tag{6.110}$$

and the step change mechanism is

$$h_{\text{new}} = h_{\text{old}}\left(\frac{\tau}{E_n}\right)^{1/(p+1)}. \tag{6.111}$$

In practice we are normally *conservative* and choose to take the new step as 80–90% of the calculated value.

Stability Regions of the Adams Methods

Both Adams–Bashforth and Adams–Moulton methods have limited stability regions. Typical stability regions are smaller than the equivalent order Runge–Kutta methods. This makes the methods inappropriate for the solution of "stiff" problems. Figure 6.7 shows the stability regions in the *hλ-plane* for several different order Adams methods.

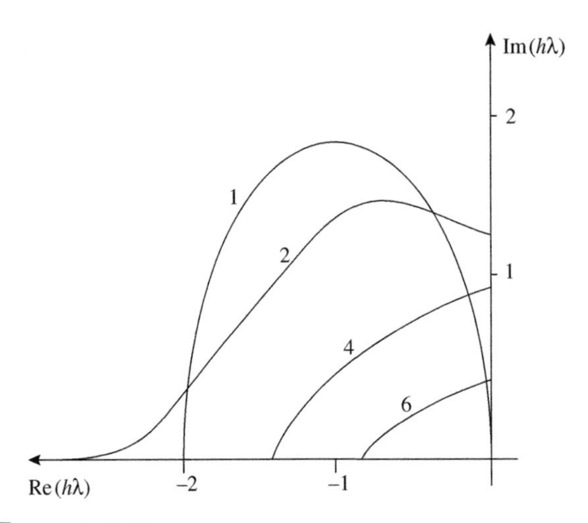

FIGURE 6.7 Stability regions for Adams methods.

A Class of Linear Multi-step Methods for Stiff Problems

If the problem being solved is "stiff", then we need to find alternatives to the classic Adams methods which have larger stability regions.

One class of implicit LMM called "the backward differentiation formulae (BDF)" has been used very successfully in solving stiff problems. They are the basis of many implementations to be found in major computer libraries such as IMSL, NAG and more recently in the MATLAB ODE solver suite (ODE15S).

From the general LMM, we can generate the following pth backward difference formula.

$$\alpha_o y_n + \alpha_1 y_{n-1} + \cdots + \alpha_p y_{n-p} = h\beta_o f(y_n, t_n). \tag{6.112}$$

That is, the derivative $f\left(y_{n,} t_n\right)$ at the nth point is represented by back values of the solutions $y_{n-1}, y_{n-2}, \ldots$.

These methods are normally implemented in variable step-variable order mode so that the computational effort is optimized. They incorporate an explicit predictor to provide an initial starting value $y_n^{(0)}$ for the unknown variable y_n, and use a Newton-like iteration for solution of the implicit formulae. Their stability regions are shown in the diagram seen in Fig. 6.8. Note that for $p = 1$ we have the backward Euler method.

Advantages of BDF methods

- Efficient on large problems with complex right-hand side due to the lower number of evaluations of the equations compared to single-step methods.
- Increasing the accuracy demanded does not increase the cost proportionately.

Problems with BDF methods

- Poor behaviour on certain oscillatory problems.
- Uncertain behaviour on unstable problems.
- Large overhead cost within the integrator.

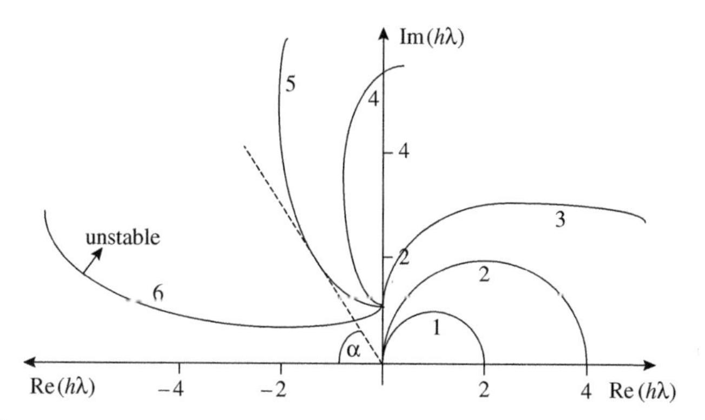

FIGURE 6.8 Stability regions for the BDF methods.

Stability regions of the BDF methods are substantially larger than the Adams methods as seen in Fig. 6.8. Note that the stable region around the imaginary axis becomes substantially less as the order increases. Higher order BDF methods approach $A(\alpha)$-*stability* regions and are less attractive when the solutions are oscillatory and thus $h\lambda$ would lie near the imaginary axis.

6.7. DIFFERENTIAL-ALGEBRAIC EQUATION SOLUTION TECHNIQUES

Differential-algebraic equation solution techniques are an extension of standard ordinary differential ODE solvers. However, there are some very significant differences between the two types of solution procedures, namely index problems and consistent initialization issues. Recall that the general form of a DAE system is

$$y' = f(y, z, t), \quad y(0) = y_0,$$
$$0 = g(y, z, t).$$

6.7.1. Substitute for Algebraic Variables

One simplistic approach, which can be used on small systems with the appropriate structure, is to substitute the algebraic variables z into the right-hand sides of the differential equations f and eliminate z completely. This procedure results in a set of ODEs with no algebraic constraints. The following example shows how this can be done on a simple tank problem.

EXAMPLE 6.7.1 (Elimination of algebraic variables to get an ODE set). Consider the equations which describe the simple tank dynamics given in Example 6.1.2. These are given as one ODE and four algebraic constraints. The equations are

$$\frac{dh}{dt} = (F_1 - F_2)/A, \tag{6.113}$$

$$F_1 = C_V\sqrt{P_1 - P_2}, \tag{6.114}$$

$$F_2 = C_V\sqrt{P_2 - P_3}, \tag{6.115}$$

$$P_2 = P_0 + \rho g h. \tag{6.116}$$

In this case, we can observe that the expression for P_2 in Eq. (6.116) can be substituted into Eqs (6.114) and (6.115) to eliminate P_2. Following this, we can substitute the expressions for F_1 and F_2 into the ODE to get the final single ODE:

$$\frac{dh}{dt} = \frac{1}{A}\left[C_V(P_1 - P_0 - \rho g h)^{\frac{1}{2}} - C_V(P_0 + \rho g h - P_3)^{\frac{1}{2}}\right]. \tag{6.117}$$

■ ■ ■

This approach appears to be useful, however there are two major drawbacks:

(a) It is often not possible to solve for z in terms of the differential variables y,
(b) When there are many algebraic variables, the substitution task becomes almost impossible.

It has also been shown that in special circumstances it is not wise to substitute to remove the algebraic constraints as information is lost.

6.7.2. Explicit ODE Solver

Here we can use an explicit ODE solver and solve the algebraic equation set each time the right-hand side of the ODEs are evaluated.

This approach attempts to decouple the algebraic and differential systems, using a simple explicit (and hence low computational cost) solver for the ODEs and then iterate to solve the accompanying algebraic system. Consider the general DAE system given by

$$y' = f(y, z, t), \tag{6.118}$$

$$0 = g(y, z, t). \tag{6.119}$$

In this case, the differential variables y at time t_n would be known by the explicit integrator and hence are usually passed to the function evaluation routine to obtain the function value f_n. Before attempting to evaluate f_n it will be necessary to evaluate z_n by solving Eqs (6.119) for the unknowns z_n. Having both y_n and z_n allows the evaluation of f_n. This is then used to advance to the new value of y_{n+1}.

Again we can illustrate this on the previous example.

EXAMPLE 6.7.2 (Using an explicit ODE solver and independent algebraic solver). If we consider the simple tank problem again, we can see that the only state or differential variable is h. The unknown values in the right-hand side of the ODE are F_1 and F_2 and these come from the solution of the three algebraic equations. Hence, having the value of the liquid height h at time t_n from the explicit ODE solver, we can proceed in the following steps:

- Solve the algebraic system of three equations for F_1, F_2 and P_2.
- Evaluate the right-hand side of the differential equation.
- Return to the integrator to update the value of liquid height at time t_{n+1}.
- Go back to the first step.

■ ■ ■

The problems with this approach are

(a) The need to solve a separate set of algebraic equations at each step.
(b) The decoupling of differential and algebraic variables.
(c) The increased load when using higher order methods like the Runge–Kutta techniques. This is a lot of work per step due to the number of internal stage calculations, which require the solution of the algebraic subsystem.
(d) The technique is very inefficient for "stiff" problems.

6.7.3. Using Implicit Differential-Algebraic Equation Solvers

Here, we use an implicit solver which simultaneously solves the ODEs and the algebraic equations in a coupled manner. This approach is the preferred option to cover most circumstances. It requires a modification of the standard implicit ODE code.

For example any implicit method, be it Runge–Kutta or Backward Differentiation Formulae (BDF) can be written as

$$y = h\gamma f(y) + \psi, \tag{6.120}$$

where h is the current steplength, γ is a constant related to the method and ψ is a vector of known information. Here, y is being solved to advance to the next solution point. Using this formula, we can solve both the states and algebraic variables simultaneously using a typical Newton iteration to give

$$\mathbf{\Gamma}^k \begin{pmatrix} \Delta y \\ \Delta z \end{pmatrix}^{k+1} = -f(y, z, t)^k, \tag{6.121}$$

where the Jacobian $\mathbf{\Gamma}^k$ is

$$\mathbf{\Gamma}^k = \begin{bmatrix} \mathbf{I} - h\gamma\,\partial f/\partial y & -h\gamma\,\partial f/\partial z \\ -h\gamma\,\partial g/\partial y & -h\gamma\,\partial g/\partial z \end{bmatrix} \tag{6.122}$$

representing the combined ODE and algebraic system. The right-hand side function value in the Newton iteration shown in Eq. (6.121) is given by

$$-f = \begin{bmatrix} \psi + h\gamma f(y^k, z^k, t) - y^k \\ g(y^k, z^k, t) \end{bmatrix} \tag{6.123}$$

and the increment vectors Δy and Δz for differential and algebraic variables are defined as

$$\Delta y^{k+1} = y^{k+1} - y^k, \tag{6.124}$$

$$\Delta z^{k+1} = z^{k+1} - z^k. \tag{6.125}$$

This set of algebraic equations is solved at each step, requiring the calculation of a Jacobian matrix and the subsequent iteration of the Newton method to some convergence criterion.

This approach (and slight variations of it) is the most widely used technique for coupled solutions. A number of well known codes are available to handle DAE systems in this way. The most prominent of the BDF based routines is the DASSL code by Petzold [46] and the suite of routines by Barton and co-workers at MIT [47]. Other codes based on variants of Runge–Kutta methods are also available such as DIRK (Cameron [48]) and Radau5 (Hairer *et al.* [49]) and ESDIRK [50]. The widely available MATLAB system has a suite of routines suitable for solving DAE systems using codes like ODE15S.

6.7.4. Exploiting Equation Structure

The final technique exploits the algebraic equation structure and seeks to reduce the problem to the smallest possible coupled DAE set or to reduce the problem to a series of simple sequential algebraic calculations before evaluating the ODEs

Here, we analyse the equations to find a partition and precedence order in the algebraic set. This technique is used extensively to decompose large sparse problems into smaller more tractable subproblems which can be solved sequentially. The same technique is applicable to the algebraic subset of the DAE problem with the intention to find partitions and a precedence order which makes the algebraic system a fully sequential calculation. After carrying out this analysis, it might be possible to structure the original model in sequential form, so that each algebraic variable is a simple explicit substitution, knowing only the differential variables y and any previously calculated algebraic variables. The calculation sequence could be

$$z_1 = q_1(y), \tag{6.126}$$

$$z_2 = q_2(y, z_1), \tag{6.127}$$

$$\vdots \tag{6.128}$$

$$z_n = q_n(y, z_{n-1}, z_{n-2}, \ldots, z_1), \tag{6.129}$$

with the ODEs being

$$y' = f(y, z, t). \tag{6.130}$$

In some cases, certain algebraic equations form coupled sets and must be solved simultaneously. This is a function of the original problem structure. This technique often shares the advantages of the simultaneous DAE solver with some efficiency in solution due to the algebraic structuring.

Partitioning and Precedence Ordering in Algebraic Systems

The key to carrying out this structural analysis is contained in two principal steps:

- establishing an output assignment for the equation system,
- generating the partitions and precedence order.

This analysis is carried out on the algebraic subset of the DAEs given by Eqs (6.118) and (6.119). The first step seeks to assign to each algebraic equation an "output" variable which is calculated knowing all other variables in the equation. This is done for all equations.

The basis of the analysis is a structural representation of the equation set which identifies which variables occur in each equation. To do this, we can construct an incidence matrix **J** for the algebraic system as defined by

$$\mathbf{J}_{ij} = \begin{cases} 1 & \text{if } z_j \text{ exists in } g_i, \\ 0 & \text{otherwise.} \end{cases} \tag{6.131}$$

In this case, we assume that the differential variables y are known values in evaluating the algebraic equations. This would be the case if a pure ODE solver were to be used.

The assignment problem of an output variable to an equation can be regarded as a permutation of the rows of **J** to produce a diagonal with non-zero elements. If the assignment does not exist then the incidence matrix is structurally singular. Many algorithms exist for finding an assignment. The method by Duff [51] is widely used.

The method then finds a permutation matrix **R** which transforms **J** into a matrix with entries on all diagonals. That is,

$$\mathbf{B} = \mathbf{RJ}. \tag{6.132}$$

If this cannot be achieved, then the matrix **J** is singular.

EXAMPLE 6.7.3 (Assignment of variables for the simple tank problem). The algebraic equations for the tank problem are given by

$$(g_1) = F_1 - C_V\sqrt{P_1 - P_2}, \tag{6.133}$$

$$(g_2) = F_2 - C_V\sqrt{P_2 - P_3}, \tag{6.134}$$

$$(g_3) = P_2 - P_0 - \rho g h. \tag{6.135}$$

In this case, the unknown algebraic variables are P_2, F_1 and F_2. The differential variable is the liquid height h.

The incidence matrix is given by

$$\mathbf{J} = \begin{matrix} \\ g_1 \\ g_2 \\ g_3 \end{matrix} \begin{matrix} P_2 & F_1 & F_2 \\ \end{matrix} \begin{pmatrix} 1 & 1 & 0 \\ 1 & 0 & 1 \\ 1 & 0 & 0 \end{pmatrix}. \tag{6.136}$$

The assignment can be achieved by the following row permutations to give:

$$\mathbf{B} = \begin{matrix} g_3 \\ g_1 \\ g_2 \end{matrix} \begin{pmatrix} \boxed{1} & 0 & 0 \\ 1 & \boxed{1} & 0 \\ 1 & 0 & \boxed{1} \end{pmatrix} = \begin{pmatrix} 0 & 0 & 1 \\ 1 & 0 & 0 \\ 0 & 1 & 0 \end{pmatrix} \begin{pmatrix} 1 & 1 & 0 \\ 1 & 0 & 1 \\ 1 & 0 & 0 \end{pmatrix} = \mathbf{RJ}. \tag{6.137}$$

Hence, the system of algebraic equations is non-singular because an assignment is possible and we have assigned the variables P_2, F_1 and F_2 to be the output variables from the equations g_3, g_1 and g_2.

It is worth noting that assignments for small systems, say up to 15 variables can be easily done by hand. For hundreds of equations it is necessary to use computer based techniques.

Having established an assignment for the algebraic system, it is now necessary to partition the equations and also obtain a precedence order for the calculations. This phase can be done using a manual directed graph (digraph) representation of the equation system when the problem is small or again using matrix manipulations. This process identifies the *strong components* of the equation system which correspond to subsystems of equations which must be solved simultaneously rather than separately. The partitioning and ordering can be performed using matrix methods. The most well known is that due to Tarjan [52] which effectively finds permutations of the assignment matrix to produce a block diagonal lower matrix **M**, given by

$$\mathbf{M} = \mathbf{D}\mathbf{B}\mathbf{D}^{\mathrm{T}} \tag{6.138}$$

These algorithms are commonly used within sparse matrix packages for improving solution efficiency.

An alternative to this is the digraph technique which is found in Sargent and Westerberg [53]. Here we construct a digraph of the equation system, which consists of nodes and edges. The nodes represent the equations and the edges of the directed graph represent the variable dependencies in the equations. The construction of the digraph consists of drawing an edge from g_i to g_k, if the assigned output variable of g_i appears in g_k. This can be seen in the following example of the simple tank.

EXAMPLE 6.7.4 (Construction of digraph for tank problem). Consider the output assignments shown in Eq. (6.137). In this case, there are three nodes. For node g_1 the output variable is F_1 and this does not appear in any other equation. Hence, there is no edge from node g_1. For g_3 the output variable is P_2 and this appears in both g_1 and g_2, hence there are two edges out of g_3. This simple digraph is given in Fig. 6.9. ■ ■ ■

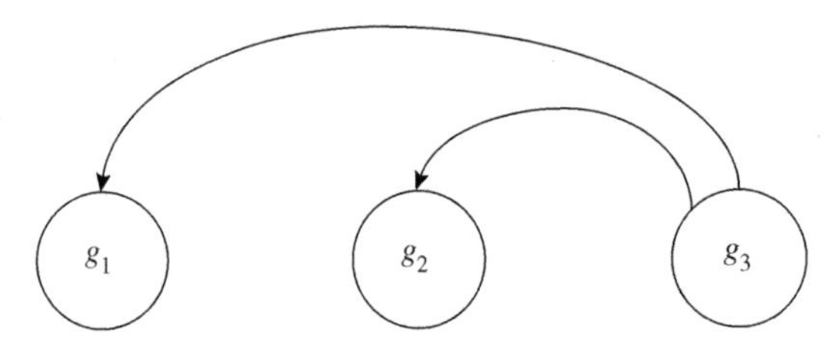

FIGURE 6.9 Digraph of equation system.

The analysis of the digraph to obtain the partitions and precedence order proceeds by

- starting at any node and trace edges until:
  – There is a node with no output. Delete that node from the digraph and add it to a list. Rearrange the digraph after deleting the node and edges going to it.
  – A former node in the path is reached. Merge all the nodes between the occurrences, into a composite node, since these represent an information loop. Rearrange the digraph and continue.
- when all nodes have been removed to the list then the partitioning is given by the composite nodes and the precedence order is the reverse order of the list.

The next example illustrates this.

EXAMPLE 6.7.5 (Partitioning and precedence order from the digraph). Starting with the g_3 node in the digraph in Fig. 6.9, we trace along an edge to g_2. This has no output, hence we can delete it from the graph and put it on our list of terminal nodes as LIST= g_1. The digraph is then seen in Fig. 6.10

Clearly the next edge tracing starts at g_3 and ends at g_1 where there are no output edges. Hence, we remove g_1 from the graph and add it to the list which becomes LIST $= (g_1, g_2)$. The digraph is left with a single node g_3, which can also be added to the list to give the final list sequence of LIST $= (g_1, g_2, g_3)$.

The list shows three separate partitions, of one equation each with a calculation sequence commencing at g_3 and going to g_2 and g_1. Thus, we would solve g_3 for P_2 followed by g_2 for F_2 and finally g_1 for F_1.

In this case, there are no strong components in this graph and hence no sets of algebraic equations which require simultaneous solution. Also, it needs to be mentioned that the precedence order is not unique. Neither is the assignment.

In many cases, problems which appear to be complex DAE sets can be structured in such a way that all the algebraic variables can be calculated sequentially given the values of the differential variables, as would be the case in using an explicit ODE solver.

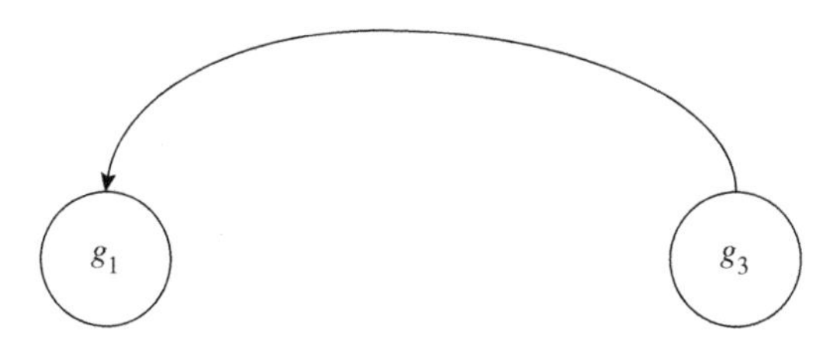

FIGURE 6.10 Reduced digraph.

6.8. SUMMARY

In this chapter, we have considered the numerical solution of lumped parameter models arising from conservation and constitutive equations. In many cases, these equation systems lead to couple DAEs. This chapter dealt with the key numerical concepts of error propagation and the stability of the solution. These were complemented by a discussion of the various generic types of numerical techniques available to the modeller.

The importance of matching the characteristics of the solver to the characteristics of the mathematical problem were emphasised. This is vitally important if efficient and reliable solutions are to be obtained.

The final section considered the solution of DAE systems. Here, there are two basic approaches worth considering. One considers exploiting the underlying structure of the coupled algebraic subsystem so as to improve robustness and ease of solution. The other solves the DAE set as a fully coupled system using implicit methods.

6.9. REVIEW QUESTIONS

Q6.1. Describe the characteristics of the two principal classes of problems which arise from lumped parameter modelling. (Section 6.1)

Q6.2. How do you handle higher order differential systems with the currently available numerical routines for first ODEs or DAEs? (Section 6.2.1)

Q6.3. Outline the major categories of ODE solvers and describe their basic characteristics. When is it appropriate to use explicit numerical methods? (Section 6.3.1)

Q6.4. What factors contribute to the global error of a numerical solution? What is the "local error" associated with a numerical solution? What are the key issues in controlling error? (Section 6.4)

Q6.5. Describe what is meant by the stability of the numerical solution and show how this is related to both the problem and to the particular numerical method being used. (Section 6.5)

Q6.6. Describe what is the truncation error of a numerical method and a procedure by which it can be estimated. (Section 6.4.6)

Q6.7. How is absolute stability of a numerical method described? What role does it play in determining the step size of the numerical routine? (Section 6.5.2)

Q6.8. Write down the general formula which describes the Runge–Kutta methods. How is this displayed as a Butcher Block? (Section 6.6.1)

Q6.9. What three principal types of Runge–Kutta methods can be derived? Describe the characteristics of these methods in terms of accuracy, work per step and stability. (Section 6.6.1)

Q6.10. What methods are typically used for local error control in Runge–Kutta and LMMs? What advantages and disadvantages exist for the different approaches? (Section 6.6.3)

Q6.11. Give four ways by which DAE systems could be solved and discuss the advantages and disadvantages of each. (Section 6.7)

6.10. APPLICATION EXERCISES

A6.1. Consider the following sets of ODEs and convert them to sets of first-order equations, using the transformations given in Section 6.2.1.

(a)

$$\frac{d^2u}{dt^2} + \frac{du}{dt}\frac{dv}{dt} = \sin t, \qquad \frac{dv}{dt} + v + u = \cos t.$$

(b)

$$\frac{d^2y}{dt^2} - (1 - y^2)\frac{dy}{dt} + y = 0.$$

(c)

$$\frac{d^2y}{dt^2} + \frac{dy}{dt} - \frac{dv}{dt} = 0, \qquad \frac{d^2y}{dt^2} + \frac{dv}{dt} + 3y + v = 0.$$

A6.2. Consider the following normalized differential equations which come from a reaction engineering application [45].

$$\frac{dy_1}{dt} = k_1 y_1 + k_2 y_2 y_3,$$

$$\frac{dy_2}{dt} = k_1 y_1 - k_2 y_2 y_3 - k_3 y_2^2,$$

$$\frac{dy_3}{dt} = k_3 y_2^2.$$

The initial conditions are given by $y(0) = [1, 0, 0]$. The values of the reaction rate coefficients are $k_1 = 0.04$, $k_2 = 10^4$, $k_3 = 3 \times 10^7$. These equations are known to be ill-conditioned or "stiff".

(a) Try solving these equations using an explicit Runge–Kutta method such as ODE23 or ODE45 within MATLAB. Solve these to local error tolerances of 1×10^{-3}, 1×10^{-4} and 1×10^{-6}. Observe the number of steps taken, the number of times the function subroutine is called and the steplength used by the solver throughout the time interval (0, 1000).

(b) Now solve the same equations with the MATLAB implicit linear multistep solver ODE15S, which is an implementation of the backward differentiation methods and hence suitable for stiff problems. Using the same local error tolerances solve the problem and note the same statistics as in part (a).

What differences are there between the two methods and why?

A6.3. The Trapezoidal Rule is a well-known numerical method where the new solution y_n is given by the formula:

$$y_n = y_{n-1} + \frac{h_n}{2}(f_{n-1} + f_n),$$

which means that the method combines both explicit and implicit behaviour, because of the f_{n-1} and f_n terms. The method has some particularly interesting stability properties.

(a) Examine the Trapezoidal Rule to determine the truncation error of the method as outlined in Section 6.4.6. What is the principal truncation error term?

(b) Calculate the form of the stability function K for the method by applying it to the linear test problem $y' = \lambda y$ as was done in Section 6.5.2. Sketch the absolute stability region in the complex $h\lambda$ plane. What features are important from the region and particularly its behaviour on stiff or unstable problems?

A6.4. Solve the simple ODE given by

$$\frac{dy}{dt} = 5(y - t^2),$$

where $y(0) = 0.08$ and $t = (0, 5)$.

(a) Use a simple fixed step Euler method written in MATLAB using steplengths $h = 0.1$ and $h = 0.01$. Comment on the behaviour of the solutions and deduce why the behaviour occurs.

(b) Use a simple Backward Euler method written in MATLAB and again solve the problem for the same fixed steplengths used in part (a). Comment on the solutions and also comparisons of the solution between the two methods. Examine the stability characteristics of the problem in attempting to explain the numerical behaviour.

A6.5. Determine the maximum stable steplength that could be taken with Euler's method when solving the following ODE problem.

Solve the following stiff ODE system given by

$$\frac{dy_1}{dt} = Ky_1y_3 + Ky_2y_4,$$

$$\frac{dy_2}{dt} = -Ky_1y_4 + Ky_2y_3,$$

$$\frac{dy_3}{dt} = 1 - y_3,$$

$$\frac{dy_4}{dt} = -y_4 - 0.5y_3 + 0.5.$$

The initial conditions are given by

$$y = [1.0, 1.0, -1.0, 0.0]^T$$

(a) Solve this set of equations for $K = 10^4$ and $t = (0, 2)$ using the BDF method such as ODE15S in MATLAB. Use a local error tolerance of 1×10^{-4}. Plot the value of $(y_1^2 + y_2^2)^{0.5}$ versus time t generated by the solver.

(b) Now solve the problem again under the same conditions but use an explicit solver such as ODE23 or ODE45 from MATLAB. Calculate the value of $(y_1^2 + y_2^2)^{0.5}$ versus time t and compare the trend with the solution from part (a). Is there any significant difference? If so, explain the difference in terms of the underlying characteristics of the problem.

[Hint: Consider the eigenvalues of the system equations as time proceeds from 0 to 2.0. Is there a fundamental change in model behaviour through the time interval as reflected in the eigenvalues? Note that you can obtain an analytical expression for y_3 and y_4 and then substitute them into the first two equations to get the Jacobian as a function of time.]

A6.6. Consider the set of DAEs which arise from the modelling of the open tank shown in Example 6.1.2. The equations are given by

$$\frac{dz}{dt} = \frac{(F_1 - F_2)}{A}, \quad F_1 = C_{v_1}\sqrt{P_1 - P_2},$$

$$F_2 = C_{v_2}\sqrt{P_2 - P_3}, \quad P_2 = P_0 + \frac{\rho g z}{1000}.$$

A possible set of parameters and constants for the equations is given by

$$A = 12.5\,\text{m}^2, \quad C_{v_1} = 3.41\,\text{m}^3/\text{kPa}^{1/2}\text{h}, \quad C_{v_2} = 3.41\,\text{m}^3/\text{kPa}^{1/2}\text{h},$$

$$\rho = 1000\,\text{kg/m}^3, \quad g = 9.81\,\text{m/s}^2.$$

The specified system variables are

$$P_0 = 100\,\text{kPa} \quad P_1 = 400\,\text{kPa} \quad P_3 = 100\,\text{kPa}.$$

(a) Solve the equation set by converting it to a pure ODE problem and then use a simple Euler method or explicit Runge–Kutta method to solve the equation. Solve for local error tolerances of 1×10^{-3}, 1×10^{-4} and 1×10^{-6} and note the execution time and number of function evaluations in MATLAB.

(b) Set up the problem by calling a nonlinear equation solver each time the differential equations need to be evaluated. This will take the values of the differential variables and solve for the unknown values of the algebraic variables.

(c) Use an implicit DAE solver such as the MATLAB routine ODE15S which implements the BDF method. Alternatively use a simple Backward Euler method. Note the performance over a range of local error tolerances.

(d) Carry out the structural analysis of the algebraic equation set and then solve this sequential followed by the evaluation of the differential equations.

Comment on the performance of the methods used and their complexity of setup. For this problem what is the best solution approach?

A6.7. Consider the model of the evaporator system considered in Application Exercise A5.9. The system is shown in Fig. 6.11.

The model equations for the forced circulation evaporator are given by

$$\rho A_S \frac{dL_2}{dt} = F_1 - F_4 - F_2, \quad M\frac{dX_2}{dt} = F_1X_1 - F_2X_2, \quad C\frac{dP_2}{dt} = F_4 - F_5,$$

$$T_2 = a_2P_2 + b_2X_2 + c_2, \quad T_3 = a_3P_2 + c_3, \quad F_4 = \frac{\left[Q_{100} + F_1C_{p1}(T_1 - T_2)\right]}{\lambda},$$

$$T_{100} = a_TP_{100} + c_T, \quad Q_{100} = a_q(F_1 + F_3)(T_{100} - T_2), \quad F_{100} = \frac{Q_{100}}{\lambda_S},$$

$$Q_{200} = \frac{UA_2(T_3 - T_{200})}{[1 + UA_2/(2C_PF_{200})]}, \quad T_{201} = T_{200} + \frac{Q_{200}}{C_PF_{200}}, \quad F_5 = \frac{Q_{200}}{\lambda}.$$

The following seven (7) parameters are specified:

$$\rho A_S, \ M, \ C, \ C_P, \ \lambda, \ \lambda_S, \ UA_2.$$

All constants a_i, b_j, c_k in the thermo-physical correlations are known.

(a) Using the model equations and the evaporator schematic, analyse the equations to see if it is possible to solve the algebraic equations in a sequential manner. Make use of the digraph analysis to do this.

(b) Give the solution sequence for this structured set of algebraic equations.

(c) Give values for the parameters, and solve the model using an appropriate solver.

A6.8. Consider the DAE set which was derived from the modelling of the enclosed tank with adiabatic compression of gas which was given in Application Exercise A4.1. Look at possible solution techniques for this problem and compare two principal methods—one based on structuring of the equation set, the other on a direct DAE solution.

A6.9. Solve the Application Exercise of A5.7 for various values of the overall heat transfer coefficient. Investigate a numerical scheme which solves the problem as a coupled DAE set as well as considering the structuring of the equations and the use of an ODE solver. Compare the performance of the solution methods.

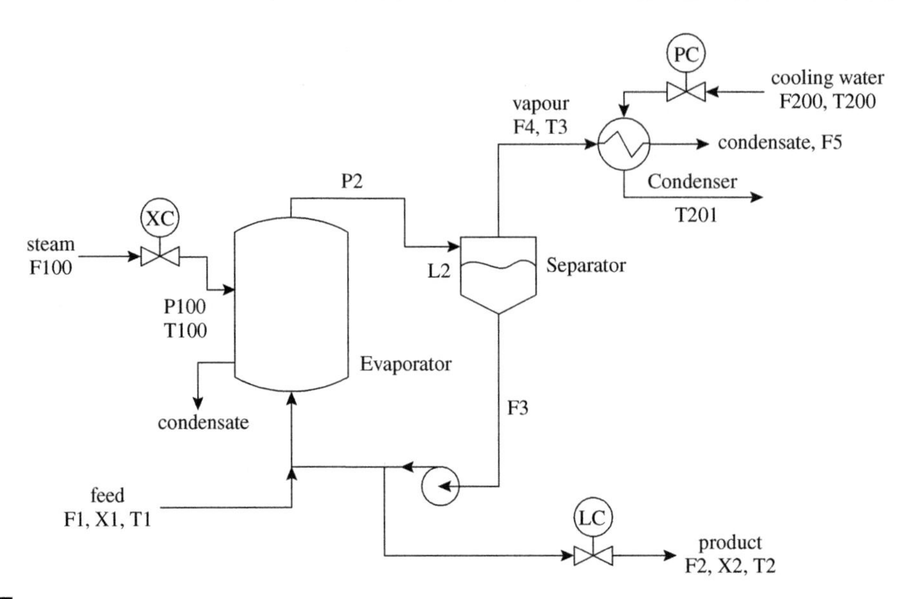

FIGURE 6.11 Forced circulation evaporator system.

A6.10. (a) Consider the cascade of CSTRs modelled in Application Exercise A5.8 and solve this problem for the given data, starting from a situation where coolant flow has been established but no reactant flow has commenced. Formulate the numerical solution by structuring the equations and using an ODE solver.

(b) Investigate alternative startup procedures in order to reach the high conversion condition.

As a guide to the system behaviour, the following approximate steady-state operating points have been found from operational experience using the counter-current cooling scheme.

Low Conversion Point

The following data were obtained:

Data	Reactor 1	Reactor 2	Reactor 3
T_{ss} [°C]	34.5	45	62
C_{ss} [kgmol/m^3]	0.576	0.537	0.462
$T_{C_{ss}}$ [°C]	25.9	25.6	25

High Conversion Point

The following data were obtained:

Data	Reactor 1	Reactor 2	Reactor 3
T_{ss} [°C]	171	174	161
C_{ss} [kgmol/m^3]	0.0775	0.009	0.00117
$T_{C_{ss}}$ [°C]	30.2	27.6	25

A6.11. Consider the dynamic behaviour of a CSTR with a recycle stream. The equations for dimensionless concentration and temperature are given by

$$\frac{dy}{dt} = -\lambda y + Da(1-y)\exp\left(\frac{\theta}{1+\theta/\gamma}\right),$$

$$\frac{d\theta}{dt} = -\lambda\theta + Da \cdot B(1-y)\exp\left(\frac{\theta}{1+\theta/\gamma}\right) - \beta(\theta - \theta_c),$$

where y is the concentration, θ the temperature, Da the Damköhler number, b the adiabatic temperature rise, β the heat transfer coefficient and λ is the recirculation coefficient.

Given that $\lambda = 1,\ Da = 0.1618,\ \beta = 3.0,\ B = 14.0,\ \theta_c = 0.0$ and $\gamma = 10^4$.

(a) Solve the coupled set of equations for two different initial conditions.

(b) Analyse the equations to check their stability at various points in the solution space.

A6.12. The following algebraic equations are the result of a modelling exercise:

$$x_1x_2 + 3x_3^2 - 5 = 0,$$

$$x_2x_3x_7 + 2 = 0,$$

$$5x_4 - x_1 - 5 = 0,$$

$$x_3x_6 + 2x_7 - 2x_1 = 0,$$

$$-2x_5 - 4x_6 + 34 = 0.$$

(a) Analyse the equations to develop a structured approach to their solution and show clearly the analysis steps you took to achieve this task.

7 DYNAMIC MODELS—DISTRIBUTED PARAMETER SYSTEMS

Distributed parameter models, as their name suggests, incorporate the spatial variation of states within the balance volume. They account for situations where the scalar field of the intensive quantities is both a function of time and position. This could be a concentration, volume-specific internal energy or temperature. What this means is that the application of the general conservation equations leads to models which are represented by partial differential equations (PDEs) in one, two or three spatial dimensions. This has already been introduced in Chapter 3, which introduced the general conservation balance over a 3D volume $\mathcal{V}$ with a closed boundary surface $\mathcal{F}$.

In this chapter, we develop the basic principles of distributed parameter system (DPS) modelling by considering the origin of such models, their general form and the approaches to their construction. A vital part of the model is represented by the constitutive equations which have been dealt with in Chapter 4.

Another very important aspect of these models is the initial and boundary conditions which apply to the particular system under consideration. These can be difficult issues to resolve in some modelling applications.

Following this chapter, we consider the task of solving such systems using a range of numerical methods.

7.1. DEVELOPMENT OF DPS MODELS

The development of DPS models is illustrated with the simple example of a packed bed catalytic reactor. The problem statement of the modelling task of the packed bed tubular catalytic reactor is given as follows:

EXAMPLE 7.1.1 (DPS model of a simple packed bed tubular catalytic reactor).

Process system

Consider a tubular reactor completely filled with catalyst and with ideal plug flow. A pseudo-first-order catalytic reaction $A \rightarrow P$ takes place in an incompressible fluid phase.

The process is shown in Fig. 7.1.

Modelling goal

Describe the behaviour of the reactor for temperature control purposes.

From the problem description above, we may extract the following modelling assumptions which will be generally valid for any further model we derive in this chapter for this example problem.

Assumptions

$\mathcal{A}1$. Plug flow is assumed in the reactor, which has constant volume.

$\mathcal{A}2$. A first-order reaction $A \rightarrow P$ takes place.

$\mathcal{A}3$. An incompressible (liquid) phase is present as a bulk phase.

$\mathcal{A}4$. A solid phase catalyst is present which is uniformly distributed.

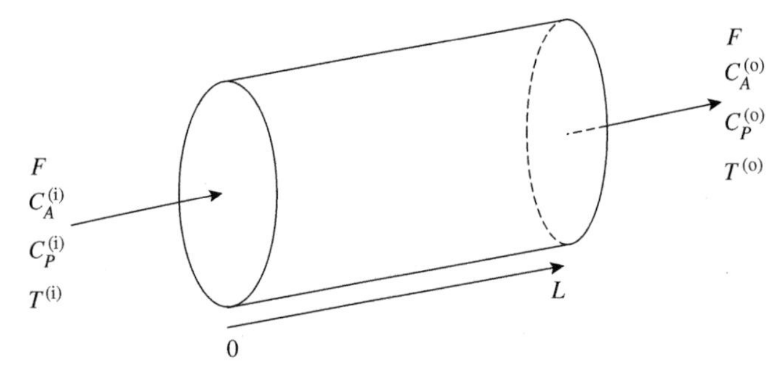

FIGURE 7.1 Simple packed bed tubular catalytic reactor.

7.1.1. Origin of DPS Models

All process engineering models based on first principles originate from the general differential form of the conservation equations given as Eq. (3.36). This general form can be expressed in various co-ordinate systems leading to a variety of different mathematical forms of the conservation equations. For example, Eq. (3.37) is expressed in 3D rectangular (Cartesian) co-ordinate system assuming the diffusion coefficient D to be constant.

This joint origin of all process models appears in the mathematical properties of their model equations. In cases when we do not neglect the spatial variations of the volume-specific intensive variables related to the conserved extensive quantities we necessarily have

- first-order spatial derivative in the directions we have convection,
- second-order spatial derivative in the directions we have diffusion,
- first-order time derivative if we have a dynamic model.

Therefore, in the general case, when no special additional assumptions are given, the mathematical form of the conservation balances is a second order (in spatial co-ordinates) partial differential equation. *Therefore, DPS models are the most general and natural form of process models.*

It has been emphasized in Chapter 4 that constitutive equations are also needed for making the model complete. These include potential variable relations and source relations are also needed to make the model complete.

7.1.2. Modelling of Distributed Parameter Systems

Balance Volumes

The modelling of a process system starts with the *balance volume definitions*. As discussed in Chapter 3, the key to modelling is to define clearly our balance regions or volumes. In the case of DPS, we define a local co-ordinate system and then carry out balances over a representative element of that system.

In the case of co-ordinates, we apply three commonly used approaches, which *are normally dictated by the geometry of the system*. These are:

- *Rectangular co-ordinates*; typically Cartesian co-ordinates in terms of the directions x, y and z.
- *Cylindrical co-ordinates*; where the key dimensions are radius (r), angle (θ) and length (z).
- *Spherical co-ordinates*; where two angles (θ and ϕ) are given, plus a radius r.

These co-ordinate systems are discussed further in Section 3.1.1.

It is up to the modeller to decide which geometry suits the actual physical system. Under certain circumstances, a 3D situation may be simplified to a 1D approach by the assumptions made about the relative effects in the other two co-ordinates. The desired accuracy, dependent on the application, may also dictate the final geometry.

It is usually the case that most physical systems assume one of these geometric shapes, or the region of interest approximates to a slab, cylinder or sphere. The co-ordinate systems and the balance volumes with their geometric spaces are given in Figs. 7.2–7.4.

Conservation Balances

Once the geometry is decided, then it is necessary to carry out the balance of mass, energy or momentum over a representative infinitesimal volume of the space. This can take the form of a "slice" or "shell" in the region of interest. The conservation balance has the following general form:

$$\left\{\begin{matrix}\text{net change}\\ \text{of quantity in time}\end{matrix}\right\} = \left\{\begin{matrix}\text{flow in}\\ \text{through boundary}\end{matrix}\right\} - \left\{\begin{matrix}\text{flow out}\\ \text{through boundary}\end{matrix}\right\} + \left\{\begin{matrix}\text{net}\\ \text{generation}\end{matrix}\right\} - \left\{\begin{matrix}\text{net}\\ \text{consumption}\end{matrix}\right\}.$$

The differential form of the conservation balances has been derived in Chapter 3 and given in a form independent of co-ordinate system in Eq. (3.36). In order to

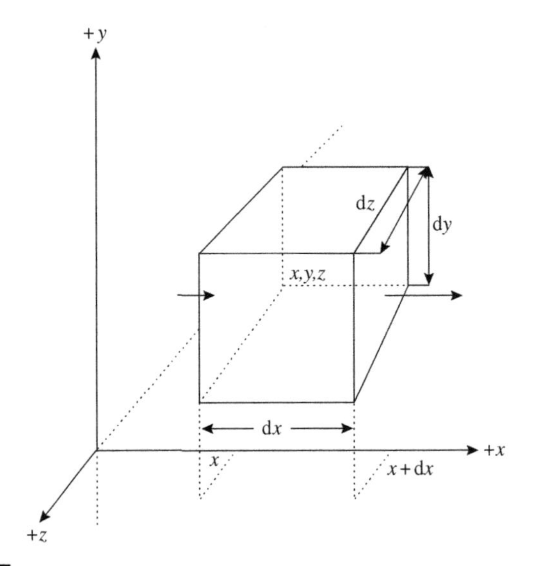

FIGURE 7.2 Balance volume in rectangular co-ordinates.

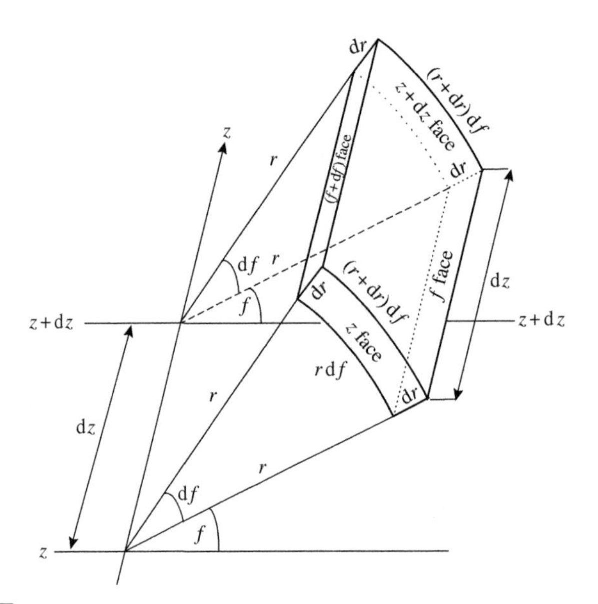

FIGURE 7.3 Balance volume in cylindrical co-ordinates.

develop the balance equations of the DPS model, we have to derive a special case of this general equation taking into account

- the geometry of the system to select the co-ordinate system,
- modelling assumptions that define the general terms in the equation.

Thereafter, the constitutive equations should also be given, based on the modelling assumptions. The following example of the packed bed plug flow catalytic reactor will illustrate how a distributed parameter model can be derived.

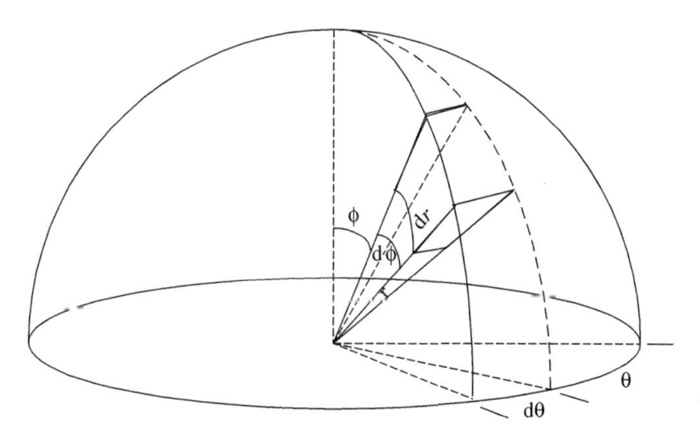

FIGURE 7.4 Balance shell for spherical co-ordinates.

EXAMPLE 7.1.2 (Derivation of the DPS model equations for the simple packed bed tubular catalytic reactor (Example 7.1.1 continued)). In order to simplify the form of the equations, we add some more simplifying assumptions.

Assumptions *additional to the ones listed as 1–4* are the following:

$\mathcal{A}5$. The reactor is uniformly distributed in its cross-section, i.e. no radial diffusion or convection takes place.
$\mathcal{A}6$. Constant physico-chemical properties.
$\mathcal{A}7$. The heat transfer through the reactor wall is considered, but the outer temperature is assumed to be constant, $T_{\rm W}$.

Balance volumes
A single volume encapsulating the whole reactor. This is now a *distributed* system. The rectangular co-ordinate system is chosen.

Model equations
Variables

$$0 \le x \le L, \quad 0 \le t, \quad C_A(x,t), \quad T(x,t), \quad \hat{U}(x,t), \tag{7.1}$$

where x is the spatial coordinate in axial direction, L the length of the reactor, t the time, C_A the reactant concentration (i.e. volume-specific component mass), $\hat{U}$ the volume-specific internal energy and T is the temperature.

The balance equations are special forms of Eq. (3.37) with one spatial co-ordinate only and with the sources being the chemical reaction and the heat transfer through the wall.

Component mass balances

$$\frac{\partial C_A}{\partial t} = D\frac{\partial^2 C_A}{\partial x^2} - F\frac{\partial C_A}{\partial x} - r_A, \tag{7.2}$$

where F is the mass flowrate of the inert incompressible fluid, D the diffusion coefficient and r_A is the reaction rate.

Energy balance

$$\frac{\partial \hat{U}}{\partial t} = \kappa \frac{\partial^2 \hat{U}}{\partial x^2} - F \frac{\partial \hat{U}}{\partial x} - \Delta H r_A - q_{\text{tr}}, \tag{7.3}$$

where q_{tr} is the heat transfer rate, κ the heat diffusion coefficient and ΔH is the reaction enthalpy.

Constitutive equations

Potential relation:

$$\hat{U} = c_P \rho T, \tag{7.4}$$

where c_P is the constant specific energy and ρ is the density.

Reaction rate relation:

$$r_A = k_0 \mathrm{e}^{-E/RT} C_A, \tag{7.5}$$

where k_0 is the pre-exponential factor, E is the activation energy, and R is the universal gas constant.

Transfer rate relation:

$$q_{\text{tr}} = K(T - T_{\text{w}}), \tag{7.6}$$

■ ■ ■ where K is the heat transfer coefficient.

It is important to note that in most cases the potential and source relations are directly substituted in the conservation equations and not shown separately as given above. If the overall mass and volume of any phases in the system are constant and the physico-chemical properties (density and specific heat) are also constant, then the energy conservation equation is directly given as a PDE for the temperature. This is, however, very poor engineering practice because the underlying assumptions are not directly shown and they may not even be valid.

Conservation Balances in Spherical and Cylindrical Co-ordinate Systems

The co-ordinate system dependent form of the conservation balance equations can be derived from the operator form of the differential balance equations in Eq. (3.36). Observe that only the convection and diffusion terms are dependent on the co-ordinate system. These forms can also be obtained using the transformation in Eqs (3.1)–(3.4).

A simple and often used example is the co-ordinate system dependent form of the diffusion term in the spherical co-ordinate system assuming uniform dispersion in all directions:

$$\frac{1}{x^2} \frac{\partial}{\partial x} \left(x^2 \frac{\partial T}{\partial x} \right). \tag{7.7}$$

7.1.3. Deriving DPS Models from Microscopic Balances

Distributed parameter system (DPS) models can be derived from considering the conservation balances to be applied at a particular point in the system. The procedure normally considers an arbitrary finite volume in the macroscopic space, develops the mass, energy or momentum conservation equations for this finite volume and then takes the limit as this element goes to a point. In this case, the principal dimensions are driven to zero. This microscopic approach can be applied to all principal geometries or developed in one chosen geometry such as in rectangular co-ordinates and then transformed to cylindrical or spherical co-ordinates by the transformations shown in Section 3.1.1. The following example illustrates the application of a microscopic balance for total mass within a 3D system.

EXAMPLE 7.1.3 (Application of mass balance in 3D space using rectangular co-ordinates). Consider the conservation of mass in the 3D region as shown in Fig. 7.2. The dimensions of the 3D volume are given by $\mathrm{d}x$, $\mathrm{d}y$ and $\mathrm{d}z$. We can denote the flow of mass per unit area in the x direction which passes into and out of the y–z faces as

$$\rho v_x, \tag{7.8}$$

$$\rho v_{x+\mathrm{d}x}, \tag{7.9}$$

where ρ is the fluid volume-specific mass (or density), and v_x, $v_{x+\mathrm{d}x}$ are the flow velocities into and out of the y–z faces. Similar flows in the y and z directions can be defined.

We can carry out the normal balance for the conservation of mass over this finite volume to give

$$\begin{aligned}\frac{\partial(\rho \cdot \mathrm{d}x \cdot \mathrm{d}y \cdot \mathrm{d}z)}{\partial t} &= (\rho v_x - \rho v_{x+\mathrm{d}x}) \cdot \mathrm{d}y \cdot \mathrm{d}z \\ &\quad + (\rho v_y - \rho v_{y+\mathrm{d}y}) \cdot \mathrm{d}x \cdot \mathrm{d}z + (\rho v_z - \rho v_{z+\mathrm{d}z}) \cdot \mathrm{d}x \cdot \mathrm{d}y.\end{aligned} \tag{7.10}$$

We can now divide both sides by the volume $\mathrm{d}x \cdot \mathrm{d}y \cdot \mathrm{d}z$ of the element and take the limit as the volume shrinks to zero (to a point), giving

$$\frac{\partial \rho}{\partial t} = -\frac{\partial}{\partial x}(\rho v_x) - \frac{\partial}{\partial y}(\rho v_y) - \frac{\partial}{\partial z}(\rho v_z), \tag{7.11}$$

since

$$\lim_{di \to 0} \frac{(\rho v_i - \rho v_{i+di})}{di} = \frac{\partial}{\partial i}(\rho v_i), \quad i = x, y, z. \tag{7.12}$$

In vector form, we can write

$$\frac{\partial \rho}{\partial t} = -\nabla \cdot \rho v, \tag{7.13}$$

which is known as the *equation of continuity* and expresses the time rate of change in fluid density in the 3D space. If the density does not change and is thus incompressible, then we have

$$\rho \nabla \cdot v = 0, \tag{7.14}$$

$$\nabla \cdot v = 0. \tag{7.15}$$

■ ■ ■

Using the approach to DPS model development as shown in Example 7.1.3, it is possible to develop general conservation equations for component mass, energy and momentum. The microscopic approach to the development of distributed parameter models is covered extensively in the book by Bird *et al.* [13] and the reader is referred to this reference for the detailed balance equations in their various forms.

7.1.4. Boundary Conditions and Initial Conditions

The specification of boundary conditions and initial conditions is extremely important in DPS models. There are several different types of boundary conditions which are applied to these systems. In order to have a well-posed model, we need to develop the governing equations as well as specifying initial conditions for time dependent problems as well as appropriate boundary conditions for the system. The following sections discuss these issues.

The derivation of the initial and boundary conditions starts with the investigation of *the domain of the PDEs in our DPS model*. For model equations that are parabolic the domain is usually semi-infinite in the time domain, and there is usually a specific initial time but no final time given. In the spatial co-ordinate directions, we usually have a finite domain, where the region of the model is a closed finite volume in space. This means, that *we most often have a semi-infinite multi-dimensional rectangle as the domain of our PDS model.*

Boundary Conditions

Boundary conditions are specified for all time on each of the boundaries of the problem. In a heat exchange problem, this might mean setting the incoming fluid temperature T and concentration C_A at $x = 0$. It can be a function of time $T(0, t)$ and $C_A(0, t)$ hence act as forcing functions or disturbances on the system.

Boundary conditions appear in three major forms which are important to consider. These forms are commonly called first or Dirichlet, second or Neumann and Robbins or third type conditions. Consider the following 2D solution domain D with boundary $\sum$ shown in Fig. 7.5

For the general equation in (x, y), we can identify three distinct boundary conditions.

(a) *The Dirichlet problem*, where the value of the function is specified on the boundary, i.e.

$$\Phi = f(x, y) \quad \text{on } \Sigma_1. \tag{7.16}$$

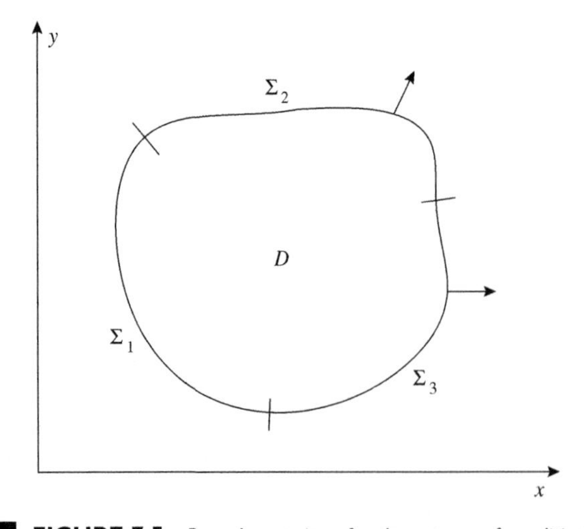

FIGURE 7.5 Boundary regions for three types of conditions.

(b) *The Neumann problem*, where the normal derivative is specified, i.e.

$$\frac{\partial \Phi}{\partial n} = g(x, y) \quad \text{on } \Sigma_2 \tag{7.17}$$

Here $\partial \Phi / \partial n$ refers to differentiation along the normal to $\sum_2$ directed away from the interior of D.

(c) *The Robbins problem, third type or mixed condition*, where we have

$$\alpha(x, y)\Phi + \beta(x, y)\frac{\partial \Phi}{\partial n} = \gamma(x, y) \quad \text{on } \Sigma_3 \tag{7.18}$$

with

$$\alpha(x, y) > 0, \quad \beta(x, y) > 0 \text{ for } (x, y) \in D. \tag{7.19}$$

Assumptions Relevant to the Boundary Conditions

Conditions on the physical boundaries of the system
For a mass transfer situation, we could interpret the Dirichlet condition as setting a concentration on the boundary,

$$C(0, t) = C_0. \tag{7.20}$$

The Neumann condition would be equivalent to a flux at the boundary. For example, for a perfectly isolated boundary at $x = 0$ orthogonal to the x co-ordinate direction the mass flux should be equal to zero, so that

$$\frac{\partial C_A}{\partial x}(0, t) = 0 \tag{7.21}$$

for the component concentration C_A.

The third type of boundary condition would be equivalent to convective mass transfer across the boundary. If again we assume a boundary orthogonal to the x co-ordinate direction and assume component mass transfer driven by the difference between the concentration within the system on the boundary at $x = x_M$ with $C_A(x_M, t)$ and a given fixed outer concentration C^* we have

$$\frac{\partial C_A}{\partial x}(x_M, t) = K\left(C^* - C_A(x_M, t)\right). \tag{7.22}$$

Number of boundary conditions
The number of independent boundary conditions along a given co-ordinate direction depends on the order of the partial derivative operator in that direction. *The number of independent boundary conditions along a co-ordinate direction should be equal to the order of the corresponding partial derivative operator.* Therefore, if we have no diffusion but only convection in a direction we should specify one condition; otherwise, with diffusion we must have two boundary conditions. The two boundary conditions can be on one or both sides of the interval of the co-ordinate direction. Care needs to be taken that two or more boundary conditions are not set simultaneously for a specified boundary region!

The effect of the balance volume shape

Note that not every boundary of the domain in the given co-ordinate system represents a real physical boundary. An example is a sphere where the real physical boundary is at $r = R$, R being the radius of the sphere, but we have five virtual co-ordinate boundaries at $r = 0, \theta = 0, \theta = 2\pi, \varphi = 0$ and $\varphi = 2\pi$. Virtual boundary conditions express the continuity of the processes and the distributions should be constructed accordingly in these cases. We shall see an instance of such a construction in the case of a spherical catalyst in Example 7.2.4.

Infinitely large balance volumes

Special boundary conditions are used to express the fact that the balance volume is very large or infinitely large in a co-ordinate direction. We may set the size of the balance volume in that direction and

- set the concentration of the reactants to zero using Dirichlet conditions at the boundary;
- specify that all convective and diffusive flow is zero at that boundary using Neumann conditions.

Boundary conditions of the above types are shown in the examples given in this chapter. The initial and boundary conditions of the packed bed plug flow catalytic reactor are given below.

EXAMPLE 7.1.4 (Derivation of the initial and boundary conditions for the simple packed bed tubular catalytic reactor (Example 7.1.1 continued)). In order to simplify the form of the initial and boundary conditions, we add some more simplification assumptions.

Assumptions *additional to the ones listed as 1–7 are:*

$\mathcal{A}$8. The initial distribution of component A and the initial temperature in the reactor is uniformly constant.

$\mathcal{A}$9. The reactor is very long in the x co-ordinate direction to enable full conversion.

Initial and boundary conditions

Initial conditions

$$C_A(x, 0) = C_A^*, \quad T(x, 0) = T^*. \tag{7.23}$$

Boundary conditions

$$\frac{\partial C_A}{\partial x}(L, 0) = 0, \quad C_A(0, t) = C_A^{(i)}, \tag{7.24}$$

$$\frac{\partial T}{\partial x}(L, 0) = 0, \quad T(0, t) = T^{(i)}, \tag{7.25}$$

where $C_A^{(i)}$ and $T^{(i)}$ are the inlet concentration and the inlet temperature, while C_A^* and T^* are the initial concentrations and temperature, respectively.

Note that two boundary conditions are needed because of the presence of the diffusion in both of the balance equations. The first condition at the outlet of the reactor expresses the fact that the reactor is very long and the reaction is fully completed before the flow exits.

Initial Conditions

Initial conditions set the values of the states at the initial time (typically $t = 0$) for the *whole* of the region of interest.

For example, in the DPS model of the tubular catalytic reactor we needed to set the initial fluid temperature and concentration along the length of the reactor. These conditions can be given as

$$T(x, 0) = f_1(x), \tag{7.26}$$

$$C_A(x, 0) = f_2(x), \tag{7.27}$$

where f_1 and f_2 are given functions in space.

These conditions are typical of the initial values which must be given.

7.1.5. Non-dimensional Forms

Where possible we should express the equations in terms of non-dimensional variables. This procedure helps to reduce the arithmetic involved with the numerical method. It also provides a useful scaling of the variables used in the equations which improves the numerical solution procedure. It is also the case that all problems with the same non-dimensional formulation can be dealt with by means of one solution. It is often the case that *different physical problems become identical mathematical problems* when the models are non-dimensionalized.

EXAMPLE 7.1.5 (Non-dimensional forms). Consider the steady-state reaction and diffusion in a porous medium:

$$\frac{\mathrm{d}}{\mathrm{d}x}\left(D_{\mathrm{e}}\frac{\mathrm{d}c}{\mathrm{d}x}\right) + R(c) = 0,$$

where

$$-D_{\mathrm{e}}\frac{\mathrm{d}c}{\mathrm{d}x} = 0 \quad \text{at } x = 0,$$

$$c = c_1 \quad \text{at } x = L,$$

and

$$R(c) = -kc^2.$$

We define

$$\hat{c} = \frac{c}{c_1}; \quad \hat{x} = \frac{x}{L}.$$

Then substitute to get

$$\frac{D_{\mathrm{e}}c_1}{L^2}\frac{\mathrm{d}^2\hat{c}}{\mathrm{d}\hat{x}^2} - kc_1^2(\hat{c})^2 = 0,$$

$$-\frac{D_{\mathrm{e}}c_1}{L}\frac{\mathrm{d}\hat{c}}{\mathrm{d}\hat{x}} = 0 \quad \text{at } \hat{x}L = 0,$$

$$c_1\hat{c} = c_1 \quad \text{at } \hat{x}L = L.$$

Rearrange by multiplying throughout by $\left(L^2/D_e c_1\right)$ to obtain

$$\frac{d^2\hat{c}}{d\hat{x}^2} = \frac{kc_1L^2}{D_e}(\hat{c})^2 \equiv \phi^2(\hat{c})^2$$

with

$$\frac{d\hat{c}}{d\hat{x}} = 0 \quad \text{at } \hat{x} = 0, \qquad \hat{c} = 1 \quad \text{at } \hat{x} = 1,$$

■ ■ ■ where ϕ^2 = Thiele modulus squared.

7.1.6. Modelling Assumptions and their Effect on the Model Equations

The most important modelling assumptions which affect the model equations of a DPS are briefly summarized here. The assumptions are grouped according to the model element or property they reflect.

- *Shape of the balance volumes*
  The shape of the balance volumes determines the co-ordinate system adopted. This affects the mathematical form of the conservation balances in their convection and diffusion terms.
- *Size of the balance volumes*
  Very large balance volumes (in any of their co-ordinate directions) may call for special boundary conditions.
- *Phases in the process system*
  Solid phase in a balance volume implies the absence of convection.
- *Flow conditions*
  Plug flow conditions imply convection in the direction of the flow with uniform flowrate in every other direction. The flow conditions determine the vector field describing the convective flow.
- *Mixing conditions*
  Perfect mixing in any of the co-ordinate direction implies no diffusion and uniform distribution of the intensive properties in that direction.

7.2. EXAMPLES OF DISTRIBUTED PARAMETER MODELLING

This section outlines some of the major areas which have benefited from the application of DPS modelling. Most of the examples are drawn from the process industries and show where DPS modelling has been used.

7.2.1. Heat Transfer Analysis

In most cases, heat transfer is inherently a distributed process. This has been a very common application area spanning 1D, 2D and 3D modelling. Applications range from double-pipe heat exchangers of one spatial variable through to irregular shaped non-homogeneous 3D bodies with complex boundary conditions.

DPS Model of a Double-pipe Heat Exchanger

The mathematical model of a double-pipe heat exchanger shows a lot of similarities to the model of the tubular catalytic rector we have seen before in this chapter. The process system consists of a double pipe heated from outside by condensing saturated steam. This situation is described by Roffel and Rijnsdorp [54]. The DPS model of the heat exchanger is described below.

EXAMPLE 7.2.1 (*Double-pipe heat exchanger*).
Process system
Consider a simple 1D double-pipe exchanger where a liquid stream is being heated by condensing saturated steam at a temperature, T_S. This is shown in Fig. 7.6. Fluid enters at temperature $T_L(0, t)$ and exits ($z = L$) at $T_L(L, t)$. Heat transfer takes place between the steam (T_s) and the wall (T_w) and then to the fluid ($T_L(z, t)$). The spatial variation is related to the fluid temperature whilst the steam temperature is a "lumped" variable, T_S.

Assumptions

$\mathcal{A}1$. The overall mass (volume) of the liquid as well as that of the wall is constant.
$\mathcal{A}2$. No diffusion takes place.
$\mathcal{A}3$. Steam temperature reacts instantaneously to supply changes.
$\mathcal{A}4$. Heat transfer coefficients are constant.
$\mathcal{A}5$. Specific heats and densities are constant.
$\mathcal{A}6$. Time delays are negligible for fluid.
$\mathcal{A}7$. Liquid is in plug flow.

Balance volumes
Two volumes ($\Sigma^{(HEL)}$ and $\Sigma^{(HEW)}$) for the fluid and the wall, respectively, both of them are *distributed*. The rectangular co-ordinate system is chosen.

Model equations
Variables

$$0 \leq z \leq L, \quad 0 \leq t, \quad T_L(z, t), \quad T_W(z, t), \tag{7.28}$$

where z is the axial co-ordinate along the tube and L is the tube length.

Energy balance for the liquid

$$\rho_L A_L c_f \frac{\partial T_L}{\partial t} = -u \rho_L A_f c_L \frac{\partial T_L}{\partial z} + h_L A_L (T_w - T_L). \tag{7.29}$$

Energy balance for the wall

$$M_W c_W \frac{\partial T_W}{\partial t} = h_s A_s (T_s - T_w) - h_L A_L (T_w - T_L), \tag{7.30}$$

where

ρ_L fluid density [kg/m^3]
A_f flow area [m^2]
c_L fluid heat capacity [J/kg K]
u fluid velocity [m/s]
h_L fluid heat transfer coefficient [W/m^2 K]

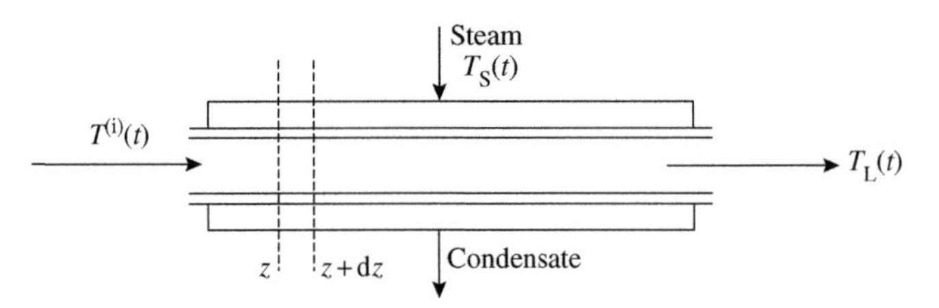

FIGURE 7.6 Double-pipe heat exchanger.

A_L internal tube area per length of tube [m^2/m]
M_w mass of tube per length [kg/m]
c_w metal heat capacity [J/kg K]
A_s external tube area [m^2/m]
h_s steam side heat transfer coefficient [W/m^2 K]

Initial conditions

$$T_L(z, 0) = f_1(z), \tag{7.31}$$

$$T_w(z, 0) = f_2(z). \tag{7.32}$$

Boundary condition

$$T_L(0, t) = T^{(i)}(t) \quad \text{for all t} \geq 0, \tag{7.33}$$

where $T^{(i)}$ is the inlet concentration of the liquid.

We can *simplify the above PDEs* by defining some time constants for the system. For example,

- $\tau_L = \frac{\rho_L c_L A_f}{h_L A_L}$ (characteristic fluid heating time),
- $\tau_{sw} = \frac{M_w c_w}{h_s A_s}$ (steam to wall thermal constant),
- $\tau_{wL} = \frac{M_w c_w}{h_L A_s}$ (wall to fluid thermal constant).

Modified equations are then

$$\frac{\partial T_L}{\partial t} = -u\frac{\partial T_L}{\partial z} + \frac{1}{\tau_L}(T_w - T_L), \tag{7.34}$$

$$\frac{\partial T_w}{\partial t} = \frac{1}{\tau_{sw}}(T_s - T_w) - \frac{1}{\tau_{wL}}(T_w - T_L). \tag{7.35}$$

■ ■ ■

There are some interesting key steps in developing the DPS model above which are as follows:

1. Only one spatial co-ordinate is needed because of the plug flow assumption.
2. We have substituted the constitutive equations into the balance equations. The transfer rate relation was simply substituted as the only source term. The

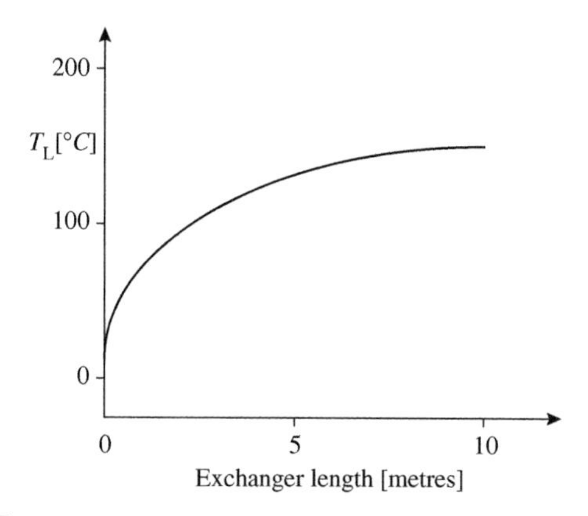

FIGURE 7.7 Steady-state temperature profile.

potential relation $\hat{U} = c_P \rho T$ with constant physico-chemical properties has been substituted into the energy conservation equations (7.29) and (7.30).
3. Note that T_s is a function of time only and acts as a forcing function on the system.
4. The model equations (7.34) and (7.35) are *coupled PDEs* since the wall temperature (T_W) depends on the fluid temperature (T_L).
5. A particular steady-state solution profile to the above model is shown in Fig. 7.7.

Three-dimensional Heat Transfer

It is quite common that multi-dimensional problems arise where heat transfer is important. Some interesting and important examples are for example:

(i) Temperature distribution in a carcass of frozen meat being thawed.
(ii) Temperature profiles in walls of engines.
(iii) Temperature profiles in high temperature reactor systems where refractory materials are used.

The following example deals with the modelling of 3D heat transfer in a solid body.

EXAMPLE 7.2.2 (Three-dimensional heat transfer in a solid body).
Process system
Consider a solid body of a rectangular shape of finite length in each co-ordinate direction. Let us assume heat generation (or loss in the solid) depending on the spatial location.

Modelling goal
Describe the dynamic behaviour of the solid body.

Assumptions

$\mathcal{A}1$. The mass of the body and its volume are constant.
$\mathcal{A}2$. The specific heat c_P and the density ρ are constant.
$\mathcal{A}3$. The heat generation or loss is a given function $q(x, y, z, t)$.

$\mathcal{A}4$. Assume uniform initial conditions.
$\mathcal{A}5$. The temperature outside of the body is the same all over the surface and does not change in time.

Balance volumes
A single volume encapsulating the whole solid body but this time this balance volume is a *distributed* one.

Model equations
Variables

$$0 \leq x \leq L_x, \quad 0 \leq y \leq L_y, \quad 0 \leq z \leq L_z, \quad 0 \leq t, \quad T(x, y, z, t), \tag{7.36}$$

where x, y, z are the spatial coordinates and L_x, L_y, L_z are the lengths of the body in the respective direction. The Cartesian co-ordinate system is selected.

Energy balance

$$c_P\rho\frac{\partial T}{\partial t} = \frac{\partial}{\partial x}\left(K_x\frac{\partial T}{\partial x}\right) + \frac{\partial}{\partial y}\left(K_y\frac{\partial T}{\partial y}\right) + \frac{\partial}{\partial z}\left(K_z\frac{\partial T}{\partial z}\right) + q, \tag{7.37}$$

where K_x, K_y, K_z are thermal conductivities in the x, y and z directions, respectively. T is a function $T(z, y, z, t)$ and c_P, ρ are heat capacity and density, respectively.

Initial and boundary conditions
Initial conditions

$$T(x, y, z, 0) = T^*, \tag{7.38}$$

where T^* is the given uniform initial condition.

Boundary conditions

$$T(0, y, z, t) = T_E, \quad T(L_x, y, z, t) = T_E, \tag{7.39}$$

$$T(x, 0, z, t) = T_E, \quad T(x, L_y, z, t) = T_E, \tag{7.40}$$

$$T(x, y, 0, t) = T_E, \quad T(x, y, L_z, t) = T_E, \tag{7.41}$$

■ ■ ■ where T_E is the given uniform and constant temperature of the environment.

There are again some interesting key steps in developing the above DPS model that are worth mentioning:

1. We have substituted the potential relation $\hat{U} = c_P\rho T$ with constant physico–chemical properties into the energy conservation equation.
2. Note that q is a function of time and space and acts as a forcing function on the system.

7.2.2. Reaction Engineering

Modelling of a Pollutant in a River

As an interesting example of deriving a distributed parameter model, we consider a modification of a problem posed by Roffel and Rijnsdorp [54].

EXAMPLE 7.2.3 [(Modelling of a pollutant in a river).]

Process system

Here, we are concerned about a pollutant discharged to a river and its subsequent concentration as it flows downstream. Figure 7.8 shows the geometry of the situation, showing the cross section of the river and the representative slice over which the conservation balances can be performed. Note that the axial distance is the co-ordinate length z.

Modelling goal

To describe the distribution of the pollutant in the river.

Assumptions

$\mathcal{A}1$. The river has a constant cross-sectional area and the material is ideally mixed over the cross section.
$\mathcal{A}2$. The pollutant disappears via a first-order reaction.
$\mathcal{A}3$. Axial dispersion is present.
$\mathcal{A}4$. No radial dispersion.
$\mathcal{A}5$. River flow is constant.
$\mathcal{A}6$. The river water is considered isothermal.
$\mathcal{A}7$. No pollution is in the river in the beginning of the process.

Balance volumes

A single distributed parameter volume encapsulating the whole river.

Model equations

Variables

$$0 \leq x \leq L, \quad t \geq 0, \quad c(x,t), \tag{7.42}$$

where x is the spatial coordinate in axial direction, L the length of the river, t the time, and c the pollutant concentration.

Component mass balance

$$\frac{\partial c}{\partial t} = -u\frac{\partial c}{\partial z} + D\frac{\partial^2 c}{\partial z^2} - kc, \tag{7.43}$$

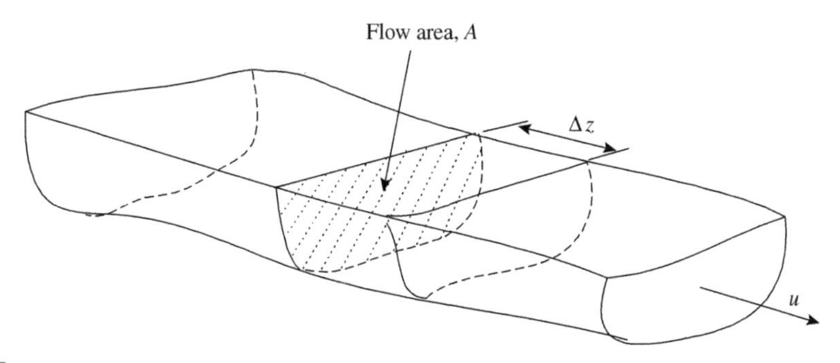

FIGURE 7.8 Pollutant dispersion in a river.

with

c concentration [kg/m^3]
A flow area [m^2]
u flow velocity [m/s]
D dispersion coefficient [m^2/s]
k reaction rate [s^{-1}]

Initial and boundary conditions

$$c(x, 0) = 0, \quad c(0, t) = C_A^{(\mathrm{i})}, \tag{7.44}$$

where $C_A^{(\mathrm{i})}$ is the inlet concentration of the pollutant.

Normalization of the equations
We can take the governing equation and define some non-dimensional groups as follows:

$$\theta = \frac{tu}{L}; \quad Pe = \frac{uL}{D}; \quad Z = \frac{z}{L}; \quad C = \frac{c}{c_0}; \quad K = \frac{kL}{u}$$

and rearrange to get

$$\frac{\partial C}{\partial \theta} = -\frac{\partial C}{\partial Z} + \frac{1}{Pe}\frac{\partial^2 C}{\partial Z^2} - KC, \tag{7.45}$$

where Pe is the Peclet number which indicates the ratio of convective to dispersive transport. As the magnitude of $Pe \to \infty$, dispersion is negligible and we obtain plug flow behaviour. As the value of $Pe \to 0$ we approach a well mixed situation. ■ ■ ■

The key steps in developing the DPS model above were as follows:

1. We have substituted the reaction rate equation as the only source into the component mass conservation equation.
2. We assumed just one source of pollution at the very beginning of the river and gave it as an initial condition. If multiple pollution sources are present then we have to add appropriate spatially located additional source terms into the component mass balance equation.

Non-isothermal Reaction in a Catalyst Pellet

Finlayson [55] describes the modelling of reactions in a spherical catalyst pellet where diffusion and reaction take place ($x = 0$ is at the centre of the catalyst particle). The process system is the same as in Section 3.2.2 but now we give all of the ingredients of its PDS model.

EXAMPLE 7.2.4 (Non-isothermal reaction in a spherical catalyst pellet).
Process system
Consider a spherical catalyst pellet where diffusion and a first-order chemical reaction take place.

Modelling goal
To describe the concentration profile of the reactant in the pellet.

Assumptions

$\mathcal{A}1$. The overall mass and the volume of the pellet are constant.
$\mathcal{A}2$. No convection takes place.
$\mathcal{A}3$. Only a single first-order reaction takes place.
$\mathcal{A}4$. Constant physico-chemical properties.
$\mathcal{A}5$. The pellet is uniform in all directions, i.e. all variables are functions of the distance from the centre and the time only.
$\mathcal{A}6$. Given initial conditions.
$\mathcal{A}7$. Heat and component mass transfer occurs on the pellet surface.

Balance volumes
A single volume is a *distributed* one encapsulating the whole pellet.

Model equations
Variables

$$0 \leq x \leq 1, \quad 0 \leq t, \quad c(x,t), \quad T(x,t), \tag{7.46}$$

where x is the normalized spatial coordinate, t is time, c is the reactant concentration and T is the temperature. The spherical co-ordinate system is selected.

Component mass and energy balances

$$M_1 \frac{\partial T}{\partial t} = \frac{1}{x^2}\frac{\partial}{\partial x}\left(x^2 \frac{\partial T}{\partial x}\right) + \phi^2 \beta c \exp\left[\gamma\left(1 - \frac{1}{T}\right)\right], \tag{7.47}$$

$$M_2 \frac{\partial c}{\partial t} = \frac{1}{x^2}\frac{\partial}{\partial x}\left(x^2 \frac{\partial c}{\partial x}\right) - \phi^2 c \exp\left[\gamma\left(1 - \frac{1}{T}\right)\right], \tag{7.48}$$

where M_1, M_2, ϕ^2, β, γ are dimensionless groups and the following boundary and initial conditions apply:

Initial and boundary conditions
Boundary conditions at the centre ($x = 0$)

$$\frac{\partial c}{\partial x} = \frac{\partial T}{\partial x} = 0. \tag{7.49}$$

Boundary conditions at the pellet surface ($x = 1$)

$$-\frac{\partial T}{\partial x} = \frac{Nu}{2}(T - g_1(t)), \tag{7.50}$$

$$-\frac{\partial c}{\partial x} = \frac{Sh}{2}(c - g_2(t)). \tag{7.51}$$

These are energy and mass transfer relations driven by the bulk conditions given by g_1, g_2; in turn, functions of time.

Initial conditions

$$T\,(x,0) = h_1(x); \quad c(x,0) = h_2(x), \tag{7.52}$$

■ ■ ■ which are the initial profiles.

There are key steps in developing the PDS model above which are as follows:

1. Only one spatial co-ordinate is needed because of the spherical geometry and the uniform distribution in all directions.
2. We have substituted the constitutive equations into the balance equations. The reaction rate relation was simply substituted as the only source term. The potential relation $\hat{U} = c_P \rho T$ with constant physico-chemical properties has been substituted into the energy conservation equation.
3. The model equations are *coupled PDEs*.
4. Observe the virtual boundary conditions at the centre ($x = 0$) describing no flow conditions there.

Problems such as these can display very complex behaviour and show such phenomena as multiple steady states. In this case, the parameter ϕ is the Thiele modulus, which represents the reaction rate to diffusion rate in the catalyst.

Because of the nonlinear nature of these equations, the transient behaviour can also be very complex.

This type of problem is one of the more complex modelling applications with coupled mass and heat transfer, nonlinear terms and multiple solutions.

7.3. CLASSIFICATION OF DPS MODELS

The classification of DPS models has great practical importance when solving the model equations numerically. Numerical solution techniques should be chosen according to the type of model equations. Moreover the number and type of the boundary conditions depend also on the classification result.

7.3.1. Classification of the Partial Differential Part

The classification of the PDEs in a DPS model is performed on the basis of the partial differential operator present in the equations. The general form of the conservation expression expressed in Cartesian co-ordinates (see Eq. (3.37)) will serve as the example equation for the classification:

$$\frac{\partial \hat{\Phi}}{\partial t} = D\left(\frac{\partial^2 \varphi}{\partial x^2} + \frac{\partial^2 \varphi}{\partial y^2} + \frac{\partial^2 \varphi}{\partial z^2}\right) - \left(\frac{\partial \hat{\Phi} v_x}{\partial x} + \frac{\partial \hat{\Phi} v_y}{\partial y} + \frac{\partial \hat{\Phi} v_z}{\partial z}\right) + \hat{q}. \tag{7.53}$$

Let us *associate an algebraic equation with the PDE* as follows:

1. Replace $\hat{q}$ by 0.
2. Replace every term containing a partial derivative by another term composed as follows. The partial derivative factor is replaced by the independent variable with power equal to the order of the partial derivative with respect to that variable. All the other factors (model parameters and other variables) in the term remain unchanged.

For the conservation equation (7.53) above we get

$$t = Dx^2 + Dy^2 + Dz^2 - v_x x - v_y y - v_z z. \tag{7.54}$$

The geometry of the second order multi-dimensional curve $t = t(x, y, z)$ given in Eq. (7.54) determines the class of the PDE. Let us illustrate how this classification is performed by assuming that we only have a single space co-ordinate, so that Eq. (7.54) is in the form:

$$t = Dx^2 - v_x x. \tag{7.55}$$

We can then define the following special cases:

- $D \neq 0$ gives a parabola then the original PDE is of *parabolic form*,
- $D = 0,\ v_x \neq 0$ gives a line which we may regard as a degenerated hyperbola giving rise to a *hyperbolic* PDE.

A typical example of a parabolic PDE is the time dependent diffusion problem as given in Example 7.3.1.

EXAMPLE 7.3.1 (Time dependent diffusion problem). The unsteady-state problem is given by

$$\frac{\partial c}{\partial t} = D\left(\frac{\partial^2 c}{\partial x^2} + \frac{\partial^2 c}{\partial y^2}\right)$$

with boundary conditions

$$c(x, 0) = c_1; \quad c(x, 1) = c_2; \quad 0 < x < 1,$$
$$c(0, y) = c_3 : \quad c(1, y) = c_4; \quad 0 < y < 1,$$

and initial condition

$$c = f(x, y) \quad \text{at } t = 0.$$

Hyperbolic models generally arise from vibration problems or where discontinuities persist in time. These can be shock waves, across which there are discontinuities in speed, pressure and density. Example 7.3.2 gives a typical example.

EXAMPLE 7.3.2 (Discontinuities in time). One-dimensional wave equation

$$\frac{\partial^2 u}{\partial t^2} = c^2 \frac{\partial^2 u}{\partial x^2},$$

where u is the transverse displacement at a distance x from one end of a system whose length is z.

Boundary conditions

$$u(0, t) = u_1 \quad u(z, t) = u_2 \quad \text{all } t.$$

Initial conditions

$$u(x, 0) = g(x) \quad \text{at } t = 0,$$

Generally, hyperbolic equations in two independent variables are solved numerically or by special techniques such as the method of characteristics.

7.3.2. The Effect of Steady-State Assumptions

The previous development has lead to time varying distributed parameter models. We can, however, modify these models through the use of steady-state conditions which effectively eliminate the time-varying terms in the model.

In order to see what effect the steady-state assumption has on the form of the DPS model, let us take the steady-state version of the general form of the conservation expression expressed in Cartesian co-ordinates as the example equation:

$$0 = D\left(\frac{\partial^2 \varphi}{\partial x^2} + \frac{\partial^2 \varphi}{\partial y^2} + \frac{\partial^2 \varphi}{\partial z^2}\right) - \left(\frac{\partial \hat{\Phi} v_x}{\partial x} + \frac{\partial \hat{\Phi} v_y}{\partial y} + \frac{\partial \hat{\Phi} v_z}{\partial z}\right) + \hat{q}. \tag{7.56}$$

The associated algebraic equation reads

$$0 = Dx^2 + Dy^2 + Dz^2 - v_x x - v_y y - v_z z. \tag{7.57}$$

The above equation describes an elliptic curve. In the case of two spatial variables, we have

$$0 = Dx^2 + Dy^2 - v_x x - v_y y. \tag{7.58}$$

Therefore, the steady-state assumption leads to models which are elliptic in form and which can be solved more robustly and at less cost compared with full dynamic models. A typical example is given by the steady-state diffusion problem as given in Example 7.3.3.

EXAMPLE 7.3.3 (Steady-state diffusion problem).

$$\frac{\partial^2 c}{\partial x^2} + \frac{\partial^2 c}{\partial y^2} = 0, \quad 0 < x, y < 1,$$

$$c(x, 0) = c_1; \quad c(x, 1) = c_2; \quad 0 < x < 1,$$

$$c(0, y) = c_3; \quad c(1, y) = c_4; \quad 0 < y < 1,$$

where this is defined on a rectangular co-ordinate system.

Only a few special types of elliptic equations have been solved analytically due to the problem of describing the boundary by equations. In the case of difficult boundary shapes it is essential to use an approximation method. It should be realized that the approximation method, whether it be finite difference, finite element or method of weighted residuals will normally provide numerical solutions as accurate as the original model equations justify. In the real world, the modelling process and associated data contain numerous approximations and errors.

Setting boundary conditions for such problems needs to be done with care. This has been discussed in Section 7.1.4.

7.3.3. Assumptions Leading to the DPS Model Classes

In summary, DPS models can be of different forms dictated by the underlying engineering assumptions as follows:

1. *With no additional assumptions they are PDEs in parabolic form.*

2. *With a steady-state assumption they are elliptic PDEs.*
3. *With no diffusion they are first-order hyperbolic PDEs.*

7.4. LUMPED PARAMETER MODELS FOR REPRESENTING DPSs

Lumping of process models of DPS is a widely used technique to transform the set of partial differential equations in the model into a set of ordinary differential equations. The lumped model is a finite approximation of the DPS model in the space co-ordinate directions whereas the time variable remains the only independent variable in a lumped model. Most often models of originally DPS systems developed for dynamic analysis, control or diagnostic purposes are lumped in space.

The conceptual steps in lumping a DPS system model are as follows:

1. Divide the distributed parameter balance volume of the process system into a finite number of subvolumes.
2. *Lump* each subvolume into a perfectly mixed subvolume (call it *lump*) with the variables averaged.
3. Describe convection in the original DPS system as in- and outflows of the connected set of neighbouring lumps using the appropriate direction.
4. Describe diffusion affecting all the neighbouring lumps as in- and outflows of neighbouring lumps.
5. Use the same sources for the lumps as for the original DPS model.
6. Develop the balance equations for every lump.
7. Respect boundary conditions at the lumps which coincide with the overall boundaries of the process system.

In the case of a single spatial co-ordinate x the above steps reduce to:

1. Divide the length of the process system L into a finite number of intervals N, usually of equal size $\Delta x = L/N$.
2. *Lump* each subvolume into a perfectly mixed subvolume of equal size $V^{(k)} = V/N$.
3. Describe convection in the original DPS system as in- and outflows of the connected set of neighbouring lumps using the appropriate direction, i.e. $v\varphi^{(k-1)}$ being the inflow and $v\varphi^{(k)}$ the outflow for the kth lump with v being the flowrate.
4. Describe diffusion affecting all the neighbouring lumps, such that

$$D\frac{(\varphi^{(k)} - \varphi^{(k-1)})}{\Delta x} \quad \text{and} \quad D\frac{(\varphi^{(k+1)} - \varphi^{(k)})}{\Delta x}$$

for the kth lump.

Note that some of the solution methods do a kind of lumping when solving PDE models, e.g. the method of lines (see details in Chapter 8).

As an illustration of the above procedure, a lumped model of the double-pipe heat exchanger described in Example 7.2.1 is given.

EXAMPLE 7.4.1 (Lumped model of the double-pipe heat exchanger in Example 7.2.1).

Process system
Consider the simple double pipe heat exchanger where the liquid stream is being heated by condensing saturated steam. The system is the same as in Example 7.2.1.

Additional Assumptions *to the ones listed as 1–7.*

$\mathcal{A}$8. The heat exchanger is described as a sequence of three well mixed volumes.

Balance volumes
We consider two lots of three balance volumes with equal holdups.

Model equations

Variables

$$\left(T_{\rm L}^{(k)}(t), T_{\rm W}^{(k)}(t),\ k = 1, 2, 3\right), \quad 0 \leq t, \tag{7.59}$$

where $T_L^{(k)}(t)$ and $T_{\rm W}^{(k)}(t)$ is the liquid and the wall temperature in the kth tank pair respectively and t is the time. The model parameters are the same as in the Example 7.2.1.

Energy balances for the liquid

$$\frac{{\rm d}T_{\rm L}^{(k)}}{{\rm d}t} = 3u\left(T_{\rm L}^{(k-1)} - T_{\rm L}^{(k)}\right) + \frac{1}{\tau_{\rm L}}\left(T_{\rm W}^{(k)} - T_{\rm L}^{(k)}\right), \tag{7.60}$$

$$k = 1, 2, 3, \quad T_{\rm L}^{(0)}(t) = T_{\rm L}^{({\rm i})}(t). \tag{7.61}$$

$T_{\rm L}^{({\rm i})}$ is the inlet fluid temperature to the whole reactor.

Energy balances for the wall

$$\frac{{\rm d}T_{\rm W}^{(k)}}{{\rm d}t} = \frac{1}{\tau_{\rm SW}}\left(T_{\rm S} - T_{\rm W}^{(k)}\right) - \frac{1}{\tau_{\rm WL}}\left(T_{\rm W}^{(k)} - T_{\rm L}^{(k)}\right), \quad k = 1, 2, 3. \tag{7.62}$$

Initial conditions

$$T_{\rm L}^{(k)}(0) = f_1^{(k)}, \quad k = 1, 2, 3, \tag{7.63}$$

$$T_{\rm W}^{(k)}(0) = f_2^{(k)}, \quad k = 1, 2, 3 \tag{7.64}$$

where the values of $f_1^{(k)}$, $k = 1, 2, 3$ and $f_2^{(k)}$, $k = 1, 2, 3$ are taken from the functions f_1 and f_2 at the appropriate spatial locations z_k, $k = 1, 2, 3$.

7.5. SUMMARY

The chapter has introduced the principles underlying the development of distributed parameter models. The key driver for DPS modelling is the need to capture the variation in key states as a function of spatial position. In many cases, we simplify our approach by adopting a lumped parameter view of the system. This can be most useful but there are clearly occasions when that assumption is not valid. In those cases we are forced to model the spatial variations.

In modelling process systems as DPS models, the level of complexity rises significantly. This has implications on the solution techniques as well as for the development of the model. Only when the lumped parameter system assumption is invalid, should we consider the need to treat the problem as a distributed system. Chapter 8 deals with a variety of techniques for solving the distributed parameter models.

7.6. REVIEW QUESTIONS

Q7.1. Describe why distributed parameter models arise in process engineering applications. (Sections 7.1 and 7.2)

Q7.2. What are the principal geometries used in the development of distributed parameter models? Give a process engineering example of each approach describing how it arises. (Section 7.1)

Q7.3. What are the roles of initial conditions and boundary conditions in the formulation of distributed parameter models? Describe the three principal types of boundary conditions and give some examples of their use in practical modelling problems. (Section 7.1)

Q7.4. What are the advantages and disadvantages of non-dimensional forms of the governing PDEs? (Section 7.1.5)

Q7.5. Describe the classes of modelling assumptions which affect the form of the DPS model equations. Give examples illustrating the possible effects. (Section 7.1)

Q7.6. Describe the three main forms of partial differential systems which arise from distributed parameter modelling. What are the most common forms in process engineering? (Section 7.3)

Q7.7. Describe the process of representing the behaviour of partial differential equation models in terms of lumped parameter models. What are the key characteristics of the resultant equation systems in terms of problem size and accuracy? (Section 7.4)

7.7. APPLICATION EXERCISES

A7.1. Develop the general mass and energy balance equations for a tubular plug flow reactor, where the reaction kinetics are given by the following general expression:

$$r_i = k_i C_i^n = k_0 \, \mathrm{e}^{-E/(RT)} C_i^n .$$

Assume that the inlet flowrate is F_{in}, inlet concentration is $C_{A,in}$ and temperature T_0. The reactor will be cooled or heated depending on whether the reaction is exothermic or endothermic.

Give the initial conditions and boundary conditions for such a system.

A7.2. For the application example in A7.1 develop a lumped model representation for the general problem.

A7.3. Consider the modelling of a liquid–liquid contactor or extraction column where a solute A contained in a liquid B is extracted using a solvent S in countercurrent flow. Assume that the liquid rates are L_B and L_S and that a simple mass transfer relation holds for the transfer of solute from B to S.

Show the fundamental geometry to be used in developing the balance equations.

A7.4. Consider the development of a dynamic model to predict the behaviour of a packed bed adsorber. In this case, a gas stream contains a hydrocarbon which must be adsorbed onto an activated carbon adsorbent. Assume that the bulk concentration of the species is C_b and that the adsorbed concentration is C_a. The relationship which relates C_a to C_b is given by $C_a = f(C_b)$.

Assume that the bed voidage is ε, the bed cross-sectional area is A and that the interpellet gas velocity is u. Develop the dynamic mass balance which describes the behaviour of the adsorbed species with time and length along the adsorber.

Define appropriate initial and boundary conditions for such a situation.

A7.5. Consider a well-mixed batch reactor for the production of octyl phthalate in which an endothermic reaction takes place between octanol and phthalic anhydride. The temperature profile of the batch is controlled by the use of a circulating hot oil system which consists of a heat transfer coil located inside the reactor and an external gas fired heater which maintains a fixed inlet temperature of oil to the reactor.

Model the temperature profile in the hot-oil coil within the reactor and use it to predict the dynamics of the energy transfer to the batch. Clearly state all assumptions in the model.

A7.6. A hot rectangular slab of steel is hardened by immersing it vertically in a tank of oil using an overhead crane. Develop a dynamic model of the system, stating clearly the assumptions used in developing the model.

A7.7. Consider the problem of freezing of a lake due to subzero air temperatures during winter. Develop a one dimensional model of the temperature as a function of time and depth in the lake as the freezing occurs. Consider how the ice layer on top of the cold water increases in thickness as the process proceeds. State the assumptions used in developing the model as well as the necessary boundary and initial conditions needed for the solution.

Consider how the moving ice–water interface can be tracked as time proceeds.

A7.8. Consider the problem of the double pipe heat exchanger given in Example 7.2.1. Derive the governing energy balance equations using the microscopic balance approach as set out in section 7.1.3. Compare the final derived balance equations with those in the example.

A7.9. For the 3D heat transfer problem in a rectangular slab given in Example 7.2.2, derive the governing energy balance equation from the application of a microscopic balance. Compare your model with that shown in the example.

A7.10. Derive, using the microscopic balance approach, the model described in Example 7.2.3 which considers the concentration changes in time for flow of a pollutant in a river. Check your development against that stated in the example.

A7.11. For the double-pipe heat exchanger problem given in Example 7.2.1, consider the case where the outer tube side is now a counter-current liquid process stream which provides heating to the colder inner tube stream. Derive the distributed parameter model which describes the temperature dynamics for both the hot stream, the cold stream and the tube wall.

A7.12. Derive the mass and energy balance equations related to the mass and heat transfer in a spherical catalyst pellet given in Example 7.2.4. Use a microscopic balance approach to do this and compare your model with the governing equations given in the example.

8 SOLUTION STRATEGIES FOR DISTRIBUTED PARAMETER MODELS

Partial differential equations (PDEs) occur in many process engineering applications from fluid flow problems to heat transfer. Generally, most problems lead to parabolic equations which are time dependent and hence represent unsteady-state processes. In contrast, steady-state problems lead to elliptic PDEs.

Many numerical techniques have been developed to solve PDEs. The most widely used are still the finite difference methods, which are generally much simpler to set up compared with other available methods. However, in the last 20 years, and especially the last 10 years the finite element method has been used to solve many problems with irregular shaped boundaries as well as a diverse range of problems. In chemical reactor analysis, the orthogonal collocation technique has also been extensively used because of reasons of accuracy and efficiency. Another popular method often used is the method of lines, which converts a PDE to a set of ODEs subsequently using efficient ODE codes to solve the sets of equations. The choice between methods for a particular problem is not always an easy one. It is often a matter of experimentation, since many factors have to be considered. These include the complexity of the problem, ease of setting up the solution method, accuracy required, stability problems and computational effort.

8.1. AREAS OF INTEREST

The three major areas of interest from the point of view of process modelling are

(a) parabolic equations in one dimension,

(b) parabolic equations in two dimensions,
(c) elliptic equations.

The application of finite difference methods to these problems will be considered first in Section 8.2. The problems of accuracy and stability of these methods are also covered. Secondly, in Section 8.3, the method of lines will be introduced as a means of reducing a parabolic PDE to a set of ODEs as an initial value problem (ODE-IVP). Thirdly, the method of weighted residuals (MOWR) will be covered in Section 8.4 and, in particular, the method of orthogonal collocation (OCM) for application to both boundary value problems in ordinary differential equation (ODE-BVP) and then for the numerical solution of PDEs in Sections 8.5 and 8.6.

The application of numerical methods to PDEs, ordinary differential equation, or to boundary value problems converts the original problem into:

- linear or nonlinear algebraic equations in the case of:
  – finite difference methods applied to elliptic or parabolic PDEs (PPDEs),
  – orthogonal collocation methods applied to elliptic PDEs and to steady-state BVPs;
- linear or nonlinear differential or DAEs in the case of:
  – method of lines applied to parabolic PDEs
  – orthogonal collocation applied to PPDEs.

8.2. FINITE DIFFERENCE METHODS

Finite difference methods still remain a popular general method for the solution of all types of PDEs. The aim of the method *is to replace the derivatives in the governing equation by finite difference approximations* (FDAs).

Hence, we convert the problem to a set of algebraic equations. That is, we establish a mesh over the interval of interest in, for example, the domain of x as shown in Fig. 8.1, so that

$$\left.\begin{array}{ll} x_i = a + i \cdot \Delta x, & 0 \le i \le N \\ \Delta x = (b-a)/N & \end{array}\right\} \quad \text{uniform mesh.} \tag{8.1}$$

Solution of the algebraic equations gives the solution values of the original equation at the mesh points, x_i.

Note that we have $N+1$ total points, $N-1$ internal points.

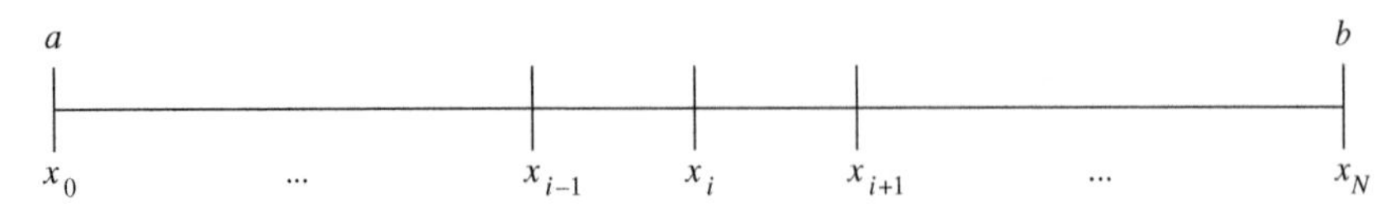

FIGURE 8.1 Finite difference mesh.

8.2.1. Finite Difference Approximations

Consider a continuous function $u(x)$, where we write a Taylor series expansion about x_i so that

$$u(x_i + \Delta x) = u(x_i) + \Delta x u'(x_i) + \tfrac{1}{2}\Delta x^2 u''(x_i) + \tfrac{1}{6}\Delta x^3 u'''(x_i) + \cdots, \tag{8.2}$$

$$u(x_i - \Delta x) = u(x_i) - \Delta x u'(x_i) + \tfrac{1}{2}\Delta x^2 u''(x_i) - \tfrac{1}{6}\Delta x^3 u'''(x_i) + \cdots. \tag{8.3}$$

From these expansions or combinations of them we can obtain the following FDAs. Using the notation $u_i = u(x_i)$, we have the following first- and second-order approximations, formed using forward, backward or central differences.

First-order approximations:

$$u_i' = \frac{\mathrm{d}u(x_i)}{\mathrm{d}x} \simeq \frac{u_{i+1} - u_i}{\Delta x} + O(\Delta x) \quad \text{(forward difference)}, \tag{8.4}$$

$$u_i' = \frac{\mathrm{d}u(x_i)}{\mathrm{d}x} \simeq \frac{u_i - u_{i-1}}{\Delta x} + O(\Delta x) \quad \text{(backward difference)}. \tag{8.5}$$

Second-order approximations:

$$u_i' = \frac{\mathrm{d}u(x_i)}{\mathrm{d}x} \simeq \frac{u_{i+1} - u_{i-1}}{2\Delta x} + O(\Delta x^2) \quad \text{(central difference)}, \tag{8.6}$$

$$u_i'' = \frac{\mathrm{d}^2u(x_i)}{\mathrm{d}x^2} \simeq \frac{u_{i+1} - 2u_i + u_{i-1}}{\Delta x^2} + O(\Delta x^2) \quad \text{(central difference)}. \tag{8.7}$$

It is obvious that higher order derivatives can be estimated as well as FDAs of higher accuracy. These FDAs form the basis of the finite difference method. Detailed coverage of finite difference methods is given by Smith [56].

8.2.2. Parabolic Equations in One Dimension

These are evolutionary problems since they are not only dependent on a spacial variable x but are transient. Hence, the solution is a function of x and t.

EXAMPLE 8.2.1 We consider an explicit method for the problem of unsteady diffusion in one dimension,

$$\frac{\partial u}{\partial t} = k\frac{\partial^2 u}{\partial x^2},$$

with boundary conditions

$$u(0, t) = 1, \quad u(1, t) = 0$$

and initial conditions

$$u(x, 0) = 2x\colon\ 0 \le x \le \tfrac{1}{2} \text{ and } 2(1 - x)\colon\ \tfrac{1}{2} \le x \le 1,$$

where we assume

$$k = \text{constant}.$$

An explicit finite difference method

From our previous FDAs, we can write

$$\frac{u_{i,j+1} - u_{i,j}}{\Delta t} = k\left(\frac{u_{i+1,j} - 2u_{i,j} + u_{i-1,j}}{\Delta x^2}\right).$$

If we let $k = 1$ we can write

$$u_{i,j+1} = u_{i,j} + r\left(u_{i-1,j} - 2u_{i,j} + u_{i+1,j}\right),$$

where

$$r = \frac{\Delta t}{\Delta x^2}.$$

Hence, it is possible to calculate the unknown mesh values $u_{i,j+1}$ at the $(i, j+1)$ mesh point in terms of known values along the jth row. This is called an *explicit method* and is the finite difference scheme shown in Fig. 8.2.

Note that we have accuracy of $O(\Delta t)$ and $O(\Delta x^2)$.

We can now solve this PDE by choosing values of Δt and Δx and then perform the explicit calculations. We consider three cases which differ in the value $r = \Delta t/\Delta x^2$ as follows:

Case 1

$$\Delta x = 0.1, \quad \Delta t = 0.001 \implies r = 0.1, \tag{8.8}$$

$$u_{i,j+1} = \tfrac{1}{10}\left(u_{i-1,j} + 8u_{i,j} + u_{i+1,j}\right). \tag{8.9}$$

Case 2

$$\Delta x = 0.1, \quad \Delta t = 0.005 \implies r = 0.5, \tag{8.10}$$

$$u_{i,j+1} = \tfrac{1}{2}(u_{i-1,j} + u_{i+1,j}). \tag{8.11}$$

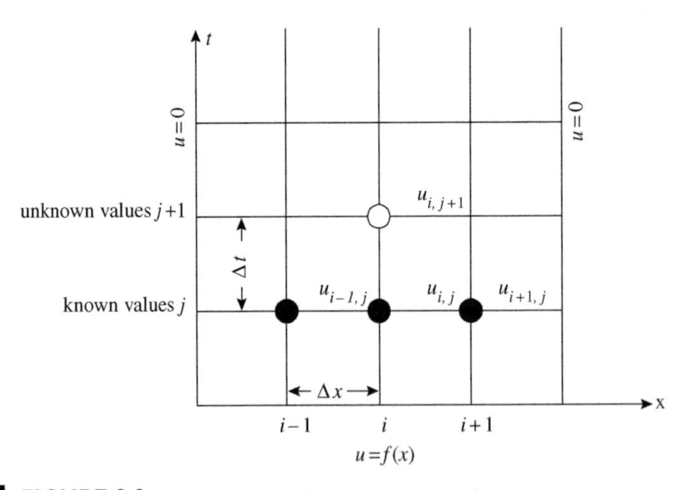

FIGURE 8.2 Explicit finite difference mesh and solution points.

Case 3

$$\Delta x = 0.1, \quad \Delta t = 0.01 \implies r = 1.0, \tag{8.12}$$

$$u_{i,j+1} = u_{i-1,j} - u_{i,j} + u_{i+1,j}. \tag{8.13}$$

When we solve the 1D problem using these three explicit methods we observe the following.

Cases 1 and 2 produce quite reasonable results as seen in the table below compared with the analytic solution. Case 1 results are slightly more accurate than those of Case 2. Case 3 produces *totally unacceptable* results with the solution becoming negative.

t	True solution (at x = 0.3)	Solutions		Relative Error (%)	
		Case 1	Case 2	Case 1	Case 2
0.005	0.5966	0.5971	0.6000	0.08	0.57
0.010	0.5799	0.5822	0.6000	0.4	3.5
0.020	0.5334	0.5373	0.5500	0.7	3.1
0.100	0.2444	0.2472	0.2484	1.1	1.6

The results clearly indicate that the value r is critical to the production of a stable solution. In this case, *the explicit method is valid only when* $0 < r \leq \frac{1}{2}$.

Crank–Nicholson Implicit Method

The explicit finite difference method was simple but had a serious limitation on the time step (Δt) since $0 < r \leq \frac{1}{2}$. That is, $\Delta t \leq \Delta x^2/2$ and Δx *must be small to obtain reasonable spatial accuracy.*

We would like to obtain a method where the stability limit on r is greatly increased. We can obtain this if we evaluate the term $\partial^2 u/\partial x^2$ at some weighted average of its value at the time levels j and $j+1$.

EXAMPLE 8.2.2 (Crank–Nicholson implicit method). For the equation

$$\frac{\partial u}{\partial t} = \frac{\partial^2 u}{\partial x^2},$$

we have

$$\frac{u_{i,j+1} - u_{i,j}}{\Delta t} = \beta \left(\frac{u_{i+1,j+1} - 2u_{i,j+1} + u_{i-1,j+1}}{\Delta x^2} \right) + (1-\beta) \left(\frac{u_{i+1,j} - 2u_{i,j} + u_{i-1,j}}{\Delta x^2} \right).$$

For

$$\beta = 0, \quad \text{Explicit method (Euler);} \quad 0(\Delta t), 0(\Delta x^2),$$

$$\beta = \tfrac{1}{2}, \quad \text{Crank–Nicholson;} \quad 0(\Delta t^2), 0(\Delta x^2),$$

$$\beta = 1, \quad \text{Implicit method (Backward Euler);} \quad 0(\Delta t), 0(\Delta x^2).$$

The mesh is shown in Fig. 8.3.

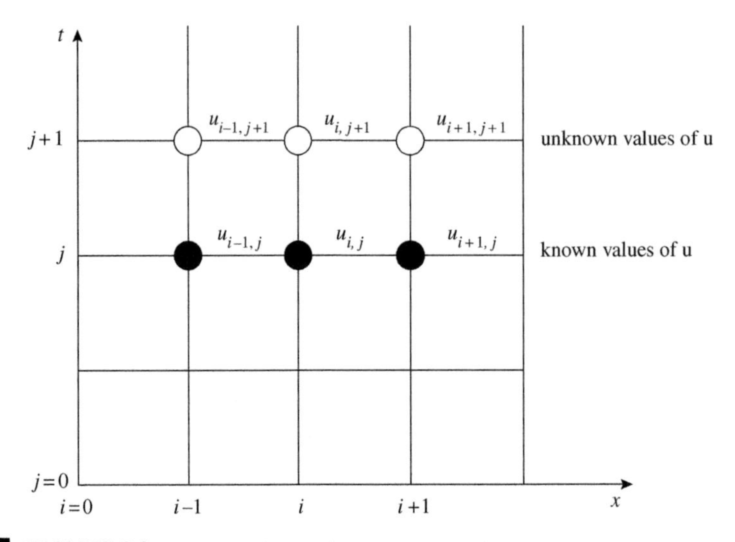

FIGURE 8.3 Crank–Nicholson finite difference scheme.

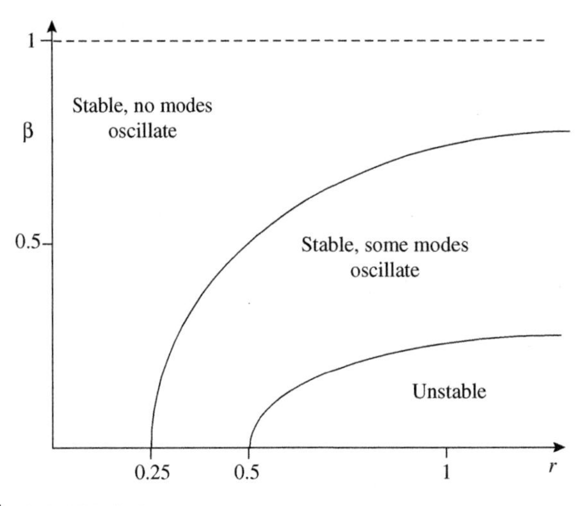

FIGURE 8.4 Stability regions for the Crank–Nicholson method.

The stability of these equations is given by

$$r = \frac{\Delta t}{\Delta x^2} \leq \frac{0.5}{1 - 2\beta}. \tag{8.14}$$

We can plot r versus β to obtain the stability properties of the FDA for $\partial u/\partial t = \partial^2 u/\partial x^2$. This is shown in Fig. 8.4.

Solution of the equations

The solution of the Crank–Nicholson method as well as the fully implicit method means that it is necessary to solve a tridiagonal system of equations to find the unknowns.

Suppose we have $N-1$ internal mesh points at each time row (j); then the equations are

$$-r\beta u_{i-1,j+1} + (1+2r\beta)u_{i,j+1} - r\beta u_{i+1,j+1} \quad (8.15)$$
$$= u_{i,j} + r(1-\beta)(u_{i+1,j} - 2u_{i,j} + u_{i-1,j}), \quad i = 1, \ldots, N-1,$$

where the unknowns are

$$u_{i-1,j+1}, \quad u_{i,j+1}, \quad u_{i+1,j+1}. \quad (8.16)$$

We can write the simultaneous equations in the matrix form:

$$\mathbf{A}u_{j+1} = f(u_j) \quad j = 0, 1, \ldots. \quad (8.17)$$

We obtain the solution at each timestep j by using a modified elimination and back substitution algorithm, such as Thomas' algorithm which accounts for the tridiagonal structure. Fixed or Dirichlet boundary conditions are imposed at $j = 0$ and N before the final matrix equations is solved.

EXAMPLE 8.2.3 (Crank–Nicholson method). Consider using Crank–Nicholson ($\beta = 1/2$) for our test problem with $\Delta x = 0.1$ and $\Delta t = 0.01$ and so $r = 1$. The equations at each time point $j+1$ are

$$-u_{i-1,j+1} + 4u_{i,j+1} - u_{i+1,j+1} = u_{i-1,j} + u_{i+1,j}.$$

We obtain the matrix equations:

$$\begin{pmatrix} 4 & -1 & & & \\ -1 & 4 & -1 & & \\ & \ddots & \ddots & \ddots & \\ & & \ddots & \ddots & \ddots \\ & & & \ddots & \ddots & \ddots \\ & & & & \ddots & \ddots & \ddots \\ & & & & & \ddots & \ddots & \ddots \\ & & & & & \ddots & \ddots & \ddots \\ & & & & & & -1 & 4 & -1 \\ & & & & & & & -1 & 4 \end{pmatrix} \begin{pmatrix} u_{1,j+1} \\ u_{2,j+1} \\ \vdots \\ \vdots \\ \vdots \\ u_{i,j+1} \\ \vdots \\ \vdots \\ \vdots \\ u_{9,j+1} \end{pmatrix} = \begin{pmatrix} u_{0,j} + u_{2,j} \\ u_{1,j} + u_{3,j} \\ \vdots \\ \vdots \\ \vdots \\ \vdots \\ \vdots \\ \vdots \\ \vdots \\ u_{8,j} + u_{10,j} \end{pmatrix}$$

for the mesh.

For this problem, it is possible to reduce the amount of work since the initial conditions are symmetric about $x = 0.5$. Applying the method gives the following results at $x = 0.5$.

t	True solution (at $x = 0.5$)	Numerical solution	Relative Error
0.01	0.7743	0.7691	−0.7
0.02	0.6809	0.6921	+1.6
0.10	0.3021	0.3069	+1.6

This shows approximately the same error as the explicit method (but this method uses 10 times the number of steps). The results show the increased accuracy of Crank–Nicholson $\mathcal{O}(\Delta t^2)$ over the Euler method $\mathcal{O}(\Delta t)$ and the ability of the Crank–Nicholson method to take large steps in Δt.

8.2.3. Parabolic Equations in Two Dimensions

Here, we are concerned about the solution of the equation

$$\frac{\partial u}{\partial t} = k\left(\frac{\partial^2 u}{\partial x^2} + \frac{\partial^2 u}{\partial y^2}\right) \tag{8.18}$$

over the rectangular region $0 < x < a,\ 0 < y < b$, where u is initially known at all points within and on the boundary of the rectangle, and is known subsequently on the boundary for all time. That is, we have initial conditions for the whole region as well as boundary conditions. The mesh is shown in Fig. 8.5.

We define the co-ordinates of the mesh as (x, y, t) with

$$x_i = i\Delta x; \quad y_j = j\Delta y; \quad t_n = n\Delta t, \tag{8.19}$$

where i, j, n are positive integers and we denote the value of u at the mesh points as

$$u(i\Delta x, j\Delta y, n\Delta t) = u_{i,j,n}. \tag{8.20}$$

The simplest finite difference method would be an explicit method. Let us consider this approach to our problem, as well as some others.

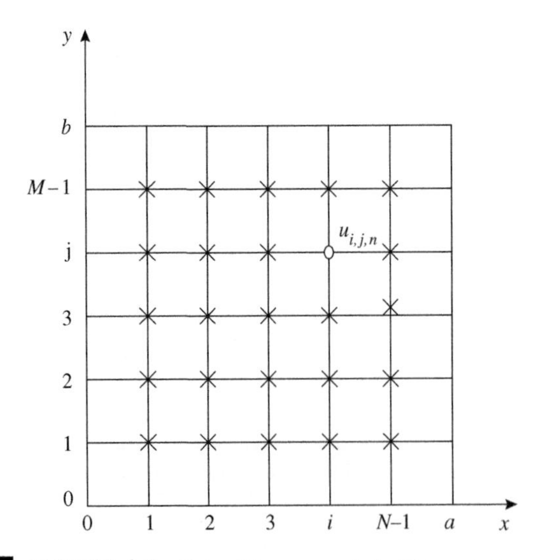

FIGURE 8.5 Finite difference mesh for 2D parabolic equations.

Explicit Finite Difference Method

We would have from (8.18)

$$\frac{u_{i,j,n+1} - u_{i,j,n}}{\Delta t} = k\left(\frac{u_{i,j,n} - 2u_{i,j,n} + u_{i+1,j,n}}{\Delta x^2}\right) + k\left(\frac{u_{i,j-1,n} - 2u_{i,j,n} + u_{i,j+1,n}}{\Delta y^2}\right). \quad (8.21)$$

This is obviously a simple formulation but the stability limit shows that

$$k\left(\frac{1}{\Delta x^2} + \frac{1}{\Delta y^2}\right)\Delta t \le \frac{1}{2}, \quad (8.22)$$

which implies very small values of Δt. Hence, it is generally *not* a practical method.

Crank–Nicholson Method

In this case, we introduce some implicit nature into the discretization to help overcome the stability issue.

EXAMPLE 8.2.4 (Crank–Nicholson method). We would have

$$\frac{u_{i,j,n+1} - u_{i,j,n}}{\Delta t} = \frac{k}{2}\left(\left(\frac{\partial^2 u}{\partial x^2} + \frac{\partial^2 u}{\partial y^2}\right)_{i,j,n} + \left(\frac{\partial^2 u}{\partial x^2} + \frac{\partial^2 u}{\partial y^2}\right)_{i,j,n+1}\right).$$

This would be valid for all values of Δx, Δy and Δt but requires the solution of $(M-1) \times (N-1)$ simultaneous algebraic equations at each timestep, where $N = a/\Delta x, M = b/\Delta y$.

The equations must normally be solved by iterative means rather than via the simple tridiagonal system algorithm.

Alternating-Direction Implicit (ADI) Method

This method was originally developed to overcome the problems associated with the previous methods. For a typical problem, the method is about 25 times less work than the explicit method and about seven times less work than the Crank–Nicholson method.

Assume we know the solution at $t = n \cdot \Delta t$, we replace *only one* of the second-order terms, say $\partial^2 u/\partial x^2$ by an implicit difference approximation at the unknown $(n+1)$ time level. The other term $\partial^2 u/\partial y^2$ is replaced by an explicit representation. We have

$$\frac{u_{i,j,n+1} - u_{i,j,n}}{k \cdot \Delta t} = \frac{u_{i-1,j,n+1} - 2u_{i,j,n+1} + u_{i+1,j,n+1}}{\Delta x^2} + \frac{u_{i,j-1,n} - 2u_{i,j,n} + u_{i,j+1,n}}{\Delta y^2}. \quad (8.23)$$

We can use this to advance from the nth to $(n+1)$th timestep by solving the $(N-1)$ equations in each row parallel to the x axis. Hence, to solve for all the variables we need to solve $(M - 1)$ sets of $(N - 1)$ equations. This is much easier than solving $(M - 1)(N - 1)$ equations.

The advancement of the solution to the $(n + 2)$th time level is then achieved by replacing $\partial^2 u/\partial y^2$ by an implicit approximation and $\partial^2 u/\partial x^2$ by an explicit approximation, so that

$$\frac{u_{i,j,n+2} - u_{i,j,n+1}}{k \cdot \Delta t} = \frac{u_{i-1,j,n+1} - 2u_{i,j,n+1} + u_{i+1,j,n+1}}{\Delta x^2} + \frac{u_{i,j-1,n+2} - 2u_{i,j,n+2} + u_i}{\Delta y^2}. \tag{8.24}$$

This involves solving $(N-1)$ sets of $(M-1)$ equations and we are solving in columns parallel to the y axis. The time interval Δt must be the same for each advancement.

The method generally works well for simple, ideal problems. For nonlinear problems or problems with complex geometry it may fail completely, when using the standard iterative techniques. An alternative is to use an implicit iterative method.

8.2.4. Elliptic Equations

Elliptic equations arise from the consideration of steady state behaviour. Hence there is no time derivative term and we solve the equation over the spatial domain subject to the chosen boundary conditions.

The following example illustrates an approach to the solution of an elliptic problem.

EXAMPLE 8.2.5 (Elliptic PDE). Consider the 2D heat conduction problem given by:

$$\frac{\partial^2 T}{\partial x^2} + \alpha \frac{\partial^2 T}{\partial y^2} = Q$$

where Q is an internal heat generation term and α is a thermal conductivity. Boundary conditions are:

$$T = 0 \quad \text{on } x = 0, x = 1$$
$$T = 0 \quad \text{on } y = 0$$
$$\frac{\partial T}{\partial y} = -T \quad \text{on } y = 1.$$

A second order FDA for the elliptic problem could be written as:

$$\frac{(T_{i+1,j} - 2T_{i,j} + T_{i-1,j})}{\Delta x^2} + \alpha \frac{(T_{i,j+1} - 2T_{i,j} + T_{i,j-1})}{\Delta y^2} = Q_{i,j}$$

which becomes on rearrangement:

$$2(1 + \alpha)T_{i,j} = T_{i+1,j} + T_{i-1,j} + \alpha(T_{i,j+1} + T_{i,j-1}) - h^2 Q_{i,j}.$$

This FDA of the elliptic PDE can be solved for $T_{i,j}$ by iterating point by point or solving for all T_i at a given level j. Fixed value (Dirichlet) boundary conditions at $x = 0, 1$ and $y = 0$ are easily imposed. The derivative of Neumann boundary condition needs special treatment as the next section shows. ■ ■ ■

8.2.5. Modification for Boundary Conditions

The elliptic problem had a boundary condition in terms of $\partial T/\partial y = -T$ on $y = 1$.

It is possible to handle this condition by using the "false boundary" approach. We introduce an $(N + 1)$th line of points in the y-direction. Schematically, we see this illustrated in Fig. 8.6.

We can write the boundary condition as

$$\frac{T_{i,N+1} - T_{i,N-1}}{2\Delta y} = -T_{i,N} \quad \text{(central difference)}. \tag{8.25}$$

The FDA for the Nth row is given by

$$2(1+\alpha)T_{i,N} = T_{i+1,N} + T_{i-1,N} + \alpha\left(T_{i,N+1} + T_{i,N-1}\right) - h^2 Q_{i,N}. \tag{8.26}$$

We can then replace $T_{i,N+1}$ by the boundary condition (rearranged)

$$T_{i,N+1} = T_{i,N-1} - 2\Delta y T_{i,N} \tag{8.27}$$

to obtain the expression for the Nth row as

$$2(1+\alpha+\alpha h)T_{i,N} = T_{i+1,N} + T_{i-1,N} + 2\alpha T_{i,N-1} - h^2 Q_{i,N}. \tag{8.28}$$

Curved boundaries can be handled by special interpolation functions [56].

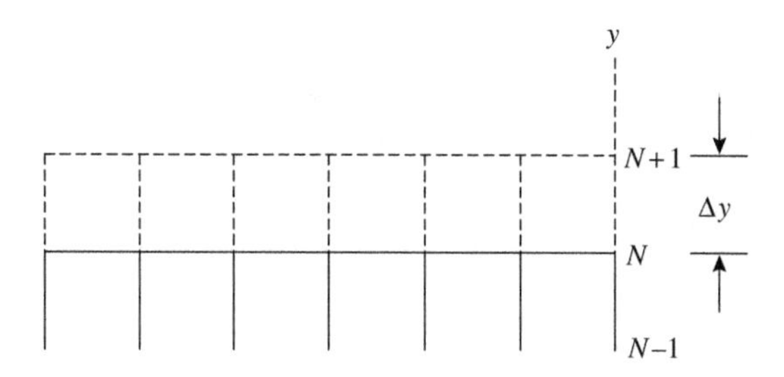

FIGURE 8.6 False boundary finite difference scheme.

8.3. METHOD OF LINES

This is very similar to the finite difference method except we *do not discretize the time variable*. Hence, we produce a set of ODEs, which can be integrated using the standard packages. This technique is closely related to the "lumping" procedure for PDEs.

Thus, we convert the equation

$$\frac{\partial u}{\partial t} = \frac{\partial^2 u}{\partial x^2} \tag{8.29}$$

to

$$\frac{du_i}{dt} = f_i(u), \quad i = 0, \ldots, N. \tag{8.30}$$

EXAMPLE 8.3.1 (Packed bed reactor with radial dispersion). Consider a packed bed reactor with radial dispersion:

$$\frac{\partial c}{\partial t} = \alpha \nabla^2 c + \beta R(c),$$
$$\frac{\partial c}{\partial r} = 0 \quad \text{at } r = 0; \qquad \frac{\partial c}{\partial r} = Bi_w \left(c(1, z) - c_w(z)\right).$$

We have, after expanding $\nabla^2 c$ the discretized equations:

$$\frac{dc_i}{dt} = \alpha \left[\frac{c_{i-1} - 2c_i + c_{i+1}}{\Delta r^2} + \left(\frac{a-1}{r_i} \right) \frac{c_{i+1} - c_{i-1}}{2\Delta r} \right] + \beta R_i,$$

where $a = 1, 2, 3$ depending on geometry.

For the derivative boundary conditions we use a false boundary to give

$$\frac{c_1 - c_{-1}}{2\Delta r} = 0; \qquad \frac{c_{N+1} - c_{N-1}}{2\Delta r} = Bi_w \left(c_N - c_w\right)$$

and the equations for $i = 0$ and N are found from substituting the boundary conditions into the governing equations.

Thus, we have a set of ODEs

$$\frac{dc_0}{dt} = \frac{2a}{\Delta r^2} (c_1 - c_0) + \beta R_0,$$
$$\frac{dc_i}{dt} = \alpha \left[\frac{c_{i-1} - 2c_i + c_{i+1}}{\Delta r^2} + \left(\frac{a-1}{r_i} \right) \frac{c_{i+1} - c_{i-1}}{2\Delta r} \right] + \beta R_i \quad i = 1, \ldots, N-1,$$
$$\frac{dc_N}{dt} = \alpha \left[\frac{2c_{N-1} - Bi_w 2\Delta r (c_{N-1}) - 2c_N}{\Delta r^2} - (a-1) Bi_w (c_w - c_N) \right] + \beta R_N.$$

A common problem with the method of lines is that the resulting set is often "stiff" and thus implicit ODE methods are normally needed to solve these problems.

EXAMPLE 8.3.2 (A dispersion–convection problem). Consider a dispersion–convection problem given by

$$\frac{\partial y}{\partial t} = \frac{\partial^2 y}{\partial x^2} - c \frac{\partial y}{\partial x},$$

$c = \text{constant} > 0.$

Boundary conditions and initial condition are

$$y(0,t) = 1, \quad t > 0,$$
$$\frac{\partial y}{\partial x}(1,t) = 0, \quad t > 0,$$
$$y(x,0) = 0, \quad 0 < x < 1.$$

Applying the method of lines gives

$$\frac{\mathrm{d}y_i}{\mathrm{d}t} = \frac{y_{i-1} - 2y_i + y_{i+1}}{(1/N)^2} - c\left(\frac{y_{i+1} - y_{i-1}}{(2/N)}\right), \quad i = 1, \ldots, N,$$

$$y_i(0) = 0, \quad i = 1, \ldots, N, \quad y_0(t) = 1, \quad y_{N+1}(t) = y_{N-1}(t).$$

In this case, the stiffness of the problem is proportional to the square of the discretization value N.

8.4. METHOD OF WEIGHTED RESIDUALS

In this section, we consider the numerical solution of general boundary value problems (ODE-BVP) as a prelude to the solution of PDEs in Section 8.6. The key concept is that we substitute a trial function, typically a power series function into the governing equation which creates a "residual" or error since the chosen function is rarely the correct solution. We then minimize the "weighted residual" by applying various weighting criteria which distribute the error in different ways over the spatial domain. Hence, the term method of weighted residuals (MOWR). A comprehensive treatment of these techniques in chemical engineering is given by Finlayson [55,57].

EXAMPLE 8.4.1 (A simple heat conduction problem). Consider a simple heat conduction problem which we will use as a test problem [57]. Here, we have the BVP:

$$\frac{\mathrm{d}}{\mathrm{d}x}\left(c(u)\frac{\mathrm{d}u}{\mathrm{d}x}\right) = 0 \quad \text{where } c(u) = 1 + \alpha u$$

with boundary conditions

$$u(0) = 0; \quad u(1) = 1.$$

The exact solution is given by:

$$u(x) = -1 + \sqrt{3x+1} \quad \text{for } \alpha = 1,$$
$$u(0.5) = 0.5811$$

We will now develop and compare MOWR approaches with the analytical solution.

8.4.1. Basic Steps in the Method of Weighted Residuals

The procedure for applying the MOWR is as follows:

Step 1. Choose a trial function and expand the function.

Step 2. Try and fit boundary conditions to the trial function.

Step 3. Substitute the trial function into the problem and form a residual.
Step 4. Make the residual zero by using one of the following criteria:
Collocation, Galerkin, Least squares, Subdomain or Moments.

These criteria distribute the error in the solution in different ways and can be used for a variety of applications.

EXAMPLE 8.4.2 (MOWR approach to the simple heat conduction problem). We can now apply the MOWR approach to our simple heat conduction problem.

Step 1: Choose $\phi_N = \sum_{i=0}^{N+1} c_i x^i$ as the trial function (or test solution).
Step 2: Now fit the boundary conditions to the trial function:

$$\phi_N(0) = 0 \quad \text{implies that} \quad c_0 = 0,$$

$$\phi_N(1) = 1 \quad \text{implies that} \quad \sum_{i=1}^{N+1} C_i = 1$$

or

$$c_{N+1} = 1 - \sum_{i=1}^{N} c_i$$

or

$$c_1 = 1 - \sum_{i=2}^{N+1} c_i$$

so that we can modify and write the trial function as

$$\phi_N = \left(1 - \sum_{i=2}^{N+1} c_i\right) x + \sum_{i=2}^{N+1} c_i x^i,$$

$$\phi_N = x + \sum_{i=2}^{N+1} c_i \left(x^i - x\right).$$

Finally, we have

$$\phi_N = x + \sum_{i=1}^{N} A_i (x^{i+1} - x),$$

which we will now use as our approximate solution. Notice that we have the unknown coefficients A_i. The higher the order of the polynomial the more A_i terms we have. It should be noted that regardless of the values of A_i the boundary conditions are satisfied by the trial function. Put the values of $x = 0$ and $x = 1$ in and see.

Step 3: Form the residual by substituting the trial function into the expanded BVP to obtain

$$R(x, \phi_N) = (1 + \alpha\phi_N)\phi_N'' + \alpha(\phi_N')^2,$$

where ϕ_N' and ϕ_N'' are the first and second derivations of ϕ_N.

Then, the weighted residual is

$$\int_0^1 w_k R(x, \phi_N)\,\mathrm{d}x = 0, \quad k = 1, \ldots, N.$$

Step 4: Apply a weighting criterion to reduce the residual to zero. That is, we choose a method of distributing the error in a particular way. We consider a collocation approach in this case, although other approaches can be used.

Collocation method

Here, we force ϕ_N to satisfy the residual at a given set of N points. This leads to N algebraic equations whose solution gives the value of the coefficients A_i. At the collocation points, the residual $R(x, \phi_N) = 0$. As more collocation points are used, we force the approximation to become more accurate.

Hence, for the collocation method, the weight (w_k) is

$$w_k = \delta(x - x_k) \quad \text{(dirac delta)}, \tag{8.31}$$

meaning that $w_k = 1$ when $x = x_k$, otherwise $w_k = 0$.

Hence,

$$\int_0^1 w_k R(x, \phi_N)\,\mathrm{d}x = R(x_k, \phi_N) = 0. \tag{8.32}$$

EXAMPLE 8.4.3 (A fixed point collocation example). Collocate at $x = 0.5$. Here we want the residual to be zero at the mid-point.

From the trial function ϕ_N we obtain

$$\phi_1 = x + A_1(x^2 - x), \quad \phi_1' = 1 + A_1(2x - 1), \quad \phi_1'' = 2A_1.$$

The residual becomes

$$R(x, \phi_1) = \left\{1 + \alpha\left[x + A_1\left(x^2 - x\right)\right]\right\} 2A_1 + \alpha\left[1 + A_1(2x - 1)\right]^2.$$

Now let $x = \frac{1}{2}$ (collocation point), $\alpha = 1$ giving the residual equation

$$\left[1 + \tfrac{1}{2} + A_1\left(\tfrac{1}{4} - \tfrac{1}{2}\right)\right] 2A_1 + 1 + A_1(1 - 1) = 0,$$

$$-\tfrac{1}{2}A_1^2 + 3A_1 + 1 = 0, \quad A_1 = -0.3166.$$

The other root is not feasible, so the approximate solution is

$$\phi_1 = x - 0.3166(x^2 - x).$$

and

$$\phi_1(0.5) = 0.5795$$

which compares well with the analytic solution 0.5811.

8.5. ORTHOGONAL COLLOCATION

In this case, the choice of collocation points is not arbitrary. The trial functions are chosen as sets of orthogonal polynomials and the collocation points are the roots of these polynomials. Also, we can formulate the solution so that the dependent variables are the solution values at the collocation points, rather than the coefficients of the expansion [57]. The use of orthogonal polynomials means that the solution error decreases faster as the polynomial order increases in comparison to arbitrarily placed collocation points.

Instead of any trial functions, $\phi_i(x)$, let us use orthogonal polynomials of the form

$$P_m(x) = \sum_{i=0}^{m} c_i x^i. \tag{8.33}$$

We can define the polynomials by

$$\int_a^b w(x) P_n(x) P_m(x)\, \mathrm{d}x = 0, \quad n = 0, 1, \ldots, m-1. \tag{8.34}$$

A particular sequence of orthogonal polynomials leads to the *Legendre polynomials*:

$$P_0 = 1, \quad P_1 = x, \quad P_2 = 1 - 3x^2, \quad \ldots \tag{8.35}$$

with the polynomial $P_m(x)$ having m roots, which become the collocation points. Other orthogonal polynomials such as the Jacobi polynomials can also be used [58]. In the next two sections, we obtain the key polynomial expansions suitable for problems with symmetry and those without symmetry.

8.5.1. Orthogonal Collocation with Symmetry

Consider problems on the domain $0 \leq x \leq 1$ and symmetric about $x = 0$. These can typically be reactor or catalyst problems. We can exploit the geometry of the problem by considering polynomials expanded in powers of x^2 [57]. For problems with Dirichlet boundary conditions, we could use

$$y(x) = y(1) + (1 - x^2) \sum_{i=1}^{N} a_i P_{i-1}(x^2), \tag{8.36}$$

and the orthogonal polynomial can be defined as

$$\int_0^1 w(x^2) P_m(x^2) P_n(x^2) x^{a-1}\, \mathrm{d}x = 0, \quad n = 0, 1, \ldots, m-1, \tag{8.37}$$

where

$$a = \begin{cases} 1 & \text{planar,} \\ 2 & \text{cylindrical,} \\ 3 & \text{spherical.} \end{cases}$$

specify the geometry of the problem.

In the above, we know $P_{i-1}(x^2)$ and the expression automatically satisfies the boundary condition since the presence of the $(1-x^2)$ term ensures it. Thus, we build the symmetry information into our trial function. We can substitute the trial solution into our differential equation to form the residual, which is then set to zero at the N *interior* collocation points x_j. These points are the roots of the Nth order polynomial $P_N(x^2) = 0$ at x_j. This gives N equations for the N coefficients a_i.

An equivalent form of the trial function is given by

$$y(x) = \sum_{i=1}^{N+1} d_i x^{2i-2}. \tag{8.38}$$

At the collocation points we have

$$y(x_j) = \sum_{i=1}^{N+1} x_j^{2i-2} d_i, \tag{8.39}$$

and thus the derivative term is

$$\left.\frac{dy}{dx}\right|_{x_j} = \sum_{i=1}^{N+1} \left.\frac{dx^{2i-2}}{dx}\right|_{x_j} d_i \tag{8.40}$$

with the Laplacian

$$\nabla^2 y = \sum_{i=1}^{N+1} \nabla^2 (x^{2i-2})_{x_j} d_i \tag{8.41}$$

given for each of the geometries, planar, cylindrical and spherical.

These can be written in matrix form as

$$y - \mathbf{Q}d, \quad \frac{dy}{dx} = \mathbf{C}d, \quad \nabla^2 y = \mathbf{D}d, \tag{8.42}$$

where

$$Q_{ji} = x_j^{2i-2}, \quad C_{ji} = \left.\frac{dx^{2i-2}}{dx}\right|_{x_j}, \quad D_{ji} = \left.\nabla^2(x^{2i-2})\right|_{x_j}. \tag{8.43}$$

We can obtain $d = \mathbf{Q}^{-1}y$ and then get the derivatives in terms of the values of the function at the collocation points.

$$\frac{dy}{dx} = \mathbf{CQ}^{-1}y \equiv \mathbf{A}y, \tag{8.44}$$

$$\nabla^2 y = \mathbf{DQ}^{-1}y \equiv \mathbf{B}y. \tag{8.45}$$

The matrices $(\mathbf{A}, \mathbf{B})$ are $(N+1)^*(N+1)$ and the collocation points are shown in Fig. 8.7.

The symmetry condition is imposed at $x = 0$. It is important to note that there are N interior collocation points, with the $N+1$ point at $x = 1$.

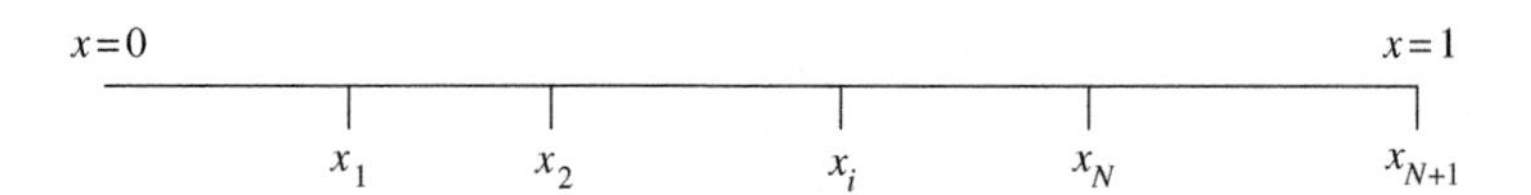

FIGURE 8.7 Location of collocation points for symmetric polynomials.

Quadrature
To accurately evaluate integrals we can use:

$$\int_0^1 f(x^2)x^{a-1}\,\mathrm{d}x = \sum_{j=1}^{N+1} W_j f\left(x_j^2\right). \tag{8.46}$$

To find the quadrature weights, evaluate the definite integral for $f_i = x^{2i-2}$. Thus,

$$\int_0^1 x^{2i-2}x^{a-1}\,\mathrm{d}x = \sum_{j=1}^{N+1} W_j x_j^{2i-2} = \frac{1}{2i-2+a} \equiv f_i, \tag{8.47}$$

$$W\mathbf{Q} = f \quad \text{giving} \quad W = f\mathbf{Q}^{-1}. \tag{8.48}$$

Tables 8.1 and 8.2 give information on the collocation points and collocation matrices for the symmetric polynomials [55].

Note that these tables are to be *used only for symmetric polynomials*. The following examples show how the numerical solution is formulated for problems with symmetry.

TABLE 8.1 Roots of Symmetric Polynomials

Points	$a = 1$, planar geometry	$a = 2$, cylindrical geometry	$a = 3$, spherical geometry
	$w = 1 - x^2$	$w = 1 - x^2$	$w = 1 - x^2$
$N = 1$	0.44721 35955	0.57735 02692	0.65465 36707
$N = 2$	0.28523 15165	0.39376 51911	0.46884 87935
	0.76505 53239	0.80308 71524	0.83022 38963

EXAMPLE 8.5.1 (Diffusion and reaction in a catalyst pellet). A model of diffusion and reaction in a catalyst pellet, can be described by a BVP [55].

Consider

$$\frac{\mathrm{d}^2 y}{\mathrm{d}x^2} = \phi^2 R(y)$$

with

$$\frac{\mathrm{d}y}{\mathrm{d}x} = 0 \quad \text{at } x = 0,$$

$$y(1) = 1.$$

TABLE 8.2 Matrices for Orthogonal Collocation Using Symmetric Polynomials

Points	Weights
Planar geometry $(a = 1)$	
$N = 1$	$W = \begin{pmatrix} 0.8333 \\ 0.1667 \end{pmatrix}$ $\mathbf{A} = \begin{pmatrix} -1.118 & 1.118 \\ -2.500 & 2.500 \end{pmatrix}$ $\mathbf{B} = \begin{pmatrix} -2.5 & 2.5 \\ -2.5 & 2.5 \end{pmatrix}$
$N = 2$	$W = \begin{pmatrix} 0.5549 \\ 0.3785 \\ 0.0667 \end{pmatrix}$ $\mathbf{A} = \begin{pmatrix} -1.753 & 2.508 & -0.7547 \\ -1.371 & -0.6535 & 2.024 \\ 1.792 & -8.791 & 7 \end{pmatrix}$ $\mathbf{B} = \begin{pmatrix} -4.740 & 5.677 & -0.9373 \\ 8.323 & -23.26 & 14.94 \\ 19.07 & -47.07 & 28 \end{pmatrix}$
Cylindrical geometry $(a = 2)$	
$N = 1$	$W = \begin{pmatrix} 0.375 \\ 0.125 \end{pmatrix}$ $\mathbf{A} = \begin{pmatrix} -1.732 & 1.732 \\ -3 & 3 \end{pmatrix}$ $\mathbf{B} = \begin{pmatrix} -6 & 6 \\ -6 & 6 \end{pmatrix}$
$N = 2$	$W = \begin{pmatrix} 0.1882 \\ 0.2562 \\ 0.0556 \end{pmatrix}$ $\mathbf{A} = \begin{pmatrix} -2.540 & 3.826 & -1.286 \\ 1.378 & -1.245 & 2.623 \\ 1.715 & -9.715 & 8 \end{pmatrix}$ $\mathbf{B} = \begin{pmatrix} -9.902 & 12.30 & -2.397 \\ 9.034 & -32.76 & 23.73 \\ 22.76 & -65.42 & 42.67 \end{pmatrix}$
Spherical geometry $(a = 3)$	
$N = 1$	$W = \begin{pmatrix} 0.2333 \\ 0.1 \end{pmatrix}$ $\mathbf{A} = \begin{pmatrix} -2.291 & 2.291 \\ -3.5 & 3.5 \end{pmatrix}$ $\mathbf{B} = \begin{pmatrix} -10.5 & 10.5 \\ -10.5 & 10.56 \end{pmatrix}$
$N = 2$	$W = \begin{pmatrix} 0.0949 \\ 0.1908 \\ 0.0476 \end{pmatrix}$ $\mathbf{A} = \begin{pmatrix} -3.199 & 5.015 & -1.816 \\ -1.409 & -1.807 & 3.215 \\ 1.697 & -10.70 & 9 \end{pmatrix}$ $\mathbf{B} = \begin{pmatrix} -15.67 & 20.03 & -4.365 \\ 9.965 & -44.33 & 34.36 \\ 26.93 & -86.93 & 60 \end{pmatrix}$

Evaluate the residuals at N interior collocation points x_j to give

$$\sum_{i=1}^{N+1} B_{ji} y_i = \phi^2 R(y_j), \quad j = 1, \ldots, N,$$

and the outer boundary condition gives

$$R(y) = y^2, \quad N = 1.$$

We have from the residual equations

$$B_{11}y_1 + B_{12}y_2 = \phi^2 R(y_1) = \phi^2 y_1^2$$

with

$$y_2 = 1.$$

For planar geometry $a = 1$ and for $w = 1 - x^2$, we have in Table 8.2

$$B_{11} = -2.5 \quad W_1 = 5/6,$$

$$B_{12} = 2.5 \quad W_2 = 1/6.$$

The residual equation becomes

$$-2.5y_1 + 2.5y_2 = \phi^2 y_1^2, \quad y_2 = 1.$$

Solve for $y_1 = f(\phi^2)$ giving

$$y_1 = \frac{-2.5 + \sqrt{6.25 + 10\phi^2}}{2\phi^2}.$$

■ ■ ■ This provides a good solution upto $\phi = 2$.

EXAMPLE 8.5.2 (A reactor problem). Consider the following reactor problem given by

$$\frac{1}{x^{a-1}}\frac{\mathrm{d}}{\mathrm{d}x}\left(x^{a-1}\frac{\mathrm{d}y}{\mathrm{d}x}\right) + g(x^2)y = f(x^2)$$

with $\mathrm{d}y/\mathrm{d}x = 0$ at $x = 0$; $y(1) = 1, a = 1$.

The residuals at the collocation points are

$$\sum_{i=1}^{N+1} B_{ji}y_i + g\left(x_j^2\right)y_j = f\left(x_j^2\right), \quad y_{N+1} = 1.$$

If at $x = 1$ we had $-\mathrm{d}y/\mathrm{d}x = ay + b$ as the boundary condition, then we would write the additional residual equation:

$$-\sum_{i=1}^{N+1} A_{N+1\ i}\ y_i = ay_{N+1} + b.$$

If we need to solve for another point (not a collocation point), then we can use

$$y(x) = \sum_{i=1}^{N+1} d_i x^{2i-2}$$

and

$$d = \mathbf{Q}^{-1}y.$$

■ ■ ■

Note that we do not expand the Laplacian $\nabla^2 y$ when using the trial functions with symmetry built into them. Tables 8.1 and 8.2 give the matrices **A** and **B** for the three types of geometry. When $a = 1$ we still use the matrices **B** when writing the residuals, since the geometry information is built into the polynomials.

8.5.2. Orthogonal Collocation without Symmetry

Many problems in process engineering do not have symmetry properties and hence we need unsymmetric polynomials to handle these cases. We can of course always use unsymmetric polynomials on symmetric problems by imposing the symmetry condition as a boundary condition.

Hence, we now have polynomials in even and odd powers of x, given by [55]

$$y(x) = b + cx + x(1-x)\sum_{i=1}^{N} a_i P_{i-1}(x) \tag{8.49}$$

with the polynomials defined by

$$\int_a^b w(x)P_m(x)P_n(x)\,\mathrm{d}x = 0 \quad n = 0, 1, \ldots, m-1. \tag{8.50}$$

This gives N interior collocation points, with two boundary conditions ($x = 0,\ x = 1$) as seen in Fig. 8.8.

In a similar manner to Subsection 8.5.1, which developed the method for symmetric problems, the collocation points and the respective collocation matrices can be defined for the unsymmetric case. It should be noted that if unsymmetric polynomials are to be used then the Laplacian must be handled in *expanded form.* That is,

$$\nabla^2 u \equiv \frac{\mathrm{d}^2 u}{\mathrm{d}x^2} + \frac{a-1}{x}\frac{\mathrm{d}u}{\mathrm{d}x}. \tag{8.51}$$

Tables 8.3 and 8.4 give information on the collocation points and the corresponding matrices.

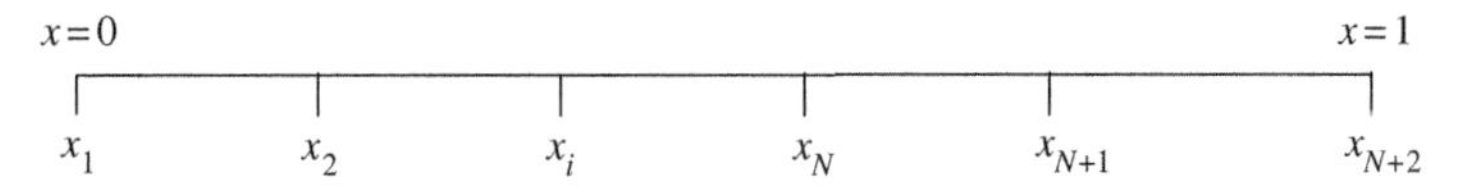

FIGURE 8.8 Location of collocation points for unsymmetric polynomials.

TABLE 8.3 Roots of Unsymmetric Polynomials

$N = 1$	0.5000 00000
$N = 2$	0.21132 48654
	0.78867 51346

EXAMPLE 8.5.3 (Diffusion and reaction problem). Again for the diffusion and reaction problem

$$\frac{\mathrm{d}^2 y}{\mathrm{d}x^2} = \phi^2 R(y); \quad \left.\frac{\mathrm{d}y}{\mathrm{d}x}\right|_{x=0} = 0, \quad y(1) = 1.$$

TABLE 8.4 Matrices for Orthogonal Collocation using Unsymmetric Polynomials

$N=1$	$W=\begin{pmatrix}\frac{1}{6}\\ \frac{2}{3}\\ \frac{1}{6}\end{pmatrix}$	$\mathbf{A}=\begin{pmatrix}-3 & 4 & -1\\ -1 & 0 & 1\\ 1 & -4 & 3\end{pmatrix}$
		$\mathbf{B}=\begin{pmatrix}4 & -8 & 4\\ 4 & -8 & 4\\ 4 & -8 & 4\end{pmatrix}$
$N=2$	$W=\begin{pmatrix}0\\ \frac{1}{2}\\ \frac{1}{2}\\ 0\end{pmatrix}$	$\mathbf{A}=\begin{pmatrix}-7 & 8.196 & -2.196 & +1\\ -2.732 & 1.732 & 1.732 & -0.7321\\ 0.7321 & -1.732 & -1.732 & 2.732\\ -1 & 2.196 & -8.196 & 7\end{pmatrix}$
		$\mathbf{B}=\begin{pmatrix}24 & -37.18 & 25.18 & -12\\ 16.39 & -24 & 12 & -4.392\\ -4.392 & 12 & -24 & 16.39\\ -12 & 25.18 & -37.18 & 24\end{pmatrix}$

Even though this is a symmetric problem, let us use the unsymmetric polynomials to get the following residual equations:

$$\sum_{i=1}^{N+2} B_{ji} y_i = \phi^2 R(y_j), \quad j = 2, \ldots, (N+1)$$

plus the boundary conditions

$$\sum_{i=1}^{N+2} A_{1i} y_i = 0, \quad j = 1,$$

$$y_{N+2} = 1, \quad j = N+2.$$

We can solve these using Newton's method.

EXAMPLE 8.5.4 (A heat transfer problem). Apply to the heat transfer problem

$$\frac{\mathrm{d}}{\mathrm{d}x}\left[k(T)\frac{\mathrm{d}T}{\mathrm{d}x}\right] = 0,$$

$$T(0) = 0 \quad T(1) = 1,$$

$$k(T) = 1 + T.$$

Upon expanding, we get

$$(1+T)\frac{\mathrm{d}^2T}{\mathrm{d}x^2} + \left(\frac{\mathrm{d}T}{\mathrm{d}x}\right)^2 = 0.$$

This is an unsymmetric problem and we have to use unsymmetric polynomials to get the residual equations

$$\left(1+T_j\right)\sum_{i=1}^{N+2} B_{ji}T_i + \left[\sum_{i=1}^{N+2} A_{ji}T_i\right]^2 = 0, \quad j = 2, \ldots, (N+1)$$

together with the boundary conditions

$$T_1 = 0, \quad T_{N+2} = 1.$$

Once we specify the number of internal collocation points, we can set up the equations and solve for the unknowns.

For $N = 1$, we have

$$(1+T_2)\sum_{i=1}^{3} B_{2i}T_i + \left(\sum_{i=1}^{3} A_{2i}T_i\right)^2 = 0, \quad j = 2$$

$$(1+T_2)\,(4T_1 - 8T_2 + 4T_3) + (-T_1 + T_3)^2 = 0$$

with

$$T_1 = 0 \quad T_3 = 1,$$

giving

■ ■ ■ $$T_2 = 0.579.$$

8.5.3. Orthogonal Collocation on Elements

In cases like the diffusion-reaction problem where the profiles can be very steep, it can be of advantage to divide the region into elements and use orthogonal collocation on separate elements rather than one polynomial over the whole region (global collocation).

It is then possible to reduce the number of total collocation points. This can significantly reduce computation time.

To do this, we use trial functions plus continuity conditions across the element boundaries. We use *unsymmetric* polynomials since we do not want to impose the symmetry condition on each element.

Setting up the problem

We divide the global x domain into a number of elements. A typical element exists between x_k and x_{k+1}. This situation is shown in the Fig. 8.9.

Hence,

$$\Delta x_k = x_{k+1} - x_k \tag{8.52}$$

and NE = total number of elements.

At the end points of each element, we want the function and the first derivative to be equal to the adjacent element and continuous across the element boundary.

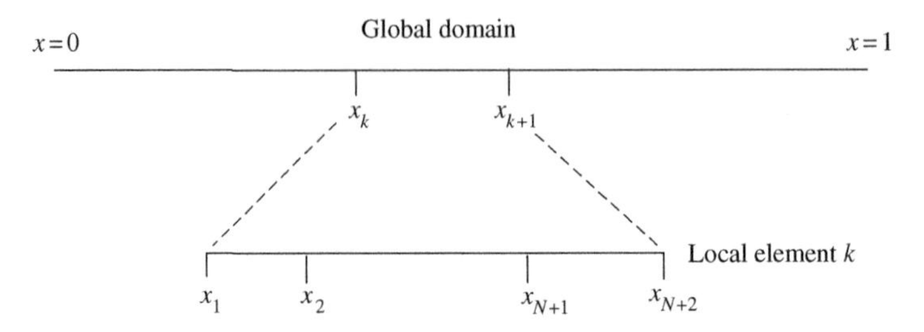

FIGURE 8.9 Layout of collocation on finite elements.

EXAMPLE 8.5.5 (Reaction and diffusion equation). Consider the 1D reaction and diffusion equation [57]:

$$\frac{\mathrm{d}^2 c}{\mathrm{d}x^2} + \frac{a-1}{x}\frac{\mathrm{d}c}{\mathrm{d}x} = \phi^2 f(c)$$

with boundary conditions

$$\left.\frac{\mathrm{d}c}{\mathrm{d}x}\right|_{x=0} = 0; \qquad -\left.\frac{\mathrm{d}c}{\mathrm{d}x}\right|_{x=1} = Bi_m(c(1)-1).$$

For the kth element we can define

$$s = \frac{x - x_k}{x_{k+1} - x_k} = \frac{x - x_k}{\Delta x_k},$$

so that s is between 0 and 1 on each element.

This gives the "local" BVP as

$$\frac{1}{\Delta x_k^2}\frac{\mathrm{d}^2 c}{\mathrm{d}s^2} + \frac{a-1}{x_k + s\cdot\Delta x_k}\cdot\frac{1}{\Delta x_k}\frac{\mathrm{d}c}{\mathrm{d}s} = \phi^2 f(c).$$

We can now apply collocation without symmetry on the element equation to give:

$$\frac{1}{\Delta x_k^2}\sum_{j=1}^{N+2} B_{ij}c_j^k + \frac{a-1}{x_k + s_i\Delta x_k}\cdot\frac{1}{\Delta x_K}\sum_{j=1}^{N+2} A_{ij}c_j^k = \phi^2 f\left(c_i^k\right),$$

where

$$k = 1, 2, \ldots, \mathrm{NE}, \quad i = 2, 3, \ldots, N+1.$$

Continuity conditions

Let the concentration c at the end of the $(k-1)$th element be the same as the concentration c at the start of the kth element (c_i^k = *concentration at* ith *point in* kth *element*).

Thus, we have

$$\left.\frac{\mathrm{d}c}{\mathrm{d}x}\right|_{x_k^-} = \left.\frac{\mathrm{d}c}{\mathrm{d}x}\right|_{x_k^+}, \tag{8.53}$$

and so the residual equation is given by

$$\frac{1}{\Delta x_{k-1}} \sum_{i=1}^{N+2} A_{N+2,\, i} c_i^{k-1} = \frac{1}{\Delta x_k} \sum_{i=1}^{N+2} A_{1,\, i} c_i^k, \tag{8.54}$$

and this is applied at *each* point between elements. We must now deal with the global boundary conditions of the problem.

Boundary conditions
These are only applied in the *first* and *last* element. We have for the first element:

Condition 1

$$\frac{1}{\Delta x^1} \sum_{i=1}^{N+2} A_{1,\, i} c_i^1 = 0. \tag{8.55}$$

For the last element:
Condition 2

$$-\frac{1}{\Delta x^{\mathrm{NE}}} \sum_{i=1}^{N+2} A_{N+2,\, i} c_i^{\mathrm{NE}} = Bi_m \left[C_{N+2}^{\mathrm{NE}} - 1 \right]. \tag{8.56}$$

We must remember that $c_{N+2}^{k-1} = c_1^k$ at each element. The matrix equations resulting from gathering all the elements together have a block diagonal structure.

EXAMPLE 8.5.6 (A solution to the diffusion equation using collocation on elements). Consider the solution to

$$\frac{\mathrm{d}^2 c}{\mathrm{d}x^2} = \phi^2 c$$

with

$$\frac{\mathrm{d}c}{\mathrm{d}x} = 0; \quad c(1) = 1.$$

We use 2 elements and 2 internal collocation points per element.

We use *unsymmetric* polynomials, and the original BVP can be written as

$$\frac{1}{\Delta x_k^2} \sum_{i=1}^{N+2} B_{ji} c_i^k = \phi^2 c_j^k, \qquad j = 2(1)N+1 \quad k = 1, 2.$$

Continuity gives

$$\frac{1}{\Delta x_1} \sum_{i=1}^{N+2} A_{N+2,\, i} c_i^1 - \frac{1}{\Delta x_2} \sum_{i=1}^{N+2} A_{1,\, i} c_1^2 = 0$$

with

$$c_{N+2}^1 = c_1^2.$$

The boundary conditions are as follows:

$$x = 0, \quad \frac{1}{\Delta x_1} \sum_{i=1}^{N+2} A_{1i} c_i^1 = 0,$$

$$x = 1, \quad c_{N+2}^2 = 1.$$

Hence, the matrix equations are

$$\begin{pmatrix} A_{11} & A_{12} & A_{13} & A_{14} & & & \\ B_{21} & B_{22} & B_{23} & B_{24} & & & \\ B_{31} & B_{32} & B_{33} & B_{34} & & & \\ A_{41} & A_{42} & A_{43} & (A_{44}-A_{11}) & -A_{12} & -A_{13} & -A_{14} \\ & & & B_{21} & B_{22} & B_{23} & B_{24} \\ & & & B_{31} & B_{32} & B_{33} & B_{34} \\ & & & & & & 1 \end{pmatrix} \begin{pmatrix} c_1^1 \\ c_2^1 \\ c_3^1 \\ c \\ c_2^2 \\ c_3^2 \\ c_4^2 \end{pmatrix}$$

$$= \begin{pmatrix} 0 \\ \Delta x^2 \phi^2 c_2^1 \\ \Delta x^2 \phi^2 c_3^1 \\ 0 \\ \Delta x^2 \phi^2 c_2^2 \\ \Delta x^2 \phi^2 c_3^2 \\ 1 \end{pmatrix},$$

where

$$c = c_{N+2}^1 = c_1^2.$$

■ ■ ■

The matrix equation can then be solved using a Newton-like iteration to solve for the unknown concentrations $c_1^{(1)}$ to $c_4^{(2)}$.

8.6. ORTHOGONAL COLLOCATION FOR PARTIAL DIFFERENTIAL EQUATIONS

Orthogonal collocation can be easily applied to transient problems to produce a set of ODE initial value problems. This follows the development by Finlayson [57].

For the general PPDE, we can use a similar development of the non-symmetrical trial function but now the coefficients are functions of t. That is,

$$y(x, t) = b(t) + c(t)x + x(1 - x) \sum_{i=1}^{N} a_i(t) P_{i-1}(x), \tag{8.57}$$

which can be written as

$$y(x, t) = \sum_{i=1}^{N+2} d_i(t) x^{i-1}. \tag{8.58}$$

At the collocation points x_j, we would have

$$y(x_j, t) = \sum_{i=1}^{N+2} d_i(t) x_j^{i-1} \Longrightarrow y(t) = Qd(t), \tag{8.59}$$

with

$$\frac{\partial y(x_j, t)}{\partial x} = \sum_{i=1}^{N+2} d_i(t)(i-1)x_j^{i-2} \Longrightarrow \frac{\partial y}{\partial x} = Cd(t), \tag{8.60}$$

$$\nabla^2 y = Dd(t). \tag{8.61}$$

Hence, for the transient problem in one dimension, the formulation is identical to the boundary value problem except that the coefficients are changing with time. This means we will have sets of ODEs to solve.

EXAMPLE 8.6.1 (Nonlinear diffusion and reaction problem). Consider the following one dimensional problem given by:

$$\frac{\partial u}{\partial t} = \frac{\partial}{\partial x}\left(D(u)\frac{\partial u}{\partial x}\right) + R(u)$$

with boundary conditions:

$$u(0, t) = 1$$
$$u(1, t) = 0$$

and initial conditions:

$$u(x, 0) = 0$$

For this problem, we can write the collocation equations as

$$\frac{\mathrm{d}u_j}{\mathrm{d}t} = \sum_{i=1}^{N+2} A_{ji} D(u_i) \sum_{z=1}^{N+2} A_{iz} u_z + R(u_j), \quad j = 1, \ldots, N \tag{8.62}$$

or

$$\frac{\mathrm{d}u_j}{\mathrm{d}t} = D(u_j) \sum_{i=1}^{N+2} B_{ji} u_i + \frac{\mathrm{d}D(u_j)}{\mathrm{d}u}\left(\sum_{i=1}^{N+2} A_{ji} u_i\right)^2 + R(u_j), \quad j = 1, \ldots, N \tag{8.63}$$

with the usual boundary conditions:

$$u_1(t) = 1, \quad u_{N+2}(t) = 0. \tag{8.64}$$

■ ■ ■ We can solve these ODEs using standard packages.

8.7. SUMMARY

This chapter has considered a wide range of numerical methods for the solution of various types of PDEs. The most well known of the methods are those based on FDAs for the terms in the PDEs. These are conceptually simple and relatively easy to implement. A range of classes were discussed which covered both explicit and implicit methods. The implicit methods allow larger discretizations to be made without breaching stability bounds.

The method of lines is a useful technique which converts PPDEs into sets of ODEs. The problem can then be easily solved using standard integrators.

Finally, we considered a polynomial approximation method based on collocation. This is a powerful technique and easily applied to all classes of PDEs. It is conceptually more difficult but implementation is straightforward. It is widely used in the process industries for the solution of these types of problems.

8.8. REVIEW QUESTIONS

Q8.1. What are the three classes of PDEs encountered in process engineering applications? Give some examples of when and where these equation classes arise. (Section 8.1)

Q8.2. Describe the principal categories of numerical methods for the solution of PDE systems. (Section 8.1)

Q8.3. What are the steps needed in applying the finite difference method to the solution of PDE models? What key forms of finite difference methods exist and what are the characteristics of each in terms of stability and complexity of solution? (Section 8.2)

Q8.4. What is the advantage of using some form of implicit method in the solution of PDEs using finite difference methods? When is it necessary? (Section 8.2.2)

Q8.5. What is the method of lines and what properties of the method does one need to consider in using it? (Section 8.3)

Q8.6. What are the key concepts behind the method of weighted residuals? Describe the steps in applying the technique to a problem. (Section 8.4)

Q8.7. What is special about orthogonal polynomials that make them useful for applying the method of weighted residuals to a problem? (Section 8.5)

Q8.8. Describe the approach needed in applying the orthogonal collocation technique to the general solution of PDEs? (Section 8.6)

Q8.9. Describe some techniques for handling the solution of 2D and 3D PDE models.

8.9. APPLICATION EXERCISES

A8.1. The following equation arises from the modelling of a 1D process engineering problem:

$$\frac{\partial u}{\partial t} = \frac{\partial^2 u}{\partial x^2} \quad 0 < x < 1$$

with the following initial conditions:

$$u = 1 \quad 0 \leq x \leq 1, \quad t = 0,$$

and boundary conditions set as

$$\frac{\partial u}{\partial x} = 0, \quad x = 0, \; t > 0,$$

$$\frac{\partial u}{\partial x} = -u, \quad x = 1, \; t > 0.$$

(a) Devise a finite difference solution using second-order differences in the spatial variable and then solve the differences equations using a range of discretization levels in Δx.

A8.2. Devise an explicit finite difference method for the following model

$$\frac{\partial c}{\partial t} = \frac{\partial}{\partial x}\left(D\frac{\partial c}{\partial x}\right) + \frac{\partial}{\partial y}\left(D\frac{\partial c}{\partial y}\right),$$

where

$$D = D(c).$$

(a) Determine the truncation error in Δx and Δt.
(b) Give an approximate guide to a first choice of Δt for a stable solution when

$$D = 1 + \lambda c, \quad \lambda = 2.$$

A8.3. Consider the following problem of reaction in a spherical catalyst pellet [57], where the mass balance gives the following governing equation:

$$\frac{\partial c}{\partial t} = \frac{1}{r^2}\frac{\partial}{\partial r}\left(r^2\frac{\partial c}{\partial r}\right) - \phi^2 f(c)$$

with boundary conditions:

$$-\frac{\partial c}{\partial r}(0) = 0, \quad -\frac{\partial c}{\partial r}(1) = Bi_m(c(1) - 1),$$

where Bi_m is the Biot number for mass and the initial condition is

$$c(r) = 0 \quad t = 0.$$

Solve this problem to steady state from the initial conditions using 2 internal collocation points for three cases:

$$f = c, \quad \phi^2 = 1, \quad Bi_m = 100,$$

$$f = c^2, \quad \phi^2 = 1, \quad Bi_m = 100,$$

$$f = \frac{c}{(1+20c)^2}, \quad \phi = 32, \quad Bi_m = 100.$$

A8.4. Consider the governing equations for mass and energy arising from the modelling of a packed bed reactor, where r is the radial dimension of the bed and z is the axial dimension. The coupled PDEs for this system are given by

$$\frac{\partial c}{\partial z} = \frac{1}{r}\frac{\partial}{\partial r}\left(\gamma r \frac{\partial c}{\partial r}\right) + \beta R(c, T),$$

$$\frac{\partial T}{\partial z} = \frac{1}{r}\frac{\partial}{\partial r}\left(\hat{\gamma} r \frac{\partial T}{\partial r}\right) + \hat{\beta} R(c, T),$$

with the following boundary conditions:

$$\frac{\partial c}{\partial r} = 0, \quad -\frac{\partial T}{\partial r} = Bi(T - T_{\mathrm{W}}), \quad \text{for } r = 1,$$

$$\frac{\partial c}{\partial r} = 0, \quad \frac{\partial T}{\partial r} = 0, \quad \text{for } r = 0$$

and initial conditions

$$c(r, 0) = C_0, \quad T(r, 0) = T_0.$$

(a) Formulate a finite difference solution to this problem which is of second order in r.

(b) Apply an orthogonal collocation solution method to this problem.

A8.5. Solve the convection–dispersion problem

$$\frac{\partial c}{\partial t} = \frac{\partial}{\partial x}\left(D(c)\frac{\partial c}{\partial x}\right)$$

with the conditions:

$$c(0, t) = 0 \quad c(1, t) = \sin(t),$$

$$c(x, 0) = 2x, \quad 0 \le x \le \frac{1}{2},$$

$$c(x, 0) = 2(1 - x) \quad \frac{1}{2} \le x \le 1,$$

$$D(c) = 1 - 0.1c^2$$

using the method of lines. Use central differences for the spatial derivatives and use a discretization of $N = 20$. Integrate the resulting equation set from $t = 0$ to $t = 0.1$. Plot the profiles at $t = 0.01, 0.03, 0.05, 0.1$.

A8.6. Consider the following problem which arises from the modelling of axial and radial diffusion in a tubular reactor. The energy balance gives

$$\frac{\partial T}{\partial z} = \frac{\alpha}{r}\frac{\partial}{\partial r}\left(r\frac{\partial T}{\partial r}\right) + \frac{1}{Pe}\frac{\partial^2 T}{\partial z^2} + \beta R.$$

The radial dimension boundary conditions are given by

$$\frac{\partial T}{\partial r} = 0 \quad \text{at } r = 0, \qquad -\frac{\partial T}{\partial r} = Bi_{\text{wall}}(T(1, z) - T_{\text{wall}}) \quad \text{at } r = 1$$

and the axial dimension conditions could be

$$\frac{\partial T}{\partial z} = Pe(T - 1) \ \text{ at } z = 0, \qquad \frac{\partial T}{\partial z} = 0 \ \text{ at } z = 1.$$

(a) Devise an orthogonal collocation solution to this problem.
(b) Apply a finite difference solution to this problem.

A8.7. For the general nonlinear parabolic diffusion equation in one dimension is given by

$$\frac{\partial c}{\partial t} = \frac{\partial}{\partial x}\left(D(c)\frac{\partial c}{\partial x}\right) + R(c)$$

with initial condition

$$c(x, 0) = 0$$

and boundary conditions

$$c(0, t) = 1, \quad c(1, t) = 0.$$

(a) Devise a finite difference scheme of second order in the spatial variable for the solution. Solve it for a range of spatial discretization levels.
(b) Develop a global orthogonal collocation solution to this problem and solve it for two different numbers of internal points.
(c) Consider the formulation of the orthogonal collocation solution on several elements.

A8.8 For the model of the double pipe heat exchanger given in Example 7.2.1.

(a) Develop a numerical solution based on a finite difference approach. Test the method by solving the problem over the exchanger length of 10 m for a series of discretizations in Δz ranging from 0.1 to 1.0. Comment on the solution accuracy.
(b) Try solving the problem using orthogonal collocation and compare the results to those of the finite difference methods. Can a few collocation points produce an acceptable solution?
Assume reasonable values for all unspecified parameters. Use physical properties relevant to water for the fluids.

A8.9. For the model of the pollutant dispersion in a river given in Example 7.2.3.

(a) Set up a numerical solution of the problem using a finite difference approach and solve it for a range of spatial discretizations. Comment on the behaviour of the solution as a function of discretization and timestep.

(b) Use the method of lines to set up the ODEs to simulate the behaviour of the model. Investigate the effect of spatial discretization on the solution as well as the integration accuracy. Use MATLAB to carry out the solution.

Assume reasonable values for all unspecified parameters.

9 PROCESS MODEL HIERARCHIES

For even a moderately complicated process system, there is usually a set of different models for the same purpose that describe the system in different detail and/or at different levels. These models are not independent of each other but are related through the process system they describe and through their related modelling goals. One may arrange these models in a hierarchy:

- Starting from a static model with only the input and output variables of the process system through to the most detailed dynamic distributed parameter model which includes all the controlling mechanisms. This is a *hierarchy driven by the level of detail.*
- Starting from the molecular level, describing elementary molecular kinetics to the most detailed dynamic distributed parameter model with all the controlling mechanisms described. This is a *hierarchy driven by characteristic sizes* which is used in multi-scale modelling. It spans the nano to the macro scale.
- Starting from the models describing the slowest mode of the process system to the models incorporating the fastest modes. This is a *hierarchy driven by characteristic times.*

Hierarchies are driven by basic or fundamental modelling assumptions affecting

- the balance volumes, their number and type (lumped or distributed),
- the system boundaries or the focus of modelling,
- one or more basic controlling mechanisms, such as diffusion, present in most of the balance volumes in the process system.

These are all general composite assumptions specified for the overall system.

Figure 9.1 shows schematically the model space defined by such multi-scale modelling concepts. Any model $M(d, s, t)$ can be located within the multi-scale domain and families of models can then be developed by moving along one of the characteristic co-ordinate axes.

Together with the hierarchies various ways of obtaining one model in the hierarchy from another one are also described. These methods can also be seen as model simplification or model enrichment techniques.

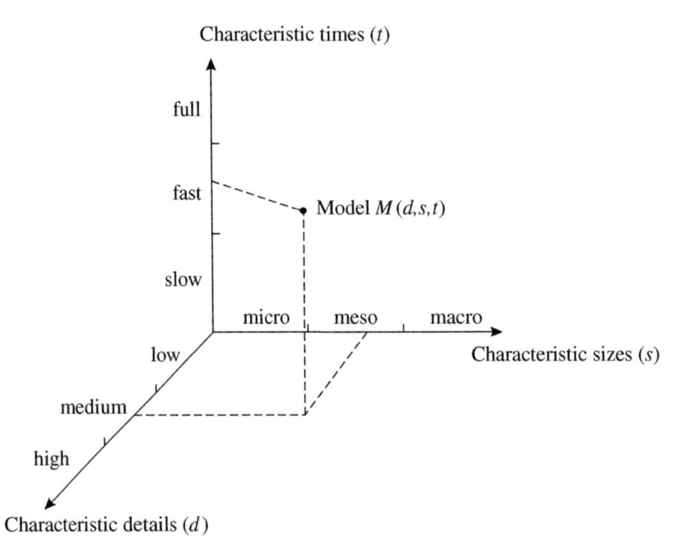

FIGURE 9.1 Multi-scale modelling domain.

These hierarchies and the transfer of information between levels is described in this chapter and illustrated on the example of a packed bed catalytic reactor. The problem statement of the modelling task of the packed bed tubular catalytic reactor is given in Example 9.0.1.

Note that the same process system was used in Chapter 7 in Example 7.1.1 to illustrate the development of distributed parameter models.

EXAMPLE 9.0.1 (Modelling of a simple packed bed tubular catalytic reactor).

Process system

Consider a tubular reactor completely filled with catalyst and with ideal plug flow. A pseudo-first-order catalytic reaction $A \rightarrow P$ takes place in an incompressible fluid phase.

The process is shown in Fig. 9.2.

Modelling goal

Describe the behaviour of the reactor for design purposes.

From the problem description above, we may extract the following modelling assumptions which will be generally valid for any further model we derive in this chapter.

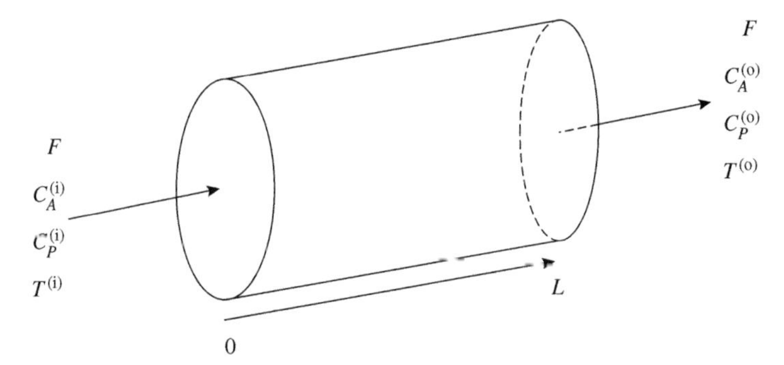

FIGURE 9.2 Simple tubular catalytic plug flow reactor.

Assumptions

$\mathcal{A}1$. Plug flow is assumed in the reactor, which has constant volume.
$\mathcal{A}2$. A pseudo-first-order reaction $A \rightarrow P$ takes place.
$\mathcal{A}3$. An incompressible (liquid) phase is present as a bulk phase.
$\mathcal{A}4$. A solid phase catalyst is present which is uniformly distributed.

9.1. HIERARCHY DRIVEN BY THE LEVEL OF DETAIL

In order to capture the idea of how different models describing the same process system with different level of detail relate to each other, let us imagine a process system which we view from different (spatial) distances. In principle, we are able to see finer detail as we get closer to the system. We shall arrange the models corresponding to different levels of magnification, so that the different granularity reflects *hierarchy levels driven by detail*. The more detail we take into account, the lower the level of model we consider. The number of levels varies according to the complexity of the process system and according to the modelling goal.

The sequence of models in this hierarchy is naturally developed if one performs a process engineering design of a system in a top–down approach. First, the process functionality and its connection to the environment is defined as a top level model. Therefore, the flowsheet is developed with the necessary operating units and their connection forming a middle level model. Finally, the detailed design is performed with a bottom level model with all the necessary details.

9.1.1. Hierarchy levels driven by detail

In the following, we shall illustrate the characteristics and relationship of models at different hierarchy levels which are driven by detail. This will be illustrated on a simple case with only three levels. Here, we have a *top, middle* and *bottom level*. Figure 9.3 shows schematically the model hierarchy in relation to one of the characteristic co-ordinates.

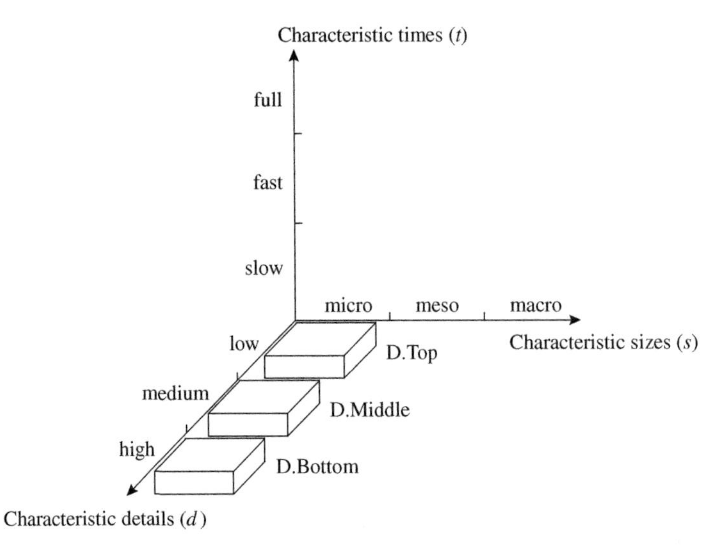

FIGURE 9.3 Model hierarchy driven by detail.

1. *Top hierarchy level* (D.Top)
   If we look at the process from a long distance, we can only see its major characteristics without much detail. From the process modelling point of view, we only consider the static or averaged long term material and energy balances dependent on the process inputs and outputs.
   The *modelling goal* could be to design the overall structure to give the basic flowsheet or structure of the process system and its connections to the environment.
   As far as the *balance volumes* are concerned, there is only a single balance volume which is coincident for mass and energy balances encapsulating the whole process system.
   An example of a top hierarchy level model of the simple packed bed tubular catalytic reactor introduced in Example 9.0.1 is given below.

EXAMPLE 9.1.1 (Top level model of a simple packed bed tubular catalytic reactor).

Process system
Consider a plug flow tubular reactor completely filled with catalyst. A pseudo-first-order catalytic reaction $A \rightarrow P$ takes place in an incompressible fluid phase. The flowsheet with the balance volume and process variables is shown in Fig. 9.4.

Modelling goal
To describe the overall behaviour of the reactor for design purposes.

Assumptions *additional to the ones listed as 1–4 are the following*:

$\mathcal{A}5$. The reactor is at steady state.
$\mathcal{A}6$. Adiabatic conditions apply.
$\mathcal{A}7$. There are constant physico-chemical properties.

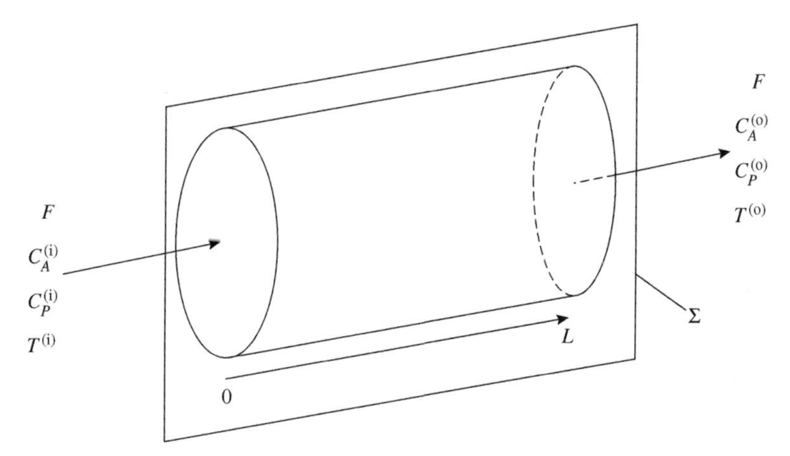

FIGURE 9.4 Top level model of a simple tubular catalytic plug flow reactor.

Balance volume
A single volume (Σ) encapsulating the whole reactor.

Model equations

Mass balances

$$FC_A^{(i)} + FC_P^{(i)} - FC_A^{(o)} - FC_P^{(o)} = 0, \tag{9.1}$$

where F is the mass flowrate of the reaction mixture consisting of the reactant, product and an inert incompressible fluid. $C_A^{(i)}$ and $C_P^{(i)}$ are the inlet concentrations of the reactant and product, respectively, and $C_A^{(o)}$ and $C_P^{(o)}$ are the outlet concentrations.

Energy balance

$$FT^{(i)}\left(c_{PA}C_A^{(i)} + c_{PP}C_P^{(i)}\right) - FT^{(o)}\left(c_{PA}C_A^{(o)} - Fc_PC_P^{(o)}\right) = F\left(C_A^{(i)} - C_A^{(o)}\right)\Delta H_R, \tag{9.2}$$

where c_{PA} and c_{PP} are the specific heats of the reactant and product, respectively and ΔH_R is the heat of reaction. ■ ■ ■

2. *Middle hierarchy level* (D.Middle)
Middle hierarchy level models are normally used when we want to describe our process system in terms of its subsystems or subprocesses. For example, a process plant can be seen as a process system consisting of operating units as subsystems. The flowsheet defines the connections between these subsystems. For a single operating unit, such as a distillation column, we may wish to consider parts of the unit with a definite subfunction. For example the stripping, feed and rectification sections of the column, as subsystems. For distributed parameter process systems, a middle hierarchy level model can serve as the lumped version of the process model with a moderate number of lumped balance volumes.

The *modelling goal* is usually to perform the design of the process subsystems and establish their connections. For "lumped" middle level models, the goal is often to describe the dynamic input–output behaviour of the process system.

The *balance volume* set is usually driven by the subsystem decomposition. Here, we allocate at least one balance volume for each of the subsystems. For distributed parameter models we define perfectly mixed balance volumes for each lumped subsystem. *It is important to note that the union of all the balance volumes in the middle level model is the balance volume of the top level model.* We only consider convection and transfer without time delay between the balance volumes.

As an example of a middle hierarchy level model driven by detail, we describe a lumped version of the distributed bottom level model of the simple packed bed tubular catalytic reactor introduced in Example 9.0.1.

EXAMPLE 9.1.2 (Lumped middle level model of a simple packed bed tubular catalytic reactor).

Process system
Consider a plug flow tubular reactor completely filled with catalyst and with a pseudo first-order $A \rightarrow P$ catalytic reaction. The flowsheet with the balance volumes and process variables is shown in Fig. 9.5.

Modelling goal
To describe the approximate dynamic input–output behaviour of the reactor for design purposes.

Assumptions *additional to the ones listed as 1–4 are the following:*

$\mathcal{A}5$. The reactor is described as a sequence of three CSTRs.
$\mathcal{A}6$. Adiabatic conditions apply.
$\mathcal{A}7$. Constant physico-chemical properties.
$\mathcal{A}8$. Uniform initial conditions in the reactor.

Balance volumes
We consider three balance volumes ($\Sigma_1^{(D.Middle)}$, $\Sigma_2^{(D.Middle)}$ and $\Sigma_3^{(D.Middle)}$) with equal holdups and

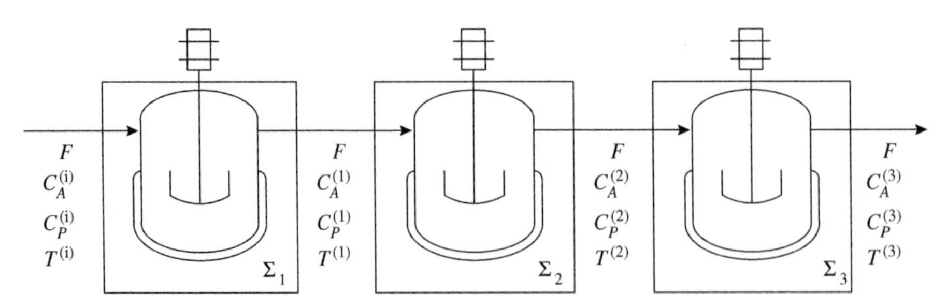

FIGURE 9.5 Lumped model of the catalytic plug flow reactor.

$$\Sigma = \Sigma_1^{(\text{D.Middle})} \bigcup \Sigma_2^{(\text{D.Middle})} \bigcup \Sigma_3^{(\text{D.Middle})}. \tag{9.3}$$

Model equations

Variables

$$\left(C_A^{(k)}(t),\ T^{(k)}(t),\ k = 1, 2, 3\right), \qquad 0 \le t, \tag{9.4}$$

where $C_A^{(k)}(t)$ and $T^{(k)}(t)$ are the reactant concentration and the temperature in the kth tank respectively, and t is the time.

Component mass balances

$$\frac{\mathrm{d}C_A^{(k)}}{\mathrm{d}t} = \frac{3F}{V}\left(C_A^{(k-1)} - C_A^{(k)}\right) - r_A, \tag{9.5}$$

$$k = 1,\ 2,\ 3, \quad C_A^{(\mathrm{o})}(t) = C_A^{(\mathrm{i})}(t) \tag{9.6}$$

with V being the volume of the reactor, F is the flowrate, r_A the reaction rate and $C_A^{(\mathrm{i})}$ is the inlet reactant concentration to the whole reactor.

Energy balances

$$\frac{\mathrm{d}T^{(k)}}{\mathrm{d}t} = \frac{3F}{V}(T^{(k-1)} - T^{(k)}) - \frac{\Delta H}{\rho c_P} r_A, \tag{9.7}$$

$$k = 1, 2, 3, \quad T^{(\mathrm{o})}(t) = T^{(\mathrm{i})}(t), \tag{9.8}$$

where ρ is the density, c_P is the specific heat of the material in the reactor, ΔH the heat of reaction and $T^{(\mathrm{i})}$ is the inlet reactant concentration to the whole reactor.

Constitutive equations

$$r_A = k_0 \mathrm{e}^{-E/(RT)} C_A, \tag{9.9}$$

where k_0 is the pre-exponential factor, E the activation energy, and R is the universal gas constant.

Initial conditions

$$C_A^{(k)}(0) = C_A^*, \quad k = 1, 2, 3, \tag{9.10}$$

$$T^{(k)}(0) = T^*, \quad k = 1, 2, 3, \tag{9.11}$$

■ ■ ■ where C_A^* and T^* are the initial concentration and temperature, respectively.

3. *Bottom hierarchy level* (D.Bottom)
   If we consider all the process details, then we have arrived at the bottom level model. From the process modelling point of view, we consider all relevant material and energy balances of the process system with the most detailed set of constitutive equations.
   The *modelling goal* is usually to perform final detailed design of the process system.

Here, we have the finest *balance volume* set. For process systems with only lumped (perfectly mixed) subsystems, we have balance volumes for every perfectly mixed holdup and for every phase in the system. For distributed parameter process systems, we set up balance volumes for every holdup of the same type of spatial distribution and for every phase or we lump the system into a high number of perfectly mixed balance volumes. *Here again the union of all the balance volumes in the bottom level model is the balance volume of the top level model as driven by detail.* We consider only convection and transfer without time delay between the balance volumes.

An example of a distributed, bottom hierarchy level model of the simple packed bed tubular catalytic reactor introduced in Example 9.0.1 is as follows.

EXAMPLE 9.1.3 (Distributed bottom level model of a simple packed bed tubular catalytic reactor).

Process system
Consider a plug flow tubular reactor completely filled with catalyst and a pseudo first-order catalytic reaction $A \to P$. The flowsheet is the same as shown in Fig. 9.2.

Modelling goal
To describe the overall behaviour of the reactor for final design purposes.

Assumptions *Additional to the ones listed as 1–4 are the following:*

$\mathcal{A}5$. The reactor is uniformly distributed in its cross section, and no radial diffusion or convection takes place.
$\mathcal{A}6$. Adiabatic conditions apply.
$\mathcal{A}7$. Constant physico-chemical properties.
$\mathcal{A}8$. No diffusion in the axial direction.
$\mathcal{A}9$. The initial distribution of component A and the temperature in the reactor is uniformly constant.

Balance volumes
A single volume ($\Sigma^{(\text{D.Bottom})}$) encapsulating the whole reactor but this time this balancing volume is a *distributed* one.

Model equations

Variables

$$0 \leq x \leq L, \quad t \geq 0, \quad C_A(x,t), \quad T(x,t), \tag{9.12}$$

where x is the spatial coordinate in axial direction, L the length of the reactor, t the time, C_A the reactant concentration and T is the temperature.

Component mass balances

$$\frac{\partial C_A}{\partial t} = -F\frac{\partial C_A}{\partial x} - r_A, \tag{9.13}$$

where F is the mass flowrate of the inert incompressible fluid and r_A is the reaction rate.

Energy balance

$$\frac{\partial T}{\partial t} = -F\frac{\partial T}{\partial x} - \frac{\Delta H}{c_P} r_A \tag{9.14}$$

where c_P is the specific heat of the material in the reactor and ΔH is the heat of reaction.

Constitutive equations

$$r_A = k_0 \, \mathrm{e}^{-E/RT} C_A, \tag{9.15}$$

where k_0 is the pre-exponential factor, E the activation energy, and R is the universal gas constant.

Initial and boundary conditions

$$C_A(x, 0) = C_A^*, \quad C_A(0, t) = C_A^{(\mathrm{i})}, \tag{9.16}$$

$$T(x, 0) = T^*, \quad T(0, t) = T^{(\mathrm{i})}, \tag{9.17}$$

where $C_A^{(\mathrm{i})}$ and $T^{(\mathrm{i})}$ are the inlet concentration and the inlet temperature, while C_A^* and T^* are the initial concentration and temperature, respectively. ■ ■ ■

9.1.2. The relation between models of different levels driven by detail

Modelling goal and modelling assumptions

As only the level of detail varies with changing the level in this hierarchy, the modelling assumptions on the system boundaries and input and output relations with its environment are the same for the models in all levels. The basic assumptions on the presence or absence of controlling mechanisms are also the same, such as the absence of diffusion in the system.

The assumptions on the balance volumes are the ones characterizing a particular level in the hierarchy driven by detail. The set of modelling assumptions on a lower level is larger than that on a higher level. That is because additional modelling assumptions are needed to incorporate more detail into a lower level model.

The "precision" of the models on various levels can be considerably different and the modelling goal is the one which dictates which level to choose for the given purpose.

Balance volumes

From the above examples, we can conclude that *a balance volume of a higher hierarchy level driven by detail is obtained as a union of some balance volumes in the lower level.* Moreover,

$$\Sigma = \bigcup_{k=1}^{N_\Sigma^{(\mathrm{D.Middle})}} \Sigma_k^{(\mathrm{D.Middle})} = \bigcup_{k=1}^{N_\Sigma^{(\mathrm{D.Bottom})}} \Sigma_k^{(\mathrm{D.Bottom})} \tag{9.18}$$

should also hold. That is, the union of all balance volumes in any of the hierarchy levels ($\Sigma_k^{(\mathrm{D.Middle})}$ or $\Sigma_k^{(\mathrm{D.Bottom})}$) should encapsulate the whole process system.

Variables and model equations

This relation of the balance volumes gives a guideline as to how the variables in the models of the higher and lower hierarchy level relate to each other. The *input and output variables of the higher level model appear in the lower level model as input and output variables of some of the lower level submodels.* However, the majority of the lower level variables affect the higher level variables only in an indirect way through the model parameters.

The state variables are computed as weighted averages of the corresponding set of lower level state variables. For example, a particular temperature $T_k^{(D.Middle)}$ of a balance volume $\Sigma_k^{(D.Middle)}$ on the middle level is usually a weighted average of the temperatures $T_j^{(D.Bottom)}$, $j = j_1, \ldots, j_k$, where the index j runs on the bottom level balance volumes which form the middle level balance volume $\Sigma_k^{(D.Middle)}$:

$$\Sigma_k^{(D.Middle)} = \bigcup_{j=j_1,\ldots,j_k} \Sigma_j^{(D.Bottom)}.$$

However, there is most often a clear correspondence between the initial and boundary conditions of the models of different levels driven by detail.

EXAMPLE 9.1.4 (Relation between different level models driven by detail for a simple packed bed tubular catalytic reactor.)

Balance volumes
The volume and boundaries of the single bottom level balance volume ($\Sigma^{(D.Bottom)}$) is the same as the overall balance volume for the top level model encapsulating the whole reactor but this balancing volume is a *distributed* one. Moreover, the balance volume relation of the 3-CSTR middle level model in Eq. (9.3) is an example of the relation (9.18).

Model equations
Variables
The overall input ($C_A^{(i)}$, $C_P^{(i)}$ and $T^{(i)}$) and output variables ($C_A^{(o)}$, $C_P^{(o)}$ and $T^{(o)}$) to the process system are given the same identifier in all the Examples 9.1.1–9.1.3.

The middle level model variables (9.4) can be interpreted as values of the bottom level variables (9.12) at given spatial points (x_1, x_2, x_3), i.e.

$$C_A^{(k)}(t) = C_A(x_k, t), \quad T^{(k)}(t) = T(x_k, t), \quad k = 1, 2, 3. \tag{9.19}$$

Initial and boundary conditions
The same or the corresponding initial conditions have been given to the middle and bottom level models in Eqs. (9.10)–(9.11) and Eqs. (9.16)–(9.17) using the same initial values C_A^* and T^*.

Model Simplification and Enrichment

In a model hierarchy driven by the level of detail, one can define more than one middle level, varying, for example the number of balance volumes in the middle level lumped model. Even the top level model can be seen as a single lumped model. In this case, the model on a higher hierarchy level can be seen as a simplified version of the one in the lower hierarchy level. In this way, adding more detail to obtain a lower level model is regarded as model enrichment.

In the case of hierarchy driven by the level of detail, model simplification is performed by lumping already existing balance volumes together to form just one lump. Model enrichment on the other hand is done by dividing a volume into parts forming several new volumes. More about model simplification and enrichment of dynamic process models can be found in Section 13.3.

9.2. HIERARCHY DRIVEN BY CHARACTERISTIC SIZES

The relationship between different models describing the same process system with different levels of characteristic sizes is best understood if we imagine the crudest *macro level* model and then zoom in to get to its lower-level dependents. Zooming means that we not only see finer detail but we also focus on a particular part of the process system. Other aspects are not within the scope of the modelling. The more we zoom in on the original macro level model the lower the level of model we obtain. We could consider three levels defined as *macro, meso* and *micro levels* driven by the characteristic sizes. These are illustrated in Fig. 9.6 which shows the size hierarchy.

The models arranged in this hierarchy naturally arise when a bottom–up modelling approach is adopted for a process system. This is often the case in practical modelling projects of industrial relevance. Here, one should combine a large set of specific information and data often obtained for different purposes. The laboratory scale measured data and estimated parameters are collected to develop models of the key controlling mechanisms that are at the lowest micro level driven by characteristic sizes. Thereafter, the critical plant equipment is described using pilot plant scale

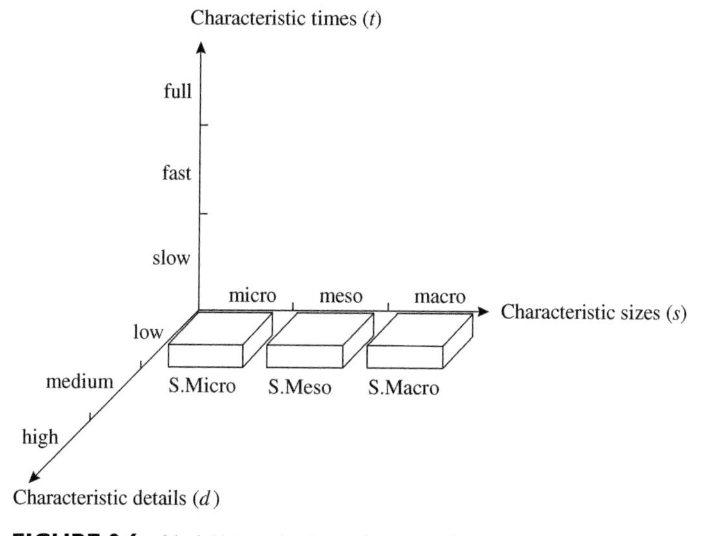

FIGURE 9.6 Model hierarchy driven by size scale.

data which can be regarded as a meso level model. Finally, all data and models are integrated to the overall plant model on the highest macro level.

From the zooming analogy it follows that we can obtain a set of different lower level models as we focus on different parts or processes of the original system. Therefore, the relationship between models of different levels driven by characteristic sizes is much more complicated than was the case for hierarchies driven by detail.

There is a clear correspondence between the models in *multi-scale modelling* and the models in a hierarchy driven by characteristic sizes. We shall show in Section 9.2.3 that the related models obtained by multi-scale modelling form a model hierarchy driven by characteristic sizes.

9.2.1. Hierarchy Levels Driven by Characteristic Sizes

The characteristics and relationships between models of different levels driven by characteristic sizes will be illustrated again on the simple packed bed tubular catalytic reactor introduced in Example 9.0.1.

1. *Macro hierarchy level* (S.Macro)

The model of a process system on the macro hierarchy level can be any of the models from the hierarchy driven by detail or driven by characteristic times. That is, from the levels (D.Top)–(D.Bottom) or (T.Full)–(T.Slow).

The *modelling goal* has no special characteristics in this case.

The *balance volumes* are chosen according to the modelling goal but it is important that the *union of all the balance volumes of the macro level should encapsulate the whole process system*, or

$$\Sigma = \bigcup_{k=1}^{N_\Sigma^{(\text{S.Macro})}} \Sigma_k^{(\text{S.Macro})}. \tag{9.20}$$

2. *Meso hierarchy level* (S.Meso)

The models of the meso hierarchy level usually correspond to models of subsystems of the overall process system. Therefore there is a *set of different meso level models* for complex processes of multiple operating units and/or phases and for processes with holdups with different spatial distributions.

The *modelling goal* is usually to understand transport and transfer processes for equipment design purposes (i.e. (D.Middle)).

The *balance volume* set is established according to the actual modelling goal. *It is important to note that the union of all the balance volumes of a particular meso hierarchy level model encapsulates only the corresponding subsystem.* This is usually only a part of the whole process system, or

$$\bigcup_{k=1}^{N_\Sigma^{(\text{S.Meso})}} \Sigma_k^{(\text{S.Meso})} \subset \Sigma. \tag{9.21}$$

As an example of a *meso* hierarchy level model, we describe a model of a catalyst particle of the simple packed bed tubular catalytic reactor introduced in Example 9.0.1.

EXAMPLE 9.2.1 (Model of a catalyst particle as a meso level model of a simple packed bed tubular catalytic reactor.)

Process system
The process system consists of a catalyst particle and its surrounding bulk fluid phase layer. The flowsheet with the balance volumes and process variables is shown in Fig. 9.7.

Modelling goal
To describe the main distributed transport and transfer processes near a catalyst particle for design purposes.

Assumptions

Ames1. There is no convection in the process system.
Ames2. Only molecular diffusion takes place.
Ames3. Constant physico-chemical properties.
Ames4. Uniform initial conditions in the reactor.
Ames5. There is a pseudo-first-order reaction $A \to P$ on the surface of the particle.
Ames6. Isothermal conditions.
Ames7. No spatial variations along the surface.

Balance volumes
There is a single distributed parameter balance volume $\Sigma^{(S.Meso)}$ which encapsulates part of the fluid phase near the surface. One of its boundaries is the catalyst surface, the other is the homogeneous bulk of the fluid phase. The balance volume is distributed only in one spatial direction normal to the surface.

Model equations
Variables

$$0 \le x \le L_L, \quad t \ge 0, \quad C_A(x,t), \tag{9.22}$$

where x is the spatial coordinate, L_L the width of the surface layer, t time and C_A is the reactant concentration.

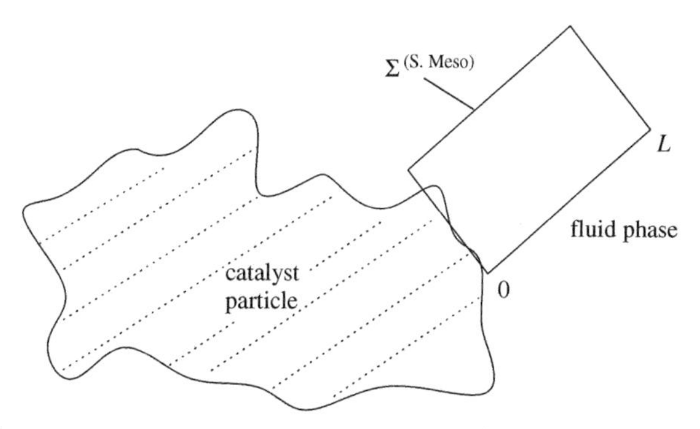

FIGURE 9.7 Catalyst particle as a meso level model.

Component mass balances

$$\frac{\partial C_A}{\partial t} = D\frac{\partial^2 C_A}{\partial x^2}, \tag{9.23}$$

where D is the molecular diffusion coefficient.

Initial and boundary conditions

$$C_A(x,0) = C_A^*, \tag{9.24}$$

$$\frac{\partial C_A}{\partial t}(0,t) = -k^{(\mathrm{S.Meso})}C_A(0,t), \tag{9.25}$$

$$C_A(L_\mathrm{L},t) = C_A^{(F)}, \tag{9.26}$$

where $C_A^{(\mathrm{F})}$ is the reactant concentration in the bulk fluid, C_A^* the initial concentration and $k^{(\mathrm{S.Meso})}$ is the reaction rate constant. Equation (9.25) indicates that the reaction only takes place on the catalyst surface.

3. *Micro hierarchy level* (S.Micro)
The models of processes on the molecular level form the set of bottom level models. From the process modelling point of view, we consider all relevant processes with balance volumes small enough to encapsulate only the molecular level process and its environment. From this description, it is clear that there is a set of different micro level models for processes with multiple phases and/or for processes with holdups with different spatial distributions.

The *modelling goal* is usually to design the reactions and/or transport processes on which the process system is built.

There is usually a single *balance volume* which is the same for mass and energy balances but small enough compared with the sizes of the molecules in order to encapsulate only a relatively small and homogeneous set of them. The balance volume can be either perfectly mixed or distributed. For transport and transfer processes through, or on, an interphase surface we usually use a distributed parameter balance volume.

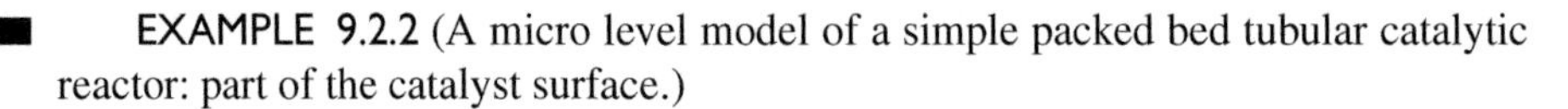

EXAMPLE 9.2.2 (A micro level model of a simple packed bed tubular catalytic reactor: part of the catalyst surface.)

Process system
Consider part of the catalyst surface where the catalytic reaction $A \to P$ takes place. The balance volume $\Sigma^{(\mathrm{S.Micro})}$ with the phases and surface considered is shown in Fig. 9.8.

Modelling goal
To describe the behaviour of the catalytic reaction for final design purposes.

Assumptions

Amic1. The process system is perfectly mixed.
Amic2. Isothermal conditions.
Amic3. The considered reaction steps are as follows:
(a) reversible adsorption of the reactant: $A^{(\mathrm{F})} \leftrightarrow A^{(\mathrm{AD})}$,

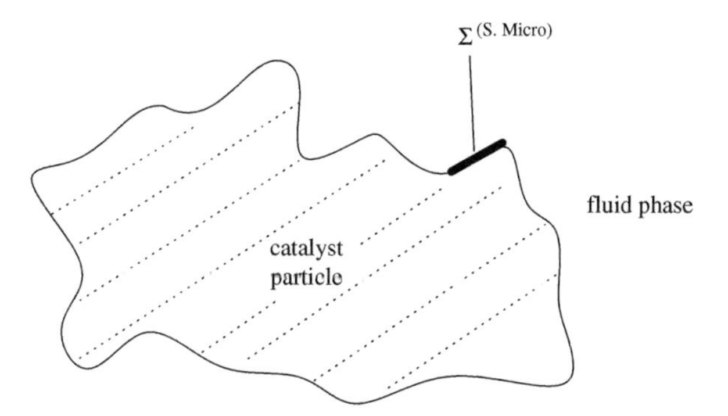

FIGURE 9.8 Reaction kinetic model as micro level model.

(b) first-order irreversible reaction on the surface: $A^{(AD)} \to P^{(AD)}$,
(c) irreversible desorption of the product: $P^{(AD)} \to P^{(F)}$.

Balance volumes
There is a single perfectly mixed balance volume $\Sigma^{(S.Micro)}$ which is part of the catalyst surface together with its fluid layer.

Model equations

Variables

$$C_A^{(F)}(t), \quad C_A^{(AD)}(t), \quad C_P^{(F)}(t), \quad C_P^{(AD)}(t), \; 0 \leq t, \tag{9.27}$$

where t is time, $C_A^{(F)}$, $C_A^{(AD)}$ are the reactant and $C_P^{(F)}$, $C_P^{(AD)}$ are the product concentration in the bulk fluid phase and adsorbed on the surface respectively.

Component mass balances

$$\frac{dC_A^{(F)}}{dt} = -k_{ad} C_A^{(F)} \left(1 - C_A^{(AD)}\right) + k_{ad-} C_A^{(AD)}, \tag{9.28}$$

$$\frac{dC_A^{(AD)}}{dt} = -k_{ad-} C_A^{(AD)} - k_R C_A^{(AD)}, \tag{9.29}$$

$$\frac{dC_P^{(AD)}}{dt} = k_R C_A^{(AD)} - k_{de} C_P^{(AD)}, \tag{9.30}$$

$$\frac{dC_P^{(F)}}{dt} = k_{de} C_P^{(AD)}, \tag{9.31}$$

where k_{ad}, k_{ad-} are the adsorption and desorption coefficients for the reactant, k_R the reaction rate coefficient and k_{de} is the desorption coefficient of the product.

Initial conditions

$$C_A^{(F)}(0) = C_A^*, \quad C_A^{(AD)}(0) = 0, \tag{9.32}$$

$$C_P^{(F)}(0) = 0, \quad C_P^{(AD)}(0) = 0, \tag{9.33}$$

where C_A^* is the initial reactant concentration in the bulk fluid phase.

9.2.2. The Relation between Models of Different Levels Driven by Characteristic Sizes

Modelling Goal and Modelling Assumptions

The modelling goal as well as the assumptions on the system boundaries and on the input–output relations of the system and its environment are drastically and characteristically different on the various levels. The same applies for the assumptions on the balance volumes.

Because the modelled process system is different on the various levels, the set of modelling assumptions on the controlling mechanisms is also different. There is, however, a relationship connecting models on various levels driven by characteristic sizes seen also in their modelling assumptions. The modelling assumptions on a lower level concern the detailed assumptions on a particular subprocess or controlling mechanism made at the higher levels.

Balance Volumes

Because the lower-level models driven by characteristic sizes are related to their higher-level parent through zooming, the union of balance volumes of the same spatial distribution at any level does not cover the union of that on a higher level. This can be stated as

$$\bigcup_{k=1}^{N_{\Sigma}^{(S.Lower)}} \Sigma_k^{(S.Lower)} \subset \bigcup_{k=1}^{N_{\Sigma}^{*}} \Sigma_k^{(S.Higher)}, \tag{9.34}$$

where $N_{\Sigma}^{*} \leq N_{\Sigma}^{(S.Higher)}$. Usually, $N_{\Sigma}^{*} = 1$ when we focus only on a particular single balance volume at the lower-level and we build up a more detailed model of it as a lower level model.

Variables and Model Equations

The above relation of the balance volumes gives a guideline on how the variables of the higher and lower hierarchy level models relate to each other. In the general case, a higher level variable is obtained as some kind of average value of the lower level variable of the same kind. This is averaged over the lower level balance volumes which in turn constitute the higher level balance volume of the corresponding variable.

It is important to note, however, that only some of the variables and controlling mechanisms of a higher level model are represented in a particular lower level model. That is because different and even non-intersecting lower level models exist for a given higher level model. In this case some of the higher level variables originate from a lower level model and others are computed from another lower level model by averaging.

Moreover, there is most often a relationship between the initial and boundary conditions of the models of different levels driven by characteristic sizes.

EXAMPLE 9.2.3 (Relation between different level models driven by characteristic sizes of a simple packed bed tubular catalytic reactor). The (S.Macro) macro hierarchy level model of the reactor is chosen to be the same as its (D.Bottom) bottom hierarchy level model described in Example 9.1.3. The catalytic pellet model as a

(S.Meso) meso level model is shown in Example 9.2.1 and the surface kinetic model as the (S.Micro) micro level model is in Example 9.2.2.

Balance volumes

We have the inclusion relation between the balance volumes of different hierarchy levels driven by characteristic sizes:

$$\Sigma^{(S.\mathrm{Micro})} \subset \Sigma^{(S.\mathrm{Mezo})} \subset \Sigma^{(D.\mathrm{Bottom})} \tag{9.35}$$

Moreover the micro level balance volume $\Sigma^{(S.\mathrm{Micro})}$ can be seen as part of the boundary surface for the meso level balance volume $\Sigma^{(S.\mathrm{Meso})}$.

Model equations

Variables

The micro level model variable (9.27) can be interpreted as the value of the meso level variable (9.22) at the boundary, i.e.

$$C_A(0, t) = C_A^{(\mathrm{F})}. \tag{9.36}$$

Initial and boundary conditions

The overall reaction kinetic expression $k^{(S.\mathrm{Meso})}C_A$ which appears in the boundary condition (9.25) of the meso level component mass balance can be regarded as a solution of the micro level model for $\mathrm{d}C_A^{(\mathrm{F})}/\mathrm{d}t$ as its depends on its variable $C_A^{(\mathrm{F})}$. ■ ■ ■

9.2.3. Multi-scale Modelling and the Hierarchy Driven by Characteristic Sizes

Multi-scale modelling is an exciting and emerging field which covers modelling aspects from the atomic to the continuum through the molecular and various meso-scales. It includes but is not limited to such interesting and important topics as surface reaction, formation of molecular and macromolecular aggregates, estimation of mechanical properties of materials and fluid media at intermediate scales.

The work in multi-scale modelling is mainly performed in the area of solid-state physics [59], physics of multi-phase systems [60], as well as multi-phase flows [61], diffusion [62] and transport processes [63].

The set of models developed for multi-scale modelling can be arranged into a hierarchy driven by characteristic sizes. The micro level model of this hierarchy corresponds to the molecular level multi-scale model, and the other models are arranged according to the smallest species they take into account.

The model hierarchy driven by the level of detail for the simple packed bed catalytic reactor described earlier in this section can be seen as a set of multi-level models developed in the area of surface kinetics.

9.3. HIERARCHY DRIVEN BY CHARACTERISTIC TIMES

The relation between models of different hierarchy levels driven by characteristic times is best understood if we imagine a detailed *dynamic model*. We can take a model from the level (D.Bottom) or (D.Middle) with its full dynamics described as a

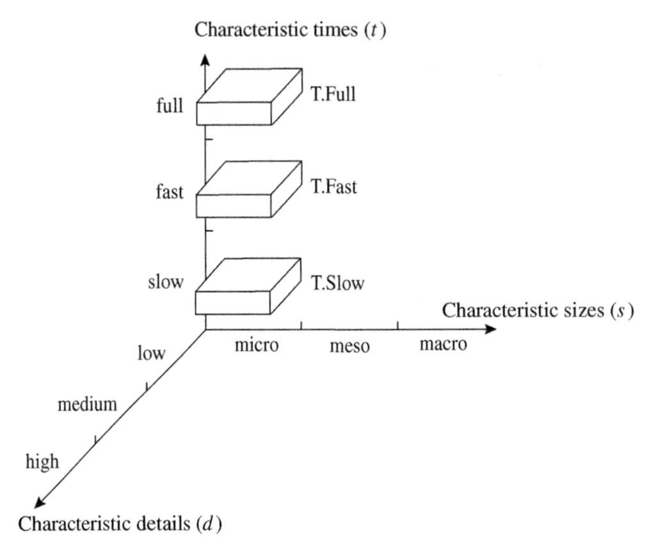

FIGURE 9.9 Model hierarchy driven by timescales.

full timescale hierarchy level model and then focus the modelling attention on the slow time response model to get the *slow hierarchy level model.* By focusing on the fast time responses we obtain the *fast hierarchy level model* counterpart. This hierarchy based on characteristics times is shown in Fig. 9.9.

9.3.1. Hierarchy Levels Driven by Characteristic Times

The relation between the dynamic models of different hierarchy levels driven by characteristic times is illustrated on the example of the simple tubular catalytic reactor introduced in Example 9.0.1.

1. *Full timescale hierarchy level* (T.Full)
   Full timescale models are detailed dynamic models which describe the dynamic behaviour of the whole process system. Therefore, they are usually the same as bottom or middle level dynamic models from the hierarchy driven by details. These are (D.Bottom) or (D.Middle) level models.

   The usual *modelling goal* is to design a control or diagnostic system for a process operation.

   The *balance volume* set is chosen to encapsulate the whole process system and to model it with the necessary detail.

   The middle level lumped dynamic model as described in Example 9.1.2 is used as the full timescale (T.Full) hierarchy level model to illustrate the models of different hierarchy levels driven by characteristic times.
2. *Fast hierarchy level* (T.Fast)
   If we want to obtain a model describing only the fast mode dynamic response of a process system then we should neglect every phenomenon which is "slow".

This is usually done by imposing *quasi steady-state assumptions* in the form of

$$\frac{\mathrm{d}\varphi}{\mathrm{d}t} = 0, \tag{9.37}$$

on a dynamic variable φ with slow time variation compared with the variation in other variables of the same model. In this case, all "slow" mode variables are considered as constants, $\varphi \simeq$ constant.

It is important to remember that "slow" and "fast" are relative notions within a particular dynamic model because slow variations in one model could well be fast in others. The notion of "slow" and "fast" may also depend on the operating range of the process system.

Usually dynamic variables related to energy balances can be considered as slow variables compared to the concentrations. However, some of the system concentrations may vary slowly compared to other concentrations in complex reaction networks.

The *balance volume* set is the same as for the full timescale level model. Moreover, the set of variables is the same but some of the differential variables are transformed to algebraic variables due to the quasi steady-state assumptions.

The concepts above are illustrated on the example of the simple catalytic tubular reactor.

EXAMPLE 9.3.1 (Fast response time model derived from the lumped middle level model of a simple packed bed tubular catalytic reactor.)

Process system
Consider a plug flow tubular reactor completely filled with catalyst, and with a pseudo first-order catalytic reaction $A \rightarrow P$ exactly the same as for Example 9.1.2. The flowsheet with the balance volumes and process variables is shown in Fig. 9.5.

Modelling goal
To describe the approximate *fast* dynamic input-output behaviour of the reactor for design purposes.

Assumptions additional to the ones listed as 1–8 is:

$\mathcal{A}$9. *Neglect slow dynamics associated with temperatures.*

Balance volumes
We consider three balance volumes ($\Sigma_1^{(\mathrm{D.Middle})}$, $\Sigma_2^{(\mathrm{D.Middle})}$ and $\Sigma_3^{(\mathrm{D.Middle})}$) with equal holdups and

$$\Sigma = \Sigma_1^{(\mathrm{D.Middle})} \bigcup \Sigma_2^{(\mathrm{D.Middle})} \bigcup \Sigma_3^{(\mathrm{D.Middle})}. \tag{9.38}$$

Model equations

Variables

$$\left(C_A^{(k)}(t),\ T^{(k)},\quad k = 1, 2, 3\right),\quad 0 \le t, \tag{9.39}$$

where $C_A^{(k)}(t)$ and $T^{(k)}$ are the reactant concentration and the temperature in the kth tank respectively and t is the time.

Component mass balances

$$\frac{\mathrm{d}C_A^{(k)}}{\mathrm{d}t} = \frac{3F}{V}\left(C_A^{(k-1)} - C_A^{(k)}\right) - r_A, \tag{9.40}$$

$$k = 1, 2, 3, \quad C_A^{(0)}(t) = C_A^{(\mathrm{i})}(t) \tag{9.41}$$

with V being the volume of the reactor, F is the flowrate, r_A the reaction rate and $C_A^{(\mathrm{i})}$ is the inlet reactant concentration to the whole reactor.

Constitutive equations
Reaction rate equation

$$r_A = k_0\,\mathrm{e}^{-E/(RT)}C_A, \tag{9.42}$$

where k_0 is the pre-exponential factor, E is the activation energy and R is the universal gas constant.

Derived from static energy balances

$$0 = \frac{3F}{V}\left(T^{(k-1)} - T^{(k)}\right) - \frac{\Delta H}{\rho c_P} r_A \tag{9.43}$$

$$k = 1, 2, 3, \quad T^{(0)} = T^{(\mathrm{i})}, \tag{9.44}$$

where ρ is the density, c_P is the specific heat of the material in the reactor, ΔH the reaction enthalpy and $T^{(\mathrm{i})}$ is the inlet temperature.

Initial conditions

$$C_A^{(k)}(0) = C_A^*, \quad k = 1, 2, 3, \tag{9.45}$$

■ ■ ■ where C_A^* is the initial concentration.

3. *Slow hierarchy level* (T.Slow)
We can proceed in a similar fashion if we want to obtain a model describing only the slow mode dynamic response of a process system. Here, we can assume that the fast processes occur instantaneously. Interestingly enough it is also done by imposing *quasi steady-state assumptions* in the form of

$$\frac{d\varphi}{\mathrm{d}t} = 0 \tag{9.46}$$

on dynamic variables φ but now with fast time variation compared with the variation in other variables of the same model. The reason behind the quasi steady-state assumption above is that the fast mode variable φ is regarded as adjusting instantaneously to its new value.

It is important to remember again that "slow" and "fast" are relative notions within a particular dynamic model because slow variations in one model could well be fast in others.

The *balance volume* set is the same as for the full timescale level model. Moreover the set of variables is the same but some of the differential variables are transformed to algebraic variables due to the quasi steady-state assumptions.

The concepts above are shown on the lumped middle level model of the simple catalytic tubular reactor.

EXAMPLE 9.3.2 (Slow response time model derived from the lumped middle level model of a simple packed bed tubular catalytic reactor.)

Process system
Consider a plug flow tubular reactor completely filled with catalyst, and with a pseudo first-order catalytic reaction $A \to P$ exactly the same as for Example 9.1.2. The flowsheet with the balance volumes and process variables is shown in Fig. 9.5.

Modelling goal
To describe the approximate *slow* dynamic input–output behaviour of the reactor for design purposes.

Assumptions *additional to the ones listed as 1–8:*

$\mathcal{A}9$. *Neglect fast dynamics in component concentrations.*

Balance volumes
We consider three balance volumes ($\Sigma_1^{(\text{D.Middle})}$, $\Sigma_2^{(\text{D.Middle})}$ and $\Sigma_3^{(\text{D.Middle})}$) with equal holdups and

$$\Sigma = \Sigma_1^{(\text{D.Middle})} \bigcup \Sigma_2^{(\text{D.Middle})} \bigcup \Sigma_3^{(\text{D.Middle})}. \tag{9.47}$$

Model equations

Variables

$$\left(C_A^{(k)},\ T^{(k)}(t),\quad k = 1, 2, 3\right),\qquad 0 \le t, \tag{9.48}$$

where $C_A^{(k)}$ and $T^{(k)}(t)$ are the reactant concentration and the temperature in the kth tank, respectively and t is the time.

Energy balances

$$\frac{\mathrm{d}T^{(k)}}{\mathrm{d}t} = \frac{3F}{V}\left(T^{(k-1)} - T^{(k)}\right) - \frac{\Delta H}{\rho c_P} r_A, \tag{9.49}$$

$$k = 1, 2, 3,\quad T^{(0)}(t) = T^{(\mathrm{i})}(t) \tag{9.50}$$

where ρ is the density, c_P is the specific heat of the material in the reactor, ΔH the heat of reaction, V the volume of the reactor, F the flowrate, r_A the reaction rate and $T^{(\mathrm{i})}$ is the inlet reactant concentration to the whole reactor.

Constitutive equations
Reaction rate equation

$$r_A = k_0\,\mathrm{e}^{-E/(RT)} C_A, \tag{9.51}$$

where k_0 is the pre-exponential factor, E the activation energy, and R is the universal gas constant.

Derived from static component mass balances

$$0 = \frac{3F}{V}\left(C_A^{(k-1)} - C_A^{(k)}\right) - r_A, \tag{9.52}$$

$$k = 1, 2, 3, \quad C_A^{(\mathrm{o})}(t) = C_A^{(\mathrm{i})}(t), \tag{9.53}$$

where $C_A^{(\mathrm{i})}$ is the inlet reactant concentration.

Initial conditions

$$T^{(k)}(0) = T^*, \quad k = 1, 2, 3, \tag{9.54}$$

■ ■ ■ where T^* is the initial temperature.

9.3.2. The Relation between Models of Different Levels Driven by Characteristic Times

Modelling Goal and Modelling Assumptions

The modelling goal is usually related to control or diagnosis. The type of control or diagnostic system considered affects the hierarchy level that is used for modelling. Plant wide optimizing control usually needs slow hierarchy level models, whilst regulation or on-line diagnosis needs fast ones.

The modelling assumptions not related to the timescale of the differential variables are exactly the same for all levels including the assumptions on the system boundaries and balance volumes.

Balance volumes

From the above, we can conclude that *the balance volumes of all hierarchy levels driven by characteristic times are exactly the same in spatial distribution* and the union of them encapsulates the whole process system. Hence, a

$$\Sigma = \bigcup_{k=1}^{N_\Sigma^{(\mathrm{T.Any})}} \Sigma_k^{(\mathrm{T.Any})}. \tag{9.55}$$

Variables and Model Equations

The variables in models of all hierarchy levels driven by characteristic times are the same by construction. However, *some dynamic variables affected by quasi steady-state assumptions in the fast or slow hierarchy level models, become algebraic variables*. Moreover, a dynamic variable transformed to an algebraic one by quasi steady-state assumptions in the slow model should remain dynamic in the fast model and vice versa.

The initial (and boundary) conditions of the models at different levels driven by characteristic times are exactly the same, except for the initial conditions of the dynamic balance equations affected by the fast or slow model assumptions. Of course, the variables set to algebraic ones lose their initial conditions.

EXAMPLE 9.3.3 (Relationship between different level models driven by characteristic times of a simple packed bed tubular catalytic reactor.)

Balance volumes
They are exactly the same for the three models inherited from the middle level model driven by detail:

$$\Sigma_k^{(\text{D.Middle})}, \quad k = 1, 2, 3.$$

Model equations
Variables
The variables are the same for models of all hierarchy levels driven by characteristic times as defined in Eq. (9.4):

$$\left(C_A^{(k)}(t),\ T^{(k)}(t),\ k = 1, 2, 3\right), \quad 0 \leq t,$$

but $C_A^{(k)}$, $k = 1, 2, 3$ become algebraic variables in the slow hierarchy level model and $T^{(k)}$, $k = 1, 2, 3$ are algebraic variables in the fast one.

Balance equations
By the quasi steady-state assumptions generating the fast and slow models, we transform the lumped dynamic energy balance to an algebraic set of equations (Eqs. (9.43)–(9.44)) in the fast model while the component mass balances of the slow model become algebraic equations (9.52)–(9.53).

Initial conditions
The same or the corresponding initial conditions have been given to the dynamic variables in the slow and fast hierarchy level models using the same initial values C_A^* and T^*.

The application of multi-time scale models to various process operations has received significant attention. For example, Robertson and Cameron [64,65] used spectral techniques to develop a family of models suitable for startup and shutdown problems in process systems. The models contained the appropriate timescale behaviour for the application.

9.4. SUMMARY

This chapter has presented an overview of model hierarchies. The hierarchies were based on three different but related views of the system being modelled. Inevitably we must simplify our view of the system for modelling purposes. How we do that can be in terms of detail of the balance volumes or the granularity of the model. We can also look at characteristic sizes based on a macro, meso or micro view of the phenomena occurring in the process. Finally, we can model our process based on the characteristic dynamic modes. This leads to a hierarchy of process models which contain various timescales of dynamic behaviour from very fast to the slowest modes.

The use of hierarchies is becoming an important aspect of modelling, driven by the application area. The model we require for detailed design purposes can be quite different from one used for control or for startup and shutdown studies. No one model will capture the necessary system attributes for all foreseeable applications. We need models "fit for purpose".

9.5. FURTHER READING

Model hierarches are useful not only for establishing relationships between related models but also for organizing safe and well-grounded information transfer between them. There are rapidly developing fields within process systems engineering where process model hierachies play a key role.

Multi-scale modelling
This field is driven by the basic science and engineering of materials, which are used in electronic and optielectronic device fabrication. The related field in process systems engineering has become a "challenge area" as reviewed by a recent paper in *AIChE Journal* [66]. Review papers in the field are also available, see e.g. [67,68].

9.6. REVIEW QUESTIONS

Q9.1. Describe the correspondence between the *variables* of the models of the simple packed bed tubular catalytic reactor (Example 9.0.1) of different hierarchy levels driven by detail. (Section 9.1)

Q9.2. Describe the relation between the *balance volumes* of the models of the simple packed bed tubular catalytic reactor (Example 9.0.1) of different hierarchy levels driven by characteristic sizes. (Section 9.2)

Q9.3. Compare the middle (D.Middle) level model (see in Example 9.1.2) and the meso (S.Meso) level model (see in Example 9.2.1) of the simple packed bed tubular catalytic reactor. What are the differences? Which one could be a sub-model of the other? (Section 9.2)

Q9.4. Characterize the process systems appearing on the micro (S.Micro) level in the hierarchy driven by characteristic sizes. Give examples of the controlling mechanisms you would describe on this level. (Section 9.2)

Q9.5. Characterize a potential full timescale (T.Full) level model in the hierarchy driven by characteristic times. Which controlling mechanisms can be regarded relatively slow and which ones relatively fast? (Section 9.3)

Q9.6. Compare the set of model equations in the full timescale (T.Full), fast (T.Fast) and slow (T.Slow) level models in the hierarchy driven by characteristic times. How does the number of dynamic balance equations vary with the level? (Section 9.3)

9.7. APPLICATION EXERCISES

A9.1. Give the weighted average form of the relation of the middle and top level, and that of the bottom and middle level models driven by detail of the simple packed bed tubular catalytic reactor introduced in Example 9.0.1. (Section 9.1)

A9.2. Develop a meso (S.Meso) level model of a heat exchanger describing the heat transfer through a metal wall. How would you use the modelling result on the macro level describing the lumped model of a countercurrent heat exchanger consisting of three lump pairs? (See Example 13.2.2 in Chapter 14) (Section 9.2)

A9.3. Consider a binary distillation column with 20 stages. Give the problem statement of modelling of its top, middle and bottom level models in the hierarchy driven by detail. (Section 9.1)

A9.4. Consider a binary distillation column with 20 stages. Give the problem statement of modelling of its macro, meso and micro level models in the hierarchy driven by characteristic sizes considering its middle (D.Middle) level model as its macro level model. (Section 9.2)

A9.5. Consider a complex process system familiar to you and try to develop and describe the various model hierarchy levels driven by detail, driven by characteristic sizes and driven by characteristic times.

PART II
ADVANCED PROCESS MODELLING AND MODEL ANALYSIS

10 BASIC TOOLS FOR PROCESS MODEL ANALYSIS

This chapter is mainly devoted to the presentation of mathematical and computer science background material which aids in the task of process model analysis. Concepts from systems and control theory such as nonlinear state space representations (SSRs) and their linearized counterparts are extremely important as a starting point for understanding the behaviour of models. The elementary notions behind these topics are given in appendix A.

Section 10.5 is especially important because it contains the background material for structural representation of nonlinear and linear dynamic state space equations using structure matrices and graphs. These are powerful concepts which allow the modeller to represent and analyze the model for a number of important characteristics which have direct influence on model behaviour in various applications such as control.

10.1. PROBLEM STATEMENTS AND SOLUTIONS

The analysis of process system models leads to mathematical problems of various types. It is convenient and useful to formulate these mathematical problems in a formal way specifying the inputs to the problem, the desired output or question to be solved and indicate the procedure or method of solution. Such a formal problem description can also be useful when we want to analyse the computational needs of a mathematical problem or one of its solution methods.

The formal description of a mathematical problem will be given in a similar form to *algorithmic problem statements* widely used in computer science [69]. According to this formalism, problem statements have the following attributes:

- The title or name of the problem, with its optional abbreviation, e.g. AVERAGING.
- The *Given* section where the input data of the problem are described.
- The *Question* to be answered or the output to be produced are specified under the *Find/Compute* section.
- an optional *Method or Procedure* section describing the way of solving the problem.

The following simple example of sample averaging illustrates the above concepts.

EXAMPLE 10.1.1 (Problem statement).

AVERAGING

Given:
A statistical sample of n elements, represented by a set of real numbers $\{x_1, x_2, \ldots, x_n\}$ taken as measurements of independent equally distributed random variables.

Compute:
The sample mean $\bar{x}$, where

$$\bar{x} = \frac{1}{n}\sum_{i=1}^{n} x_i.$$

Problem statements where a yes/no *Question* is to be answered are called *decision problems*, whilst problems with a *Find/Compute* section are termed *search problems*.

In the *Method or Procedure* section the key steps in solving the problem are usually given in the form of a *conceptual problem solution*. The ingredients of a conceptual problem solution are as follows:

1. *Solvability/feasibility analysis*
   Here we answer the following key questions: Do we have a solution at all? If yes, is it unique?
2. *Solution method (algorithm)*
   Here we set out the way in which the output is computed or the approach for arriving at a decision.
3. *Analysis of the problem and its solution method*
   Here we provide an analysis of how many computational steps are needed for solution and how this number of steps depends on the problem size.

The question of analysing the number of computational steps of a problem solution and its dependence on the problem size is the subject of algorithmic complexity theory within the theory of algorithms. The size of a problem is measured by the number of digits or characters needed to describe an *instance* of the problem. This is the set of problem ingredients specified in the *Given* section. There are fundamental complexity classes which can be used to classify a particular problem. Problems with polynomial dependence of their computational steps on their problem size belong to the class *P* for decision problems or class *FP* for search problems. The problem classes P and FP are regarded to be computationally easy, while all the other problem classes are computationally hard.

It is surprising but a fundamental fact in algorithmic complexity theory that the problem class to which a particular problem belongs is not dependent on the solution method but only on the problem statement itself. This is explained by the definition of the computational steps needed to solve a problem. It is the largest number of steps required to solve the problem for the worst possible *instance* of the problem using the best available method.

10.2. BASIC NOTIONS IN SYSTEMS AND CONTROL THEORY

10.2.1. The Notion of a System, Linear and Time-Invariant Systems

A "system" can be regarded as a basic notion in system theory. It is a general concept under which a particular application can be classed. Such particular instances include a "process system", a "financial system" or an "environmental system". We understand the system to be part of the real world with a boundary between the system and its environment. The system interacts with its environment only through its boundary. The effects of the environment on the system are described by time dependent *input* functions $u(t)$ from a given set of possible inputs $u \in \mathcal{U}$, while the effect of the system on its environment is described by the *output* functions $y(t)$ taken from a set of possible outputs $y \in \mathcal{Y}$. The schematic *signal flow diagram* of a system $\mathbf{S}$ with its input and output signals is shown in Fig. 10.1.

We can look at the signals belonging to a system as the input causing the time-dependent behaviour of the system that we can observe through the output of the system. Therefore, the system can be described as an operator $\mathbf{S}$ which maps inputs $u(t)$ into outputs $y(t)$, expressed as

$$\mathbf{S} : \mathcal{U} \rightarrow \mathcal{Y}, \quad y = \mathbf{S}[u].$$

For process control applications, we often distinguish between manipulated input variables $u(t)$ and disturbance variables $d(t)$ within the set of input variables to the system. Both manipulated input and disturbance variables act upon the system to produce the system behaviour. Disturbance variables are often regarded as uncontrolled being determined by the environment in which the system resides. These disturbances could be weather variations, process feed quality variations or variations in utility systems

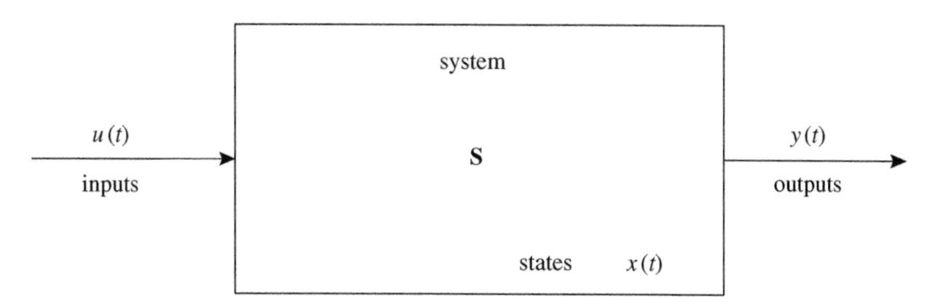

FIGURE 10.1 Signal flow diagram of a system.

such as steam or cooling water temperatures. In contrast, the manipulated input variables can be directly affected by a controller or a person belonging to the environment of the system.

There are systems with special properties which are especially interesting and easy to handle from the viewpoint of their analysis and control. Here we investigate a number of these systems.

Linear Systems

The first property of special interest is linearity. A system **S** is called *linear* if it responds to a linear combination of its possible input functions with the same linear combination of the corresponding output functions. Thus, for the linear system we note that

$$\mathbf{S}[c_1 u_1 + c_2 u_2] = c_1 y_1 + c_2 y_2 \tag{10.1}$$

with $c_1, c_2 \in \mathcal{R}$, $u_1, u_2 \in \mathcal{U}$, $y_1, y_2 \in \mathcal{Y}$ and $\mathbf{S}[u_1] = y_1$, $\mathbf{S}[u_2] = y_2$.

Time-Invariant Systems

The second interesting class of systems are time-invariant systems. A system **S** is *time-invariant* if its response to a given input is invariant under time shifting. Loosely speaking, time-invariant systems do not change their system properties in time. If we were to repeat an experiment under the same circumstances at some later time, we get the same response as originally observed.

The notion of *time invariance* is illustrated in Fig. 10.2, where we see two identical inputs to the system separated by a time Δt and note that the time shifted outputs are also identical. In many process system applications, this can be a reasonable assumption over a short time frame. In other cases, phenomena such as catalyst deactivation or heat transfer fouling lead to non time-invariant systems. An in-depth knowledge of the system mechanisms as well as the time frame of the intended analysis often resolves the validity of the time-invariant assumption. There are a number of implications which flow from the previous discussion.

Time-invariant systems have constant or time-independent parameters in their system models.

Linear and time-invariant systems are termed LTI systems.

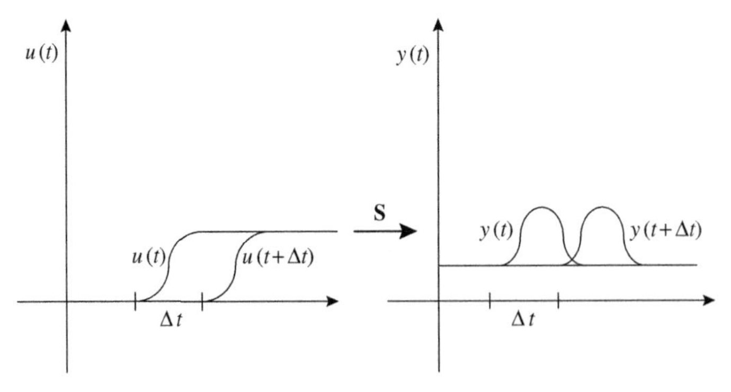

FIGURE 10.2 Notion of time invariance.

Continuous and Discrete Time Systems

Further, we may classify systems according to the time variable $t \in \mathcal{T}$ we apply to their description. There are *continuous time* systems where time is an open interval of the real line ($\mathcal{T} \subseteq \mathcal{R}$). Discrete time systems have an ordered set $\mathcal{T} = \{\ldots, t_0, t_1, t_2, \ldots\}$ as their time variable set.

Single-Input Single-Output and Multiple-Input Multiple-Output Systems

Finally, the class of *SISO* systems is often distinguished from the class of *MIMO systems* because of the relative simplicity in their system models.

In the most general and abstract case we describe the system by an *operator* **S**. However, in most of the practical cases outlined in the subsequent sections we give a concrete form of this operator. For example, **S** can be a linear differential operator of order n and thus leads to either sets of linear ODEs or to sets of linear PDEs. Nonlinear counterparts also exist. Also, the operator **S** can be characterized by a set of parameters p which are called *system parameters*.

10.2.2. Different Descriptions of Linear Time-Invariant Systems

The system **S** can be described in alternative ways [70]:

- in the time domain,
- in the operator domain,
- in the frequency domain.

The operator and frequency domain description of systems is only used for linear systems, most frequently for LTI systems. These descriptions can be obtained by using Laplace transformation or Fourier transformation of the time domain description of systems to obtain the operator domain or frequency domain description respectively.

Process models are naturally and conventionally set up in the time domain. Therefore, the various forms of *time domain description* will only be treated here. The different forms are explained and compared with the example of a continuous time SISO LTI system.

Continuous time LTI systems may be described in the time domain by

- input–output models,
- state space models.

Input–output models are further subdivided into linear differential equation models and impulse response models.

Linear Differential Equations with Constant Coefficients

If we consider the system input and output and their possibly higher order derivatives as

$$u(t), \frac{\mathrm{d}u}{\mathrm{d}t}, \frac{\mathrm{d}^2u}{\mathrm{d}t^2}, \ldots; \quad y(t), \frac{\mathrm{d}y}{\mathrm{d}t}, \frac{\mathrm{d}^2y}{\mathrm{d}t^2}, \ldots,$$

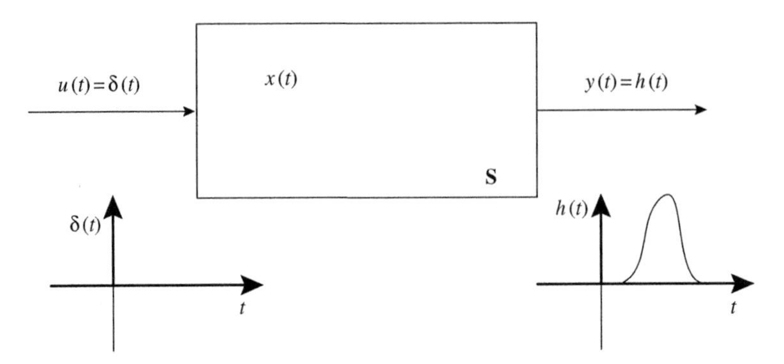

FIGURE 10.3 Impulse response function.

the general form of the input–output model for an LTI SISO system is given by the following higher order linear differential equation with constant coefficients:

$$a_n \frac{d^n y}{dt^n} + a_{n-1} \frac{d^{n-1} y}{dt^{n-1}} + \cdots + a_1 \frac{dy}{dt} + a_0 y = b_0 u + b_1 \frac{du}{dt} + \cdots + b_r \frac{d^r u}{dt^r}. \quad (10.2)$$

This equation has the following initial conditions:

$$y(t_0) = y_0, \quad \frac{d^i y}{dt^i} = y_i(t_0), \quad i = 1, \ldots, n.$$

Note that, in this case, the system parameters p consist of the set of constant coefficients:

$$p = [a_0, a_1, \ldots, a_n, b_0, b_1, \ldots, b_r]^T. \quad (10.3)$$

Impulse Response Representation

The impulse response function is the response of an LTI SISO system to a special test input signal called the *Dirac delta function* or unit impulse function. The Dirac delta function $\delta(t)$ is a function or more precisely a distribution in the mathematical sense, which is equal to zero everywhere except at $t = 0$ such that

$$\int_{-\infty}^{\infty} \delta(t)\, dt = 1.$$

The concept of impulse response representation is illustrated in Fig. 10.3, where $h(t)$ is the impulse response function and $\delta(t)$ is the Dirac delta function.

The output of **S** can be written as

$$y(t) = \int_{-\infty}^{\infty} h(t - \tau) u(\tau)\, d\tau. \quad (10.4)$$

This equation describes the *convolution* of the functions $h(\cdot)$ and $u(\cdot)$ in the time domain. Here, $h(\cdot)$ is the impulse response of the system and τ is the variable of integration.

Because $u(t)$ is a function $\mathcal{R}^+ \to \mathcal{R}$, that is u maps from the set of positive real numbers to the set of real numbers, we start the integration at time $t = 0$. The upper bound for the integration is t because the system is *causal*, so that

$$y(t) = \int_0^t h(t-\tau)u(\tau)\,\mathrm{d}\tau. \tag{10.5}$$

Note that the *system parameters* of this representation are the parameters of the impulse response function $h(\cdot)$. Impulse response functions can be parametrized or represented by either the parameters of the input–output differential Eq. (10.3) or by the parameters of the state space representation (SSR) so there is no need to specify it as a function.

Unit Step Responses

The Dirac delta function used to generate the impulse response from a system is physically not realizable, nor even a function but a so-called distribution in the mathematical sense. Therefore, one usually applies another special test function, the so-called *unit step function* to investigate the dynamic response characteristics of LTI systems.

The unit step function denoted by $1(t)$ is identically zero till $t = 0$ where it jumps to 1 and remains identically 1 as $t \to \infty$. The unit step function is the primitive function of the Dirac impulse function:

$$1(t) = \int_{-\infty}^t \delta(\tau)\,\mathrm{d}\tau \tag{10.6}$$

and has a discontinuity (a jump) at $t = 0$.

The response of a system to the unit step function is called the *unit step response function*. The unit step response function $h^*(t)$ *is the primitive function of the impulse response function*

$$h^*(t) = \int_0^t h(\tau)\,\mathrm{d}\tau \tag{10.7}$$

because integration and the response generation of a LTI system are both linear operations and thus commutative.

Note that a unit step response can be defined and generated experimentally for every input–output pair of a MIMO LTI system to give a full characterization of its multivariate dynamical responses.

It is important to note that a step response function will converge to a final value h_∞ in the limit (when $t \to \infty$) if the impulse response function has a finite integral in the limit. That is

$$h_\infty = \int_0^\infty h(\tau)\,\mathrm{d}\tau < \infty. \tag{10.8}$$

This is exactly the condition for a LTI SISO system to be bounded input bounded output (BIBO) stable (refer to Section 13.1.3). The value h_∞ is called the *static gain* of the system. Note again that one can define static gains separately for every input–output pair of a MIMO LTI system and arrange them into a matrix.

State Space Representation

Input–output representations describe the system with zero initial conditions. Given the assumption of zero initial condition, we needed the impulse response function $h(t)$ or its Laplace transform, the *transfer function* $H(s)$, a time instant t_0 and the past and future history of the input $u(t),\ 0 \leq \tau < t_0 \leq t < \infty$ to compute $y(t)$. Let us introduce new information which is called the *state of the system at* t_0 which contains all past information on the system up to time t_0.

Now, to compute $y(t)$ for $t \geq t_0$ (all future values) we only need $u(t),\ t \geq t_0$ and the state at $t = t_0$.

The development of a state space model of a process system normally involves identifying a number of classes of variables in the system. These include:

- state variable vector $x \in \mathcal{R}^n$;
- (manipulable) input variable vector $u \in \mathcal{R}^r$ which is used to manipulate the states;
- disturbance variable vector $d \in \mathcal{R}^v$;
- output variable vector $y \in \mathcal{R}^m$ which is usually the measurements taken on the system. These can be directly related to the states.
- system parameter vector $p \in \mathcal{R}^w$.

It can be shown that *the general form of SSR or state space model* of a MIMO LTI system without considering disturbances separately from manipulated inputs is in the following form:

$$\begin{aligned} \dot{x}(t) &= \mathbf{A}x(t) + \mathbf{B}u(t), \quad \text{(state equation)}, \\ y(t) &= \mathbf{C}x(t) + \mathbf{D}u(t), \quad \text{(output equation)} \end{aligned} \tag{10.9}$$

with given initial condition $x(t_0) = x(0)$ and

$$x(t) \in \mathcal{R}^n, \quad y(t) \in \mathcal{R}^m, \quad u(t) \in \mathcal{R}^r \tag{10.10}$$

being vectors of finite-dimensional spaces and

$$\mathbf{A} \in \mathcal{R}^{n \times n}, \quad \mathbf{B} \in \mathcal{R}^{n \times r}, \quad \mathbf{C} \in \mathcal{R}^{m \times n}, \quad \mathbf{D} \in \mathcal{R}^{m \times r} \tag{10.11}$$

being matrices. Note that $\mathbf{A}$ is called a *state matrix*, $\mathbf{B}$ is the *input matrix*, $\mathbf{C}$ is the *output matrix* and $\mathbf{D}$ is the *input-to-output coupling matrix*.

The parameters of a state space model consist of the constant matrices

$$p = \{\mathbf{A}, \mathbf{B}, \mathbf{C}, \mathbf{D}\}.$$

The SSR of an LTI system is the quadruplet of the constant matrices $(\mathbf{A}, \mathbf{B}, \mathbf{C}, \mathbf{D})$ in Eq. (10.9). *The dimension of an SSR* is the dimension of the state vector: $\dim x(t) = n$. The state space $\mathcal{X}$ is the set of all states:

$$x(t) \in \mathcal{X}, \quad \dim \mathcal{X} = n. \tag{10.12}$$

Simple Examples of LTI System Models

Some simple LTI system models are introduced in this subsection in order to illustrate how to construct the canonical form (10.9) of their SSR and to show their step response functions generated by MATLAB.

Having identified the state space realization matrices $(\mathbf{A}, \mathbf{B}, \mathbf{C}, \mathbf{D})$ of an LTI system, there is a simple formula to compute the static gain h_∞ of the system from the system parameters in the case of SISO systems:

$$h_\infty = -\mathbf{C}\mathbf{A}^{-1}\mathbf{B} + \mathbf{D}. \tag{10.13}$$

Note that the static gain related to an input–output pair (i, j) of a MIMO system can also be computed by a similar extended version of the above formula:

$$h_\infty(i, j) = -\mathbf{C}^{(j_{\mathrm{row}})}\mathbf{A}^{\ 1}\mathbf{B}^{(i_{\mathrm{col}})} + \mathbf{D}_i^{(j_{\mathrm{row}})}, \tag{10.14}$$

where the upper index $^{(j_{\mathrm{row}})}$ refers to the jth row, and $^{(i_{\mathrm{col}})}$ to the ith column of the particular matrix.

EXAMPLE 10.2.1 (A simple stable SISO LTI system). Consider a simple LTI SISO system model in the form:

$$\frac{dx_1}{dt} = -4x_1 + 3x_2 + 7u_1, \tag{10.15}$$

$$\frac{dx_2}{dt} = 5x_1 - 6x_2 + 8u_1, \tag{10.16}$$

$$y = x_1. \tag{10.17}$$

1. Construct the state space model representation matrices.
2. Compute the unit step response function.
3. Compute the static gain of the system.

The standard matrix-vector form of the system model above is in the form of Eq. (10.9) with the following vectors and matrices

$$x = \begin{bmatrix} x_1 \\ x_2 \end{bmatrix}, \quad u = [u_1], \quad y = [y],$$

$$\mathbf{A} = \begin{bmatrix} -4 & 3 \\ 5 & -6 \end{bmatrix}, \quad \mathbf{B} = \begin{bmatrix} 7 \\ 8 \end{bmatrix}, \quad \mathbf{C} = [\ 1\ 0\ ], \quad \mathbf{D} - [0].$$

The unit step response is computed using the MATLAB procedure `step(A,B,C,D)` to obtain the step response shown in Fig. 10.4.

Note that the steady state gain is the asymptotic value of this function as time goes to infinity. The static gain can also be computed separately from the formula (10.13) to get 7.333 in this case, which is also seen from the figure.

The next example shows a similar system extended with a new input variable to form a MISO model.

EXAMPLE 10.2.2 (A simple stable MISO LTI system). Consider a simple LTI MISO model in the form:

$$\frac{dx_1}{dt} = -4x_1 + 3x_2 + 7u_1, \tag{10.18}$$

$$\frac{dx_2}{dt} = 5x_1 - 6x_2 + 8u_2, \tag{10.19}$$

$$y = x_1. \tag{10.20}$$

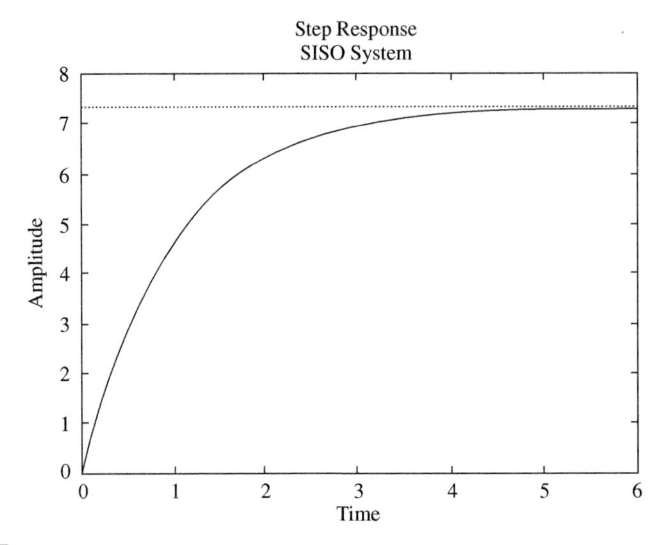

FIGURE 10.4 Unit step response for a stable SISO LTI system.

Construct the state space model representation matrices, compute the unit step response function and compute the static gain of the system.

Observe that the system above is almost the same as that in Example 10.2.1 but a new input variable has been added.

The standard matrix-vector form of the above system model is in the form of Eq. (10.9) with the following vectors and matrices:

$$x = \begin{bmatrix} x_1 \\ x_2 \end{bmatrix}, \quad u = \begin{bmatrix} u_1 \\ u_2 \end{bmatrix}, \quad y = [y],$$

$$\mathbf{A} = \begin{bmatrix} -4 & 3 \\ 5 & -6 \end{bmatrix}, \quad \mathbf{B} = \begin{bmatrix} 7 & 0 \\ 0 & 8 \end{bmatrix}, \quad \mathbf{C} = \begin{bmatrix} 1 & 0 \end{bmatrix}, \quad \mathbf{D} = [0].$$

Observe, that both the vector u and also the input matrix $\mathbf{B}$ have changed.

The unit step response is again computed using MATLAB for the input–output pair $(1, 1)$ to obtain the step response shown in Fig. 10.5.

Note that the steady state gain is the asymptotic value of this function as time goes to infinity. The static gain can also be computed separately from the multi input multi output formula (10.14) to get 4.667 in this case which is also seen from the figure.

The next example shows a MISO system with the same dimension of the system variables and matrices but with the state matrix $\mathbf{A}$ changed. This example shows an unstable system and illustrates how it can be discovered from the unit step response.

EXAMPLE 10.2.3 (An unstable MISO LTI system). Consider a simple LTI MISO system model in the form:

$$\frac{\mathrm{d}x_1}{\mathrm{d}t} = 7u_1, \tag{10.21}$$

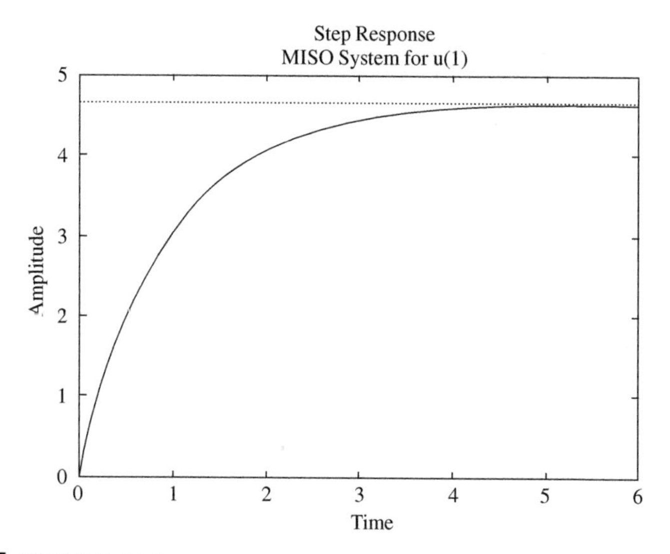

FIGURE 10.5 Unit step response for a stable MISO LTI system.

$$\frac{\mathrm{d}x_2}{\mathrm{d}t} = 5x_1 - 6x_2 + 8u_2, \tag{10.22}$$

$$y = x_1. \tag{10.23}$$

Construct the state space model representation matrices, compute the unit step response function and compute the static gain of the system.

Observe that the system above is almost the same as that in Example 10.2.2 but the first state equation has been changed to make the *first state variable an integrator*, where x_1 does not appear on the right hand side of the state equation.

The standard matrix-vector form (10.9) of the above system model contains the following vectors and matrices:

$$x = \begin{bmatrix} x_1 \\ x_2 \end{bmatrix}, \quad u = \begin{bmatrix} u_1 \\ u_2 \end{bmatrix}, \quad y = [y],$$

$$\mathbf{A} = \begin{bmatrix} 0 & 0 \\ 5 & -6 \end{bmatrix}, \quad \mathbf{B} = \begin{bmatrix} 7 & 0 \\ 0 & 8 \end{bmatrix}, \quad \mathbf{C} = [\ 10\ ], \quad \mathbf{D} = [0].$$

Observe, that only the state matrix **A** has changed compared with Example 10.2.2.

If we have an unstable system, then we cannot expect the unit step response computed with respect to any input to converge to a finite value. This is seen on the Fig. 10.6 of the unit step for the first ($i = 1$) input variable u_1. ■ ■ ■

Transformation of States

It is important to note that SSRs are not unique but we can find infinitely many equivalent state space representations with the same dimension giving rise to the same input–output description of a given system. It should be noted that *two SSRs which describe the same system are equivalent if they correspond to the same impulse response function.*

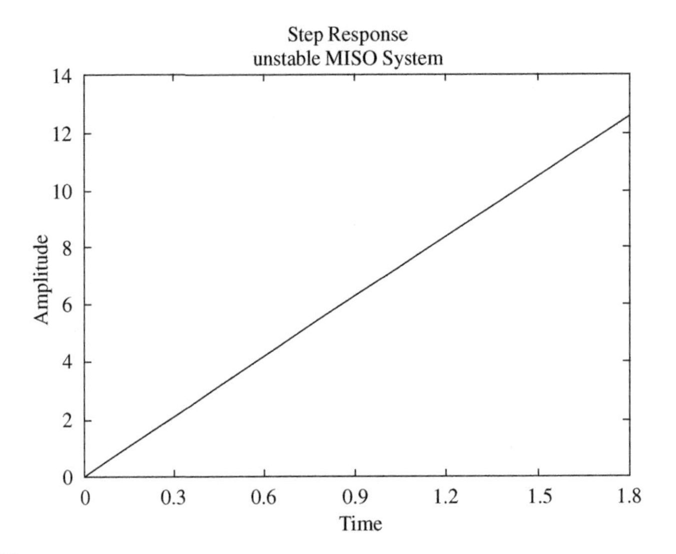

FIGURE 10.6 Unit step resonse for an unstable SISO LTI system.

In order to get another equivalent SSR from a given one, it is possible to transform the state variables using a non-singular transformation matrix $\mathbf{T}$. Then the following relations hold between two possible equivalent SSRs:

$$\begin{aligned} \dot{x}(t) &= \mathbf{A}x(t) + \mathbf{B}u(t), & \dot{\bar{x}}(t) &= \bar{\mathbf{A}}\bar{x}(t) + \bar{\mathbf{B}}u(t), \\ y(t) &= \mathbf{C}x(t) + \mathbf{D}u(t), & y(t) &= \bar{\mathbf{C}}\bar{x}(t) + \bar{\mathbf{D}}u(t), \end{aligned} \tag{10.24}$$

which are related by the transformation

$$\mathbf{T} \in \mathcal{R}^{n \times n}, \quad \det \mathbf{T} \neq 0. \tag{10.25}$$

That is,

$$\bar{x} = \mathbf{T}x \Rightarrow x = \mathbf{T}^{-1}\bar{x}. \tag{10.26}$$

If we transform the state space model equations accordingly, we get

$$\dim \mathcal{X} = \dim \bar{\mathcal{X}} = n \tag{10.27}$$

and

$$\mathbf{T}^{-1}\dot{\bar{x}} = \mathbf{A}\mathbf{T}^{-1}\bar{x} + \mathbf{B}u. \tag{10.28}$$

Finally,

$$\dot{\bar{x}} = \mathbf{T}\mathbf{A}\mathbf{T}^{-1}\bar{x} + \mathbf{T}\mathbf{B}u, \quad y = \mathbf{C}\mathbf{T}^{-1}\bar{x} + \mathbf{D}u. \tag{10.29}$$

So the transformed equivalent SSR for the same system is

$$\bar{\mathbf{A}} = \mathbf{T}\mathbf{A}\mathbf{T}^{-1}, \quad \bar{\mathbf{B}} = \mathbf{T}\mathbf{B}, \quad \bar{\mathbf{C}} = \mathbf{C}\mathbf{T}^{-1}, \quad \bar{\mathbf{D}} = \mathbf{D}. \tag{10.30}$$

State Space Models with Disturbance Variables

If the distinction between manipulated variables $u(t)$ and disturbance variables $d(t)$ is to be taken into account then we consider an extended state space model form instead of Eq. (10.9) as:

$$\begin{aligned} \dot{x}(t) &= \mathbf{A}x(t) + \mathbf{B}u(t) + \mathbf{E}d(t), \quad \text{(state equation)}, \\ y(t) &= \mathbf{C}x(t) + \mathbf{D}u(t) + \mathbf{F}d(t), \quad \text{(output equation)} \end{aligned} \tag{10.31}$$

with given initial condition $x(t_0) = x(0)$, and an additional disturbance variable

$$d(t) \in \mathcal{R}^{v} \tag{10.32}$$

with new state space matrices

$$\mathbf{E} \in \mathcal{R}^{n \times v}, \quad \mathbf{F} \in \mathcal{R}^{m \times v}. \tag{10.33}$$

Note that in this case the parameters of the state space models are

$$p = \{\mathbf{A}, \mathbf{B}, \mathbf{C}, \mathbf{D}, \mathbf{E}, \mathbf{F}\}. \tag{10.34}$$

Linear State Space Models with Time Varying Coefficients

The first step to generalize the state space models for LTI systems is to allow the parameters to depend on time. It is important to note that the system described by this extended state space model will no longer be time invariant. The resulting *linear form with time varying coefficients* is given by

$$\frac{\mathrm{d}x(t)}{\mathrm{d}t} = \mathbf{A}(t)x(t) + \mathbf{B}(t)u(t) + \mathbf{E}(t)d(t), \tag{10.35}$$

$$y(t) = \mathbf{C}(t)x(t) + \mathbf{D}(t)u(t) + \mathbf{F}(t)d(t), \tag{10.36}$$

where $\mathbf{A}(t)$ is an $n \times n$ state matrix, $\mathbf{B}(t)$ is an $n \times r$ input matrix, $\mathbf{C}(t)$ is an $m \times n$ output matrix, $\mathbf{D}(t)$ is an $m \times r$ input-to-output coupling matrix, $\mathbf{E}(t)$ is an $n \times v$ disturbance input matrix and $\mathbf{F}(t)$ is an $m \times v$ disturbance output matrix.

10.2.3. Nonlinear State Space Models

If the state of a nonlinear system can be described at any time instance by a finite-dimensional vector, then we call the system a *concentrated parameter system*. Process systems described by lumped system models where the model is constructed from a finite number of perfectly mixed balance volumes are *concentrated parameter systems* from a system theoretical viewpoint.

Having identified the relevant input, output, state and disturbance variables for a concentrated parameter nonlinear system, the general nonlinear state space equations can be written in matrix form as

$$\frac{\mathrm{d}x(t)}{\mathrm{d}t} = f(x(t), u(t), d(t), p), \tag{10.37}$$

$$y(t) = h(x(t), u(t), d(t), p), \tag{10.38}$$

or in expanded form as

$$\begin{pmatrix} \frac{dx_1}{dt} \\ \frac{dx_2}{dt} \\ \vdots \\ \frac{dx_n}{dt} \end{pmatrix} = \begin{pmatrix} f_1(x_1, \ldots, x_n, u_1, \ldots, u_r, d_1, \ldots, d_v, p_1, \ldots, p_w) \\ f_2(x_1, \ldots, x_n, u_1, \ldots, u_r, d_1, \ldots, d_v, p_1, \ldots, p_w) \\ \vdots \\ f_n(x_1, \ldots, x_n, u_1, \ldots, u_r, d_1, \ldots, d_v, p_1, \ldots, p_w) \end{pmatrix}$$

and

$$\begin{pmatrix} y_1 \\ y_2 \\ \vdots \\ y_m \end{pmatrix} = \begin{pmatrix} h_1(x_1, \ldots, x_n, u_1, \ldots, u_r, d_1, \ldots, d_v, p_1, \ldots, p_w) \\ h_2(x_1, \ldots, x_n, u_1, \ldots, u_r, d_1, \ldots, d_v, p_1, \ldots, p_w) \\ \vdots \\ h_m(x_1, \ldots, x_n, u_1, \ldots, u_r, d_1, \ldots, d_v, p_1, \ldots, p_w) \end{pmatrix}.$$

The nonlinear vector–vector functions f and h in Eq. (10.38) characterize the nonlinear system. Their parameters constitute the system parameters.

10.3. LUMPED DYNAMIC MODELS AS DYNAMIC SYSTEM MODELS

The development of lumped dynamic models for process systems has been described in detail in Section 5.2.

Lumped dynamic models are in the form of DAE systems where the

- differential equations originate from conservation balances of conserved extensive quantities (mass, component masses, energy or momentum) for every balance volume;
- algebraic equations are of mixed origin derived from constitutive relations. These are normally nonlinear algebraic equations. They are often of full structural rank indicating that the differential index of the DAE model is 1. Higher index models result when this system is rank deficient.

Therefore, lumped dynamic models can be often transformed to set of explicit first-order NLDEs with given initial conditions. This form shows that lumped dynamic models can be viewed as nonlinear state space models described in Section 10.2.3.

In this section, we show how the above elements of a lumped dynamic model of a process system appear in the nonlinear state space model. This correspondence gives special properties to those nonlinear state space models which originate from lumped dynamic models. These special properties can aid in the application of *grey-box* methods for dynamic simulation and process control.

10.3.1. System Variables for Process Systems

In order to develop a state space model of a process system, we need to identify the system variables. This includes the state, input, disturbance and output variables of the system. For process systems these choices are dictated by the development of the lumped parameter dynamic model of the process and by the setup of the measurements and any control system. More precisely, *state variables are fixed by the lumped parameter dynamic model, while input, disturbance and output variables are fixed by the design, instrumentation and purpose of the model.* We now consider these variable categories.

State variables $x(t)$
The differential equations in a lumped dynamic model originate from conservation balances for the conserved extensive quantities over each balance volume. As the state variables are the differential variables in these balance equations the natural set of state variables of a process system is the set of conserved extensive variables or their intensive counterparts for each balance volume. This fact fixes the number of state variables. Hence, the dimension of the state space model is equal to the number of conserved extensive quantities. If there are c components in the system, then in general we can write c component mass balances for each species plus an energy balance for each phase. Thus, the total number of states n is obtained by multiplying the states per phase $(c + 1)$ by the number of balance volumes n_Σ: $n = (c + 1)n_\Sigma$. Moreover the physical meaning of the state variables is also fixed by this correspondence.

Manipulated input and disturbance variables $u(t)$, $d(t)$
In order to identify potential input variables and disturbances one should look carefully at the process to identify all dynamic effects from the environment which act upon the system to affect its behaviour. Those potential input variables which can be influenced by a device such as a control valve through an instrumentation system or changed manually form the set of potential manipulated input variables. The actual purpose of the modelling decides which variables will be regarded as actual disturbances and manipulated input variables from the overall set. Typically, the following type of signals are used as manipulated input variables in a process system:

- flowrates,
- split and recycle ratios,
- utility flowrates and temperatures,
- pressures,
- current and voltage controlling heaters, shaft rotation, motors or valves.
- switches.

Output variables, $y(t)$
The set of potential output variables is fixed by the measurement devices and each measured variable can be regarded as an output of the system. The modelling or control goal is the one which finally determines which variables are actually used for a given purpose from the possible set. Typically, the following types of signals are used as output variables in a process system:

- temperatures,
- pressures,
- concentration related physical quantities like mole or mass fraction,
- level related physical quantities like weight, head or total mass,
- flowrate related physical quantities.

10.3.2. State Equations for Process System Models

Chapter 5 has considered in some detail the derivation of lumped parameter models of process systems through the application of conservation principles.

In order to illustrate how the state equation of a nonlinear state space model can be obtained from the lumped dynamic model of a process system consider the following simple example.

EXAMPLE 10.3.1 (Nonlinear state space model form of a CSTR model). Consider the nonlinear model describing the dynamics of a non-isothermal CSTR. The reaction is first order, $A \longrightarrow B$ and is exothermic. The reaction rate constant k is given by $k_0\,\mathrm{e}^{-E/(RT)}$. The reactor is cooled by coolant at temperature T_C. The feed is at temperature T_i. It is assumed that the physical properties remain constant and that inlet and outlet flows are equal.

The mass and energy balances lead to the following equations:

$$V\frac{\mathrm{d}C_A}{\mathrm{d}t} = FC_{A_i} - FC_A - kVC_A, \tag{10.39}$$

$$V\rho c_P\frac{\mathrm{d}T}{\mathrm{d}t} = F\rho c_P(T_i - T) + kVC_A(-\Delta H_R) - UA(T - T_C) \tag{10.40}$$

with the volume V being constant.

These can be rearranged to give

$$\frac{\mathrm{d}C_A}{\mathrm{d}t} = \frac{F}{V}(C_{A_i} - C_A) - kC_A, \tag{10.41}$$

$$\frac{\mathrm{d}T}{\mathrm{d}t} = \frac{F}{V}(T_i - T) + \frac{kC_A}{\rho c_P}(-\Delta H_R) - \frac{UA}{V\rho c_P}(T - T_C). \tag{10.42}$$

Now the *state variables* of the nonlinear state space model of the CSTR are

$$x = [C_A, T]^T,$$

which are in fact the intensive counterparts of the conserved extensive variables, these being the mass of component A and the total energy E. Let us select the flowrate and the coolant temperature as manipulated input variables:

$$u = [F, T_C]^T$$

Note that other choices for u are also possible.

Equations (10.41) and (10.42) are exactly the nonlinear state equations given in (10.37) with $n = 2$ and f_1 being the right-hand side function of the first and f_2 is that of the second state equation.

10.3.3. An Example of Constructing a Nonlinear State Space Model from Lumped Dynamic Model

The following example illustrates how one can use an already developed lumped dynamic process model in a systematic way to construct the nonlinear state space model with all of its ingredients. The construction is done in the following steps:

1. *Transform the model equations*
   Take the lumped process model with its balance and constitutive equations. Making use of the equation structure of the constitutive algebraic part, we substitute the algebraic equations into the differential ones.
2. *State equations and state variables*
   The transformed differential balance equations will form the set of state equations with the differential variables being the state variables of the system.
3. *Potential input variables*
   The potential input variables are the time-dependent variables on the right-hand side of the transformed model equations (state equations) which are *not* state variables and affect the variation of the state variables.
4. *Manipulable input variables and disturbances*
   From the set of input variables, we can select those which can be directly manipulable based on the instrumentation diagram of the flowsheet and based on the modelling goal. The rest are regarded as disturbance variables.
5. *Output variables and output equations*
   Using the instrumentation diagram, we can find out the quantities we measure for the system and their relationship to the state variables. These will be the output variables. Their relationships to state (and sometimes to input) variables form the output equations.

EXAMPLE 10.3.2 (State space model of three jacketted CSTRs in series). Consider a process system consisting of three jacketted CSTRs in series with cooling provided in a countercurrent direction. The cascade is shown in Fig. 10.7. Develop the nonlinear state space model of the system from its lumped dynamic system model.

Assumptions

$\mathcal{A}1$. Perfect mixing in each of the tanks and in their jackets.

$\mathcal{A}2$. There is a single first-order $A \rightarrow B$ endothermic reaction with the reaction rate

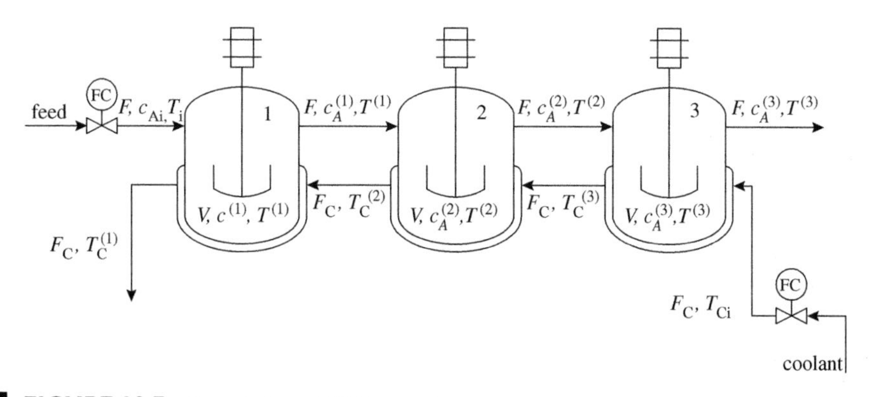

FIGURE 10.7 Three jacketted CSTRs in series.

$$r_A = kC_A \quad \left[\frac{\text{kmol}}{\text{s}\,\text{m}^3}\right]$$

in each of the tanks.

$\mathcal{A}3$. The reaction rate coefficient k obeys Arrhenius law, i.e.

$$k = k_0\,\mathrm{e}^{-E/(RT)}.$$

$\mathcal{A}4$. The volume of the mixture in the reactors and that in the cooling jackets is constant.

$\mathcal{A}5$. The corresponding volumes and heat transfer area in the three CSTRs are the same.

$\mathcal{A}6$. Countercurrent flow of the reaction mixture and the cooling water.

$\mathcal{A}7$. The heat capacity of the wall is negligible.

$\mathcal{A}8$. The physico-chemical properties are constant.

Model equations

The dynamic model of the ith CSTR consists of the following balances:

- *Component mass balance for component A in the reactor*

$$V\frac{\mathrm{d}C_A^{(\mathrm{i})}}{\mathrm{d}t} = FC_A^{(i-1)} - FC_A^{(\mathrm{i})} - VC_A^{(\mathrm{i})}k_0\mathrm{e}^{-E/RT}. \tag{10.43}$$

- *Energy balance of the reaction mixture*

$$c_P\rho V\frac{\mathrm{d}T^{(\mathrm{i})}}{\mathrm{d}t} = c_P\rho FT^{(i-1)} - c_P\rho FT^{(\mathrm{i})} - \Delta HVC_A^{(\mathrm{i})}k_0\,\mathrm{e}^{-E/RT} + K_TA(T_C^{(\mathrm{i})} - T^{(\mathrm{i})}). \tag{10.44}$$

- *Energy balance for the water in the jacket*

$$c_{PC}\rho_C V_C\frac{\mathrm{d}T_C^{(\mathrm{i})}}{\mathrm{d}t} = c_{PC}\rho_C F_C T_C^{(i+1)} - c_{PC}\rho_C F_c T_C^{(\mathrm{i})} + K_TA(T^{(\mathrm{i})} - T_C^{(\mathrm{i})}) \tag{10.45}$$

for every CSTR, i.e. for $i = 1, 2, 3$.

The following "boundary conditions" specify the inlet and the outlet conditions for the reaction mixture and the cooling water respectively:

$$C_A^{(0)} = C_{A_\mathrm{i}}, \quad T^{(0)} = T_i, \quad T_C^{(4)} = T_{C_\mathrm{i}}. \tag{10.46}$$

Moreover, we need proper initial conditions for every differential variable at time $t = 0$

$$C_A^{(\mathrm{i})}(0), \quad T^{(\mathrm{i})}(0), \quad T_C^{(\mathrm{i})}(0), \quad i = 1, 2, 3.$$

State variables and state equations

The state variables of the system consisting of the three jacketted CSTRs are dictated by the balance equations to be

$$x = \left[\, C_A^{(\mathrm{i})}, T^{(\mathrm{i})}, T_C^{(\mathrm{i})} \mid i = 1, 2, 3\right]^\mathrm{T}. \tag{10.47}$$

The state equations are then exactly the balance equations (10.43)–(10.44) for the three CSTRs, i.e. for $i = 1, 2, 3$.

Input and output variables

The potential input variables are the non-differential variables on the right-hand side of the balance equations which affect its solution, can vary in time and possibly can be manipulated:

$$C_A^{(0)} = C_{A_i}, \quad T^{(0)} = T_i, \quad T_C^{(4)} = T_{Ci}, \quad F, \quad F_C.$$

Let us assume that the flowrates are kept constant, therefore, the vector of potential input variables is

$$u^* = \left[C_{Ai}, \; T_i, \; T_{C_i}\right]^T. \tag{10.48}$$

The set of possible output variables is the same as the vector of state variables if we assume that the concentration of component A can be directly measured, i.e.

$$y^* = x. \tag{10.49}$$

■ ■ ■

10.4. STATE SPACE MODELS AND MODEL LINEARIZATION

Almost without exception, process models are nonlinear in their form. Nonlinear models are difficult to analyse directly, since there is little nonlinear analysis theory which is easy to apply. Also many models are used at, or nearby to a particular operating point and as such a linear form of the model may be adequate for analysis purposes, providing we do not use it far from the intended point of operation.

There is also the fact that many powerful and extensive linear analysis tools are available to the process engineer. These include tools for assessing performance and designing control systems based on linear systems theory. In the area of control design, the use of linear models dominates the available techniques. Hence, it can be beneficial and even vital to develop linear approximations to the original nonlinear model.

10.4.1. Linearization of Single Variable Differential Equations

Most of the models which we develop are nonlinear, since terms in the equations are raised to a power or multiplied together. We usually refer to the model as being "nonlinear in certain variables". Some variables may occur in a linear form such as cx where c is a constant or in nonlinear form such as x_1x_2, or $\sqrt{x_1}$.

EXAMPLE 10.4.1 (Simple nonlinear model). Consider the behaviour of a simple tank which leads to the dynamic equation for height of:

$$\frac{dh}{dt} = \frac{1}{A}\left\{C_{v_1}\left(P_1 - P_0 - \rho gh\right)^{1/2} - C_{v_2}\left(P_0 + \rho gh - P_3\right)^{1/2}\right\}. \tag{10.50}$$

This is clearly nonlinear in the liquid height h. ■ ■ ■

EXAMPLE 10.4.2 (Simple linear example). In contrast to the previous example, the equation describing the dynamics of a temperature sensor is given by

$$\frac{dT_s}{dt} = \frac{1}{\tau}(T - T_s). \tag{10.51}$$

This is linear in T_s.

Having the linearized equations has several advantages including:

- The super-position theory applies. This means that if x_1 and x_2 are solutions of the model then so too are $x_1 + x_2$ or cx_1 or cx_2.
- There is a wealth of powerful and readily available analysis tools, e.g. in MATLAB and Mathematica.
- The initial conditions in the ODEs are zero when the equations are written in terms of deviation variables.
- Changes in outputs are in direct relationship to input changes. If a change of Δu in the input gives rise to a change of Δy in the output, then $2\Delta u$ will lead to an output change of $2\Delta y$.

In the following paragraphs, we describe *how linearization is done*.

Linearization of equations can be applied to both algebraic and differential equations. Linearization can be done at any convenient point in the state space but is usually done at some steady-state point of the system.

It is often the case in control applications that we want to work with *deviation (perturbation) variables* which are defined in relation to a reference point or steady-state operating point. Thus, the deviation variable $\hat{x}$ can be defined as

$$\hat{x} = x - x_0, \tag{10.52}$$

where x_0 is the particular steady-state value of x. We can also use a normalized form as

$$\hat{x} = \frac{x - x_0}{x_0} \tag{10.53}$$

to ensure that we have provided some scaling to the problem. These variables, $\hat{x}$ are then "deviations" from a particular steady-state value, x_0.

Linearization is based on the application of Taylor's expansion about a particular operating point. For the general dynamic model in one variable, x we can write:

$$\frac{dx}{dt} = f(x, t) \tag{10.54}$$

and the expansion of the nonlinear function $f(x, t)$ gives

$$f(x, t) = f(x_0, t) + \left.\frac{df}{dx}\right|_{x_0} (x - x_0) + \left.\frac{d^2f}{dx^2}\right|_{x_0} \frac{(x - x_0)^2}{2!} + \cdots, \tag{10.55}$$

where x_0 is the operating point and d^if/dx^i is the ith derivative of f with respect to x evaluated at x_0.

If we truncate after the first derivative, we obtain the linear approximation:

$$f(x,t) \simeq f(x_0,t) + \left.\frac{\mathrm{d}f}{\mathrm{d}x}\right|_{x_0} (x - x_0) \tag{10.56}$$

and so the original equation can be written as

$$\frac{\mathrm{d}x}{\mathrm{d}t} = f(x_0,t) + \left.\frac{\mathrm{d}f}{\mathrm{d}x}\right|_{x_0} (x - x_0), \tag{10.57}$$

so that

$$\frac{\mathrm{d}}{\mathrm{d}t}(\hat{x} + x_0) = f(x_0,t) + \left.\frac{\mathrm{d}f}{\mathrm{d}x}\right|_{x_0} (x - x_0), \tag{10.58}$$

since

$$\frac{\mathrm{d}x_0}{\mathrm{d}t} = f(x_0,t) \quad \text{and} \quad \hat{x} = x - x_0, \tag{10.59}$$

we can subtract this to get

$$\frac{\mathrm{d}\hat{x}}{\mathrm{d}t} = \left.\frac{\mathrm{d}f}{\mathrm{d}x}\right|_{x_0} (\hat{x}); \quad \hat{x}(0) = 0, \tag{10.60}$$

which is our final linearized state space equation in the deviation variable $\hat{x}$. Often we drop the terminology $\hat{x}$ and simply write x, noting that this is a deviation variable.

EXAMPLE 10.4.3 Consider the simple nonlinear model where

$$\frac{\mathrm{d}h}{\mathrm{d}t} = F - C\sqrt{h}, \tag{10.61}$$

here, F and C are constants.

We can linearize the problem by expanding the right-hand side as a Taylor series about h_0, giving

$$\frac{\mathrm{d}h}{\mathrm{d}t} = \left.\left(F - C\sqrt{h}\right)\right|_{h_0} + \frac{\mathrm{d}}{\mathrm{d}h}\left.\left(F - C\sqrt{h}\right)\right|_{h_0} (h - h_0), \tag{10.62}$$

$$\frac{\mathrm{d}(\hat{h} + h_0)}{\mathrm{d}t} = F - C\sqrt{h_0} - \frac{1}{2}C(h_0)^{-1/2}(h - h_0). \tag{10.63}$$

If we wish to express the linearized equation in deviation terms, we can subtract the steady-state equation given by

$$F - C\sqrt{h_0} = 0, \tag{10.64}$$

from Eq. (10.63) to give the final deviation form:

$$\frac{\mathrm{d}\hat{h}}{\mathrm{d}t} = -\frac{1}{2}(h_0)^{-1/2}(\hat{h}), \tag{10.65}$$

where $\hat{h}$ is the deviation variable $(h - h_0)$.

In general, it can be seen that for a single variable nonlinear expression given by

$$\frac{dx}{dt} = f(x), \tag{10.66}$$

the linearized equation is

$$\frac{d\hat{x}}{dt} = \frac{d}{dt}(x - x_0) = \left(\frac{df}{dx}\right) \cdot (x - x_0) = \left(\frac{df}{dx}\right)\hat{x}. \tag{10.67}$$

If there are any accompanying output equations $y(t) = h(x(t), u(t))$, then these are also linearized and represented in deviation form as

$$\hat{y}(t) = \frac{\partial h}{\partial x}\hat{x} + \frac{\partial h}{\partial u}\hat{u}, \tag{10.68}$$

where the partial derivatives are evaluated at the chosen operating point.

10.4.2. Multi-variable Linearization

We can extend the single variable linearization to the case where we have many variables in our model. Hence, consider the set of ODEs given by

$$\frac{dx}{dt} = f(x) \tag{10.69}$$

or

$$\begin{aligned}
\frac{dx_1}{dt} &= f_1(x_1, x_2, x_3, \ldots, x_n),\\
\frac{dx_2}{dt} &= f_2(x_1, x_2, x_3, \ldots, x_n),\\
&\vdots\\
\frac{dx_n}{dt} &= f_3(x_1, x_2, x_3, \ldots, x_n).
\end{aligned} \tag{10.70}$$

Linearization means expanding the right-hand sides as a multi-variable Taylor series. Consider the first equation in (10.69) which can be expanded about the point $(x_{1_0}, x_{2_0}, \ldots, x_{n_0})$ to give:

$$\begin{aligned}
\frac{dx_1}{dt} &\cong f_1(x_0) + \left.\frac{\partial f_1}{\partial x_1}\right|_{x_{1_0}} (x_1 - x_{1_0}) + \cdots + \left.\frac{\partial f_1}{\partial x_n}\right|_{x_{n_0}} (x_n - x_{n_0})\\
&= f_1(x_0) + \sum_{j=1}^{n} \left.\frac{\partial f_1}{\partial x_j}\right|_{x_{j0}} (x_j - x_{j0}).
\end{aligned} \tag{10.71}$$

If we do this for all equations $(1, \ldots, n)$ we obtain

$$\frac{dx_i}{dt} = f_i(x_0) + \sum_{j=1}^{n} \left.\frac{\partial f_i}{\partial x_j}\right|_{x_{j0}} (x_j - x_{j0}), \quad i = 1, \ldots, n. \tag{10.72}$$

Finally, writing the linearized equations in deviation variable form we get

$$\frac{d\hat{x}_i}{dt} = \sum_{j=1}^{n} \left.\frac{\partial f_i}{\partial x_j}\right|_{x_{j0}} (\hat{x}_i), \quad i = 1, \ldots, n. \tag{10.73}$$

If we collect all n equations together in matrix-vector form we obtain

$$\frac{d}{dt}\hat{x} = \mathbf{J}\hat{x}, \tag{10.74}$$

where $\mathbf{J}$ is the system Jacobian (matrix of partial derivatives) evaluated at the steady-state point x_0.

$$\mathbf{J} = \begin{pmatrix} \frac{\partial f_1}{\partial x_1} & \frac{\partial f_1}{\partial x_2} & \cdots & \frac{\partial f_1}{\partial x_n} \\ \vdots & \vdots & \cdots & \vdots \\ \frac{\partial f_n}{\partial x_1} & \frac{\partial f_n}{\partial x_2} & \cdots & \frac{\partial f_n}{\partial x_n} \end{pmatrix}. \tag{10.75}$$

The vector $\hat{x}$ is given by the deviation variables:

$$\hat{x} = (x_1 - x_{1_0}, x_2 - x_{2_0}, \ldots, x_n - x_{n_0})^{\mathrm{T}}. \tag{10.76}$$

To see how this works, consider the following example:

EXAMPLE 10.4.4 (Nonlinear CSTR model). Consider a CSTR of constant volume with first-order reaction and fixed rate constant.

The nonlinear ODE is given by

$$\frac{V dC_A}{dt} = F_1 C_{A_1} - F_2 C_A - VkC_A. \tag{10.77}$$

Each term on the right-hand side is nonlinear and we can expand in deviation variable form as

$$\frac{V d\hat{C}_A}{dt} = F_{1_0}\hat{C}_{A_1} + C_{A_{1_0}}\hat{F} - F_{2_0}\hat{C}_A - C_{A_0}\hat{F}_2 - Vk(\hat{C}_A), \tag{10.78}$$

where the subscript 0 indicates the reference point for the linearization, which could be an operating steady state of the process.

10.4.3. Linearized State Space Equation Forms

When we develop models of process systems we usually end up with a nonlinear set of differential equations accompanied by a set of algebraic equations. In most control work, we use either nonlinear or linear state space forms of the system model in the form of Eq. (10.37) or Eq. (10.9).

If we regard the state space model equations of an Eq. (10.37) LTI system as being written in deviation variable form, then the state space matrices are the partial derivatives of the state and output equations with respect to the state and the input variables as follows:

$$\mathbf{A} = \left.\frac{\partial f}{\partial x}\right|_{x_0,u_0} = \begin{pmatrix} \frac{\partial f_1}{\partial x_1} & \frac{\partial f_1}{\partial x_2} & \cdots & \frac{\partial f_1}{\partial x_n} \\ \frac{\partial f_2}{\partial x_1} & \frac{\partial f_2}{\partial x_2} & \cdots & \frac{\partial f_2}{\partial x_n} \\ \vdots & \vdots & \cdots & \vdots \\ \frac{\partial f_n}{\partial x_1} & \frac{\partial f_n}{\partial x_2} & \cdots & \frac{\partial f_n}{\partial x_n} \end{pmatrix} = \begin{pmatrix} a_{11} & a_{12} & \ldots & a_{1n} \\ a_{21} & a_{22} & \ldots & a_{2n} \\ \vdots & \vdots & \ldots & \vdots \\ a_{n1} & a_{n2} & \ldots & a_{nn} \end{pmatrix}, \tag{10.79}$$

$$\mathbf{B} = \left.\frac{\partial f}{\partial u}\right|_{x_0,u_0} = \begin{pmatrix} \frac{\partial f_1}{\partial u_1} & \frac{\partial f_1}{\partial u_2} & \cdots & \frac{\partial f_1}{\partial u_r} \\ \frac{\partial f_2}{\partial u_1} & \frac{\partial f_2}{\partial u_2} & \cdots & \frac{\partial f_2}{\partial u_r} \\ \vdots & \vdots & \cdots & \vdots \\ \frac{\partial f_n}{\partial u_1} & \frac{\partial f_n}{\partial u_2} & \cdots & \frac{\partial f_n}{\partial u_r} \end{pmatrix} = \begin{pmatrix} b_{11} & b_{12} & \ldots & b_{1r} \\ b_{21} & b_{22} & \ldots & b_{2r} \\ \vdots & \vdots & \ldots & \vdots \\ b_{n1} & b_{n2} & \ldots & b_{nr} \end{pmatrix}, \tag{10.80}$$

$$\mathbf{C} = \left.\frac{\partial h}{\partial x}\right|_{x_0,u_0} = \begin{pmatrix} \frac{\partial h_1}{\partial x_1} & \frac{\partial h_1}{\partial x_2} & \cdots & \frac{\partial h_1}{\partial x_n} \\ \frac{\partial h_2}{\partial x_1} & \frac{\partial h_2}{\partial x_2} & \cdots & \frac{\partial h_2}{\partial x_n} \\ \vdots & \vdots & \cdots & \vdots \\ \frac{\partial h_m}{\partial x_1} & \frac{\partial h_m}{\partial x_2} & \cdots & \frac{\partial h_m}{\partial x_m} \end{pmatrix} = \begin{pmatrix} c_{11} & c_{12} & \ldots & c_{1n} \\ c_{21} & c_{22} & \ldots & c_{2n} \\ \vdots & \vdots & \ldots & \vdots \\ c_{m1} & c_{m2} & \ldots & c_{mn} \end{pmatrix}, \tag{10.81}$$

$$\mathbf{D} = \left.\frac{\partial h}{\partial u}\right|_{x_0,u_0} = \begin{pmatrix} \frac{\partial h_1}{\partial u_1} & \frac{\partial h_1}{\partial u_2} & \cdots & \frac{\partial h_1}{\partial u_r} \\ \frac{\partial h_2}{\partial u_1} & \frac{\partial h_2}{\partial u_2} & \cdots & \frac{\partial h_2}{\partial u_r} \\ \vdots & \vdots & \cdots & \vdots \\ \frac{\partial h_m}{\partial u_1} & \frac{\partial h_m}{\partial u_2} & \cdots & \frac{\partial h_m}{\partial u_r} \end{pmatrix} = \begin{pmatrix} d_{11} & d_{12} & \ldots & d_{1r} \\ d_{21} & d_{22} & \ldots & d_{2r} \\ \vdots & \vdots & \ldots & \vdots \\ d_{m1} & d_{m2} & \ldots & d_{mr} \end{pmatrix}. \tag{10.82}$$

These entries are evaluated at the chosen operating point of x_0, u_0.

In the case where the model explicitly includes the disturbance variables $d(t)$, it will be necessary to compute the matrices **E** and **F** defined by

$$\mathbf{E} = \left.\frac{\partial f}{\partial d}\right|_{x_0,u_0,d_0} \tag{10.83}$$

and

$$\mathbf{F} = \left.\frac{\partial h}{\partial d}\right|_{x_0, u_0, d_0}. \tag{10.84}$$

EXAMPLE 10.4.5 (Linearization of a CSTR model). Consider the nonlinear model describing the dynamics of a non-isothermal CSTR. The reaction is first order, $A \longrightarrow B$ and is exothermic. The reaction rate constant k is given by $k_0\,\mathrm{e}^{-E/(RT)}$. The reactor is cooled by coolant at temperature T_C. It is assumed that the physical properties remain constant and that inlet and outlet flows are equal.

The mass and energy balances lead to the following equations:

$$V\frac{\mathrm{d}C_A}{\mathrm{d}t} = FC_{A_i} - FC_A - kVC_A, \tag{10.85}$$

$$V\rho c_P\frac{\mathrm{d}T}{\mathrm{d}t} = F\rho c_P(T_i - T) + kVC_A(-\Delta H_R) - UA(T - T_C). \tag{10.86}$$

These can be rearranged to give

$$\frac{\mathrm{d}C_A}{\mathrm{d}t} = \frac{F}{V}(C_{A_i} - C_A) - kC_A, \tag{10.87}$$

$$\frac{\mathrm{d}T}{\mathrm{d}t} = \frac{F}{V}(T_i - T) + \frac{kC_A}{\rho c_P}(-\Delta H_R) - \frac{UA}{V\rho c_P}(T - T_C). \tag{10.88}$$

Now we linearize the equations above in the following steps:

- *First*, we can classify the variables within the nonlinear state space formulation as

$$x = [C_A, T]^T, \quad y = [C_A, T]^T, \quad u = [F, T_C]^T;$$

  other choices for u are possible.

- *Second*, convert the original nonlinear ODEs into linear state space form.

The partial derivatives with respect to the states x are given by

$$\frac{\partial f_1}{\partial x_1} = \frac{\partial}{\partial C_A}\left(\frac{F}{V}(C_{A_i} - C_A)\right) - \frac{\partial}{\partial C_A}(kC_A) \tag{10.89}$$

$$= -\frac{F}{V} - k_0\,\mathrm{e}^{-E/(RT_0)}, \tag{10.90}$$

$$\frac{\partial f_1}{\partial x_2} = \frac{\partial}{\partial T}\left(\frac{F}{V}(C_{A_i} - C_A)\right) - \frac{\partial}{\partial T}(kC_A) \tag{10.91}$$

$$= -k_0\left(\frac{E}{RT_0^2}\right)\mathrm{e}^{-E/(RT_0)}C_{A_0}. \tag{10.92}$$

$$\frac{\partial f_2}{\partial x_1} = \frac{\partial}{\partial C_A}\left(\frac{F}{V}(T_\mathrm{i} - T)\right) + \frac{\partial}{\partial C_A}\left(\frac{kC_A}{\rho c_P}(-\Delta H_\mathrm{R})\right) \tag{10.93}$$

$$- \frac{\partial}{\partial C_A}\left(\frac{UA}{V\rho c_P}(T - T_\mathrm{C})\right) \tag{10.94}$$

$$= \frac{k}{\rho c_P}(-\Delta H_\mathrm{R}) \tag{10.95}$$

$$= \frac{k_0\,\mathrm{e}^{-E/(RT_0)}}{\rho c_P}(-\Delta H_\mathrm{R}), \tag{10.96}$$

$$\frac{\partial f_2}{\partial x_2} = \frac{\partial}{\partial T}\left(\frac{F}{V}(T_\mathrm{i} - T)\right) + \frac{\partial}{\partial T}\left(\frac{kC_A}{\rho c_P}(-\Delta H_\mathrm{R})\right) \tag{10.97}$$

$$- \frac{\partial}{\partial T}\left(\frac{UA}{V\rho c_P}(T - T_\mathrm{C})\right) \tag{10.98}$$

$$= -\frac{F}{V} + k_0\left(\frac{E}{RT_0^2}\right)\mathrm{e}^{-E/(RT_0)}C_{A_0}\frac{(-\Delta H_\mathrm{R})}{\rho c_P} - \frac{UA}{V\rho c_P}. \tag{10.99}$$

The partial derivatives with respect to the inputs u are

$$\frac{\partial f_1}{\partial u_1} = \frac{\partial}{\partial F}\left(\frac{F}{V}(C_{A_\mathrm{i}} - C_A)\right) - \frac{\partial}{\partial F}(kC_A) \tag{10.100}$$

$$= \frac{(C_{A_\mathrm{i}} - C_{A_0})}{V}, \tag{10.101}$$

$$\frac{\partial f_1}{\partial u_2} = \frac{\partial}{\partial T_\mathrm{C}}\left(\frac{F}{V}(C_{A_\mathrm{i}} - C_A)\right) - \frac{\partial}{\partial T_\mathrm{C}}(kC_A) \tag{10.102}$$

$$= 0, \tag{10.103}$$

$$\frac{\partial f_2}{\partial u_1} = \frac{\partial}{\partial F}\left(\frac{F}{V}(T_\mathrm{i} - T)\right) + \frac{\partial}{\partial F}\left(\frac{kC_A}{\rho c_P}(-\Delta H_\mathrm{R})\right) \tag{10.104}$$

$$- \frac{\partial}{\partial F}\left(\frac{UA}{V\rho c_P}(T - T_\mathrm{C})\right) \tag{10.105}$$

$$= \frac{(T_\mathrm{i} - T)}{V}, \tag{10.106}$$

$$\frac{\partial f_2}{\partial u_2} = \frac{\partial}{\partial T_\mathrm{C}}\left(\frac{F}{V}(T_\mathrm{i} - T)\right) + \frac{\partial}{\partial T_\mathrm{C}}\left(\frac{kC_A}{\rho c_P}(-\Delta H_\mathrm{R})\right) \tag{10.107}$$

$$- \frac{\partial}{\partial T_\mathrm{C}}\left(\frac{UA}{V\rho c_P}(T - T_\mathrm{C})\right) \tag{10.108}$$

$$= \frac{UA}{V\rho c_P}. \tag{10.109}$$

- *Finally*, write the state space equations in deviation variable form as

$$\begin{bmatrix} \dfrac{\mathrm{d}\hat{C}_A}{\mathrm{d}t} \\ \dfrac{\mathrm{d}\hat{T}}{\mathrm{d}t} \end{bmatrix} = \begin{bmatrix} -\dfrac{F}{V} - k_0 \mathrm{e}^{-E/(RT_0)} & -k_0 \left(\dfrac{E}{RT_0^2}\right) \mathrm{e}^{-E/(RT_0)} C_{A_0} \\ +\dfrac{k_0 \mathrm{e}^{-E/(RT_0)}(-\Delta H_R)}{\rho c_P} & -\dfrac{F}{V} + k_0 \left(\dfrac{E}{RT_0^2}\right) \mathrm{e}^{-E/(RT_0)} C_{A_0} \dfrac{(-\Delta H_R)}{\rho c_P} - \dfrac{UA}{V \rho c_P} \end{bmatrix} \times \begin{bmatrix} \hat{C}_A \\ \hat{T} \end{bmatrix} + \begin{bmatrix} \dfrac{(C_{A_i} - C_{A_0})}{V} & 0 \\ \dfrac{(T_i - T_0)}{V} & \dfrac{UA}{V\rho c_P} \end{bmatrix} \begin{bmatrix} \hat{F} \\ \hat{T}_C \end{bmatrix}. \tag{10.110}$$

■■■

10.5. STRUCTURAL GRAPHS OF LUMPED DYNAMIC MODELS

As we have seen in Section 10.3, the model of a process system described by a lumped dynamic model can be easily transformed to the canonical form of concentrated parameter nonlinear state space models. The structure of a continuous-time deterministic dynamic system given in linear time-invariant or time-varying parameter state space form with the Eqs. (10.9) or (10.35) can be represented by a *directed graph* [71] as follows. The nodes of the directed graph correspond to the system variables; a directed edge is drawn from the ith vertex to the jth one if the corresponding matrix element does not vanish. Hence if the ith variable is present on the right-hand side of the jth equation then an edge exists.

10.5.1. Structure Matrices and Structure Indices

The dimensions n, m, r, v of the system variables together with the structure of the state space model matrices $\mathbf{A}, \mathbf{B}, \mathbf{C}, \mathbf{D}, \mathbf{E}, \mathbf{F}$ determine the *structure of the linear time-invariant system*. The integers n, m, r, v are called the *structure indices* of the model. The structure of a general matrix $\mathbf{W}$ is given by the *structure matrix* $[\mathbf{W}]$ whose entries are defined as

$$[w]_{ij} = \begin{cases} 0 & \text{if } w_{ij} = 0, \\ 1 & \text{otherwise.} \end{cases} \tag{10.111}$$

In other words, $[\mathbf{W}]$ is the characteristic matrix of the non-zero entries of $\mathbf{W}$.

For some structural dynamical properties, for example when we consider structural stability we need to note the sign of the matrix elements. For such purposes we apply the so-called signed structure matrices to represent the structure of the system. The signed structure of a matrix $\mathbf{W}$ is given by the *signed structure matrix* $\{\mathbf{W}\}$ whose entries are defined as

$$\{w\}_{ij} = \begin{cases} 0 & \text{if } w_{ij} = 0, \\ + & \text{if } w_{ij} > 0, \\ - & \text{if } w_{ij} < 0. \end{cases} \tag{10.112}$$

Note that the *structure of a nonlinear dynamic concentrated parameter system* can also be described similarly in the following way. If we linearize the nonlinear state space model equations (10.38) in the way it is described in Section 10.4, then we get the linearized state space model matrices **A**, **B**, **C**, **D**, **E**, **F** as in Eqs. (10.79)–(10.82). These matrices can be used to represent the structure of the nonlinear concentrated parameter state space model.

As an example we consider a simple CSTR to illustrate the concepts.

EXAMPLE 10.5.1 (Structure of a CSTR model). Determine the structure of the nonlinear state space model of the CSTR given in Examples 10.3.1 and 10.4.5.

The structure is described by

- *Structure indices*:

$$n = 2, \quad m = 2, \quad r = 2, \quad v = 0. \tag{10.113}$$

- *Signed structure matrices*:
  These matrices are derived from the matrices (10.110), which can be represented in the following form, where all variables $a_{ij}, b_{ij}, c_{ij} > 0$

$$\mathbf{A} = \begin{bmatrix} -a_{11} & -a_{12} \\ +a_{21} & -a_{22} \end{bmatrix}, \quad \mathbf{B} = \begin{bmatrix} -b_{11} & 0 \\ -b_{21} & +b_{22} \end{bmatrix}, \quad \mathbf{C} = \begin{bmatrix} 1 & 0 \\ 0 & 1 \end{bmatrix}. \tag{10.114}$$

  Thereafter, it is easy to derive the sign version of these matrices:

$$\{\mathbf{A}\} = \begin{bmatrix} - & - \\ + & - \end{bmatrix}, \quad \{\mathbf{B}\} = \begin{bmatrix} - & 0 \\ - & + \end{bmatrix}, \quad \{\mathbf{C}\} = \begin{bmatrix} + & 0 \\ 0 & + \end{bmatrix}. \tag{10.115}$$

10.5.2. Structure Graphs

The *system structure* above is represented by a *directed graph* $S = (V, \mathcal{E})$, where the vertex set V is partitioned into four disjoint parts,

$$\begin{aligned} V &= X \cup A \cup D \cup Y, \\ X \cap A &= X \cap D = X \cap Y = \emptyset, \\ A \cap D &= A \cap Y = D \cap Y = \emptyset \end{aligned} \tag{10.116}$$

with X being the set of q state variables, A the set of p input variables, D the set of r disturbances and Y the set of z output variables. All edges $e \in \mathcal{E}$ terminate either in X or in Y, by assumption:

$$v_2 \in (X \cup Y), \quad \forall\, e = (v_1, v_2) \in \mathcal{E},$$

i.e. there are no inward directed edges to actuators and disturbances. Moreover, all edges start in $(A \cup D \cup X)$, again by assumption:

$$v_1 \in (X \cup A \cup D), \quad \forall\, e = (v_1, v_2) \in \mathcal{E}.$$

That is, there are no outward directed edges from outputs of the graph. The graph S is termed the *structure graph of the dynamic system*, or *system structure graph.*

Sometimes, the set $U = A \cup D$ is called the *entrance* and Y the *exit* of the graph S.

Directed paths in the system structure graph can be used to describe the effect of a variable on other variables. A sequence $(v_1, \ldots, v_n)$, $v_i \in V$, forming a directed path in S, is an *actuator path* if $v_1 \in A$, and a *disturbance path* if $v_1 \in D$.

We may want to introduce a more general model, taking into account that the connections between the elements of a process system structure are not necessarily of the same strength. To represent such types of conditions we consider *weighted digraphs* which give full information not only on the structure matrices $[\mathbf{W}]$ but also on the entries of $\mathbf{W}$. In this way, the actual values of the elements in the state space representation matrices $(\mathbf{A}, \mathbf{B}, \mathbf{C}, \mathbf{D}, \mathbf{E}, \mathbf{F})$ can be taken into account.

For this purpose, we define a weight function w whose domain is the edge set $\mathcal{E}$ of the structure graph $S = (V, \mathcal{E})$, assigning the corresponding matrix element in $\mathbf{W}$ as weight

$$w(v_i, v_j) := w_{ij} \tag{10.117}$$

to each edge (v_i, v_j) of S.

For *deterministic state space models without uncertainty*, the edge weights w_{ij} are real numbers. In this way, a one-to-one correspondence is established between the state space model (10.31) and the weighted digraph $S = (V, \mathcal{E}; w)$.

We may put only the sign of the corresponding matrix elements $(\{\mathbf{A}\}, \{\mathbf{B}\}, \{\mathbf{C}\}, \{\mathbf{D}\}, \{\mathbf{E}\}, \{\mathbf{F}\})$ as the weights w_{ij}. In this way, a special weighted directed graph representation of the process structure results and is called the *signed directed graph (SDG)* model of the process system.

EXAMPLE 10.5.2 (The SDG of a CSTR model). The SDG model of the CSTR given in Examples 10.3.1 and 10.4.5 is constructed from the structure indices and signed structure matrices developed in Example 10.5.1 (Eqs. (10.113)–(10.115)).

Let us denote the state vertices by circles, input vertices by double circles and output vertices by rectangles. Then the SDG model of the CSTR is shown in Fig. 10.8.

The Use of Structure Graphs for Analysing Dynamic Properties

Structure graphs and SDG models are simple and transparent tools for representing the structure of process models. They represent not only a particular system they have been derived for but a whole *class of process systems* with the same structure. This property and their simplicity makes them very useful tools for analyzing dynamical properties of a class of process systems.

Most of the analysis methods based on structure graphs and SDGs will be described in Section 13.2. Here, we only give a list of the most important uses of structure graphs and SDGs:

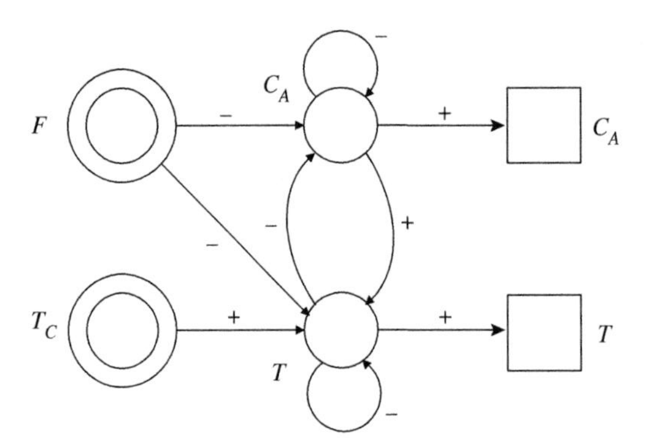

FIGURE 10.8 Structure directed graph of a CSTR.

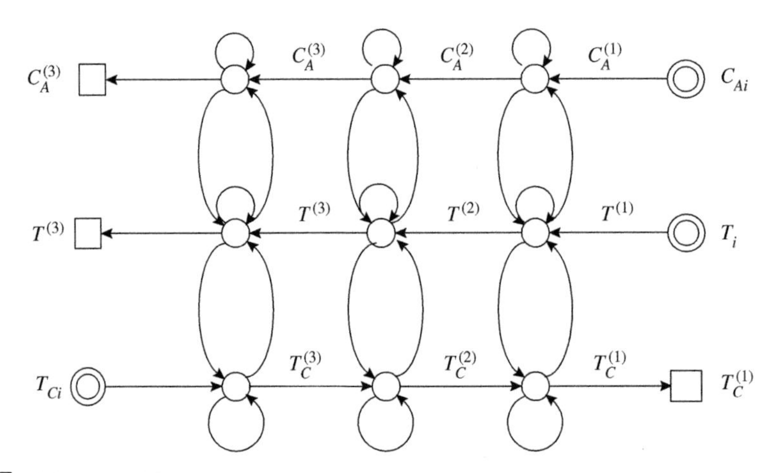

FIGURE 10.9 Structure graph of three jacketted CSTRs.

1. analysis of structural controllability and observability,
2. analysis of structural stability,
3. qualitative analysis of unit step responses.

The following example illustrates the construction of the structure graph of a process system and highlights some of the graph theoretical properties of the graph, which will be of use later on for analysing structural dynamical properties.

EXAMPLE 10.5.3 (The structure graph of the three jacketted CSTR system). Construct the structure graph of the three jacketted CSTR system given in Example 10.3.2 using the nonlinear state space model equations (10.43)–(10.45).

The structure graph of the system consisting of the three-jacketted CSTRs is shown in Fig. 10.9 with the state variables given in Eq. (10.47), with all the potential input variables u^* in Eq. (10.48) but only with a restricted set of outputs

$$y = \left[C_A^{(3)}, T^{(3)}, T_C^{(1)} \right]^T. \tag{10.118}$$

Observe that the state variable vertices span a strong component in the graph, that is a subgraph in which there is a directed path between any ordered pair of vertices ■ ■ ■ (x_i, x_j) such that the path starts at vertex x_i and ends at vertex x_j.

10.6. SUMMARY

In this chapter, we have introduced a number of key techniques which help in the model analysis task. In particular, we looked at the need to be able to represent the process system model which consists of conservation and constitutive equations in a state space form. In the general case this was nonlinear. It was important to decide about several classes of system variables which include inputs, disturbances, states and outputs.

The natural extension of this nonlinear state space form was the linearized version which allows us to apply the whole range of linear systems theory to the model. The validity of the linear model over its intended application range is an important issue. Inappropriate use of a linear model over a wide nonlinear operating space is often unjustified.

The final concept which was introduced involved the representation of the nonlinear or linear state space model in terms of a directed graph. This digraph shows the essential structural details of the system and the digraph can be used as a basis for a wide range of powerful analyses. Several distinct graph forms were discussed since they each contain certain levels of system information which can be exploited for analysis. This includes system stability, controllability and observability.

10.7. REVIEW QUESTIONS

Q10.1. Outline the key contents of a general mathematical problem statement. (Section 10.1)

Q10.2. Sketch the generalized signal flow diagram for a process system. Describe the key flows of the system. (Section 10.2)

Q10.3. What constitutes a linear system model in mathematical terms? (Section 10.2)

Q10.4. What three principal domains can be used to describe the behaviour of a system. What are the principal characteristics of each? (Section 10.2)

Q10.5. What is the significance of the impulse response function and unit step response function in determining the dynamics of a system? (Section 10.2)

Q10.6. Write down the general expressions of the SSR of a process system in both nonlinear and linear forms. For the linear form describe the principal matrices which are used and their significance. (Section 10.2)

Q10.7. Write down the means whereby you can identify states, inputs, disturbances and outputs from a knowledge of the system and the model. (Section 10.3)

Q10.8. Describe why linearization of the nonlinear state space model is often useful in model analysis and application. What role do deviation or perturbation variables play in this form of the model? (Section 10.4)

Q10.9. Describe several types of structural graphs of nonlinear and linear state space models. What are the principal characteristics of these graphs and how can they be used for further analysis of a process system? (Section 10.5)

10.8. APPLICATION EXERCISES

A10.1. Consider the simple nonlinear model where

$$\frac{dh}{dt} = F - C\sqrt{h}. \tag{10.119}$$

Here F and C are constants. Linearize the model and compare the original nonlinear equation (10.119) with the linearized form of equation to observe the effect of linearization and the error involved as we move further away from the linearization point h_0. To do this, solve both models in MATLAB and observe what happens with the deviation of the linear model as h increases. Use values of $F = 0.5$ and $C = 0.75$ with $h(0) = 0.1$. Comment on the results. (Section 10.4)

A10.2. Linearize the Arrhenius reaction rate term given by

$$k = k_0\, e^{-E/(RT)} \tag{10.120}$$

(Section 10.4)

A10.3. Linearize the expression for the flowrate through a valve given by

$$W = C_V\sqrt{P_1 - P_2}, \tag{10.121}$$

where P_1 and P_2 are variables.

Express the answer in deviation variable form. (Section 10.4)

A10.4. For the simple tank model given by the following equations:

$$\begin{aligned} \frac{dh}{dt} &= \frac{(F_1 - F_2)}{A}, \\ 0 &= F_1 - C_V\sqrt{P_1 - P_2}, \\ 0 &= F_2 - C_V\sqrt{P_2 - P_3}, \\ 0 &= P_2 - P_0 - \rho g h. \end{aligned} \tag{10.122}$$

(a) Generate the nonlinear SSR after choosing the system variables
(b) Linearize the system equations around a particular operating point. (Sections 10.4 and 10.3)

A10.5. Linearize the dynamic equations representing the substrate and oxygen dynamics of a wastewater system given by

$$\frac{dS_S}{dt} = \frac{q_F}{V}(S_{SF} - S_S) - \frac{\hat{\mu}_H}{Y_H}\left(\frac{S_S}{K_S + S_S}\right)\left(\frac{S_O}{K_{OH} + S_O}\right)X_H + (1 - f_p)b_H X_H,$$

$$\frac{dS_O}{dt} = \frac{q_F}{V}S_{OF} - \frac{q_F + q_R}{V}S_O + \frac{Y_H - 1}{Y_H}\hat{\mu}_H\left(\frac{S_S}{K_S + S_S}\right)\left(\frac{S_O}{K_{OH} + S_O}\right)X_H$$

$$+ a(1 - e^{-q_A/b})(S_{O,\text{sat}} - S_O), \tag{10.123}$$

where the states are S_S, S_O; dissolved oxygen S_O is the output and q_A is the input. All other variables are constants.

Represent the linear system in deviation variable form. (Section 10.4)

A10.6. Linearize the dynamic balance equations of the three jacketted CSTR system of Example 10.3.2, namely, Eqs. (10.43)–(10.45), with respect to the state variables and the potential input variables given by Eq. (10.48). (Section 10.4)

A10.7. Determine the structure of the nonlinear state space model of the three jacketted CSTR system of Example 10.3.2, namely, Eqs. (10.43)–(10.45) together with the output equation derived from the output variable set (10.118). Show the relationship of this structure with the structure graph seen in Fig. 10.9. (Section 10.5)

A10.8. For the tank equations given in Application Exercise A10.4 produce a structure graph of the system. (Section 10.5)

A10.9. Produce a structure graph of the wastewater dynamic equations which are given in exercise A10.5. (Section 10.5)

A10.10. Determine the structure of the underlying state space model of a system with a structure graph seen in Fig. 10.10. (Section 10.5)

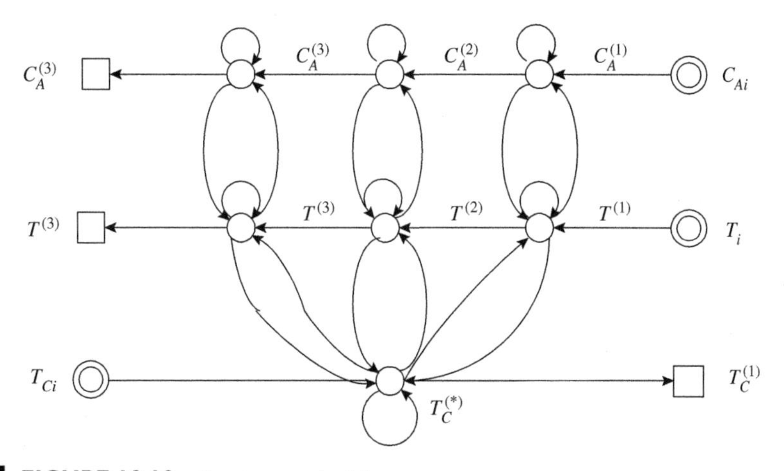

FIGURE 10.10 Structure graph of the system.

A10.11. Produce the LTI state space model matrices and the input response function of the following system model:

$$\frac{dx_1}{dt} = 7u_1, \tag{10.124}$$

$$\frac{dx_2}{dt} = 5x_1 + -6x_2 + 8u_3, \tag{10.125}$$

$$y_1 = x_1, \tag{10.126}$$

$$y_2 = x_2. \tag{10.127}$$

Produce the structure graph of the system. (Sections 10.2 and 10.5)

11 DATA ACQUISITION AND ANALYSIS

There are a number of key steps in model building, namely model calibration and validation and a number of key application areas, like process control and diagnosis which depend critically on the quality of the available measured data. This chapter gives a brief process oriented introduction to data acquisition and data analysis for modelling, control and diagnosis purposes.

The outline of the chapter is as follows:

- Often we develop a continuous time process model to be used for model calibration and validation or control. The majority of estimation and/or control methods require discrete time models. The sampling or discretization of usually continuous time process models as an important model transformation step for process parameter and structure estimation is outlined in Section 11.1.
- The quality of the estimates using model parameter and structure estimation and sometimes even the success of these procedures depends critically on the quality of the measurement data. Therefore, methods for assessing the quality of collected data, i.e. *data screening* is briefly discussed in Section 11.2.
- Most often we do not use measured data as collected, i.e. *passive data* from the process system. We usually artificially create data records specially for the estimation purposes. The procedures we use to design which experiments to be carried out and how to get proper data for the estimation are called *experimental design procedures*. We shall review experimental design for static models, i.e. design of steady-state data and experiment design for dynamic purposes in the two subsequent Sections 11.3 and 11.4.

Although it is important to review the most commonly used experiment design and data screening methods, no attempt is made in this section to give a detailed and comprehensive description of them. The most important steps and procedures are discussed from the viewpoint of process model calibration and validation. There are a number of good books that interested readers may consult [72,73].

11.1. SAMPLING OF CONTINUOUS TIME DYNAMIC MODELS

Process models are usually constructed from balance equations with suitable constitutive equations. The balance equations can be either ordinary or partial differential equations depending on the assumptions on the spatial distribution of process variables. In order to solve balance equations, we need to discretize them somehow both in space (if needed) and in time. Spatial discretization methods have been discussed in Chapter 8 on the solution of PDE models. This section discusses only discretization in time assuming that we have already lumped our PDE model as needed. In other words, *only lumped process models are considered here*.

The lumped balance equations are naturally continuous time differential equations whereas almost any known method in mathematical statistics works with discrete sets of data and uses underlying discrete time models. Therefore, the need to transform continuous time process models into their discrete time counterparts naturally arises. This type of *time discretization is called sampling*.

Almost any kind of data acquisition, data logging or control system is implemented on computers where continuous time signals are *sampled* by the measurement devices. Control signals, generated by the computer system in discrete time instances are fed back to the process system to control it. Therefore, it is convenient to generate the discrete counterpart of a process model by simply forming a composite system from the original continuous time process system, the measurement devices taking the sampled output signals and from the actuators generating the continuous time manipulated input signals to the system as shown in Fig. 11.1. The box labelled S is the original continuous time process system, the box *D/A converter* converts continuous time signals to discrete time ones and the box *A/D converter* converts discrete time signals to continuous time ones. The sampled data discrete time composite system $S^{(\mathrm{d})}$ is in the dashed line box. The discrete time input and output variables (or signals) to the discrete time system $S^{(\mathrm{d})}$ are

$$u : \{u(k) = u(t_k) \mid k = 0, 1, 2, \ldots\}, \tag{11.1}$$

$$y : \{y(k) = y(t_k) \mid k = 0, 1, 2, \ldots\}, \tag{11.2}$$

where $T = \{t_0, t_1, t_2, \ldots\}$ is the discrete time sequence.

Most often *equidistant zero-order hold sampling* is applied to the system, which means that we sample (measure) the continuous time signals and generate the manipulated discrete time signals in regular equidistant time instances, i.e.

$$t_{k+1} - t_k = h, \quad k = 0, 1, \ldots, \tag{11.3}$$

where the constant h is the sampling interval. Moreover, a zero-order hold occurs in the *D/A converter* to generate a continuous time manipulated input signal $u(t)$ from

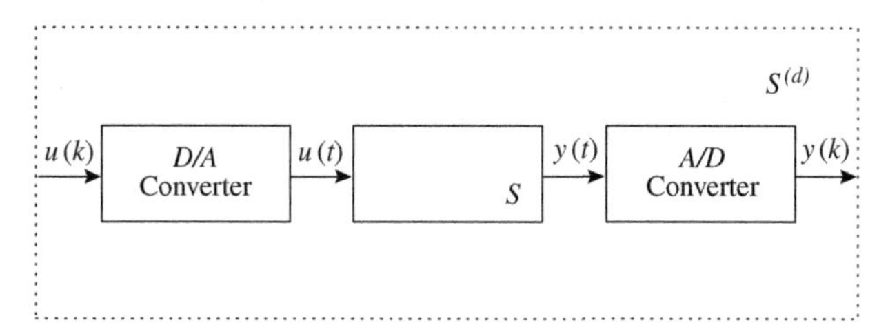

FIGURE 11.1 Signal flow diagram of a sampled data system.

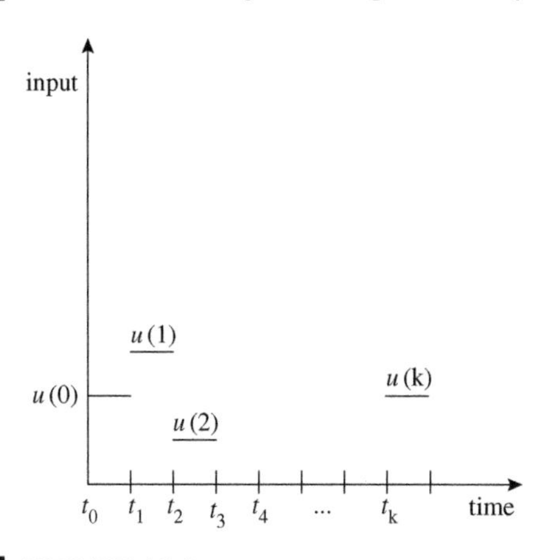

FIGURE 11.2 Equidistant zero-order hold sampling.

the discrete time one ($u(k)$):

$$u(\tau) = u(k), \quad \text{for all } \tau \in [t_k, t_{k+1}). \tag{11.4}$$

Equidistant zero-order hold sampling is illustrated in Fig. 11.2.

Now we can formulate the problem of sampling of continuous time dynamic models as follows.

Sampling Continuous Time Models

Given:
A parametrized dynamic continuous time system model in the form of

$$y = \mathcal{M}(x, u; p^{(M)}) \tag{11.5}$$

with the model parameters $p^{(M)} \in \mathcal{R}^w$ and vector-valued continuous time state and input variables $x(t) \in \mathcal{R}^n$, $u(t) \in \mathcal{R}^r$. The output variable y can also be vector-valued, i.e. $y(t) \in \mathcal{R}^m$.

Compute:
The discrete time sampled version of the model (11.5)

$$y = \mathcal{M}^{(d)}(x, u;\ p^{(Md)}) \tag{11.6}$$

with the model parameters $p^{(Md)} \in \mathcal{R}^{w_d}$ and vector-valued discrete time state and input variables $x(k) \in \mathcal{R}^n$, $u(k) \in \mathcal{R}^r$. The output variable y is a vector-valued discrete time signal, i.e. $y(k) \in \mathcal{R}^m$ such that $u(k)$ and $y(k)$ are the sampled versions of the continuous time signals $u(t)$ and $y(t)$ at any given sampling time t_k.

In the general case, we should transform a continuous time process model describing the continuous time system $\boldsymbol{S}$ to its discrete time sampled version in the following steps:

1. Take the sampled discrete time signals of the input and output signals.
2. Make a finite difference approximation (FDA) of the derivatives in the model equations by using Taylor series expansion of the nonlinear operators if needed.

For process models transformed into a LTI state space model form we have ready and explicit solution to the above problem [74].

LEMMA 11.1.1. *Let us consider a process model in the usual LTI state space form:*

$$\frac{\mathrm{d}x(t)}{\mathrm{d}t} = \mathbf{A}x(t) + \mathbf{B}u(t), \qquad y(t) = \mathbf{C}x(t) + \mathbf{D}u(t)$$

with the constant matrices $(\mathbf{A}, \mathbf{B}, \mathbf{C}, \mathbf{D})$. *Then, its sampled version using equidistant zero-order hold sampling with sampling time h is a discrete time LTI state space model in the form*

$$x(k+1) = \mathbf{\Phi}x(k) + \mathbf{\Gamma}u(k), \qquad y(k) = \mathbf{C}x(k) + \mathbf{D}u(k), \tag{11.7}$$

where the constant matrices $(\mathbf{\Phi}, \mathbf{\Gamma}, \mathbf{C}, \mathbf{D})$ *in the discrete time model are*

$$\mathbf{\Phi} = \mathrm{e}^{\mathbf{A}h} = \left(I + h\mathbf{A} + \frac{h^2}{2!}\mathbf{A}^2 + \cdots\right), \tag{11.8}$$

$$\mathbf{\Gamma} = \mathbf{A}^{-1}(\mathrm{e}^{\mathbf{A}h} - \mathbf{I})\mathbf{B} = \left(h\mathbf{I} + \frac{h^2}{2!}\mathbf{A} + \frac{h^3}{3!}\mathbf{A}^2 + \cdots\right)\mathbf{B}. \tag{11.9}$$

If numerical values of the continuous time model matrices $(\mathbf{A}, \mathbf{B}, \mathbf{C}, \mathbf{D})$ are given, then there are ready MATLAB functions to compute the matrices $(\mathbf{\Phi}, \mathbf{\Gamma}, \mathbf{C}, \mathbf{D})$. Most often, if h is small enough, it is sufficient to consider only the first order (containing h but not its higher powers) approximation of the Taylor series in Eqs. (11.8)–(11.9).

Note that the solution of the sampling problem described in Lemma 11.1.1 is just a special case of the SAMPLING CONTINUOUS TIME MODELS problem for continuous time LTI state space models with $p^{(M)} = \{\mathbf{A}, \mathbf{B}, \mathbf{C}, \mathbf{D}\}$ resulted in a discrete time LTI state space model (11.7) with the parameters $p^{(Md)} = \{\mathbf{\Phi}, \mathbf{\Gamma}, \mathbf{C}, \mathbf{D}\}$.

The sampling of a continuous time linearized state space model of a CSTR introduced in Example 10.4.5 serves as an illustration to the above.

EXAMPLE 11.1.1 (Sampled linearized state space model of a CSTR). Consider the CSTR described in Example 10.4.5 with its linearized state space model in Eq. (10.110) and assuming full observation of the state variables. Derive the sampled version of the model assuming zero-order hold equidistant sampling with sampling rate h.

The state space model matrices in symbolic form are as follows:

$$\mathbf{A} = \begin{bmatrix} -\dfrac{F}{V} - k_0 \mathrm{e}^{-E/(RT_0)} & -k_0\left(\dfrac{E}{RT_0^2}\right)\mathrm{e}^{-E/(RT_0)} C_{A_0} \\ -\dfrac{k_0 \mathrm{e}^{-E/(RT_0)}(-\Delta H_\mathrm{R})}{\rho c_P} & -\dfrac{F}{V} - k_0\left(\dfrac{E}{RT_0^2}\right)\mathrm{e}^{-E/(RT_0)} C_{A_0}\dfrac{(-\Delta H_\mathrm{R})}{\rho c_P} - \dfrac{UA}{V\rho c_P} \end{bmatrix}, \tag{11.10}$$

$$\mathbf{B} = \begin{bmatrix} \dfrac{(C_{A_\mathrm{i}} - C_{A_0})}{V} & 0 \\ \dfrac{(T_\mathrm{C} - T)}{V} & \dfrac{F}{V} + \dfrac{UA}{V\rho c_P} \end{bmatrix}, \tag{11.11}$$

$$\mathbf{C} = \mathbf{I}, \quad \mathbf{D} = 0. \tag{11.12}$$

Applying zero-order hold equidistant sampling to the model matrices above with the first-order approximation in Eqs. (11.8)–(11.9) we obtain

$$\mathbf{\Phi} = \begin{bmatrix} 1 - \left(\dfrac{F}{V} - k_0 \mathrm{e}^{-E/(RT_0)}\right)h & -hk_0\left(\dfrac{E}{RT_0^2}\right)\mathrm{e}^{-E/(RT_0)} C_{A_0} \\ -h\dfrac{k_0 \mathrm{e}^{-E/(RT_0)}(-\Delta H_\mathrm{R})}{\rho c_P} & 1 - h\left(\dfrac{F}{V} - k_0\left(\dfrac{E}{RT_0^2}\right)\mathrm{e}^{-E/(RT_0)} C_{A_0}\dfrac{(-\Delta H_\mathrm{R})}{\rho c_P} - \dfrac{UA}{V\rho c_P}\right) \end{bmatrix}, \tag{11.13}$$

$$\mathbf{\Gamma} = \begin{bmatrix} \dfrac{(C_{A_\mathrm{i}} - C_{A_0})}{V}h & 0 \\ \dfrac{(T_\mathrm{C} - T)}{V}h & \left(\dfrac{F}{V} + \dfrac{UA}{V\rho c_P}\right)h \end{bmatrix}, \tag{11.14}$$

■ ■ ■ and Eq. (11.12) remains unchanged.

11.2. DATA SCREENING

If we have collected any real data either steady-state or dynamic, we have to assess the quality and reliability of the data before using it for model calibration or validation. Data screening methods are used for this purpose assuming that we have a set of measured data

$$D[1, k] = \{d(1), d(2), \ldots, d(k)\} \tag{11.15}$$

with vector-valued data items $d(i) \in \mathcal{R}^\nu$, $i = 1, \ldots, k$ arranged in a sequence according to the time of the collection (experiment). The term "screening" indicates that this prior assessment should be as simple and as effective as possible.

Data deviations can take the following form:

- gross errors due to equipment failures, malfunctions or poor sampling,
- outliers,
- jumps or sudden changes,
- trends.

Data screening is a passive process in nature, i.e. if poor quality data are detected then it is usually better *not* to use them and preferably repeat the experiment than to try to "repair" them by some kind of filtering.

11.2.1. Data Visualization

The most simple and effective way of data screening is visual inspection. This is done by plotting the collected set of measured data against

- time or sequence number (time domain),
- frequency (frequency domain),
- one another.

When we plot data on a single signal, that is on a time dependent variable *against time or sequence number*, then we get visual information on

- trends and may associate them with shift changes, seasonal changes or slow drift due to some equipment changes,
- outliers, gross errors or jumps detected just from the pattern.

It is important to visualize the data using different scales on both the time and the magnitude axis.

If we plot several signals onto the same plot against time, then we get information about joint disturbances which affect some or all of them in the same way. This plot carries diagnostic information about the nature of the disturbances affecting the quality of the data.

We may want to transform our steady-state data to the *frequency domain* by applying Fourier transformation. The frequency plot of a signal carries information about periodic changes which may be present in the data. The time period of the change is shown as a peak in the frequency plot.

It is important to know that white noise signals, that is signals consisting of independent elements have a uniform frequency plot. Coloured noise signals, that is signals with non-zero auto-correlation coefficients, will have a smooth, peak-free non-uniform frequency plot.

If we plot *one signal against another one*, then we may discover

- cross-correlation, and/or
- linear dependence

between them.

Visualization helps in identifying quickly apparently abnormal data which does not conform to the usual patterns. There are good references to data visualization [72].

As a supplement to data visualization plots *simple statistics, such as*

- *signal mean value,*
- *auto-correlation coefficients,*
- *empirical histograms on signal value distribution*

are also usually computed and evaluated.

Example time-plots of data records of real-valued data signals are shown in Figs. 11.3 and 11.4 for data visualization purposes.

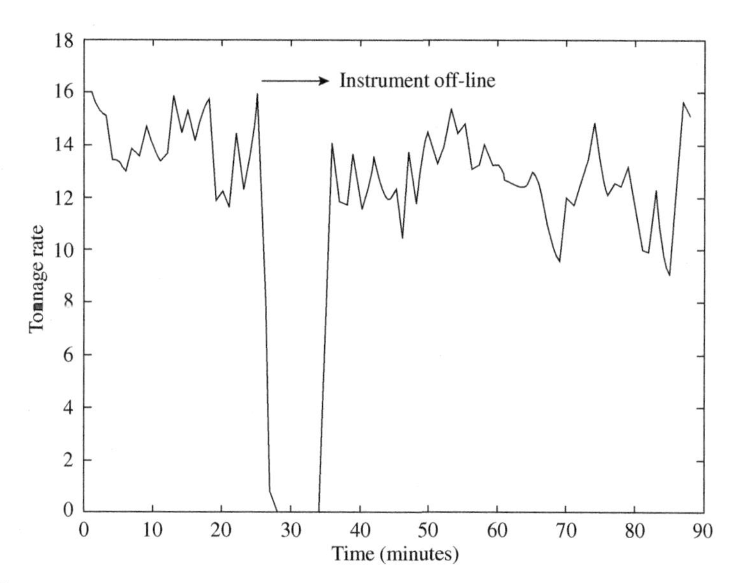

FIGURE 11.3 Data visualization example No. 1.

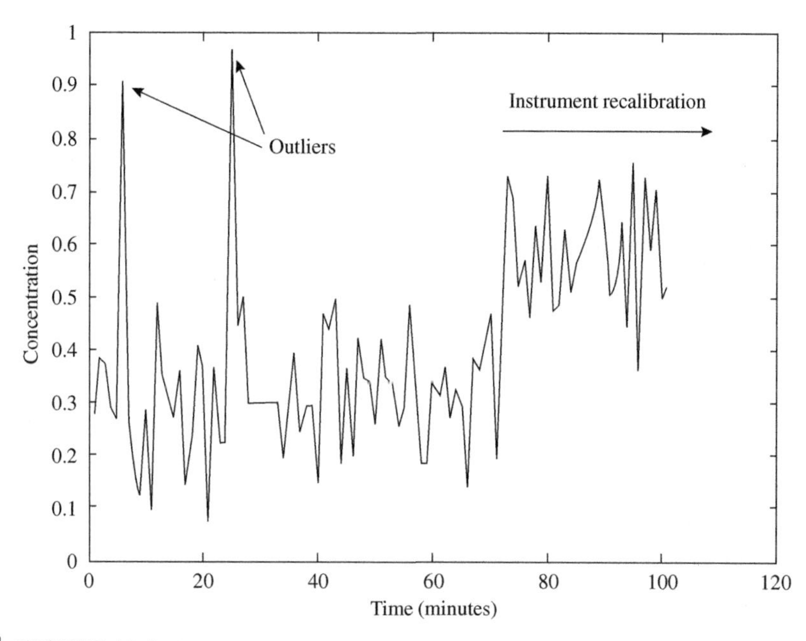

FIGURE 11.4 Data visualization example No. 2.

Two jumps forming a gross error are seen in Fig. 11.3. The reason for the gross error was a measurement device failure: the corresponding automated sensor went off-line when the first jump was observed and then went back online causing the second jump.

The first part (in time) of the plot seen in Fig. 11.4 shows two big outliers and a slow trend. These anomalies initiated a recalibration of the corresponding instrument which is seen in the form of a positive jump on the plot.

11.2.2. Outlier Tests

The notion of an outlier has no definite mathematical meaning in itself because samples taken from a random variable with finite mean and variance, such as from a Gaussian distribution may take very large value (theoretically of any magnitude) with a very small probability. Therefore, we can regard a measured data point as an outlier if its relative magnitude $\|d(i) - \overline{d}\|$ with $\overline{d}$ as the sample mean and $\| \cdot \|$ as a vector norm is much larger than the variance of the measurement errors σ_{ε}.

Apparent outliers are visible by data visualization but there are several outlier tests we may apply (see [73]).

The outlier test methods are based on two different principles: either they are looking for deviations from the "usual" distribution of the data (*statistical methods*) or they detect outlier data by *limit checking*. In both cases, we may apply the *fixed or the adaptive version* of the methods depending on whether the normal (non-outlier) data statistics are given *a priori* or they should be first estimated from test data.

1. *Statistical methods*
   In this case, the outlier detection is based on probability distribution tests. Assume we know the probability density function $p_d(x)$ of the data and have a set of measured data $D[1, k]$. If the data items are independent of each other, then we can regard the data set as a sample in a statistical sense and perform a probability distribution test to check the hypothesis "Are these data from the same underlying distribution with probability density function $p_d(x)$?".
   In the case of a Gaussian distribution the probability distribution test can be performed using the well-known χ^2-test.
2. *Limit checking*
   With these methods the signal sample mean $\overline{d}$ and the threshold for outlier detection $\Delta_d = C_{\Delta}\sigma_{\varepsilon}$ is either given *a priori* or it is estimated from time to time over a sliding data window, given a fixed large constant $C_{\Delta} > 3$.

11.2.3. Trends, Steady-State Tests

If we want to perform a parameter and/or structure estimation of static models, then we need to have steady-state data. This means no long or short term time variation in the data sequence. *Steady state or stationarity includes the absence of trends.*

There can be other circumstances when we want to have data with no trends. This can be when trends are due to slow disturbances. An example would be seasonal changes which we do not want to model.

Trend detection and removal is therefore an important data screening procedure and can be an efficient and simple process system diagnosis tool.

The trend detection or steady-state test methods can be divided into two groups: methods based on parameter estimation and methods based on other statistics.

1. *Methods based on parameter estimation*
   The simplest way to detect trends is to fit a straight line through the measured data and check whether its slope is zero. If the measurement errors are independent of each other and are normally distributed, then standard statistical hypothesis tests (see Section A.2 in the appendix) can be applied to check their constant mean.

2. *Methods based on other statistics*
   The most well known and commonly used method is the so-called CUmulative SUMmation (CUSUM) method. CUSUM is a recursive method which is based on a recursive computation of the sample mean with growing sample size:

$$s[k] = \frac{1}{k}\sum_{i=1}^{k} d(i) = \frac{1}{k}\left((k-1)\, s\,[k-1] + d(k)\right).$$

   This cumulative sum based mean is then plotted against time and inspected either by limit checking or by parameter estimation whether it has any slope different from zero. It is important to note that the statistics $s[k]$ of the CUSUM method will have decreasing variance in time assuming data with the same and constant distribution. Therefore, it becomes increasingly sensitive for even small trends as time grows. Slow periodic changes, however, are simply filtered out together with random measurement errors using these statistics. Therefore, CUSUM is applicable only if a trend of constant slope sign is expected.

More advanced methods on trend detection or testing the steady state can be found in [73].

11.2.4. Gross Error Detection

Gross errors are caused by equipment failures or malfunctions. They can be detected by different methods depending on the nature of the malfunctioning and on the presence of other "healthy" nearby signals. The available methods for gross error detection can be grouped as follows:

1. *Bias or slow trend detection*
   Gross errors can be visible bias or slow trend in the recorded signal of the sensor or sensors related to the equipment in question. Trend detection methods (see above) can then be applied to detect them.
2. *Jump detection*
   In case of sudden failure, a jump arises in the corresponding signal(s) which can be detected via bias or jump detection in time series by standard methods [75]. The simplest methods are similar to those described in Section 11.2.2 for detecting outliers. For jump detection, however, only the fixed *a priori* configured version of the methods is applicable.
3. *Balance violation detection*
   The group of related signals of the equipment subject to failure or malfunction can also be used for detecting gross error causing the violation of the causal deterministic relationship between them. Static mass and energy balances may serve for this purpose. These balances impose simple partial linear relationships between the related signals of the equipment which can be used for gross error detection [76].

11.3. EXPERIMENT DESIGN FOR PARAMETER ESTIMATION OF STATIC MODELS

The parameter estimation for static models requires steady-state (or stationary) data because of the model. Therefore, the first step in designing experiments is to ensure and test that the system is in steady state. This is done by *steady-state tests* as described in Section 11.2.3.

Thereafter, we decide the number of measurements, the spacing of the test points and the sequencing of the test points.

11.3.1. Number of Measurements

The number of measurements depends on the number of test points k_P and on the number of repeated measurements in the test points k_R. One set of repeated measurements at a particular test point is called an experiment. If we repeat measurements at the test points, then we may have a good estimate on the variances of the measurement errors which is very useful to assess the fit of the estimate (see in Section 12.2).

The number of test points depends critically on the number of state variables of the process system and on the number of estimated parameters. In general, we need to make sufficient measurements to estimate unknown parameters and possibly unknown states, i.e. the number of measurements should be significantly greater than the number of unknown parameters and states.

11.3.2. Test Point Spacing

It is important that we have a set of measurements which "span" the state space of the process system the model is valid for. This means that we have to space experimental measurement points roughly uniformly over the validity region of the process model we are going to calibrate or validate.

It is equally important that we stay within the validity region of our process model. For linear or linearized models, this validity region can be quite narrow around the nominal operating point of the model. We may wish to perform linearity tests to find out this validity region before we collect data for model calibration and validation.

11.3.3. Test Point Sequencing

For static models and process systems in steady-state, the sequence of measurements should not affect the result of parameter estimation. That is, because measurement errors of the individual measurements are usually independent and equally distributed with zero mean.

For some kinds of process systems, however, we can achieve the above nice properties of independence and zero mean only by *randomization of the measurements*. This artificially transforms systematic errors (biases, trends, etc.) into random measurement errors. These kinds of process systems include

- systems with "bio" processes which may have a "memory-effect",
- systems which are affected by drifts or slow changes due to unremovable disturbances such as daily or seasonal changes.

11.4. EXPERIMENT DESIGN FOR PARAMETER ESTIMATION OF DYNAMIC MODELS

The experimental design for parameter and structure estimation of dynamic models involves a number of additional issues compared with static models. The reason for this is that we can only design the input variable part of the independent variables in the model and then the output variables are determined by the dynamic response of the process system itself. In other words, *the items in the measurements $d(i)$, $i = 1, \ldots, k$ are partitioned into items belonging to input variables and output variables, i.e.*

$$D[1, k] = \{d(i) \mid d(i) = [y(i), u(i)]^T, \ i = 1, \ldots, k\} \tag{11.16}$$

with vector-valued data items $d(i) \in \mathcal{R}^{\nu}$, $y(i) \in \mathcal{R}^{m}$, $u(i) \in \mathcal{R}^{r}$ and $\nu = m + r$. Moreover, the sequence number of the particular measured data should be interpreted as discrete time. Therefore we cannot apply *Test point sequencing*. Moreover, *Test point spacing* can only be done in a constrained way taking into account the dynamics of the system.

The additional important issues in experiment design for parameter and/or structure estimation of dynamic models include

- sampling time selection,
- excitation.

These are now discussed.

11.4.1. Sampling Time Selection

Proper sampling of continuous time signals and process models is essential in the case of parameter and/or structure estimation of dynamic models. A detailed discussion on sampling is given in Section 11.1 where we only deal with sampling time selection.

The selection of the sampling time is closely connected with the selection of the number of measurements. *We want to have sufficiently rapid sampling for a sufficient length of time.*

Moreover, we want to have information about all time response characteristics of our dynamic process model, even about the fastest one which is usually the hydraulic effects or the component mass variations. For this reason *we have to select the sampling time to be roughly one third or one quarter of the fastest time response of the process system*, which is usually in the order of seconds for process systems.

11.4.2. Excitation

The dynamics of the process system should be sufficiently excited if we want to estimate the parameters in a dynamic model. Although disturbances and other noise sources are often present, the magnitude and frequency of their variation is usually not enough to give a good signal-to-noise ratio for real plant measurements. Therefore, *we should apply some kind of test signal on the inputs of the process system to generate sufficient excitation.* The lack of sufficient excitation may result in a highly correlated set of estimated parameters.

The most frequently applied signal is the so-called *pseudo-random binary sequence (PRBS)* which gives reasonable excitation at very modest amplitudes but requires a relatively long time period and does not disturb the normal operation of the process system too much.

The PRBS has only two signal magnitudes and jumps at randomly selected time intervals from one to the other.

In designing the PRBS test signal [77],

- the base time interval and sampling time should be about one-fifth of the smallest time constant you wish to identify;
- the sequence size should be chosen such that the sequence length is about five times larger than the major time constant of the process system.

More about sufficient excitation, parameter estimation, identifiability and test signal design can be found in [77].

11.5. SUMMARY

The chapter gives a brief introduction to the preparatory steps of statistical model calibration and validation.

The sampling of continuous time models prepares discrete time equivalents of the developed process model needed for the model calibration and validation.

The basic methods and procedures for data analysis and preprocessing such as data screening by visualization, steady-state detection, gross error and outlier detection as well as experiment design for both static and dynamic process models are outlined.

11.6. FURTHER READING

Data acquisition and analysis is an important preliminary step in model calibration and validation, and it is an important area with its own literature. Further reading in this area is therefore only directed towards the aspects important from the viewpoint of process modelling.

An interesting sub-field of data acquisition is the active learning from process data in a process modelling context which has been reported in a recent paper [78].

The calibration and re-calibration of chemical sensors is a vitally important issue in obtaining good quality and reliable process data. This sub-field of data acquisition and analysis mainly belongs to the field of analytical chemistry. An advanced calibration method based on signal processing and curve fitting is reported in [79].

Data analysis is especially important when the standard Gaussian distribution assumption of the measurement and other errors is known to be invalid. This is usually the case for various faults and failures and has a special importance for risk analysis. Special data analysis techniques are used to retrieve univariate distributions from data for risk analysis [80].

11.7. REVIEW QUESTIONS

Q11.1. Describe briefly the available data screening procedures. Which ones are applicable for a single signal and which ones for groups of signals? (Section 11.2)

Q11.2. Characterize briefly the outlier test methods. Which ones are applicable for steady-state signals and which ones for slowly changing signals? (Section 11.2)

Q11.3. Describe briefly the principles behind the available gross error detection methods. Describe the relation between them and the trend detection and outlier detection methods. (Section 11.2)

Q11.4. What are the main points or questions to be considered when designing experiments for parameter estimation for static models? (Section 11.3)

Q11.5. What are the additional main points or questions to be considered compared with the static case when designing experiments for parameter estimation for dynamic models? (Section 11.3)

11.8. APPLICATION EXERCISES

A11.1. Given a continuous time LTI system model in the form of

$$\frac{dx}{dt} = \mathbf{A}x + \mathbf{B}u, \quad y = \mathbf{C}x$$

with the matrices

$$\mathbf{A} = \begin{bmatrix} 3 & 4 \\ 5 & 6 \end{bmatrix}, \quad \mathbf{B} = \begin{bmatrix} 7 \\ 8 \end{bmatrix}, \quad \mathbf{C} = [1 \quad 0]. \tag{11.17}$$

Compute the sampled version of the above model. (Section 11.1)

A11.2. Investigate the plots in Figs. 11.3 and 11.4 by visual inspection and application of data screening methods. Mark clearly on the figure the time instance(s) and signals(s) where you detect:

1. drift (slow change),
2. outliers,
3. gross errors.

(Section 11.2.)

A11.3. Imagine the branching of a pipe in a piping system. Assume that an incompressible fluid flows in the pipe in the direction of the branch where it divides into two streams. The ratio between the cross sections in the pipes is 0.3:0.7 with the cross section of the main pipe being 1. Give the static balances you can use for gross error detection. Which sensors do you need to apply the method? What kind of malfunctioning can you detect, if

1. you know precisely the ratio between the cross section of the branched pipes as above,
2. you do not know the exact ratio between them.

(Section 11.2.4.)

12 STATISTICAL MODEL CALIBRATION AND VALIDATION

Model validation is one of the most difficult steps in the modelling process. It needs a deep understanding of modelling, data acquisition as well as basic notions and procedures of mathematical statistics. The aim of the chapter is to provide the modeller with a user-oriented and hopefully also user-friendly basic understanding of the most commonly used methods in statistical model calibration and validation.

The outline of the chapter is as follows:

- The chapter starts with a revised and extended view of grey-box modelling in section 12.1 as a motivation for paying greater attention to parameter and structure estimation in process modelling. The concept of model calibration is introduced.
- Statistical model calibration is usually done through model parameter and/or structure estimation. The basic notions about model parameter and structure estimation are described in Section 12.2. The underlaying notions and methods from mathematical statistics are given in Section A.1 in the appendix.
- Thereafter, the model parameter and structure estimation is discussed separately for static and dynamic process models in Sections 12.3 and 12.4.
- Finally, model validation methods based on either model prediction or model parameter estimation are discussed in Section 12.6.

12.1. GREY-BOX MODELS AND MODEL CALIBRATION

In practical cases, we most often have an incomplete model if we build a model from first principles according to Steps 1–4 of the SEVEN STEP MODELLING PROCEDURE described in Section 2.3. The reason for this is that we rarely have a complete model together with all the parameter values for all of the controlling mechanisms involved. An example of this is a case when we have complicated reaction kinetics, where we rarely have all the reaction kinetic parameters given in the literature or measured independently and often the form of the reaction kinetic expression is only partially known.

This section deals with incomplete or grey-box models and with the procedure called model calibration to make them solvable for dynamic simulation purposes.

12.1.1. Grey-box Models

The notion of grey-box modelling has been understood somewhat differently by various authors in the literature since the first book of Bohlin [81] appeared with this title. It is used in contrast to the so-called *empirical or black-box models when the model is built largely from measured data using model parameter and/or structure estimation techniques* (see for example [14,82]). The opposite case is the case of *white-box models when the model is constructed only from first engineering principles with all its ingredients known* as a well-defined set of equations which is mathematically solvable. In practice, of course, no model is completely "white" or "black" but all of them are "grey", since practical models are somewhere in between.

Process models developed using first engineering principles but with part of their model parameters and/or structure unknown will be termed grey-box models in the context of this book.

In order to illustrate how a grey-box model may arise consider a case of a simple CSTR.

EXAMPLE 12.1.1 (A grey-box model of a CSTR). Consider a CSTR similar to that of Example 10.3.1 with an exothermic reaction $A \longrightarrow B$. The reaction rate constant k is given by $k_0 e^{-E/(RT)}$. The reactor is cooled by coolant at temperature T_C. It is assumed that inlet and outlet flows are equal, but we do not exactly know the form of the reaction kinetic expression

$$r_A = kc(C_A). \tag{12.1}$$

The functional form of c is unknown and we assume that the physico-chemical properties depend on the temperature T but in an unspecified way, i.e.

$$c_P = c_P(T), \quad \rho = \rho(T), \quad \Delta H_R = \Delta H_R(T). \tag{12.2}$$

Describe the grey-box model of the reactor.

The rearranged mass and energy balances lead to the following model equations:

$$\frac{dC_A}{dt} = \frac{F}{V}(C_A^{(i)} - C_A) - r_A, \tag{12.3}$$

$$\frac{dT}{dt} = \frac{F}{V}(T^{(i)} - T) - \frac{r_A}{\rho c_P}(-\Delta H_R) - \frac{UA}{V\rho c_P}(T - T_C). \tag{12.4}$$

Now the *grey-box model of the CSTR* consists of the two balances above (12.3)–(12.4) with the following unknown structural elements:

1. reaction kinetic expression as in Eq. (12.1),
2. property relations as in Eqs. (12.2).

■ ■ ■

12.1.2. Model Calibration

It follows from the above that we often do not have available values of the model parameters and/or part of the model structure. Therefore, we want to obtain these model parameters and/or structural elements using experimental data from the real process. Because measured data contain measurement errors, we can only *estimate* the missing model parameters and/or structural elements. This estimation step is called *statistical model calibration.*

From the above description, we can conclude that the model calibration is performed via a model parameter and/or structure estimation using

- the developed grey-box model by the steps 1–4 of the SEVEN STEP MODELLING PROCEDURE,
- measured data from the real process system which we call *calibration data,*
- a predefined measure of fit, or *loss function* which measures the quality of the process model with its missing parameters and/or estimated structural elements.

A more precise list of the ingredients above will be given in the problem statement of model parameter and structure estimation in the next section.

Conceptual Steps of Model Calibration

In realistic model calibration there are other important steps which one should carry out besides just a simple model parameter and/or structure estimation. These steps are needed to check and to transform the grey-box model and the measured data to a form suitable for the statistical estimation and then to check the quality of the obtained model. These conceptual steps to be carried out when doing model calibration are as follows:

1. *Analysis of model specification*
   Here, we have to consider all of the ingredients of our grey-box process model to determine which parameters and/or structural elements need to be estimated to make the process model equations solvable for generating their solution as a vector for static models or as a vector-valued function of time for dynamic models. This step may involve a DOF analysis and the analysis of the non-measurement data available for the model building.
2. *Sampling of continuous time dynamic models*
   Statistical procedures use a *discrete set of measured data* and a model. To get an estimate therefore we need to discretize our grey-box process models to be able to estimate its parameters and/or structural elements (see Section 11.1 for details).

3. *Data analysis and preprocessing*
   Measurement data from a real process system are usually of varying quality. We may have data with bias, outliers or large measurement errors due to some malfunctions in the measurement devices or unexpectedly large disturbances. From the viewpoint of good quality estimates it is vital to detect and remove data of unacceptable quality. (The methods for this are described in Section 11.2).
4. *Model parameter and structure estimation*
5. *Evaluation of the quality of the estimate*
   The evaluation is done by using either empirical, usually graphical methods or by exact hypothesis testing if the mathematical statistical properties of the estimates are available. More details will be given about the measure of fit in Section 12.3.

Model calibration is usually followed by model validation when we decide on the quality of the model obtained by modelling and model calibration. Model validation is in some sense similar to model calibration because here we also use measured data, but another, independently measured data set (*validation data*) and also statistical methods. Model validation is the subject of the last Section 12.6.

12.2. MODEL PARAMETER AND STRUCTURE ESTIMATION

The estimation of some or all of the model parameters and/or parts of the model structure is the technique we use in the model calibration and in the model validation steps of the SEVEN STEP MODELLING PROCEDURE. During *model calibration* we use the developed grey-box model and measured experimental data to obtain a well-defined or solvable process model. In the *model validation* step, we again use measured experimental data (the validation data) distinct from what has been applied for model calibration. We do this in two different ways:

- to compare the predicted outputs of the model to the measured data, or
- to compare the estimated parameters of the model based on validation data to the "true" or previously estimated parameters.

12.2.1. The Model Parameter Estimation and the Model Structure Estimation Problem

The conceptual problem statement of the model parameter estimation and model structure estimation is given below:

MODEL PARAMETER ESTIMATION

Given:

- A parametrized explicit system model in the form of

$$y^{(M)} = \mathcal{M}(x;\ p^{(M)}) \tag{12.5}$$

with the model parameters $p^{(M)} \in \mathcal{R}^{\nu}$ being unknown, the vector-valued independent variable $x \in \mathcal{R}^{n}$ and vector-valued dependent variable $y^{(M)} \in \mathcal{R}^{\mu}$.

- A set of measured data

$$D[0,k] = \{(x(i), y(i)) \mid i = 0, \cdots, k\}, \tag{12.6}$$

where $y(i)$ is assumed to contain measurement error while $x(i)$ is set without any error for static models and may contain additional measurement errors for dynamic models.
- A suitable signal norm $\|\cdot\|$ to measure the difference between the model output $y^{(M)}$ and the measured independent variables y to obtain the loss function of the estimation:

$$L(p) = \|y - y^{(M)}\|. \tag{12.7}$$

Compute:
An estimate $\hat{p}^{(M)}$ of $p^{(M)}$ such that $L(p)$ is a minimum:

$$\|y - y^{(M)}\| \to \min.$$

Some of the items in the above problem statement need further discussion and explanation as follows.

1. The measured values of the dependent variable $\{y(i),\ i = 1, \ldots, k\}$ are assumed to have a zero mean measurement error. That is $y(i) = y^{(M)}(i) + \varepsilon(i)$ with ε being the measurement error while the independent variables are assumed to be set error-free for static models and may contain additional measurement error for dynamic models. In other words, $\{y(i),\ i = 1, \ldots, k\}$ is a sequence of random variables and the sequence $\{x(i),\ i = 1, \ldots, k\}$ is either deterministic in case of static models or stochastic.
2. Note that the estimate $\hat{p}^{(M)}$ itself is also a random variable. The quality of the estimate $\hat{p}^{(M)}$, or the properties of its mean value, variance and its distribution depends on
   (a) the parametrized model form, whether it is linear, nonlinear, etc.,
   (b) the signal norm selected,
   (c) the actual estimation procedure applied,
   (d) selection of the independent variable values, as given by the selection of the sequence $\{x(i),\ i = 1, \ldots, k\}$.

   The proper selection of the independent variable values is the subject of *experimental design*.
3. The general loss function defined in Eq. (12.7) has more meaningful forms. For example, in practical cases it is the sum of the squares of the difference between y and $y^{(M)}$. The notion of vector and signal norms is introduced in Section A.3 in the appendix and some simple examples are also given there.
4. It is also important to note that the MODEL PARAMETER ESTIMATION problem is an optimization problem. We want to minimize the loss function in Eq. (12.7) which depends on the model parameters $p^{(M)}$ in an implicit way through Eq. (12.5).

MODEL STRUCTURE ESTIMATION

Given:

- An explicit (but not parametrized) system model in the form

$$y^{(M)} = \mathcal{M}(x) \tag{12.8}$$

with the model $\mathcal{M}$ being an unknown form of the vector-valued independent variable $x \in \mathcal{R}^n$ and vector-valued dependent variable $y^{(M)} \in \mathcal{R}^\mu$.

- A set of measured data

$$D[0, k] = \{(x(i), y(i)) \mid i = 0, \ldots, k\} \tag{12.9}$$

where $y(i)$ is assumed to contain measurement error while $x(i)$ is set without any error for static models and may contain additional measurement errors for dynamic models.

- A $\|\cdot\|$ a signal norm to measure the difference between the model output $y^{(M)}$ and the measured variables y to obtain the loss function $L(\mathcal{M})$ in the form of Eq. (12.7).

Compute:
An estimate $\widehat{\mathcal{M}}$ of $\mathcal{M}$ such that $L(\mathcal{M})$ is minimum

$$\|y - y^{(M)}\| \to \min.$$

It can be seen from the above problem statement that the task of model structure estimation is much more complex than the task of model parameter estimation. Moreover, the problem statement above is ill-defined in the mathematical statistical sense because even non-parametric statistical methods need to have a well-defined domain of the non-parametric function in order to be estimated. This is usually a probability density function. Therefore, we usually constrain the domain, $\mathbf{Dom}_{\mathcal{M}}$ of the models $\mathcal{M}$ in the non-parametric explicit model equation (12.8) to have a finite set of explicit parametric system models (12.5). In other words, we give the *set of possible model structures* when constructing the domain $\mathbf{Dom}_{\mathcal{M}}$:

$$\mathbf{Dom}_{\mathcal{M}} = \{\mathcal{M}_1(x; p_i^{(M)}), \ldots, \mathcal{M}_M(x; p_M^{(M)})\}. \tag{12.10}$$

It follows from the above that if the model structure estimation problem is constrained by the set of possible model structures, then the conceptual solution of the model parameter estimation problem breaks down to considering the number of candidate model structures. Therefore, the conceptual solution of a model structure estimation problem consists of the following steps,

1. Construction of the set of candidate model structures (12.10).
2. Model parameter estimation for every model $\mathcal{M}_m$, $m = 1, \ldots, M$ in the set and the computation of the corresponding loss function $L_m(p_m^{(M)})$ in the form of Eq. (12.7).
3. Selection of the member $\mathcal{M}_*$ with the minimal loss $L_*(p_*^{(M)})$. Note that together with the "best-fit" model structure of $\mathcal{M}_*$, we obtain an estimate of its parameters $p_*^{(M)}$ as a side effect.

Finally, we illustrate the way the above problem specifications can be constructed by a simple example.

EXAMPLE 12.2.1 (A model structure estimation problem of a CSTR). Consider the CSTR described in Example 10.3.1, i.e. consider a CSTR with an exothermic reaction $A \longrightarrow B$. The reaction rate constant k is given by $k_0 e^{-E/RT}$. The reactor is cooled by coolant at temperature T_C. It is assumed that the physical properties remain constant and that inlet and outlet flows are equal. Assume we have several *candidate reaction kinetic expressions*:

1. $r_A = kC_A$,
2. $r_A = kC_A^{1/2}$,
3. $r_A = kC_A/(C_A^* - C_A)$.

Describe the set of candidate model structures, i.e. $\mathbf{Dom}_{\mathcal{M}}$ for model structure estimation.

The rearranged mass and energy balances lead to the following model equations:

$$\frac{\mathrm{d}C_A}{\mathrm{d}t} = \frac{F}{V}(C_A^{(\mathrm{i})} - C_A) - r_A, \tag{12.11}$$

$$\frac{\mathrm{d}T}{\mathrm{d}t} = \frac{F}{V}(T^{(\mathrm{i})} - T) - \frac{r_A}{\rho c_P}(-\Delta H_{\mathrm{R}}) - \frac{UA}{V\rho c_P}(T - T_{\mathrm{C}}). \tag{12.12}$$

Now the *set of candidate model structures* $\mathbf{Dom}_{\mathcal{M}} = \{\mathcal{M}_1, \mathcal{M}_2, \mathcal{M}_3\}$ consist of the two parametrized model equations with their parameters unknown:

1. $\mathcal{M}_1$: Eqs. (12.11)–(12.12) with $r_A = kC_A$,
2. $\mathcal{M}_2$: Eqs. (12.11)–(12.12) with $r_A = kC_A^{1/2}$,
3. $\mathcal{M}_3$: Eqs. (12.11)–(12.12) with $r_A = kC_A/(C_A^* - C_A)$

■ ■ ■

12.2.2. The Least Squares (LS) Parameter Estimation for Models Linear in Parameters

The *least squares* (LS) estimation methods receive their name from the square signal norm $\|\cdot\|_2$ which is applied for computing the loss function in Eq. (12.7):

$$L(p) = \|y - y^{(M)}\|_2 = \left(\sum_{i=0}^{k}\sum_{j=1}^{\mu}(y_j(i) - y_j^{(M)}(i))^2\right)^{1/2}.$$

As we see later in this chapter, almost every parameter or structure estimation problem for models linear in their parameters is somehow transformed to a standard least squares parameter estimation problem. Therefore, this estimation procedure and its properties are of primary importance and will be described and analysed.

Problem Statement

The LS parameter estimation problem statement for models linear in their parameters and with a single dependent variable is as follows:

LEAST SQUARES ESTIMATION (LINEAR IN PARAMETER SINGLE DEPENDENT VARIABLE MODELS)

Given:

- A linear in parameter system model in the following form:

$$y^{(M)} = x^{\mathrm{T}}p = \sum_{i=1}^{n} x_i p_i \tag{12.13}$$

with the model parameters $p \in \mathcal{R}^n$ being unknown, the vector-valued independent variable $x \in \mathcal{R}^n$ and scalar-valued dependent variable $y^{(M)} \in \mathcal{R}$.

- A set of measured data consisting of m measurements ($m \geq n$) arranged in a vector-matrix form as

$$y = \begin{bmatrix} y_1 \\ y_2 \\ \cdots \\ y_m \end{bmatrix}, \quad \mathbf{X} = \begin{bmatrix} x_{11} & x_{12} & \cdots & x_{1n} \\ x_{21} & x_{22} & \cdots & x_{2n} \\ \cdots & \cdots & \cdots & \cdots \\ x_{m1} & x_{m2} & \cdots & x_{mn} \end{bmatrix}, \tag{12.14}$$

where y_i is assumed to contain measurement error.

- A loss function L

$$L(p) = r^{\mathrm{T}}\mathbf{W}r = \sum_{i=1}^{m}\sum_{j=1}^{m} r_i W_{ij} r_j, \tag{12.15}$$

$$r_i = y_i - y_i^{(M)}, \tag{12.16}$$

where r is the *residual vector* and $\mathbf{W}$ is a suitable *positive definite symmetric weighting matrix.*

Compute:
An estimate $\hat{p}$ of p such that the loss function $L(p)$ is a minimum.

The LS Parameter Estimation Method

The LS parameter estimate for p is in the form of

$$(\mathbf{X}^{\mathrm{T}}\mathbf{W}\mathbf{X})\hat{p} = \mathbf{X}^{\mathrm{T}}\mathbf{W}y \quad \text{or} \quad \hat{p} = (\mathbf{X}^{\mathrm{T}}\mathbf{W}\mathbf{X})^{-1}\mathbf{X}^{\mathrm{T}}\mathbf{W}y. \tag{12.17}$$

Note that the two forms are mathematically equivalent but they are used for different purposes:

1. For parameter estimation, where we compute the estimate $\hat{p}$, the implicit form shown on the left-hand side is used. This allows us to use regularization and decomposition methods to solve the overdetermined linear equation resulting in a much better and numerically stable procedure even for ill-conditioned cases.
2. The explicit form shown on the right-hand side is only for theoretical analysis purposes. The statistical properties of the estimate $\hat{p}$ will be analysed using this form.

Extension to the Vector-Valued Dependent Variable Case

A linear in parameter model in the form of (12.36) can be extended to the case of vector-valued dependent variable $y^{(M)} \in \mathcal{R}^{\nu}$

$$y^{(M)} = \mathbf{P}x, \tag{12.18}$$

where the matrix of model parameters $\mathbf{P} \in \mathcal{R}^{\nu \times n}$ contains the model parameters of the submodel belonging to the jth entry of the vector $y^{(M)}$ in its rows $p^j = [p_{1j} \ldots p_{nj}]^T \in \mathcal{R}^n$.

The above model can be transformed into the form required for the parameter estimation of the single dependent variable case in (12.14) if one forms the following vectors and data matrices:

$$y^* = \begin{bmatrix} y_{1(1)} \\ \cdots \\ y_{1(\nu)} \\ \cdots \\ y_{m(1)} \\ \cdots \\ y_{m(\nu)} \end{bmatrix}, \quad p^* = \begin{bmatrix} p_{11} \\ \cdots \\ p_{1n} \\ \cdots \\ p_{\nu 1} \\ \cdots \\ p_{\nu n} \end{bmatrix}, \quad \mathbf{X}^* = \begin{bmatrix} \mathbf{X} & 0 & \cdots & 0 \\ 0 & \mathbf{X} & \cdots & 0 \\ \cdots & \cdots & \cdots & \cdots \\ 0 & 0 & \cdots & \mathbf{X} \end{bmatrix} \tag{12.19}$$

where $y_{i(j)}$ is the jth entry of the ith measured dependent data vector which is assumed to contain measurement error.

Handling Measurement Errors

There are several important remarks and observations which should be discussed concerning the LEAST SQUARES ESTIMATION (LINEAR IN PARAMETER MODELS) problem which are as follows.

1. Most often the measured dependent variable values are assumed to have an *additional measurement error with zero mean*, i.e.

$$y_i = y_i^{(M)} + \varepsilon_i, \quad i = 1, \ldots, m, \tag{12.20}$$

$$\mathbf{E}\{\varepsilon_i\} = 0. \tag{12.21}$$

where $\mathbf{E}\{.\}$ is the mean value operator. If in addition the individual *measurement errors are statistically independent of each other and are normally distributed with a diagonal covariance matrix $\mathbf{\Delta}_\varepsilon$, then we can characterize the Gaussian distribution of the measured values as*

$$y \sim \mathcal{N}(y^{(M)}, \mathbf{\Delta}_\varepsilon), \tag{12.22}$$

i.e. the measured value vector is also normally distributed. Note that in the vector-valued dependent variable case the above assumption means that the measurement errors of the vector entries are independent in addition to the independence of the measurement errors in the individual measurement points.

2. One of the most difficult problems in specifying a weighted least squares parameter estimation problem is to choose the weights properly. The weighting matrix $\mathbf{W}$ in the loss function L given by Eq. (12.15) is used to impose a different weight on the residuals of the particular measurement points. If some of the measurement points are regarded as relatively unreliable, such that their measurement error has a larger variance, then it is assigned a lower weight compared to the others. An obvious choice for the weighting matrix is to select it to be the inverse of the covariance matrix of the measurement errors if it is known or its estimate can be obtained:

$$\mathbf{W} = \mathbf{\Delta}_\varepsilon^{-1}. \tag{12.23}$$

Note that the inverse matrix above always exists because covariance matrices are positive definite.

12.2.3. Properties of LS Parameter Estimation

We can look at the explicit form of the estimation equation (12.17) as being a linear transformation of the random variable y by a constant, non-random matrix

$$\mathbf{T} = (\mathbf{X}^{\mathrm{T}}\mathbf{W}\mathbf{X})^{-1}\mathbf{X}^{\mathrm{T}} \tag{12.24}$$

in order to get the random variable $\hat{p}$. Therefore, we can compute the mean value and the covariance matrix of the estimate $\hat{p}$ following the formulae (A.13) described in section A.1.2 in the appendix as

$$\mathbf{COV}\{\hat{p}\} = \mathbf{T}\mathbf{COV}\{y\}\mathbf{T}^{\mathrm{T}}. \tag{12.25}$$

If the measurement errors are normally distributed with zero mean and covariance matrix $\mathbf{\Delta}_\varepsilon$ and the weighting matrix in the loss function Eq. (12.15) has been selected accordingly as in Eq. (12.23) then the estimate $\hat{p}$ has the following properties:

- *It is normally distributed.*
- *Its mean value is $\mathbf{E}\{\hat{p}\} = p$, hence it is unbiased.*
- *Its covariance matrix is computed as*

$$\mathbf{COV}\{\hat{p}\} = (\mathbf{X}^{\mathrm{T}}\mathbf{W}\mathbf{X})^{-1}\mathbf{\Delta}_\varepsilon. \tag{12.26}$$

- *The estimate itself is a maximum-likelihood estimate, and thus an efficient optimal estimate.*

Confidence Intervals and Regions for the Estimated Parameters

The covariance matrix of the estimate $\mathbf{COV}\{\hat{p}\}$ contains information not only on the variances of the individual entries $\hat{p}_i$ (being the diagonal elements of it) but also on the correlation between the estimates. As has been said before, if the measurement errors are statistically independent of each other and are normally distributed with zero mean and covariance matrix $\mathbf{\Delta}_\varepsilon$, with the weight matrix $\mathbf{W}$ set accordingly then the distribution of the estimate $\hat{p}$ is Gaussian with the covariance matrix in Eq. (12.26).

In order to understand what the covariance matrix means about the quality of the estimate in the case of two parameters, we can look at the shape of the 2D Gaussian probability density function shown in Figure A.2 in Section A.1 in the appendix. The elements in the covariance matrix determine the shape of the ellipsoidal cuts of the function and the direction of the main axis of the ellipses. In the case of more than two parameters *the confidence region is a multi-dimensional ellipsis* with parameters being the lengths of its axes and the direction of its main axis. Note that the lengths of the axes have nothing to do with the variances of the parameter vector entries.

Unfortunately, we can not construct confidence intervals separately in the strict mathematical sense for the normally distributed parameters in this case and give an estimate of the covariance or correlation between any pair using the off-diagonal elements. This was only possible if the estimate $\hat{p}$ contained independent entries.

In order to overcome this difficulty we may apply two different approaches to have information on the confidence region and confidence intervals for the estimated parameters.

1. *Linear transformation of the estimated parameter vector*
Similar to the standardization of vector-valued random variables described in section A.1.3 in the appendix the vector-valued transformed parameter vector

$$\hat{p}^* = \mathbf{COV}\{\hat{p}\}^{-1/2}\hat{p} - p \tag{12.27}$$

is approximately a standard Gaussian vector-valued random variable (see Section A.1.1 in the appendix). This means that its entries are regarded as independent random variables with standard Student t-distribution. This distribution can be used to construct confidence intervals for the entries independently on the given significance level. The *Student t-test* can be used for testing hypotheses independently for the mean value of the *transformed* parameter values.

Note that by this transformation we transform the multidimensional ellipse as confidence region in the space of the original parameters into a multivariable rectangle which is parallel to the co-ordinate direction in the space of transformed variables.

Thereafter, the inverse of the standardization transformation (12.27) above can be used to transform back the multivariate rectangle into the space of the original parameters. With the vectors of lower and upper confidence limits of the transformed parameters $\hat{p}_L^*$ and $\hat{p}_U^*$, we get the following lower and upper parameter vectors for the original parameters:

$$\hat{p}_{\mathrm{L}} = \mathbf{COV}\{\hat{p}\}^{1/2}\hat{p}_{\mathrm{L}}^* + \hat{p}, \qquad \hat{p}_{\mathrm{U}} = \mathbf{COV}\{\hat{p}\}^{1/2}\hat{p}_{\mathrm{U}}^* + \hat{p}. \tag{12.28}$$

2. *Individual confidence intervals*
It is possible to construct a statistically correct *joint confidence region for all of the parameters* on the $(1-\alpha)$ significance level as follows:

$$(p^{(M)} - \hat{p})^T \mathbf{X}^{\mathrm{T}} \mathbf{W} \mathbf{X} (p^{(M)} - \hat{p}) \leq \nu s^2 F(\nu, k-\nu, 1-\alpha), \tag{12.29}$$

where $F(\nu, k-\nu, 1-\alpha)$ is the $(1-\alpha)$ point ("upper α point") of the $F(\nu, k-\nu)$ distribution value for DOF $(\nu, k-\nu)$ and s^2 is the estimated (empirical) standard deviation as an estimate for the variance of the measurement errors computed from the data. The $F(\nu, k-\nu, 1-\alpha)$ value can be obtained from standard statistical tables, where ν is the numerator and $k-\nu$ is the denominator.

The inequality above is the one which provides the equation of an "elliptically shaped" contour in the space of all parameters.

In order to derive some kind of average individual confidence intervals for the parameters we first observe that the worst-case confidence intervals for the individual parameters would be the vertices of the multidimensional rectangle parallel to the co-ordinate directions and containing the whole multidimensional ellipsis. For very long and thin ellipses, however, this would be a very conservative estimate and the majority of this multi dimensional cube would not represent a reasonable parameter vector. This situation is seen on Fig. 12.1 in the case of two parameters with point E.

Note that if we construct more reasonable average confidence intervals then the same situation will also arise with the presence of non-valid parameter vectors within the volume permitted by the Descartes product of the individual confidence intervals which is a smaller cube within the conservative one. Moreover, we shall have left out some of the valid parameter vector values. See point A in Fig. 12.1.

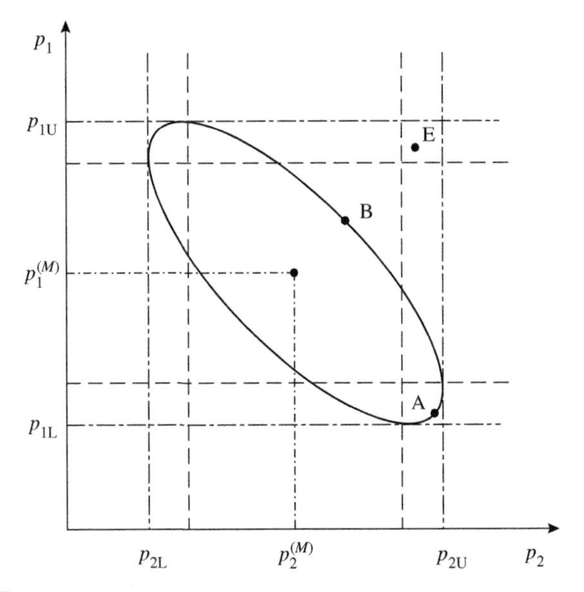

FIGURE 12.1 Confidence regions and intervals.

We obtain individual confidence intervals for the various parameters separately from the formula

$$\hat{p}_i \pm t(k, 1 - \tfrac{1}{2}\alpha)s_{b_i}, \tag{12.30}$$

where the estimated standard deviation s_{b_i} of the individual parameters p_i is the square root of the ith diagonal term of the covariance matrix $(\mathbf{X}^{\mathrm{T}}\mathbf{W}\mathbf{X})^{-1}\mathbf{\Delta}_\epsilon$.

Note that in the optimal case we would like to have uncorrelated parameter estimates, zero off-diagonal covariance elements with the minimum achievable variances. *Experimental design with the proper choice of the elements in the matrix X can be used to achieve this goal.*

Assessing the Fit

As soon as the parameter estimate $\hat{p}$ is obtained the question arises whether the quality of the estimate is good or not. We have to assess the fit of the estimate or the fit of the model with estimated parameters. There are different ways of assessing the fit based on either the statistical properties of the estimate or on the statistical properties of the residuals.

1. *Residual tests*

The simplest test to assess the fit of a parameter estimate is to check whether the residuals $r_i = y_i - y_i^{(M)}$ are randomly distributed. We can do that by *simple visual inspection plotting r_i against*

- *sequence number i,*
- *independent variables x_i,*
- *predicted values $y_i^{(M)}$*, etc.

Figure 12.2 shows typical residual patterns you may obtain.

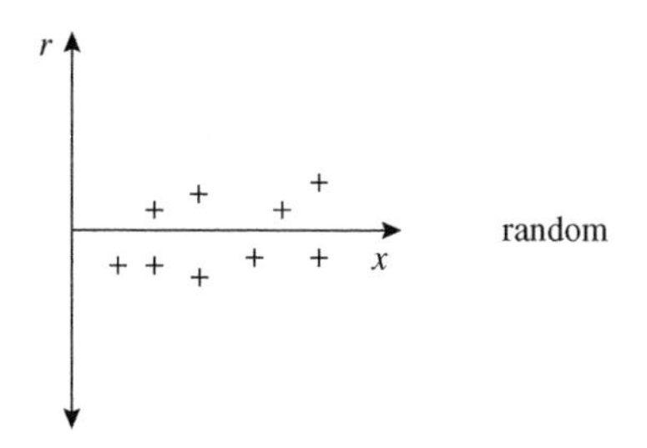

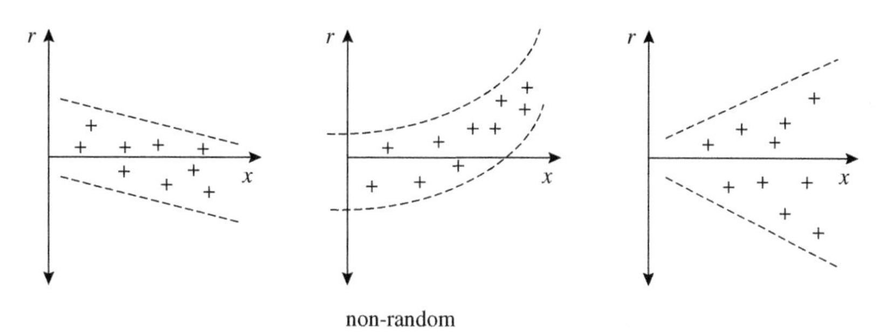

FIGURE 12.2 Residual patterns.

If we can assume that the measurement errors are statistically independent of each other and are normally distributed with zero mean covariance matrix Δ_ε then *the residuals can be regarded as a sample*

$$\mathbf{S}(\varepsilon) = \{r_1, r_2, \ldots, r_k\}$$

to test the hypothesis

$$H_0 : m = 0$$

by standard hypothesis testing (see details in Section A.2 in the appendix).

In case of a *single dependent variable* with known variance Δ_ε, we can compute the test statistics u as in Eq. (A.18) Section A.2 in the appendix. Otherwise, we compute the empirical sample variance

$$\Delta_\varepsilon \simeq \mathbf{s}^2\{\mathbf{S}(\varepsilon)\} = \frac{\sum_{i=1}^{m}(r_i - \mathbf{M}\{\mathbf{S}(\varepsilon)\})^2}{m-1} \tag{12.31}$$

with $\mathbf{M}\{\mathbf{S}(\varepsilon)\}$ being the sample mean as in Eq. (A.17)

$$\mathbf{M}\{\mathbf{S}(\varepsilon)\} = \frac{\sum_{i=1}^{k} r_i}{k}.$$

Thereafter, we can compute the test statistics

$$t = \frac{\mathbf{M}\{\mathbf{S}(\varepsilon)\} - m_0}{\sqrt{\mathbf{s}^2\{\mathbf{S}(\varepsilon)\}}} \sim t_{\nu-1}(0, 1), \tag{12.32}$$

where the test statistics t has a standard Student t-distribution $t_{\nu-1}(0, 1)$ and perform a Student t-test.

The case of vector-valued dependent variable is much more complicated and needs the full arsenal of multivariate statistics [83]. In the case of Gaussian measurement error distribution the so-called Wilks test [84] is used instead of the u test for testing hypothesis on the mean value with known covariance matrix.

If the residuals are found to be non-randomly distributed and/or the residual tests have failed then we have problems with either the model or the data or both.

2. *Correlation coefficient measures*

In order to compute the correlation coefficient measures we have to decompose the *sum of squares (SOS) about the sample mean* $l_{m\mathrm{SOS}}$ into two parts ($l_{r\mathrm{SOS}} + l_{d\mathrm{SOS}}$) as follows:

$$l_{\mathrm{SOS}} = \sum_{i=1}^{k}(y_i - \mathbf{M}\{y\})^2 = \sum_{i=1}^{k}\left(y_i - y_i^{(M)}\right)^2 + \sum_{i=1}^{k}(y_i^{(M)} - \mathbf{M}\{y\})^2 \tag{12.33}$$

with

$$\mathbf{M}\{y\} = \frac{1}{k}\sum_{i=1}^{k} y_i, \quad l_{r\mathrm{SOS}} = \sum_{i=1}^{k}\left(y_i - y_i^{(M)}\right)^2, \quad l_{d\mathrm{SOS}} = \sum_{i=1}^{k}(y_i^{(M)} - \mathbf{M}\{y\})^2$$

and $\mathbf{M}\{y\}$ being the sample mean of the measured dependent variables. In this decomposition, l_{SOS} is the total variation which is decomposed to the unexplained variation $l_{r\mathrm{SOS}}$ about regression and to the explained variation $l_{d\mathrm{SOS}}$ due to regression. The coefficient of determination

$$0 \leq \rho^2 = \frac{\sum_{i=1}^{k}(y_i^{(M)} - \mathbf{M}\{y\})^2}{\sum_{i=1}^{k}(y_i - \mathbf{M}\{y\})^2} \leq 1, \tag{12.34}$$

and its square root

$$\rho = \sqrt{\frac{\sum_{i=1}^{k}\left(y_i^{(M)} - \mathbf{M}\{y\}\right)^2}{\sum_{i=1}^{k}(y_i - \mathbf{M}\{y\})^2}}$$

measures the variation in the data explained by the model.

12.2.4. The LS Parameter Estimation for Nonlinear Models

If we have a model nonlinear in its parameters then we can apply two different techniques to estimate its parameters.

1. Transform the nonlinear model into a form which is linear in its parameters and apply LEAST SQUARES ESTIMATION (LINEAR IN PARAMETER MODELS).
2. Solve the general MODEL PARAMETER ESTIMATION problem.

If we cannot transform the model equations into a form linear in parameters then we have to solve the general nonlinear model parameter estimation problem. As it has been mentioned before the MODEL PARAMETER ESTIMATION problem is an optimization problem with the following ingredients:

- a process model nonlinear in parameters $y^{(M)} = \mathcal{M}(x;\ p)$,
- a set of measured data $D[0, k]$ as in Eq. (12.6) and a set of predicted model outputs $\{y^{(M)}(i), i = 1, \ldots, k\}$ computed from the model equation above using the measured values of the independent variables $\{x(i), i = 1, \ldots, k\}$,
- a loss function in the LS parameter estimation case in the form of SOS

$$L_{\mathrm{SOS}}(p) = \sum_{i=1}^{k} \|y(i) - y^{(M)}(i)\|_2^2, \tag{12.35}$$

which is a special case of the general loss function in Eq. (12.7) depending on the model parameters p in an implicit way through the model equation above.

The optimization problem above is usually solved by a suitable optimization algorithm, most often by the *Levenberg–Marquardt method.*

It is important to note that the *estimate $\hat{p}$ has lost its nice statistical properties mentioned in Section 12.2.3.* Therefore the lack of fit tests can only be applied with caution and their results taken as approximate in the case of models nonlinear in parameters.

Confidence Intervals and Regions for the Estimated Parameters

As has been mentioned, the LS estimate $\hat{p}$ does not have the nice statistical properties in the case of models nonlinear in their parameters as in the case of models linear in parameters.

Therefore we can only apply an *approximate approach* if we want to estimate the parameter confidence intervals for the estimated coefficients. The theory given in Section 12.2.3 is not sufficient to cover this issue as the estimated parameter confidence intervals are only applicable to linear models. The following explains a possible approach for the *single parameter case* to get the nonlinear parameter confidence interval.

This approach requires us to graph the sum of squares value (L_{SOS}) as a function of the parameter to be estimated p as seen in Figure 12.3.

Figure 12.3 shows the optimum value of p_{opt} and its LS value L_{opt}. A 95% confidence level for the sum of squares can be established using:

$$L_{\mathrm{SOS}}^{95} = r^{\mathrm{T}} r \left(1 + \frac{\nu}{k-\nu} F(\nu, k-\nu, 0.05)\right) = L_{\mathrm{opt}} \left(1 + \frac{\nu}{k-\nu} F(\nu, k-\nu, 0.05)\right),$$

where in the single-dependent variable case:

$$r^{\mathrm{T}} r = \sum_{i=1}^{k} (y(i) - y(i)^M)^2 = L_{\mathrm{opt}}$$

and k = number of data points, ν = number of parameters (now $\nu = 1$), and $F(\nu, k-\nu, 0.05)$ is the F distribution value for DOF $(\nu, k-\nu)$. This can be obtained from standard statistical tables, where ν is the numerator and $k-\nu$ is the denominator.

Once the SOS_{95} line has been drawn on the above figure, it is then possible to estimate the parameter confidence interval by simply taking the intersection points on the *SOS* line.

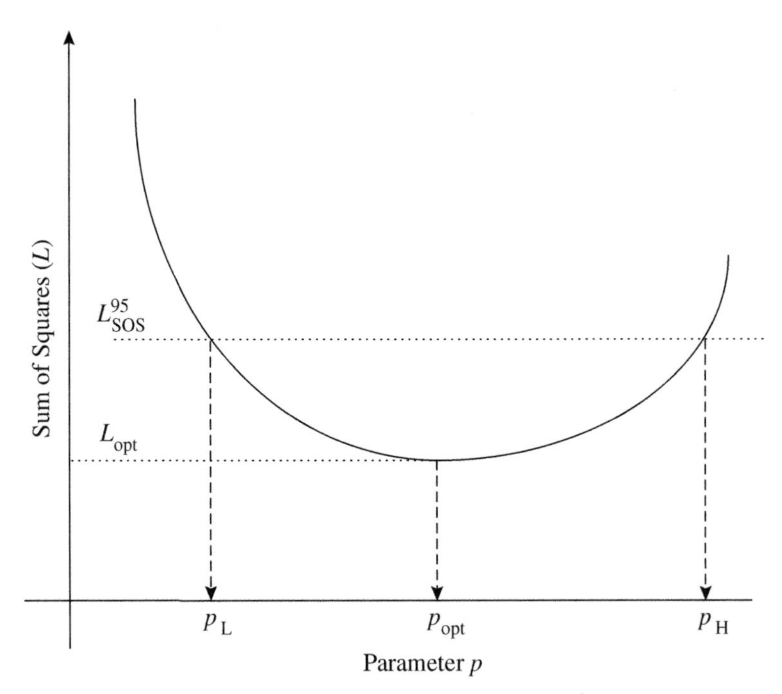

FIGURE 12.3 Nonlinear SOS variation.

Note that in the case of a nonlinear problem, the SOS *curve will not be symmetric about the* p_{opt} *value.*

We may generalize the above method for constructing confidence intervals individually in case of more than one parameter to be estimated. We have to take these confidence intervals with the same concerns as described in Section 12.2.3 for linear models. *The joint confidence region for the estimated parameters for models nonlinear in parameters, however, will not be a multidimensional ellipses. It may be non-convex of arbitrary shape.*

12.3. MODEL PARAMETER ESTIMATION FOR STATIC MODELS

An explicit parametrized system model $y^{(M)} = \mathcal{M}(x;\ p^{(M)})$ is *static* if there is no ordering between either the components x_i of the independent variable vector x or the individual independent variables in the set of measured data $x(i),\ i = 1, \ldots, k$. In other words, we can choose the measurement points for the parameter estimation arbitrarily. Moreover, the measurements of the dependent variable $y(i),\ i = 1, \ldots, k$ can often be regarded as independent random variables.

Process systems in steady state are described by static models. We only consider lumped system models in this chapter, therefore steady-state models are described by a set of usually nonlinear algebraic equations forming their static model. Grey-box static models of lumped process systems are therefore algebraic equations with unknown parameters and/or unknown structural elements such as static linear or nonlinear functions.

12.3.1. Models Linear in their Parameters

The most important class of static models is the class of static models linear in their parameters.

Examples of Static Process Models Linear in Parameters

The most general form of static explicit models linear in parameters is a special case of the parametric models in Eq. (12.5) of the MODEL PARAMETER ESTIMATION problem in the form of:

$$y^{(M)} = x^{\mathrm{T}} p^{(M)} = \sum_{i=1}^{n} x_i p_i^{(M)} \tag{12.36}$$

with the model parameters $p^{(M)} \in \mathcal{R}^{n \times \mu}$ being unknown, the vector-valued independent variable $x \in \mathcal{R}^n$ and vector-valued dependent variable $y^{(M)} \in \mathcal{R}^{\mu}$. The above simple form of a set of linear equations seems to be rare in process engineering.

There are, however, some important static models which are or can easily be transformed to the form of Eq. (12.36). These are the following.

1. *Polynomial models of one independent variable*
A static model of the form

$$y^{(M)} = \sum_{i=0}^{\nu} \xi^i p_i^{(M)} \tag{12.37}$$

with a scalar-valued independent variable $\xi \in \mathcal{R}$ and scalar-valued dependent variable $y^{(M)} \in \mathcal{R}$ can be seen as being in the form of Eq. (12.36) with

$$x = [1 \;\; \xi \;\; \xi^2 \;\; \cdots \;\; \xi^{\nu}]^{\mathrm{T}}, \quad p^{(M)} = [\, p_0^{(M)} \;\; p_1^{(M)} \;\; \cdots \;\; p_{\nu}^{(M)} \,]^{\mathrm{T}}, \tag{12.38}$$

where $\mu = 1, \; n = \nu + 1$.

2. *Polynomial models of several independent variables*
This case is an extension of the previous one and we show a simple two-variate example of a hypothetical reaction kinetic expression of the form:

$$y^{(M)} = k_1 C_A^2 - k_2 C_A C_B + k_3 C_B^{1/2} \tag{12.39}$$

with scalar-valued independent variables $C_A, \; C_B \in \mathcal{R}$ and scalar-valued dependent variable $y^{(M)} \in \mathcal{R}$. The above equation is also in the form of Eq. (12.36) with

$$x = [\, C_A^2 \;\; C_A C_B \;\; C_B^{1/2} \,]^{\mathrm{T}}, \quad p^{(M)} = [\, k_1 \;\; k_2 \;\; k_3 \,]^{\mathrm{T}}, \tag{12.40}$$

where $\mu = 1, \; n = 3$.

3. *Exponential expressions*
Temperature dependent reaction kinetic expressions can also be transformed into the form of Eq. (12.36). The simplest case is the expression

$$r_A = k_0 \mathrm{e}^{-E/RT} C_A^{\kappa}, \tag{12.41}$$

where k_0 is the pre-exponential coefficient, E the activation energy and κ is the order of the reaction. If we want to estimate the parameters we can take the logarithm of the equation above to get

$$\ln(r_A) = k_0 - E\frac{1}{RT} + C_A\kappa \tag{12.42}$$

which gives

$$y^{(M)} = \ln(r_A), \quad x = [1, \; \frac{1}{RT}, \; C_A]^{\mathrm{T}}, \quad p^{(M)} = [k_0, \; E, \; \kappa]^{\mathrm{T}}. \tag{12.43}$$

Finally, we give a simple example of a static model of a CSTR linear in its parameters.

EXAMPLE 12.3.1 (A static model of a jacketted CSTR). Consider a CSTR of constant volume, equal in and outflows and with no chemical reaction. The reactor is cooled by coolant at temperature T_{C}. It is assumed that the physical properties remain constant and that inlet and outlet flows are equal in the jacket for the coolant. Describe the static model of the reactor and show that it is linear in parameters.

The rearranged static energy balances for the reactor volume and the jacket lead to the following model equations:

$$\frac{F_{\mathrm{R}}}{V_{\mathrm{R}}}(T^{(\mathrm{i})} - T) - \frac{UA}{V_{\mathrm{R}}\rho_{\mathrm{R}}c_{P\mathrm{R}}}(T - T_{\mathrm{C}}) = 0, \tag{12.44}$$

$$\frac{F_{\mathrm{J}}}{V_{\mathrm{J}}}(T_{\mathrm{C}}^{(\mathrm{i})} - T_{\mathrm{C}}) + \frac{UA}{V_{\mathrm{J}}\rho_{\mathrm{J}}c_{P\mathrm{J}}}(T - T_{\mathrm{C}}) = 0, \tag{12.45}$$

where the flowrates F_{R}, F_{J} and the volumes V_{R}, V_{J} are assumed to be known and constant and the variables T, $T^{(\mathrm{i})}$, T_{C}, $T_{\mathrm{C}}^{(\mathrm{i})}$ can be measured. The equations above can be rearranged to give

$$T^{(\mathrm{i})} = -\frac{UA}{\rho_{\mathrm{R}}c_{P\mathrm{R}}}T_{\mathrm{C}} + \left(1 + \frac{UA}{\rho_{\mathrm{R}}c_{P\mathrm{R}}}\right)T, \tag{12.46}$$

$$T_{\mathrm{C}}^{(\mathrm{i})} = -\frac{UA}{\rho_{\mathrm{J}}c_{P\mathrm{J}}}T + \left(1 + \frac{UA}{\rho_{\mathrm{J}}c_{P\mathrm{J}}}\right)T_{\mathrm{C}}, \tag{12.47}$$

which is linear in the parameters

$$p^{(M)} = \left[\frac{UA}{\rho_{\mathrm{R}}c_{P\mathrm{R}}}, \; \frac{UA}{\rho_{\mathrm{J}}c_{P\mathrm{J}}}\right]^{\mathrm{T}} \tag{12.48}$$

with the independent variables

$$x = [T_{\mathrm{C}}, \; T]^{\mathrm{T}} \tag{12.49}$$

and dependent variables

$$y = [T^{(\mathrm{i})}, \; T_{\mathrm{C}}^{(\mathrm{i})}]^{\mathrm{T}}. \tag{12.50}$$

Note that the parameters in Eqs. (12.46) and (12.47) depend on each other linearly which makes the parameter estimation problem unusual and difficult.

The LS Parameter Estimation for Static Models Linear in Parameters

The LS parameter estimation problem in the case of static models linear in parameters is already in the same form as the LEAST SQUARES ESTIMATION (LINEAR IN PARAMETER MODELS) problem. Therefore, the LS parameter estimation method is applicable for this case in exactly the same way as it is described in Section 12.2.2.

The properties of the estimate (in Section 12.2.3), the way of computing confidence intervals (in Section 12.2.3) for the estimated parameters and assessing the fit (in Section 12.2.3) are also the same.

Checklist for Performing Weighted Least Squares Parameter Estimation for Static Models Linear in Parameters

Here, we summarize some of the most important issues one should remember and preferably check when performing weighted least squares parameter estimation for static models linear in parameters. The symptoms and consequences of the conditions *not* met are also indicated.

1. *Static model linear in parameter form*
2. *Steady-state data*
   Steady-state tests are used to test if the data set is taken from a steady state of the process as it is described in Section 11.2.3.
3. *Reliable measured data*
   Data analysis and preprocessing should be performed before parameter estimation to obtain reliable and sufficiently rich data according to Section 11.2.
4. *Measurement errors are normally distributed with zero mean*
   This is a rather difficult condition to check but it affects the quality of the estimate in a significant manner.

   Symptoms: empirical histogram of measured data is not "Gaussian-like",
   Consequences: possibly biased and non-normally distributed estimate, hypothesis test may fail.

12.3.2. Nonlinear Models

As it has been described before in Section 12.2.4 the parameter estimation of static nonlinear models is possible in two different ways.

1. The nonlinear model is transformed into a form which is linear in parameters and then the LS parameter estimation of static models linear in parameters is applied. Note that we usually estimate the *transformed parameters* which depend on the original parameters in a nonlinear way. Therefore, the statistical properties of the estimates are valid and the confidence intervals can be obtained only for these transformed parameters.

   Some simple examples of *model transformation* are described in Section 12.3.1.
2. We solve the LS model parameter estimation problem directly on the original model by optimization. Here we estimate the original model parameters in the nonlinear model the same way as it is described in Section 12.2.4.

We prefer to perform model transformation to obtain a model linear in parameters but there are static models which cannot be transformed in this way. A simple example of this kind is given below.

EXAMPLE 12.3.2 (A static model of a CSTR not linear in its parameters). Consider the CSTR described in Example 10.3.1, i.e. consider a CSTR with an exothermic reaction $A \longrightarrow B$. The reaction rate constant k is given by: $k_0 e^{-E/RT}$. The reactor is cooled by coolant at temperature T_C. It is assumed that the physical properties remain constant and that inlet and outlet flows are equal. Assume that the reaction kinetic expression is in the form

$$r_A = kC_A^{1/2}. \tag{12.51}$$

Develop a static model of the reactor and show that it is nonlinear in its parameters.

The rearranged static mass and energy balances lead to the following model equations:

$$\frac{F}{V}(C_A^{(i)} - C_A) - r_A = 0 \tag{12.52}$$

$$\frac{F}{V}(T^{(i)} - T) - \frac{r_A}{\rho c_P}(-\Delta H_R) - \frac{UA}{V\rho c_P}(T - T_C) = 0 \tag{12.53}$$

together with the constitutive reaction kinetic equation (12.51). Assume we can measure C_A, $C_A^{(i)}$, T_C, T, $T^{(i)}$ and the volume V and flowrate F are known and constant. The above model is nonlinear in the parameters

$$p^{(M)} = [k_0,\ E,\ (c_P\rho),\ \Delta H,\ (UA)]^{\mathrm{T}} \tag{12.54}$$

with the independent variables

$$x = [C_A^{(i)},\ T_C,\ T^{(i)}]^{\mathrm{T}} \tag{12.55}$$

and dependent variables

$$y = [C_A,\ T]^{\mathrm{T}}. \tag{12.56}$$

12.4. IDENTIFICATION: MODEL PARAMETER AND STRUCTURE ESTIMATION OF DYNAMIC MODELS

An explicit discrete time parametrized system model $y^{(M)} = \mathcal{M}(x;\ p^{(M)})$ is *dynamic* if

- its variables are time dependent, where time gives an ordering in the sequences $\{y(\tau), \tau = 1, \ldots, k\}$ and $\{x(\tau), \tau = 1, \ldots, k\}$;
- its independent variable vector x contains present and past values of the inputs and outputs of the process system, i.e.

$$x(\tau) = \{u(\tau),\ y(\tau - 1),\ u(\tau - 1)\ , \ldots\} \tag{12.57}$$

  up to a finite length;
- $y^{(M)}(\tau)$ is the value of the model output at any time τ.

There are significant implications for the parameter and structure estimation problem from the above conditions:

1. We should sample dynamic process models before attempting model parameter or structure estimation procedures.
2. We cannot choose the measurement points for the parameter estimation arbitrarily just the values of the input variables. The other part of the independent variable vector $x(\tau)$ is defined implicitly by the choice of the input variable sequence (see Eq. (12.57)).
3. Moreover, the measurements of the dependent variable $y(i),\ i = 1, \ldots, k$ cannot be regarded as independent random variables unless the process system itself is fully deterministic and the measurement error is caused by the measurement system itself.

Note that there is a wide and matured technology of parameter and structure estimation of dynamic models in systems and control theory (see e.g. [14,82]) where this is called *system identification*.

We only consider lumped system models in this chapter. Grey-box dynamic models of lumped process systems are therefore DAE systems with unknown parameters and/or unknown structural elements such as static linear or nonlinear functions.

12.4.1. Problem Statements and Conceptual Solutions

Model Parameter Estimation for Dynamic Models

The conceptual problem statement of the model parameter estimation and model structure estimation is given below.

MODEL PARAMETER ESTIMATION FOR DYNAMIC MODELS

Given:

- A parametrized, sampled explicit dynamic system model in the form of

$$y^{(M)} = \mathcal{M}(u;\ p^{(M)}) \tag{12.58}$$

  with the model parameters $p^{(M)} \in \mathcal{R}^{\nu}$ being unknown, the vector-valued input variable $u(\tau) \in \mathcal{R}^{r}$, a vector-valued output variable $y^{(M)}(\tau) \in \mathcal{R}^{m}$ and both the input and output variables are varying in (discrete) time, i.e. $y^{(M)} : \mathcal{P} \rightarrow \mathcal{R}^{m}$ and $u : \mathcal{P} \rightarrow \mathcal{R}^{r}$ where $\mathcal{P}$ is the set of integer numbers.
- A set of measured data which is called a *measurement record* to emphasize the time dependence of the measured data

$$D[0, k] = \{(u(\tau), y(\tau)) \mid \tau = 0, \ldots, k\}, \tag{12.59}$$

  where $y(\tau)$ is assumed to contain measurement error while $u(\tau)$ is set without any error.
- A suitable *signal norm* $\|\cdot\|$ (see Section A.3 in the appendix) to measure the difference between the model output $y^{(M)}$ and the measured independent variables y to obtain the loss function:

$$L(p) = \|r\|, \tag{12.60}$$

where r is the residual signal, or sequence

$$r(\tau) = y(\tau) - y^{(M)}(\tau), \quad \tau = 0, \ldots, k. \tag{12.61}$$

Compute:
An estimate $\hat{p}^{(M)}$ of $p^{(M)}$ such that $L(p)$ is minimum

$$\|y - y^{(M)}\| \rightarrow \min .$$

Conceptual solution method:
Observe that the MODEL PARAMETER ESTIMATION FOR DYNAMIC MODELS problem is a special case of the general MODEL PARAMETER ESTIMATION problem. Therefore, we can construct an instance of the general MODEL PARAMETER ESTIMATION problem from the instance elements given above in the *Given* section as it is described above in the introduction of Section 12.4.

Model Structure Estimation for Dynamic Models

As pointed out in Section 12.2, we usually constrain the search for a model structure by giving the *set of candidate model structures* $\mathcal{M}^{(S)}$ as

$$\mathcal{M}^{(S)} = \{\mathcal{M}_j \mid j = 1, \ldots, M\}, \tag{12.62}$$

$$y^{(Mj)} = \mathcal{M}_j(u; p^{(Mj)}) \tag{12.63}$$

with elements of explicit dynamic parametrized models $\mathcal{M}_j$, $j = 1, \ldots, M$.

It is very important to have the set of candidate models as complete as possible because the search for the best model structure is only performed over this set. If the true model structure happens not to be in the set then we shall find a structure which is nearest to the true one but not the same.

The sources for candidate model structures include but are not limited to

- process knowledge (with engineering judgement),
- known structural elements, such as reaction kinetics,
- detailed models developed for other modelling goals such as design, optimization, simulation but with suitable model simplification.

MODEL STRUCTURE ESTIMATION FOR DYNAMIC MODELS

Given:

- A set of candidate model structures $\mathcal{M}^{(S)}$ consisting of M parametrized sampled explicit dynamic system models in the form of

$$y^{(Mj)} = \mathcal{M}_j(u;\ p^{(Mj)}), \quad j = 1, \ldots, M \tag{12.64}$$

with the model parameters $p^{(Mj)} \in \mathcal{R}^{\nu_j}$ being unknown, the vector-valued input variable $u(\tau) \in \mathcal{R}^r$, a vector-valued output variable $y^{(Mj)}(\tau) \in \mathcal{R}^m$ and both the input and output variables are varying in (discrete) time, i.e. $y^{(Mj)} : \mathcal{P} \rightarrow \mathcal{R}^m$ and $u : \mathcal{P} \rightarrow \mathcal{R}^r$ where $\mathcal{P}$ is the set of integer numbers.

- A *measurement record* consisting of the measured values of inputs and outputs

$$D[0,k] = \{(u(\tau), y(\tau)) \mid \tau = 0, \ldots, k\}, \qquad (12.65)$$

where $y(\tau)$ is assumed to contain measurement error while $u(\tau)$ is set without any error.
- A suitable signal norm $\|\cdot\|$ to measure the difference between the model output $y^{(Mj)}$ and the measured independent variables y to obtain the loss function of the estimations:

$$L^{(j)}(p^{(Mj)}) = \|r^{(j)}\|, \qquad (12.66)$$

where $r^{(j)}$ is the residual of the jth model element in the set of candidate model structures

$$r^{(j)}(\tau) = y(\tau) - y^{(Mj)}(\tau), \quad \tau = 0, \ldots, k. \qquad (12.67)$$

Compute:
An estimate of the model index j with its parameter set p^{Mj} such that $L^{(j)}(p^{(Mj)})$ is minimal over $j = 1, \ldots, M$.

Conceptual solution method:
Observe that we have to solve several MODEL PARAMETER ESTIMATION FOR DYNAMIC MODELS problems to generate the solution of the MODEL STRUCTURE ESTIMATION FOR DYNAMIC MODELS problem above. Therefore, we can solve the problem in the following steps:

1. For each $\mathcal{M}_j \in \mathcal{M}^{(S)}$ perform a MODEL PARAMETER ESTIMATION FOR DYNAMIC MODELS procedure and obtain $(p^{(Mj)}, L^{(j)}(p^{(Mj)}))$, $j = 1, \ldots, M$.
2. Select the model index j^* for which the loss $L^{(j)}(p^{(Mj)})$ is the smallest, i.e.

$$j^* = \arg \min_{j=1,..,M} L^{(j)}. \qquad (12.68)$$

The above problem statement and conceptual solution has the following properties:

1. MODEL STRUCTURE ESTIMATION FOR DYNAMIC MODELS problem contains or calls M times MODEL PARAMETER ESTIMATION FOR DYNAMIC MODELS as a subroutine.
2. As a side effect the parameter set $p^{(Mj^*)}$ of the optimal model structure $\mathcal{M}_{j^*}$ is also computed.
3. We have to perform a measure of fit test to the optimal model $\mathcal{M}_{j^*}$ to evaluate not only the quality of the estimated parameters $p^{(Mj^*)}$ but also the *completeness of the set of candidate model structures. The set of candidate model structures* $\mathcal{M}^{(S)}$ *was*
   - *not complete if* $L^{(j^*)}(p^{(Mj^*)})$ *is "too large",*
   - *"too rich" if the difference between* $L^{(j^*)}(p^{(Mj^*)})$ *and other* $L^{(j)}(p^{(Mj)})$*'s is "too small"*

compared to the variance of the measurement errors.

12.4.2. Models Linear and Time Invariant in their Parameters

For *each component* $y_\ell^{(M)}$, $\ell = 1, \ldots, m$ of dynamic models linear and time invariant in their parameters we can easily construct an instance of the LEAST SQUARES PARAMETER ESTIMATION problem described in Section 12.3 with the following items in its *Given* section:

- A linear in parameter dynamic system model in the form of

$$y_\ell^{(M)}(\tau) = d^{\mathrm{T}}(\tau)p = \sum_{i=1}^{n} \mathrm{d}_i(\tau)p_i^{(\ell)}, \quad \tau = 0, 1, \ldots \tag{12.69}$$

  with the model parameters $p^{(e)} \in \mathcal{R}^n$ being unknown, the vector-valued independent variables $d(\tau) \in \mathcal{R}^n$ and scalar-valued dependent variable $y_\ell^{(M)} \in \mathcal{R}$, where the elements in the data vector $d(\tau)$ are current and past input and output values, as given by

$$\begin{aligned} \mathrm{d}_i(\tau) = \{&y_j(\tau - \theta), u_k(\tau - \eta) \mid j = 1, \ldots, m;\ k = 1, \ldots, r; \\ &1 \leq \theta \leq n_y,\ 0 \leq \eta \leq n_u\}, \quad i = 1, \ldots, n. \end{aligned} \tag{12.70}$$

  In the more general linear in parameter case the elements in the data vector $d(\tau)$ can be complicated nonlinear functions of the current and past input and output values.
- A set of measured data consisting of N measurements for the pair $(y_\ell(\tau), d(\tau))$ in $\tau = 1, \ldots, N$ $(N \geq n)$ is arranged in a vector-matrix form as

$$y = \begin{bmatrix} y_\ell(1) \\ y_\ell(2) \\ \cdots \\ y_\ell(N) \end{bmatrix}, \quad \mathbf{X} = \begin{bmatrix} d_{11} & d_{12} & \cdots & d_{1n} \\ d_{21} & d_{22} & \cdots & d_{2n} \\ \cdots & \cdots & \cdots & \cdots \\ d_{N1} & d_{N2} & \cdots & d_{Nn} \end{bmatrix}, \tag{12.71}$$

  where $y_\ell(i)$ is assumed to contain measurement error.
- A loss function L_{SOS} which is now in an SOS form

$$L_{\mathrm{SOS}} = \sum_{i=1}^{N} r_i^2, \tag{12.72}$$

$$r_i = y_\ell(i) - y_\ell^{(M)}(i), \tag{12.73}$$

  where r is the *residual vector*.

We can perform the model parameter estimation by solving the LEAST SQUARES PARAMETER ESTIMATION for each component $y_\ell^{(M)}$, $\ell = 1, \ldots, m$ of the predicted output of dynamic models linear and time invariant in their parameters.

12.4.3. Nonlinear Models

In Section 12.3.2 on static models nonlinear in parameters, we have already mentioned and used the fact that the MODEL PARAMETER ESTIMATION problem is an optimization problem with the following elements:

- We have a *quadratic* SOS *loss function* L_{SOS} in Eq. (12.72) *to be minimized as a function of model parameters* $p_i^{(\ell)}$, $i = 1, \ldots, n$ in Eq. (12.69).
- We have a set of measured data given in Eq. (12.70).

Therefore, the *conceptual solution* of the model parameter estimation problem for nonlinear dynamic models attempts to solve the optimization problem in two steps:

1. *Formulate an* L_{SOS} *minimization problem.*
2. *Combine dynamic simulation (i.e. numerical solution) and optimization* in the following substeps:
   (a) Give an initial set of estimated parameters $\hat{p}^{(0)}$ to the DAE form lumped nonlinear dynamic process models with suitable initial conditions.
   (b) Solve the model equations using the previously estimated parameters $\hat{p}^{(k)}$ to obtain the model output $\{y^{(M)}(\tau;\ \hat{p}^{(k)}),\ \tau = t_0, t_1, \ldots, t_K\}$
   (c) Compute the loss function L_{SOS} as in Eq. (12.72).
   (d) Decide on whether the minimum is found.
   (e) If yes, then END otherwise generate a new estimate of the parameters $\hat{p}^{(k+1)}$ using a suitable optimizer and go to the solution step again.

The block diagram of the steps in the conceptual solution is shown in Fig. 12.4.

12.5. THE CSTR PROBLEM: A CASE STUDY OF MODEL PARAMETER ESTIMATION

Various model parameter estimation cases are solved, examined and compared in this section in order to deepen our understanding on model parameter estimation. The following process system is considered for all three model parameter estimation problems.

12.5.1. Isotherm CSTR with Fixed Volume

Consider an isothermal continuous stirred reactor where chemical reactions take place. It is assumed that the inlet and outlet flows are equal, the reactor volume is constant and all the physical properties remain constant.

The dynamic model of the reactor consists of the component mass balances in this case in the general form of

$$V\frac{dC_A}{dt} = FC_{A_i} - FC_A - Vr_A \tag{12.74}$$

for a general component A.

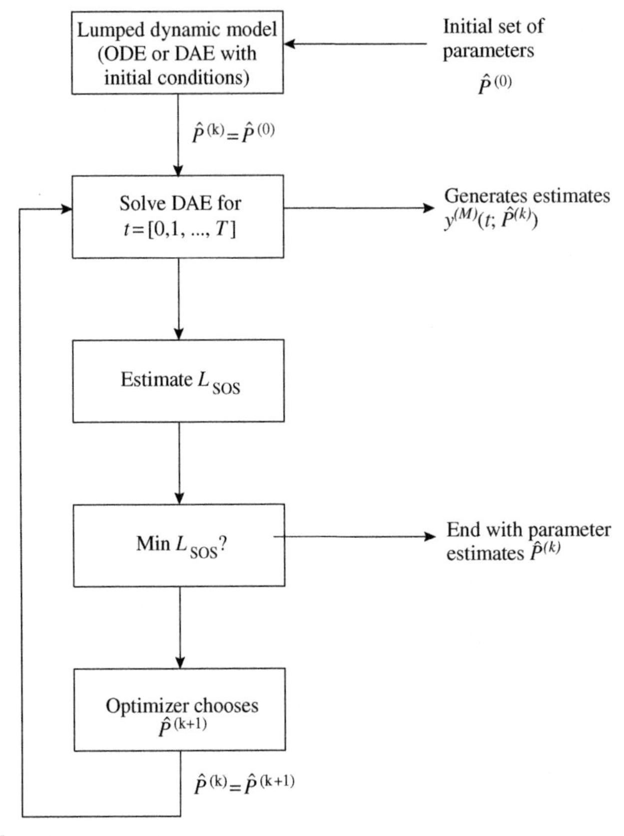

FIGURE 12.4 Parameter estimation for nonlinear models.

12.5.2. Parameter Estimation for Static Models—Linear in Parameters Single Parameter Case

Problem Statement

Given:

1. *Additional assumptions*
   - the volume of the reactor V is a known constant ($V = 3\,\text{m}^3$);
   - there is a single first order $A \rightarrow B$ reaction with the reaction rate

$$r_A = k_A C_A \quad \left[\frac{\text{kmol}}{\text{s}\,\text{m}^3}\right];$$

   - steady-state conditions.
2. *Measured variables*

   C_{A_i} inlet concentration of component A $[\text{kmol/m}^3]$,
   C_A concentration of component A in the reactor $[\text{kmol/m}^3]$
   F flowrate $[\text{m}^3/\text{sec}]$
3. *Parameters to be estimated*:

   k_A reaction rate coefficient for the reaction $A \rightarrow B$.

TABLE 12.1 Measured data for static model single parameter case

C_{A_i} [kmol/m^3]	C_A [kmol/m^3]	F [m^3/s]
0.3	0.2441	0.005
0.32	0.2659	0.005
0.34	0.2824	0.005
0.36	0.301	0.005
0.38	0.3151	0.005
0.4	0.3333	0.005
0.42	0.3556	0.005
0.44	0.3698	0.005
0.46	0.3844	0.005
0.48	0.3963	0.005
0.5	0.4164	0.005

Modelling for Parameter Estimation

Take the general dynamic conservation balance Eq. (12.74) in steady state, that is when $\mathrm{d}C_A/\mathrm{d}t = 0$ and rearrange the terms to get

$$C_{A_i} - C_A = k_A \frac{VC_A}{F}. \tag{12.75}$$

The above equation is a static model linear in its parameter k_A in the form of $y = kx$ with

$$y = C_{A_i} - C_A, \quad x = \frac{VC_A}{F}.$$

Data

The measured data are given in Table 12.1. Observe that the number of data points is $m = 11$.

The Results of LS Parameter Estimation

The parameter estimation can be performed by standard LS method using the fact that the transformed static model (12.75) is linear in its parameter.

If we perform the parameter estimation, then the following result is obtained:

$$\hat{k}_A = 5.00 \cdot 10^{-4}, \quad s_{k_A}^2 = 2.24 \cdot 10^{-10}, \quad L_{\mathrm{SOS}} = 1.38 \cdot 10^{-4}. \tag{12.76}$$

Observe that *the result is a triplet* of three numbers in the single parameter case.

We can use the estimated variance $s_{k_A}^2$ and the obtained sum of squares L_{SOS} to evaluate the quality of the estimate. The square root of $s_{k_A}^2$ indicates the quality of the estimate if compared to the mean value $\hat{k}_A$ which is roughly 5% in this case meaning that it is a good estimate. The value L_{SOS}/m is a measure of fit for the linear in parameter model itself. $L_{\mathrm{SOS}}/m = 1.25 \cdot 10^{-5}$ is in this case indicating that a good fit is obtained compared to the variance of the estimated parameters.

12.5.3. Parameter Estimation for Static Models—Linear in Parameters Two Parameter Case

Problem Statement

Given:

1. *Additional assumptions*
   - the volume of the reactor V is a known constant ($V = 3\,\text{m}^3$);
   - there are two consecutive first-order reactions $A \to B$ and $B \to C$ with the reaction rates

$$r_A = -k_A C_A, \quad r_B = -k_B C_B;$$

   - steady-state conditions.

2. *Measured variables*

   C_{A_i} inlet concentration of component A,
   C_A concentration of component A in the reactor,
   C_{B_i} inlet concentration of component B,
   C_B concentration of component B in the reactor,
   F flowrate.

3. *Parameters to be estimated*

   k_A reaction rate coefficient for the reaction $A \to B$,
   k_B reaction rate coefficient for the reaction $B \to C$.

Modelling for Parameter Estimation

In this case, the dynamic model of the CSTR consists of two component mass balance equations in the form:

$$V\frac{\mathrm{d}C_A}{\mathrm{d}t} = FC_{A_i} - FC_A - Vk_A C_A,$$

$$V\frac{\mathrm{d}C_B}{\mathrm{d}t} = FC_{B_i} - FC_B + Vk_A C_A - Vk_B C_B.$$

Taking them both in steady state, that is when $\mathrm{d}C_A/\mathrm{d}t = 0$ and $\mathrm{d}C_B/\mathrm{d}t = 0$ and rearranging the terms we get:

$$C_{A_i} - C_A = k_A\frac{VC_A}{F}, \tag{12.77}$$

$$C_{B_i} - C_B = -k_A\frac{VC_A}{F} + k_B\frac{VC_B}{F}. \tag{12.78}$$

The above equations form a static model linear in its parameters $p = [k_A,\ k_B]^{\mathrm{T}}$.

We may observe, however, that we do not need both equations to be able to estimate the parameters but just the second one (12.78). It contains both parameters

TABLE 12.2 Measured data for static model two parameter case

C_{A_i} [kmol/m^3]	C_A [kmol/m^3]	C_{B_i} [kmol/m^3]	C_B [kmol/m^3]	F [m^3/s]
0.3	0.2495	0.6	0.0311	0.005
0.32	0.2638	0.6	0.0312	0.005
0.34	0.2866	0.6	0.0313	0.005
0.36	0.3021	0.6	0.0314	0.005
0.38	0.3231	0.6	0.0319	0.005
0.4	0.3273	0.6	0.0318	0.005
0.42	0.3437	0.6	0.0322	0.005
0.44	0.3681	0.6	0.0323	0.005
0.46	0.3867	0.6	0.0328	0.005
0.48	0.3987	0.6	0.0323	0.005
0.5	0.4192	0.6	0.0329	0.005

and it is static and linear in its parameters in the form of $y = p^T x$ with

$$y = C_{B_i} - C_B, \quad p = [k_A, \ k_B]^T, \quad x = \left[-\frac{VC_A}{F}, \ \frac{VC_B}{F} \right]^T.$$

Data

The measure data set consisting of $m = 11$ items is given in Table 12.2.

The Results of LS Parameter Estimation

The parameter estimation can be performed by standard LS method using the transformed static model equation (12.78) which is linear in its parameters.

The LS parameter estimation results in the following data:

$$\hat{k}_A = 5.00 \cdot 10^{-4}, \quad \hat{k}_B = 4.50 \cdot 10^{-2}, \tag{12.79}$$

$$\mathbf{COV}\{\hat{p}\} = \begin{bmatrix} 8.23 \cdot 10^{-7} & 8.62 \cdot 10^{-6} \\ 8.62 \cdot 10^{-6} & 9.21 \cdot 10^{-5} \end{bmatrix}, \tag{12.80}$$

$$L_{SOS} = 0.297. \tag{12.81}$$

Observe that *the result is a triplet* again but now we have the vector of the estimated parameter, its covariance matrix and the sum of squares being a scalar.

We can use the following derived quantities to assess the quality of the estimate and to measure the fit:

1. *The empirical variances $s^2_{k_A}$ and $s^2_{k_A}$* to compute their square roots and compare them with the mean values, that is

$$s_{k_A} = 9.07 \cdot 10^{-4}, \quad s_{k_B} = 9.60 \cdot 10^{-3}.$$

These values indicate that our estimate is poor as these values are higher than the mean value of the estimate.

2. *The correlation coefficient* ρ *of the estimated parameters* which is computed according to the formulae:

$$\rho = \frac{[\mathbf{COV}\{\hat{p}\}]_{12}}{s_{k_A} s_{k_B}}, \tag{12.82}$$

where $[\mathbf{COV}\{\hat{p}\}]_{12}$ is the off-diagonal element of the covariance matrix. Note that $1 \geq \rho \geq -1$ is a measure of the dependence between the estimated values: if it is small compared to 1, then we have a good, uncorrelated estimate for the parameters. Unfortunately in our case $\rho = 0.99$ is another sign of a poor quality estimate.

3. *The obtained sum of squares* L_{SOS} divided by the number of measurements to assess the measure of fit $L_{SOS}/m = 2.7 \cdot 10^{-2}$ indicates that we have problems even with the fit. This may be a consequence of a poor quality parameter estimate.

Note that this poor quality estimate could have been avoided by experimental design by properly selecting the measurement points in advance.

12.5.4. Parameter Estimation for Dynamic Models—Linear in Parameters Single Parameter Case

Problem Statement

Given:

1. *Additional assumptions*
   - the volume of the reactor V is a known constant ($V = 3\,\mathrm{m}^3$);
   - the equidistant sampling interval h is a known constant ($h = 2\,\mathrm{s}$);
   - there is a single first-order $A \to B$ reaction with the reaction rate

$$r_A = -k_A C_A.$$

2. *Measured variables* measured in every sampling interval h to get N sampled data sets altogether

   $C_{A_i}(k)$ inlet concentration of component A,
   $C_A(k)$ concentration of component A in the reactor,
   $F(k)$ flowrate,

   with $k = 1, \ldots, N$.

3. *Parameters to be estimated*

   k_A reaction rate coefficient for the reaction $A \to B$.

Modelling for Parameter Estimation

We use the sampled version of the general dynamic model equation (12.74) in the form of:

$$\frac{V}{h}(C_A(k+1) - C_A(k)) = F(k)C_{A_i}(k) - F(k)C_A(k) - Vk_A C_A(k). \tag{12.83}$$

TABLE 12.3 Measured data for dynamic model single parameter case

Time [s]	C_{A_i} [kmol/m^3]	C_A [kmol/m^3]	F [m^3/s]
0	0.333	0.3333	0.005
2	0.337	0.333	0.005
4	0.341	0.3327	0.005
6	0.345	0.3324	0.005
8	0.349	0.3322	0.005
10	0.353	0.3319	0.005
12	0.3569	0.3317	0.005
14	0.3609	0.3315	0.005
16	0.3649	0.3314	0.005
18	0.3688	0.3312	0.005
20	0.3727	0.3311	0.005
22	0.3766	0.331	0.005
24	0.3805	0.3309	0.005
26	0.3844	0.3308	0.005
28	0.3883	0.3307	0.005
30	0.3921	0.3307	0.005
32	0.3959	0.3307	0.005
34	0.3997	0.3307	0.005
36	0.4035	0.3307	0.005
38	0.4072	0.3308	0.005

We can use the sampled dynamic model equation (12.83) directly for parameter estimation. The most simple form of the above model equation suitable for parameter estimation is

$$C_{A_i}(k) - C_A(k) - \frac{V}{h}\frac{C_A(k+1) - C_A(k)}{F(k)} = k_A \frac{C_A(k)V}{F(k)}, \tag{12.84}$$

which is linear in its parameters in the form of $y = kx$, where

$$y = C_{A_i}(k) - C_A(k) - \frac{V}{h}\frac{C_A(k+1) - C_A(k)}{F(k)}, \quad x = \frac{C_A(k)V}{F(k)}.$$

Note that now we have measurement errors for all measured variables, therefore the measurement errors for y and x are correlated.

Data

The measured dynamic data are given in Table 12.3. Observe that the first column indicates time.

The Results of LS Parameter Estimation

The parameter estimation can be performed by the standard LS method using the fact that the transformed model (12.84) is linear in its parameter.

The LS parameter estimate gives the following result:

$$\hat{k}_A = 5.00 \cdot 10^{-4}, \quad s^2_{\hat{k}_A} = 1.20 \cdot 10^{-10}, \quad L_{SOS} = 1.17 \cdot 10^{-3}. \qquad (12.85)$$

The results can be compared to the similar ones obtained for the same CSTR but using its static model and steady-state data given in Eq. (12.76). Comparing the corresponding elements in the two results, we come to the conclusion that the quality of the estimate is roughly the same for the dynamic and the static case.

12.6. STATISTICAL MODEL VALIDATION VIA PARAMETER ESTIMATION

The *principle of statistical model* validation is to compare by the methods of mathematical statistics either

- the (measured) system output with the model output, or
- the estimated system parameters with the model parameters.

In other words "validation" means "comparison" of

$$(y \text{ and } y_M) \quad \text{or} \quad (p \text{ and } p_M).$$

Statistical methods are needed because the measured output y is corrupted by measurement (observation) errors

$$y = y^{(M)} + \varepsilon.$$

As was emphasized in Section 12.1.2 on model calibration, model validation is similar to model calibration because here we also use measured data, but another, independently measured data set (*validation data*) and also use statistical methods. The *items of the instance of a model validation problem* are as follows:

- a developed and calibrated *process model*,
- measured data from the real process system which we call *validation data*,
- a predefined measure of fit, or *loss function* which measures the quality of the process model with its missing parameters and/or structure elements estimated.

The conceptual steps to be carried out when performing model validation are also similar to that of model calibration and include:

1. *Analysis of the process model*
   This step may involve the analysis of the uncertainties in the calibrated process model and its sensitivity analysis. The results can be applied for designing experiments for the model validation.
2. *Sampling of continuous time dynamic models*
   As before in model calibration.
3. *Data analysis and preprocessing*
   As before in model calibration. (The methods for this are described in Section 11.2.)
4. *Model parameter and structure estimation*

5. *Evaluation of the quality*
   The evaluation is done by using either empirical graphical methods or by exact hypothesis testing if the mathematical statistical properties of the estimates are available. More details are given about the measure of fit in Section 12.3.

In order to perform statistical model validation, we need to be able to solve the following mathematical problems.

1. Hypothesis testing in case of
   - *steady-state hypothesis* for single and multiple variables,
   - comparing step responses.
2. Parameter estimation in most of the cases for the linear in parameters case together with hypothesis testing on the estimated parameters.

12.7. SUMMARY

Basic methods and procedures for statistical model calibration and validation are included in this chapter based on a grey-box modelling approach.

The basic technique we use for model calibration and validation is model parameter and structure estimation for both static and dynamic models. The procedure for model parameter estimation together with the analysis of the distribution and statistics of the estimated parameters has been covered. The way of constructing confidence intervals and methods for assessing the fit have also been discussed.

12.8. FURTHER READING

The tools and techniques of model calibration and validation are not just for process models but for dynamic models of any kind. The origin of the techniques is in mathematical statistics and in systems and control theory under the name of "identification". Therefore, further reading is recommended in both of the above mentioned fields.

A recent and excellent review on model validation and model error modelling can be found in the technical report [85] which is available through the Web using the address `http://www.control.isy.liu.se`.

There are several special books and monographs available about parameter estimation in process systems such as [86–88].

12.9. REVIEW QUESTIONS

Q12.1. What is the difference between the *black-box, grey-box* and *white-box* models of the same process system. Illustrate your answer on the Example 12.1.1 in Section 12.1.

Q12.2. What are the statistical properties of the estimated parameters obtained by LS estimation for models linear in parameters and with Gaussian independent measurement errors? How can we construct confidence intervals for the individual parameters and how these intervals relate to the joint confidence region of the parameters? (Section 12.2.2)

Q12.3. What are the methods available for assessing the fit of a model linear in parameters when its parameters have been estimated by an LS method? (Section 12.2.2)

Q12.4. Give examples of process models linear in their parameters. Describe the checklist for performing LS parameter estimation for static models linear in parameters together with the possible consequences when an item in the list fails. (Section 12.3)

Q12.5. What are the methods available for performing an LS parameter estimation for models nonlinear in their parameters? How would you compute the confidence intervals for the estimated parameters? (Section 12.2.2)

Q12.6. What are the additional issues or problems to be considered for dynamic models compared with the static case when performing model parameter and/or structure estimation? (Section 12.4)

Q12.7. What are the elements in the LEAST SQUARES PARAMETER ESTIMATION problem when we apply it to model parameter estimation for dynamic models linear and time invariant in their parameters? (Section 12.4)

Q12.8. Describe the steps of the conceptual solution of the model parameter estimation for dynamic models nonlinear in their parameters. (Section 12.4)

Q12.9. Compare model calibration and model validation as a statistical procedure. What are the similarities and differences? (Sections 12.1.2 and 12.6)

12.10. APPLICATION EXERCISES

A12.1. Develop a grey-box lumped model of a simple countercurrent concentric tube heat exchanger described in Example 7.2.1. A simple 3-lump model is found in Chapter 13 as Example 13.2.2. (Section 12.1). Show how you would set up a parameter estimation problem for the heat transfer coefficients.

A12.2. Consider the CSTR described in Example 10.3.1, i.e. consider a CSTR with a first order $A \longrightarrow B$ reaction. The reaction rate expression is $r_A = kC_A^2$. It is assumed that the physical properties remain constant and that inlet and outlet flows are equal. Develop a static model for this system, show that it is linear in parameters and describe the corresponding parameter estimation problem. (Section 12.3)

A12.3. Consider a CSTR described in Example 12.2.1, i.e. consider a CSTR with a first order $A \longrightarrow B$ reaction where we do not exactly know the form of the reaction kinetic expression but we have three candidate reaction kinetic expressions

1. $r_A = kC_A$,
2. $r_A = kC_A^{1/2}$,
3. $r_A = kC_A/(C_A^* - C_A)$,

The reaction rate constant k is given by $k = k_0 e^{-E/(RT)}$. It is assumed that the physical properties remain constant and that inlet and outlet flows are equal.

(a) Develop a static model for this system.

(b) Assume that the concentration and temperature of the inlet and that of the reactor, as well as the coolant temperature are measured in time. Construct the MODEL STRUCTURE ESTIMATION problem instance for this problem.

(c) Which are the parameters belonging to each of the three related MODEL PARAMETER ESTIMATION problem? (section 12.3)

A12.4. Experiments were carried out on the melting of an ice-cube in a glass of water for a range of stirring speeds. The glass had a diameter of 0.075 m and three experiments at stirring speeds of 1, 2 and 3 revolutions per second were done. The initial conditions for each experiment were:

Speed [rev/s]	Initial ice weight [g]	Initial water weight [g]
1	12.40	207.80
2	14.28	205.00
3	14.85	205.08

The temperature profiles obtained using a thermometer were:

Time [s]	Temperature [°C]		
	1 rev/s	2 rev/s	3 rev/s
0	15.5	16.0	15.5
15	14.5	14.5	14.0
30	14.0	13.5	13.0
45	13.0	12.5	12.0
60	12.5	12.0	11.0
75	12.0	11.5	10.5
90	12.0	11.0	10.5
105	11.5	10.5	
120	11.5	10.5	
135	11.0	10.5	
150	11.0		
165	11.0		
180	11.0		

Recording was stopped when the ice had fully melted.

Develop a model for this problem.

Use this information to estimate the convective heat transfer coefficient for the melting ice by taking two data sets for calibration and then using the other for validation. Make the heat transfer coefficient a function of stirring speed.

A12.5. A series of experiments were carried out to observe the dynamics of a pool of evaporating ethanol under various conditions of radiation and windspeed. An amount of ethanol was added to a circular dish of diameter 0.15 m and the rate of evaporation was measured under four conditions which varied windspeed and radiation. Windspeed was measured using an anemometer and incident radiation levels via a pyranometer. Liquid temperature was measured with a thermocouple.

For an averaged airspeed of 3.5 m/s the following data were obtained for the two radiation scenarios.

High windspeed

Time [min]	0 W/m²		1690 W/m²	
	Weight [g]	Temp. [°C]	Weight [g]	Temp. [°C]
0	99.91	16.1	100.05	18.8
2	97.05	11.3	95.50	17.5
4	95.34	9.9	91.90	17.6
6	93.45	8.6	88.20	18.0
8	91.69	8.6	85.00	17.4
10	90.20	8.6	82.40	17.1
12	88.63	8.6	79.20	17.4
14	87.00	8.4	76.90	17.2
16	85.43	8.4	73.60	17.2

The results for an averaged windspeed of 2.2 m/s were:

Low windspeed

Time [min]	0 W/m²		1690 W/m²	
	Weight [g]	Temp. [°C]	Weight [g]	Temp. [°C]
0	99.10	15.7	79.00	16.0
2	97.10	11.2	77.26	20.5
4	95.70	10.3	74.81	22.4
6	94.50	9.6	72.44	23.0
8	93.50	9.3	69.54	23.2
10	92.30	8.9	65.48	24.5
12	91.00	8.9	62.35	24.3
14	89.80	8.8	58.07	24.3
16	88.50	8.6	—	—

(a) Review the data for quality.
(b) Use the data to help calibrate and then validate a model of an evaporating solvent pool.

13 ANALYSIS OF DYNAMIC PROCESS MODELS

Lumped dynamic process models form an important class of process models. They are in the form of DAE systems with given initial conditions. Their main application area is process control and diagnosis where they are the models which are almost extensively used.

This chapter deals with the analysis of various dynamical system properties relevant to process control and diagnosis applications. Therefore, only lumped parameter process models are treated throughout the chapter. Analytical as well as structural approaches and techniques are described.

The material presented in this chapter contains basic material for analysing dynamical properties which can then be used as model verification tools. The material is arranged in sections as follows:

- The design of model-based control and diagnostic methods usually starts with analysis of system properties. Therefore, the initial section covers the basic notions and most important analysis methods for stability, controllability and observability (Section 13.1).
- Section 13.2 deals with the analysis of structural control properties, which are dynamical properties valid for a class of process systems with the same structure.
- Dynamic process models for control and diagnostic purposes should be relatively simple, containing only a few state variables. Therefore, model simplification and reduction play a key role. This is the subject of Section 13.3.

13.1. ANALYSIS OF BASIC DYNAMICAL PROPERTIES: CONTROLLABILITY, OBSERVABILITY AND STABILITY

This section deals with the basic dynamic system properties of controllability, observability and stability. These properties are introduced for the general nonlinear SSR in the form of:

$$\dot{x} = f(x, u), \qquad y = g(x, u), \tag{13.1}$$

where f and g are given nonlinear functions.

Conditions to check controllability, observability and stability will be given for LTI systems with finite-dimensional representations in the form:

$$\dot{x} = \mathbf{A}x + \mathbf{B}u, \qquad y = \mathbf{C}x. \tag{13.2}$$

Observe that from now on we assume $\mathbf{D} = 0$ in the general form of the SSR in Eq. (13.2). Therefore, a SSR will be characterized by the triplet $(\mathbf{A}, \mathbf{B}, \mathbf{C})$. Note also that SSRs are not unique: there is infinitely many equivalent SSRs giving rise to the same input–output behaviour as shown in Section 10.2.2.

13.1.1. State Observability

The state variables of a system are *not* directly observable. Therefore, we need to determine the value of the state variables at any given time from the measured inputs and outputs in such a way that we only use functions of inputs and outputs and their derivatives together with the known system model and its parameters. We need to be able to perform this task if we want to control the states or if we want to operate a diagnostic system based on the value of state variables. *A system is called (state) observable, if from a given finite measurement record of the input and output variables, the state variable can be reconstructed at any given time.*

The problem statement of state observability can be formalized as follows:

STATE OBSERVABILITY

Given:
The form of the SSR (13.1) together with its parameters and a measurement record for inputs and outputs

$$\mathcal{D}[t_0, T] = \{u(\tau),\ y(\tau) \mid t_0 \le \tau \le T\}. \tag{13.3}$$

Question:
Is it possible to determine uniquely the value of the state variable $x(t_0)$?

Note that it is sufficient to determine the value of the state variable at a given time instance because the solution of the state equation (13.1) gives us the state at any other time instance.

LTI systems
In the case of LTI systems, the SSR matrices $(\mathbf{A}, \mathbf{B}, \mathbf{C})$ are sufficient and enough to specify a given system.

An LTI system may or may not be *state observable*, meaning that the problem above does not always have a unique solution for every $(\mathbf{A}, \mathbf{B}, \mathbf{C})$. The following theorem gives necessary and sufficient conditions for an LTI system to be state observable.

THEOREM 13.1.1. *Given an LTI with its SSR* $(\mathbf{A}, \mathbf{B}, \mathbf{C})$. *This SSR with state space* $\mathcal{X}$ *is* state observable *if and only if the* observability matrix $\mathcal{O}_n$ *is of full rank*

$$\mathcal{O}_n = \begin{bmatrix} \mathbf{C} \\ \mathbf{CA} \\ \vdots \\ \mathbf{CA}^{n-1} \end{bmatrix}, \tag{13.4}$$

that is

$$\operatorname{rank} \mathcal{O}_n = n,$$

where $\dim \mathcal{X} = n$ *is the dimension of the state vector.*

Note that *(state) observability is a realization property* and therefore does depend on the realization because the rank of the observability matrix may change if we perform any transformation of the SSR. See Eq. (10.30) for this condition.

EXAMPLE 13.1.1 (Observability matrix of a linearized CSTR model). Derive the observability matrix of the linearized CSTR model described in Examples 10.3.1 and 10.4.5 considering only the coolant temperature T_C as input and the concentration of component A (C_A) as output.

The symbolic form of the state space model matrices can be obtained from the matrices (10.114) by omitting the first column from $\mathbf{B}$ and the last row from $\mathbf{C}$:

$$\mathbf{A} = \begin{bmatrix} -a_{11} & -a_{12} \\ +a_{13} & -a_{14} \end{bmatrix}, \quad \mathbf{B} = \begin{bmatrix} 0 \\ +b_{22} \end{bmatrix}, \quad \mathbf{C} = \begin{bmatrix} 1 & 0 \end{bmatrix}. \tag{13.5}$$

Using these matrices the observability matrix in symbolic form is

$$\mathcal{O} = \begin{bmatrix} 1 & 0 \\ -a_{11} & -a_{12} \end{bmatrix}. \tag{13.6}$$

It is seen that $\mathcal{O}$ is of full rank and is a square matrix because there is only one input variable ($r = 1$).

There are ready and easy-to-use procedures in MATLAB to compute the observability matrix of an LTI system with realization $(\mathbf{A}, \mathbf{B}, \mathbf{C})$ (`obsv`) and check its rank (`rank`). Their use is illustrated by the following example.

EXAMPLE 13.1.2 (Observability matrix of simple SISO LTI model). Derive the observability matrix of the simple LTI SISO model described in Example 10.2.1.

The observability matrix will be a square matrix because we have a SISO system. We can use the MATLAB procedure `ob=obsv(A,C)` to obtain

$$\texttt{ob} = \begin{bmatrix} 1 & 0 \\ -4 & 3 \end{bmatrix}.$$

Observe the presence of the output matrix **C** in the first block being the first row of the observability matrix.

The observability matrix above has rank 2 computed by the MATLAB procedure `rank(ob)`, therefore the system is observable. ■ ■ ■

The following simple example illustrates the case when we have two outputs and an unstable system.

EXAMPLE 13.1.3 (Observability matrix of simple MIMO LTI model). Derive the observability matrix of the following unstable LTI MIMO model specified by its state space realization matrices (**A**, **B**, **C**):

$$\mathbf{A} = \begin{bmatrix} 0 & 0 \\ 5 & -6 \end{bmatrix}, \quad \mathbf{B} = \begin{bmatrix} 7 & 0 \\ 0 & 8 \end{bmatrix}, \quad \mathbf{C} = \begin{bmatrix} 1 & 0 \\ 0 & 1 \end{bmatrix}.$$

The observability matrix is a 4×2 matrix this time with the input matrix in its first block:

$$\mathcal{O} = \begin{bmatrix} 1 & 0 \\ 0 & 1 \\ 0 & 0 \\ 5 & -7 \end{bmatrix}.$$

Its rank is now 2 due to the full rank upper block being the output matrix **C** which ensures observability even when the state matrix **A** and therefore the second block **CA** in the observability matrix has rank 1. ■ ■ ■

13.1.2. State Controllability

For process control purposes over a wide operation range, we need to drive a process system from its given initial state to a specified final state. A system is called *(state) controllable* if we can always find an appropriate manipulable input function which moves the system from its given initial state to a specified final state in finite time. This applies for every given initial state final state pair.

The problem statement of state controllability can be formalized for LTI systems as follows:

STATE CONTROLLABILITY

Given:
The SSR form with its parameters as in Eq. (13.1) and the initial $x(t_1)$ and final $x(t_2) \neq x(t_1)$ states respectively.

Question:
Is it possible to drive the system from $x(t_1)$ to $x(t_2)$ in finite time?

LTI systems
For LTI systems there is a necessary and sufficient condition for state controllability which is stated in the following theorem.

THEOREM 13.1.2. *A SSR* (**A**, **B**, **C**) *is state controllable if and only if the controllability matrix* $\mathcal{C}$:

$$\mathcal{C}_n = \begin{bmatrix} \mathbf{B} & \mathbf{AB} & \cdots & \mathbf{A}^{n-1}\mathbf{B} \end{bmatrix} \tag{13.7}$$

is of full rank, that is

$$\text{rank}\, \mathcal{C}_n = n.$$

Note again that *controllability is a realization property*, and it may change if we apply state transformations to the SSR. *Joint controllability and observability*, however, *is a system property* being invariant under state transformations of SSR.

The following example illustrates the computation of the controllability matrix.

EXAMPLE 13.1.4 (Controllability matrix of a linearized CSTR model). Derive the controllability matrix of the linearized CSTR model described in Examples 10.3.1 and 10.4.5 by considering only the coolant temperature T_C as input and the concentration of component A (C_A) as output.

The symbolic form of the state space model matrices for this case has been derived in Example 13.1.1. With the matrices (13.5) the following observability matrix results:

$$\mathcal{C} = \begin{bmatrix} 0 & -a_{12}b_{22} \\ b_{22} & -a_{22}b_{22} \end{bmatrix}. \tag{13.8}$$

It is seen that $\mathcal{C}$ is of full rank and it is a square matrix because there is only one output variable ($m = 1$). ■ ■ ■

There are ready and easy-to-use procedures in MATLAB to compute the controllability matrix of an LTI system with realization (**A**, **B**, **C**) (`ctrb`) and check its rank (`rank`). Their use is illustrated by the following example.

EXAMPLE 13.1.5 (Controllability matrix of simple SISO LTI model). Derive the controllability matrix of the simple LTI SISO model described in Example 10.2.1.

The controllability matrix will be a square matrix because we have a SISO system. We can use the MATLAB procedure `co=ctrb(A,C)` to obtain

$$\texttt{co} = \begin{bmatrix} 7 & -4 \\ 8 & -13 \end{bmatrix}.$$

Observe the presence of the input matrix **B** in the first block being the first column of the controllability matrix.

The controllability matrix above has rank 2 (computed by the MATLAB procedure `rank(co)`), therefore the system is controllable. ■ ■ ■

The following simple example illustrates the case when we have two inputs and an unstable system.

EXAMPLE 13.1.6 (Controllability matrix of a simple MISO LTI model). Derive the controllability matrix of the simple LTI MISO model described in Example 10.2.3.

The controllability matrix is now a 2×4 matrix with the 2×2 input matrix **B** in its first block:

$$\mathcal{C} = \begin{bmatrix} 7 & 0 & -28 & 24 \\ 0 & 8 & 35 & -48 \end{bmatrix}.$$

Its rank is 2, therefore the system is controllable. Observe that if the input matrix **B** is of full rank in this case then the state matrix "cannot destroy" controllability. ■ ■ ■

13.1.3. Asymptotic and Bounded Input Bounded Output Stability

There are two related but different kinds of stability we investigate:

- BIBO stability which is also known as external stability,
- asymptotic stability, known as internal stability.

They are defined for both SISO and MIMO systems. In the case of LTI systems there are necessary and sufficient conditions to check both kinds of stability [70].

BIBO Stability

The system is BIBO or externally stable if it responds to any bounded-input signal with a bounded-output signal. Hence, for any

$$\{\|u(t)\| \leq M_1 < \infty \mid -T \leq t \leq \infty\} \;\Rightarrow\; \{\|y(t)\| \leq M_2 < \infty \mid -T \leq t \leq \infty\}, \tag{13.9}$$

where $\| \cdot \|$ is a suitable vector norm and M_1, M_2 are constants. For more about vector and signal norms refer to Section A.3 in Appendix A.

BIBO stability is clearly a *system property*. If we consider the system represented by its input–output operator $\mathbf{S}$ then the definition in Eq. (13.9) can be reformulated to have

$$\|\mathbf{S}\| \leq S_M < \infty,$$

where $\| \cdot \|$ is the so-called induced operator norm of the system operator induced by the signal norms applied for u and y.

Note that the definition above says nothing about the states. The variation of the state variables may or may not be bounded, given a bounded input signal to the system.

LTI systems
For SISO LTI systems there is a necessary and sufficient condition for BIBO stability.

LEMMA 13.1.1. *An SISO LTI system is BIBO or externally stable if and only if*

$$\int_0^{\infty} |h(t)\,|\,\mathrm{d}t \leq M < \infty, \tag{13.10}$$

where M is a constant and $h(t)$ is the impulse response function of the system.

Asymptotic or Internal Stability

A solution to the state equation of a system is asymptotically stable if a "neighbouring" solution, resulting from a perturbation and described by a different initial condition has the same limit as $t \to \infty$. The following defining conditions formulate the above in a rigorous mathematical way.

Consider a nonlinear lumped parameter system described by the state space model in Eq. (13.1) with the initial condition for the state vector $x(0) = x_0^0$ giving rise to the *nominal solution* $x^0(t),\ t \leq 0$. The nominal solution of the system is asymptotically stable if whenever we take another different initial condition x_0 such that $\|x_0 - x_0^0\| < \delta$ with $\delta > 0$ being sufficiently small, then

$$\|x(t) - x^0(t)\| \to 0 \quad \text{if } t \to \infty,$$

where $\| \cdot \|$ is a vector norm and $x(t)$, $t \leq 0$ is the *perturbed solution* with initial condition $x(0) = x_0$.

LTI systems
Unlike nonlinear systems which may have stable and unstable solutions *stability is a system property for LTI systems.* An LTI system with realization $(\mathbf{A}, \mathbf{B}, \mathbf{C})$ is *asymptotically or internally stable* if the solution $x(t)$ of the truncated state equation:

$$\dot{x}(t) = \mathbf{A}x(t), \quad x(t_0) = x_0 \neq 0, \ t > t_0 \tag{13.11}$$

satisfies

$$\lim_{t \to \infty} x(t) = 0. \tag{13.12}$$

We usually omit the word "asymptotic" from asymptotic stability if it does not create any confusion. We call a system "stable" if it is asymptotically stable.

Note that the above condition is similar to asymptotic stability of linear ODEs. The notion has appeared first in the context of analysing numerical stability of ODEs and DAEs in Section 5.11.5.

There is an easy-to-check necessary and sufficient condition for asymptotic stability of LTI systems.

THEOREM 13.1.3. *An LTI system is internally stable if and only if all of the eigenvalues, $\lambda_i(\mathbf{A})$ of the state matrix $\mathbf{A}$ have strictly negative real parts:*

$$\text{Re}\{\lambda_i(\mathbf{A})\} < 0, \quad \forall i.$$

Again, note that the eigenvalues of the state matrix remain unchanged by state transformations, and hence *stability is indeed a system property for LTI systems.*

The following simple examples illustrate how one can use MATLAB to check stability of LTI systems.

EXAMPLE 13.1.7 (Stability of a simple SISO LTI system). Check the asymptotic stability of the simple LTI SISO system described in Example 10.2.1.

The eigenvalues of the state matrix $\mathbf{A}$ are computed using the MATLAB procedure `eig(A)` to obtain $-1, -9$. This shows that the system is asymptotically stable.

The stability of the system implies that the unit step response function will converge to a finite value as time goes to infinity. This is seen in Fig. 10.4.

EXAMPLE 13.1.8 (Stability of a simple MISO LTI system). Check the asymptotic stability of the simple LTI MISO system described in Example 10.2.3.

The eigenvalues of the state matrix $\mathbf{A}$ are $-6, 0$. This means that the system is *not* asymptotically stable.

If we have an unstable system then we cannot expect the unit step response computed with respect to any input by `step(A,B,C,D,i)` to converge to a finite value. This is seen in Fig. 10.6 of the unit step for the first ($i = 1$) input variable u_1.

13.2. ANALYSIS OF STRUCTURAL DYNAMICAL PROPERTIES

As it has been described in Section 10.5.1 the structure of a linear or linearized state space model is given in terms of its structure indices and its (signed) structure state space model matrices ({**A**}, {**B**}, {**C**}). In this way a *class of LTI state space models* can be specified where the members in the class possess the same structure. If every member in the class, with the exception of a null measure set, has a dynamical system property, such as stability, controllability or observability, then we say that the class with the given structure has this *structural property*. We call these properties *structural stability, controllability or observability* respectively.

In some important cases, the presence of structural properties is the consequence of the phenomena taking place in the process system. This structural property, say structural stability, is then guaranteed for the class of systems defined by the phenomena. This means, that stability holds for any member of the class irrespectively of the equipment sizes, variations in the flowsheet, etc. We shall see examples of this in the subsequent subsections.

Because the structure of the process system depends mainly on the flowsheet and on the phenomena taking place, simple combinatorial methods are available to check structural properties, such as finding directed paths in a graph or computing the structural rank of a structure matrix. This makes it easy to analyse structural properties even in the early phases of process design.

The presence of a structural property in a class of process systems implies that the property holds for the members of the class but the presence of the property in a member of the class does not guarantee that property for the whole class. This means, that one may fail to show a structural property for a class, despite the fact that most members of the class may have this property. For example, continuously stirred chemical reactors (CSTRs) are not always stable, therefore we will fail to show that structural stability holds for this class of systems. But there are many CSTRs which are stable.

13.2.1. Structural Rank of the State Structure Matrix

The notion of structural rank of a structure matrix is similar to that of other structural properties. If almost every matrix of the same structure has rank n, then the structural rank of the structure matrix is equal to n. In other words, *structural rank is the maximal possible rank within the class specified by the structure matrix.*

The determination of the structural rank of a matrix is easy and uses the following property: if one can find an assignment or matching of the rows and columns determined by a non-zero entry in the structure matrix then the matrix is of full structural rank, otherwise it is the size of the maximal matching. One can find computationally easy algorithms to determine the structural rank of a structure matrix.

The following simple example highlights the relation between the structural rank of a structure matrix and the rank of a matrix with the same structure matrix.

EXAMPLE 13.2.1 (Relation between structural rank and rank of matrices). Consider the following structure matrix **Q**

$$[\mathbf{Q}] = \begin{bmatrix} \underline{x} & 0 & 0 & x & 0 \\ 0 & \underline{x} & 0 & 0 & 0 \\ x & 0 & 0 & \underline{x} & 0 \\ 0 & 0 & \underline{x} & 0 & 0 \end{bmatrix}.$$

The underlined structurally non-zero elements $\underline{x}$ show a matching between rows and columns and indicate that $s - \mathrm{rank}([\mathbf{Q}]) = 4$.

If, however, we take a special numerical matrix $\mathbf{Q}$ from the class determined by $[\mathbf{Q}]$ (that is with the same structure):

$$[\mathbf{Q}] = \begin{bmatrix} 1 & 0 & 0 & 2 & 0 \\ 0 & 2 & 0 & 0 & 0 \\ 1 & 0 & 0 & 2 & 0 \\ 0 & 0 & 5 & 0 & 0 \end{bmatrix},$$

then its rank will be smaller, actually $\text{rank}(\mathbf{Q}) = 3$. This is caused by the fact that the 4th column is twice the first. That is these two columns are linearly dependent.

Process systems whose dynamic state space model has originated from lumped dynamic models have an interesting and important special structural property which is stated below.

THEOREM 13.2.1. *The state structure matrix of a process system with convective or transfer outflow derived from lumped dynamic process models is of full structural rank.*

This fact is a consequence of the presence of the output convective terms in the conservation balance equations for the quantity which appears as a nonzero (negative) diagonal term in the linearized state equation. As none of the volumes is isolated from the others, all of them have both in- and outflows which are either convective or diffusive.

The general forms of a convective outflow in extensive variable form for the volume-specific extensive quantity φ is in the form of $-F^{(o)}\varphi$. This can be seen from Eq. (5.52) in Section 5.6. This term leads to the element $-F^{(o)}$ if one derives the right-hand sides of Eq. (5.52) to get the linearized model as is shown in Section 10.4.

The analysis of structural dynamical properties will be illustrated on the example of heat exchangers and heat exchanger networks. For this purpose, the structure matrices of a countercurrent heat exchanger are given below.

EXAMPLE 13.2.2 (Linear model structure of a heat exchanger).

Process system

Consider a countercurrent heat exchanger where the cold liquid stream is being heated by a hot liquid stream. The system is very similar to the one described in Example 7.2.1.

Modelling goal

Describe the lumped linearized model structure of the heat exchanger for structural dynamic analysis purposes.

Assumptions

- $\mathcal{A}1$. The overall mass (volume) of the liquids on both sides is constant.
- $\mathcal{A}2$. No diffusion takes place.
- $\mathcal{A}3$. No heat loss to the surroundings.
- $\mathcal{A}4$. Heat transfer coefficients are constant.
- $\mathcal{A}5$. Specific heats and densities are constant.
- $\mathcal{A}6$. Both liquids are in plug flow.
- $\mathcal{A}7$. The heat exchanger is described as a sequence of three CSTR pairs of hot and cold liquid volumes.

Balance volumes
We consider three perfectly mixed balance volumes with equal holdups for each of the hot and cold sides.

Model equations

Variables

$$\left(T_{\mathrm{h}}^{(k)}(t), T_{\mathrm{c}}^{(k)}(t),\ k = 1, 2, 3\right), \quad 0 \leq t, \tag{13.13}$$

where $T_{\mathrm{h}}^{(k)}(t)$ and $T_{\mathrm{c}}^{(k)}(t)$ is the hot and the cold side temperature in the kth tank pair respectively, and t is time.

Energy balances for the hot side

$$\frac{\mathrm{d}T_{\mathrm{h}}^{(k)}}{\mathrm{d}t} = \frac{F_{\mathrm{h}}}{V_{\mathrm{h}}^{(k)}}\left(T_{\mathrm{h}}^{(k-1)} - T_{\mathrm{h}}^{(k)}\right) - \frac{K^{(k)}A^{(k)}}{c_{P\mathrm{h}}^{(k)}\rho_{\mathrm{h}}^{(k)}V_{\mathrm{h}}^{(k)}}\left(T_{\mathrm{h}}^{(k)} - T_{\mathrm{c}}^{(k)}\right), \tag{13.14}$$

$$k = 1, 2, 3, \quad T_{\mathrm{h}}^{(0)}(t) = T_{\mathrm{h}}^{(\mathrm{i})}(t). \tag{13.15}$$

$T_{\mathrm{h}}^{(\mathrm{i})}$ is the hot liquid inlet temperature to the heat exchanger.

Energy balances for the cold side

$$\frac{\mathrm{d}T_{\mathrm{c}}^{(k)}}{\mathrm{d}t} = \frac{F_{\mathrm{c}}}{V_{\mathrm{c}}^{(k)}}\left(T_{\mathrm{c}}^{(k+1)} - T_{\mathrm{c}}^{(k)}\right) - \frac{K^{(k)}A^{(k)}}{c_{P\mathrm{c}}^{(k)}\rho_{\mathrm{c}}^{(k)}V_{\mathrm{c}}^{(k)}}\left(T_{\mathrm{c}}^{(k)} - T_{\mathrm{h}}^{(k)}\right), \tag{13.16}$$

$$k = 1, 2, 3, \quad T_{\mathrm{c}}^{(4)}(t) = T_{\mathrm{c}}^{(\mathrm{i})}(t). \tag{13.17}$$

$T_{\mathrm{c}}^{(\mathrm{i})}$ is the cold liquid inlet temperature to the heat exchanger. Note that the cold stream flows in the direction of descending volume indices.

Initial conditions

$$T_{\mathrm{h}}^{(k)}(0) = f_1^{(k)}, \quad k = 1, 2, 3, \tag{13.18}$$

$$T_{\mathrm{c}}^{(k)}(0) = f_2^{(k)}, \quad k = 1, 2, 3, \tag{13.19}$$

where the values of $f_1^{(k)},\ k = 1, 2, 3$ and $f_2^{(k)},\ k = 1, 2, 3$ are given.

Signed state structure matrix
Note that all the model parameters are positive in the model above and the balance equations (13.14)–(13.15) and (13.16)–(13.17) are the state equations of the linear state space model, therefore,

$$\{\mathbf{A}\}_{\mathrm{HE}} = \begin{bmatrix} - & + & 0 & 0 & 0 & 0 \\ + & - & 0 & + & 0 & 0 \\ + & 0 & - & + & 0 & 0 \\ 0 & 0 & + & - & 0 & + \\ 0 & 0 & + & 0 & - & + \\ 0 & 0 & 0 & 0 & + & - \end{bmatrix} \tag{13.20}$$

with the state vector

$$x_{HE} = [T_h^{(1)},\ T_c^{(1)},\ T_h^{(2)},\ T_c^{(2)},\ T_h^{(3)},\ T_c^{(3)}]^T. \qquad (13.21)$$

Note that the structure graph of a heat exchanger is shown in Fig. A.3 in Section A.5 in the appendix. ■ ■ ■

It is important to note that

1. *The state structure matrix of the countercurrent heat exchanger is indeed of full structural rank because we can easily find an assignment of the columns and rows along the main diagonal.*
2. *The diagonal elements of the state structure matrix (13.20) are all negative whereas the off-diagonal elements are either positive or zero.* This property will be important later on when analysing structural stability of countercurrent heat exchangers.
3. *Observe the diagonal blocks belonging to a volume pair connected by a heat conducting wall.* The connection is the consequence of the heat transfer between them.
4. *The other positive off-diagonal elements not belonging to the diagonal blocks indicate connections. It is easy to construct an algorithm which generates the off-diagonal connection elements from the topology of the volume pairs and their connections.*

A simple example of three jacketted CSTRs in series further illustrates the use of structure graphs to determine the structural rank of the state matrix.

EXAMPLE 13.2.3 (Structural rank of the state matrix of the three jacketted CSTR in series) (Example 10.3.2 continued).

The structure graph of the three jacketted CSTR has already been developed in Chapter 10 and is shown in Fig. 10.9. Figure 10.9 shows that there is a self-loop adjacent to each state variable vertex, that is the state structure matrix [**A**] is of full structural rank. ■ ■ ■

13.2.2. Structural Controllability and Observability

If only the structure of the system is given, then the notion of *structural controllability or observability* is used for ensuring that *all* systems with the given structure are controllable or observable. These properties can be checked using the following important result.

THEOREM 13.2.2. *[71] A linear system with structure matrices* ([**A**], [**B**], [**C**]) *is structurally controllable or observable if*

1. *the matrix* [**A**] *is of full structural rank, and*
2. *the system structure graph S is input connectable or output connectable. For controllability there should be at least one directed path from any of the input variables to each of the state variables. For observability there should be at least one directed path from each of the state variables to one of the output variables.*

As mentioned in Theorem 13.2.1, the state structure matrix of a process system is usually of full structural rank, thus input or output connectability implies structural controllability or observability for process systems in most of the cases.

Note that there is a more general similar theorem valid for systems where the structural rank of the state matrix is not full.

THEOREM 13.2.3. *[71] A linear system with structure matrices* ([**A**], [**B**], [**C**]) *is*

1. *structurally controllable if the block matrix* [**A**, **B**] *is of full structural rank:*

$$s - \text{rank}\ ([\mathbf{A}, \mathbf{B}]) = n$$

with n being the number of state variables;

2. *structurally observable if the block matrix* $[\mathbf{C}, \mathbf{A}]^{\mathrm{T}}$ *is of full structural rank:*

$$s - \text{rank}\left(\begin{bmatrix}\mathbf{C}\\ \mathbf{A}\end{bmatrix}\right) = n.$$

The use of the structure graphs to check controllability and observability of process systems is illustrated below.

EXAMPLE 13.2.4 (Structural controllability and observability of the three jacketted CSTR in series) (Example 10.3.2 continued).

The structure graph of the three-jacketted CSTR has already been developed in Chapter 10 and is shown in Fig. 10.9.

Because of the full structural rank property of the structure matrix [**A**] we only need to check input connectability to investigate structural controllability. We investigate two different cases:

1. *Rich in inputs and outputs*
   Here, we consider the full set of potential input variables u^* in Eq. (10.48) and a set of output variables y consisting of the variables at outlet as in Eq. (10.118). The structure graph of the system for this case is the one seen in Fig. 10.9.
2. *Single-input single-output*
   We pick the input and output variables belonging to the cooling water system

$$u^{\mathrm{S}} = [T_{C_{\mathrm{i}}}], \quad y^{\mathrm{S}} = [T_C^{(1)}] \qquad (13.22)$$

to form a SISO system. The structure graph of the system for this case is obtained by deleting the input and output vertices

$$C_{A_{\mathrm{i}}},\ T_{\mathrm{i}};\quad C_A^{(3)},\ T^{(3)}$$

and all of the edges adjacent to them from the graph in Fig. 10.9.

Because the state variable vertices span a strong component in the graph, and all the input variables are adjacent to a state variable from the strong component, the *system is structurally controllable in both cases.*

Because the state variable vertices span a strong component in the graph and all the output variables are adjacent to a state variable from the strong component, the *system is structurally observable in both cases.*

There is another important structural property of process systems which may enable us to perform the structural observability and controllability analysis directly on the process flowsheet. *Usually the structure graph of an operating unit is a "strong component" of the overall system structure graph.* (For the definition of a strong component see Appendix Section A.5.)

We can reduce this strong component into a single point from the viewpoint of the connectivity analysis because the vertices of a strong component should be connected to each other pairwise. *This means that we can check connectivity on the flowsheet following the direction of the streams and regard operating units as gateways between streams.* The gateway property of operating units implies that they permit to traverse in both directions between the streams they connect.

This method is illustrated on the example of a simple heat exchanger network.

EXAMPLE 13.2.5 (Structural observability and controllability analysis of heat exchanger networks). Consider a simple heat exchanger network shown in Fig. 13.1. The structure graph of a heat exchanger shown in Fig. A.3 in Section A.5 in the Appendix is strongly connected, therefore, we can perform the connectivity analysis needed for structural controllability and observability analysis directly on the flowsheet.

First, we have to determine the set of input and output variables for the system. Consider the case when there is only one input variable, the input temperature of the first hot stream $U_{\text{HEN}} = \{T_{\text{h1}}^{(\text{i})}\}$. Furthermore, the set of output variables contains all outlet temperatures from the network

$$Y_{\text{HEN}} = \{T_{\text{h1}}^{(\text{o})}, T_{\text{hm}}^{(\text{o})}; T_{\text{c1}}^{(\text{o})}, T_{\text{cm}}^{(\text{o})}\}.$$

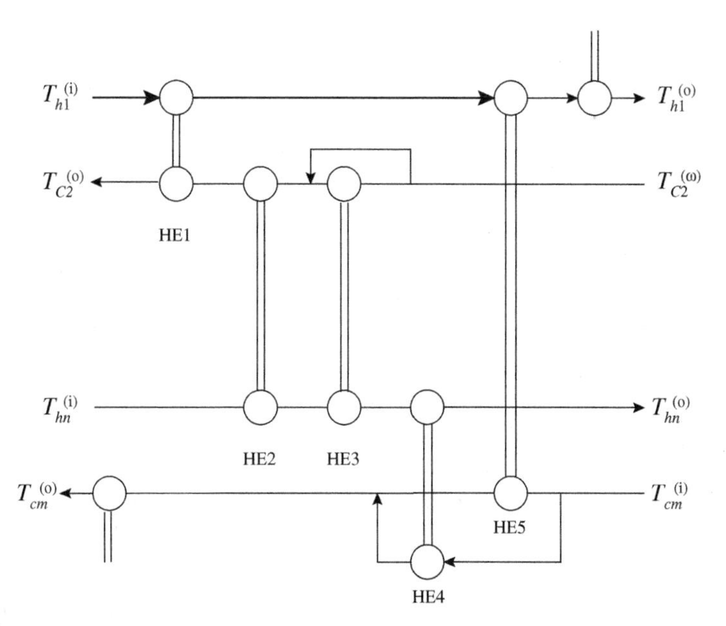

FIGURE 13.1 A simple heat exchanger network.

The set of state variables is formed from all of the temperatures of all of the lumped volumes in the system. They are represented by the heat exchanger circles because the state variables of a heat exchanger form a strong component in the overall structure graph.

For determining structural controllability, we need to check if the directed paths starting from the only input variable will reach every heat exchanger because the state structure matrix of the system is of full structural rank. A directed path connecting the input variable $T_{\mathrm{h1}}^{(\mathrm{i})}$ to the state variables in heat exchanger HE5 is shown in bold lines on the figure. It is easy to check that the system is not structurally controllable with the given input variable set U_{HEN}, because the state variables of the heat exchangers HE2, HE3 and HE4 are not input connectable.

On the other hand, the heat exchanger network is structurally observable with the given output variable set Y_{HEN} because there is at least one directed path from every heat exchanger to at least one of the output variables. ■ ■ ■

13.2.3. Structural Stability

There are two basically different methods in the literature to check structural stability of process systems:

- the method of circle families in the SDG model,
- the method of conservation matrices.

The Method of Circle Families in the SDG Model

In contrast to the analysis of structural controllability and observability where we only use the structure graph of process systems, the analysis of structural stability requires the SDG graph model of the system. For this we need to know the sign of the elements in the state space model matrices. If the SDG model of an arbitrary LTI system is given then the analysis of structural stability requires the determination of every circle family. This is a family of non-touching circles which spans the whole SDG graph, whose values need to be computed [71]. These values give the sign of the coefficients in the characteristic polynomial of the state matrix A and the Routh–Hurwitz criteria can be used to check conditions of structural stability. There are two basic problems associated with this method:

1. It is computationally hard to find all the circle families. This makes the method computationally infeasible even for a moderately large number of state variables.
2. The computation of the value of circle families involves sign addition. Often an indefinite sign results for the coefficients which makes the method non-decisive in many cases.

The Method of Conservation Matrices

Another powerful but limited method for checking structural stability of process systems is based on the notion of conservation matrices.

A real square matrix $\mathbf{F} = \{f_{ij}\}_{i,j=1}^{n}$ of order n is said to be a *column conservation matrix* or a *row conservation matrix* if it is a matrix with dominant main diagonal with respect to columns or rows. That is,

$$|f_{ii}| \geq \sum_{j \neq i} |f_{ij}| = R_i, \quad i = 1, 2, \ldots, n \tag{13.23}$$

or

$$|f_{ii}| \geq \sum_{j \neq i} |f_{ji}| = C_i, \quad i = 1, 2, \ldots, n, \tag{13.24}$$

and its elements have the following sign pattern

$$f_{ii} \leq 0, \quad f_{ij} \geq 0, \quad i \neq j. \tag{13.25}$$

In the case of proper inequality for every inequality in either Eq. (13.23) or (13.24), $\mathbf{F}$ is said to be a *strict column conservation matrix* or a *strict row conservation matrix*.

It is easy to see from the above definition that *the sum of two or more column (or row) conservation matrices is also a column (or row) conservation matrix.*

LEMMA 13.2.1. *Conservation matrices are non-singular. Furthermore, all the eigenvalues of a conservation matrix have non-positive (zero or negative) real parts, while there is no purely imaginary eigenvalue. In other words, conservation matrices are stable matrices with the real parts of all the eigenvalues being negative.*

It is important to note that the Lemma above states more than the well-known Gershgorin theorem below.

LEMMA 13.2.2. (Gershgorin). *Every eigenvalue of a square matrix* $\mathbf{F}$ *lies in at least one of the Gershgorin discs with centres* f_{ii} *and radii:*

$$\sum_{j \neq i} |f_{ij}|. \tag{13.26}$$

Because conservation matrices are non-singular and have no purely imaginary eigenvalue, they also do not have zero eigenvalues.

EXAMPLE 13.2.6 (Structural stability analysis of a simple heat exchanger). Consider the simple heat exchanger described in Example 13.1. *In order to prove structural stability of the heat exchanger we shall show that the state matrix* $\mathbf{A}_{\mathrm{HE}}$ *is a row conservation matrix.*

The signed state structure matrix $\{\mathbf{A}_{\mathrm{HE}}\}$ in Eq. (13.20) shows that it possesses the required sign pattern.

The coefficients in the energy balance equations (13.14) and (13.16) appear as elements in a row in the state matrix. The diagonal element of a hot side row is

$$-\frac{F_{\mathrm{h}}}{V_{\mathrm{h}}^{(k)}} - \frac{K^{(k)}A^{(k)}}{c_{\mathrm{Ph}}^{(k)}\rho_{\mathrm{h}}^{(k)}V_{\mathrm{h}}^{(k)}},$$

where the sum of the off-diagonal elements of the same row is

$$\frac{F_{\mathrm{h}}}{V_{\mathrm{h}}^{(k)}} + \frac{K^{(k)}A^{(k)}}{c_{\mathrm{Ph}}^{(k)}\rho_{\mathrm{h}}^{(k)}V_{\mathrm{h}}^{(k)}}.$$

It can be seen that the sum off all elements in the row is equal to zero. The same analysis can be carried out for rows belonging to the cold sides. Finally, we can conclude that all row sums are equal to zero in this case, therefore the simple heat exchanger is structurally stable. ■ ■ ■

13.3. MODEL SIMPLIFICATION AND REDUCTION

The term "model simplicity" may have different meanings depending on the context and on the set of models we consider. A process model may be more simple than another one in terms of

- *model structure*
  We can say, for example, that a linear model is simpler than a nonlinear one.
- *model size*
  For process models of the same type of structure (for example both linear) the model size can be measured in the number and dimension of model variables and parameters. Most often the number of the input and output variables are fixed by the problem statement, therefore, the number of state variables and that of parameters play a role.

We most often have the latter case, so we call a process model $^{(S)}$ simpler than another $^{(D)}$ with the same type of structure if

$$\dim(p^{(S)}) < \dim(p^{(D)}) \quad \text{and/or} \quad \dim(x^{(S)}) < \dim(x^{(D)}), \tag{13.27}$$

where $\dim(p^{(S)})$ and $\dim(p^{(D)})$ are the dimension of the parameter vector in the simple and detailed model respectively. Similarly, $\dim(x^{(S)})$ and $\dim(x^{(D)})$ stands for the dimension of the state vector in the simple and detailed models respectively.

Note that the term "model reduction" is also used for model simplification as it is stated above.

13.3.1. Problem Statement of Model Simplification

The problem statement addressed in model simplification can be stated as follows:

MODEL SIMPLIFICATION

Given:

- A detailed lumped dynamic process model $\mathcal{M}^{(D)}$ with all of its ingredients, which include model equations, parameters $p^{(D)}$ and initial conditions

$$y^{(MD)} = \mathcal{M}^{(D)}(u, x^{(D)}; \ p^{(D)}). \tag{13.28}$$

- A grey-box simplified model form $\mathcal{M}^{(S)}$, which is a process model with its equations but part of its parameters $p^{(S)}$ and/or structural elements unknown,

$$y^{(MS)} = \mathcal{M}^{(S)}(u, x^{(S)}; \ p^{(S)}). \tag{13.29}$$

- A loss function L measuring the difference between the data generated by the detailed and simplified model, usually in the form of

$$L = \|y^{(\text{MD})} - y^{(\text{MS})}\|, \tag{13.30}$$

where $|| \cdot ||$ is a suitable signal norm.

Compute:
An estimate of the simplified model and/or model structure $\widehat{\mathcal{M}}^{(\text{S})}$ such that L satisfies some stated optimality criterion.

13.3.2. Simplification of Linear Process Models

In the case of linear models, we can apply structure graphs describing the linear structure to simplify it, and obtain another linear model with less state variables.

The model simplification can be carried out in a graphical way using two basic simplification operations:

- variable removal by assuming steady state,
- variable coalescence by assuming similar dynamics.

With these two elementary transformations, we can simplify a process model structure by applying them consecutively in any required order.

Elementary Simplification Transformations

The applicability condition and the effect of the elementary simplification transformations on the structure graph are briefly described below. The elementary transformations of model structure simplification are as follows.

1. **Variable lumping**: lump(x_j, x_ℓ)
Applicability conditions: We can lump two state variables x_j and x_ℓ together to form a lumped state variable $x_{j,\ell}$ if they have "similar dynamics", that is,

$$x_j(t) = Kx_\ell(t), \quad K > 0.$$

Effect on the structure graph: We keep every edge directed to any of the original variables x_j or x_ℓ and direct them to the lumped variable, and similarly we keep every edge started from any of the original variables and start them from the lumped variable. This way *all the paths are conserved including the self-loops of the state variables.*

2. **Variable removal**: remove(x_j)
Applicability conditions: We can remove a state variable x_j from the structure graph if it is either changing much faster or much slower than the other variables. Note that a "perfect" controller forces the variable under control to follow the setpoint infinitely fast, therefore a controlled variable can almost always be removed from the structure graph. In both cases the time derivative of the variable to be removed is negligible, that is,

$$\frac{\mathrm{d}x_j(t)}{\mathrm{d}t} = 0 \;\Rightarrow\; x_j(t) = \text{constant}.$$

Effect on the structure graph: We remove the vertex corresponding to x_j from the graph and contract every edge pair (x_k, x_j), (x_j, x_ℓ) forming a directed path through the original vertex x_j into a single directed edge (x_k, x_ℓ). This way *all the paths are conserved including the self-loops of the state variables.*

The following example of the three jacketted CSTR in series shows how the composition of the two simplification transformations affect a complicated structure graph.

EXAMPLE 13.3.1 (Variable lumping and variable removal of the three jacketted CSTR in series model (Example 10.3.2 continued).)

Simplify the structure graph of the system shown in Fig. 10.9 given by:

1. lumping of all the cooling water temperatures together,
2. removing all the cooling water temperatures.

Variable lumping

We may assume that the temperature state variables belonging to the cooling water subsystem, i.e.

$$T_{\mathrm{C}}^{(1)},\ T_{\mathrm{C}}^{(2)},\ T_{\mathrm{C}}^{(3)}$$

have similar dynamical behaviour with respect to changes in the manipulated input and disturbance variables. Therefore, we can form a lumped cooling water temperature $T_{\mathrm{C}}^{(*)}$ from them by applying the variable lumping transformation twice:

$$T_{\mathrm{C}}^{(*)} = \mathrm{lump}\left(T_{\mathrm{C}}^{(1)}, \mathrm{lump}(T_{\mathrm{C}}^{(2)}, T_{\mathrm{C}}^{(3)})\right)$$

The resultant structure graph is shown in Fig. 13.2.

Variable removal

For a cooling system with large overall heat capacity, we can assume that the temperature state variables belonging to the cooling water subsystem

$$T_{\mathrm{C}}^{(1)},\ T_{\mathrm{C}}^{(2)},\ T_{\mathrm{C}}^{(3)}$$

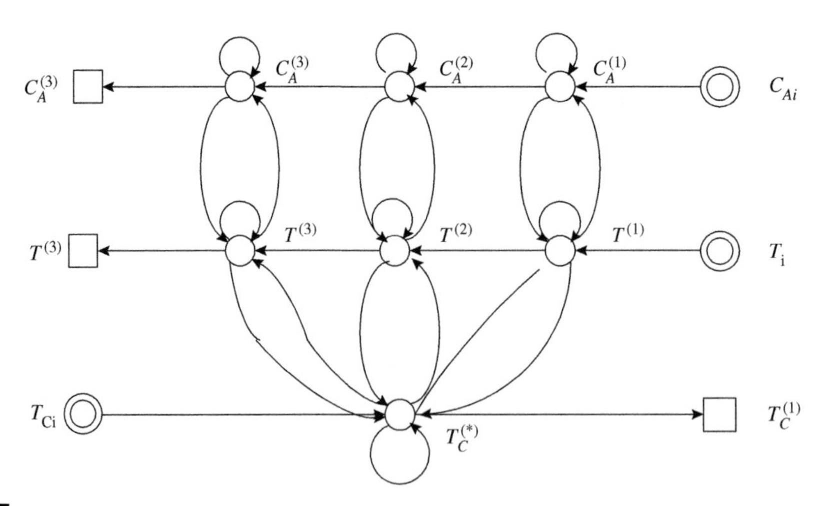

FIGURE 13.2 Simplified structure graph of three jacketted CSTRs by variable removal.

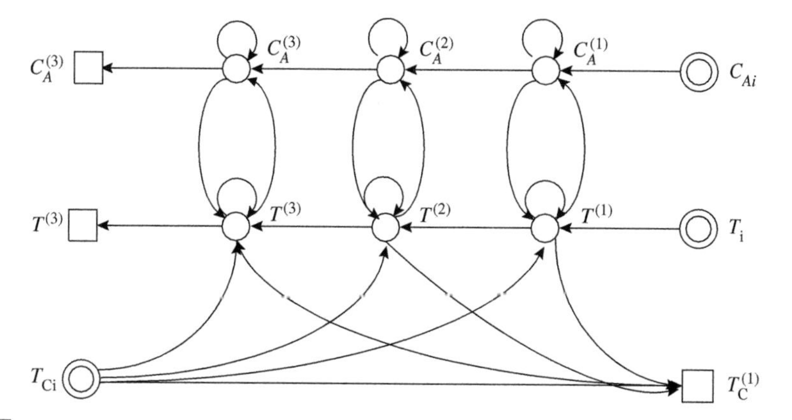

FIGURE 13.3 Simplified structure graph of three jacketted CSTRs by variable removal.

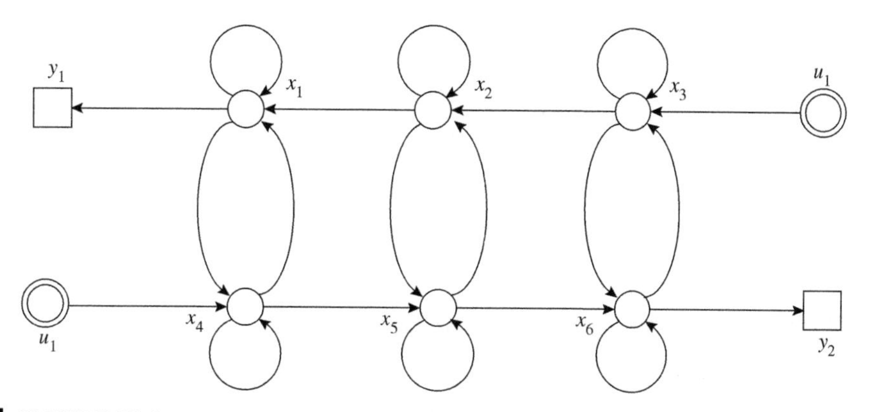

FIGURE 13.4 Structure graph of a heat exchanger.

are in quasi-steady state and are regarded as constants. Therefore, we can remove them from the structure graph by applying the variable removal transformation three times:

$$\text{remove}\left(T_C^{(1)}\right) \circ \text{remove}\left(T_C^{(2)}\right) \circ \text{remove}\left(T_C^{(3)}\right).$$

The resultant structure graph is shown in Fig. 13.3.

EXAMPLE 13.3.2 (Model simplification of the linear heat exchanger model). The structure graph of a heat exchanger model with three lumped volumes in both the cold and hot sides is shown in Fig. 13.4. Two simplified model structures are depicted in Figs. 13.5 and 13.6.

If the three variables in both the hot and cold sides are coalesced to form just a single lumped variable for the hot and cold sides then the model structure shown in Fig. 13.5 results. The lumped variables are denoted by filled circles in the figure.

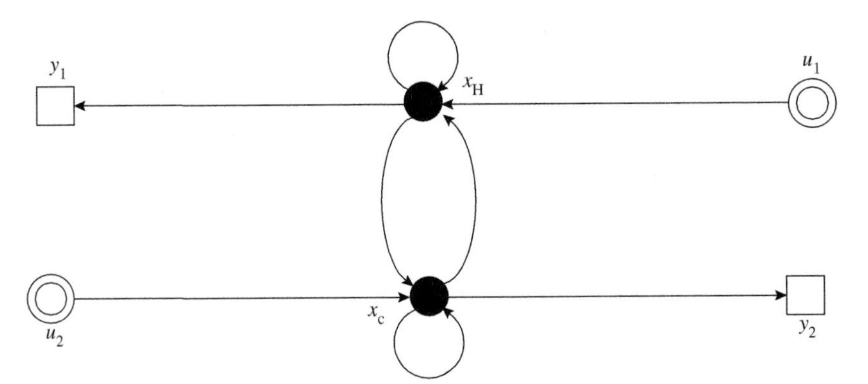

FIGURE 13.5 Simplified structure graph of a HE1.

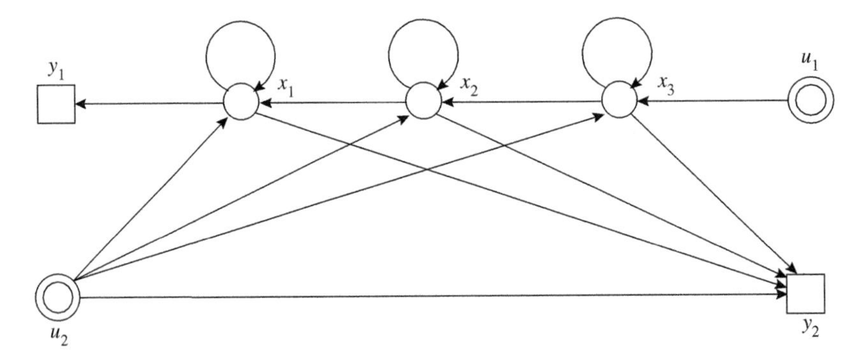

FIGURE 13.6 Simplified structure graph of a HE2.

The result of model simplification assuming all the variables in the cold side lumps are at steady state is shown in Fig. 13.6. ■ ■ ■

13.3.3. Model Simplification for Control and Diagnostic Purposes

Very often process models originating from engineering first principles are not simple enough for use in modern model-based control and diagnostic methods. These methods usually need less than ten state variables and parameters. Therefore, model simplification is an essential step before one can approach real life control and diagnostic problems.

The model simplification steps discussed in Section 13.3.2 are based on the SDG graph model of the process system. These can be conveniently combined with the analysis of structural control properties because all of the methods use combinatorial graph theoretic methods. In this way, the simplified models can be analysed if they have the same structural control properties as the detailed one.

Both the model simplification and the analysis of structural dynamic properties for LTI models are computationally easy, and they can be easily visualized. Therefore, they offer a quick and powerful tool for the model developer to adjust and analyse process models.

The Effect of Model Simplification Transformations on Structural Properties

Both variable lumping and variable removal seriously affect the structure graph of the system therefore it is important to know how these elementary transformations influence the structural properties of the underlying system model.

1. *Variable lumping*
The lumping transformation conserves all the paths including self-loops and does not lead to any direct paths from input to output vertices.
Therefore, the simplified structure graph:

- has self-loops adjacent to each of its state variable vertices, ensuring that *the state structure graph* [**A**] *has full structural rank. The state vertices again span a strong component in the structure graph*;
- remains input connectible. This property together with the fact that every input variable is adjacent to a state variable from the strong component spanned by the state variable vertices implies *structural controllability of the simplified system* in both the *"Rich in inputs and outputs"* and the *"Single input single output"* cases;
- remains output connectible. Because of the same reasons as above the *simplified system is structurally observable in both of the cases.*

2. *Variable removal*
The variable removal transformation conserves all the paths including self-loops but may lead to direct paths from input to output vertices.
Therefore, the simplified structure graph:

- has self-loops adjacent to each of its state variable vertices, ensuring that *the state structure graph* [**A**] *has full structural rank. The state vertices again span a strong component in the structure graph;*
- remains input connectible for input variables which remain adjacent to a state variable;
- remains output connectible for output variables which remain adjacent to a state variable.

13.3.4. Model Reduction

As an alternative to model structure simplification there are analytical methods to reduce the number of state variables of LTI state space models which have "similar" input–output behaviour. This alternative way is called *analytical model reduction.*

Model reduction is done in two conceptual steps:

1. construction of the so-called *balanced realization* using similarity transformations and the computation of controllability and observability gramians;
2. reducing the balanced state space realization by omitting those transformed states which do not contribute significantly to the observability and controllability gramians of the system.

It is clear from the above conceptual description that the reduced order model does not only contain fewer state variables but its state variables are transformed losing their physical meaning. That is why model reduction is considered as a black-box type model simplification technique.

Furthermore, model reduction uses a particular LTI system model, and the result may depend on the steady state linearization point in the case of nonlinear process models. Therefore, special attention is needed when this technique is applied to linearized models.

Balanced Realization

Remember that a state space realization of an LTI system model is not unique, since we can find infinitely many state space realizations giving rise to the same input–output description by applying *similarity transformations* to the state variables x by a square invertible matrix $\mathbf{T} \in \mathcal{R}^{n \times n}$, where n is the number of state variables

$$\overline{x} = \mathbf{T}x.$$

The *balanced realization* is a realization with symmetric state matrix $\mathbf{A}$, symmetric and balanced controllability and observability gramians and with an ordering of the transformed state variables.

Having a balanced realization, it is easy to calculate the contribution of each of the transformed states $\overline{x}$ to the controllability and observability of the system which is then characterized by the diagonal of the joint controllability–observability gramians arranged in a vector $g \in \mathcal{R}^n$. The elements of the gramian vector g are arranged in descending order and correspond to the elements in the transformed state variable vector. In other words, the transformed state variables are arranged in the order of their descending contribution to the joint controllability and observability of the system.

We use the MATLAB call

$$[\mathtt{AA}, \mathtt{BB}, \mathtt{CC}, \mathtt{g}, \mathtt{T}] = \mathtt{balreal}(\mathtt{A}, \mathtt{B}, \mathtt{C})$$

to obtain the balanced realization of the system. Besides the balanced realization matrices $(\mathbf{AA}, \mathbf{BB}, \mathbf{CC})$ we get the gramian in vector g and the transformation matrix $\mathbf{T}$.

This step is illustrated by the following simple example.

EXAMPLE 13.3.3 (Balanced realization for analytical model reduction of a simple LTI system). Consider a simple LTI SISO system model in the form:

$$\frac{dx}{dt} = \mathbf{A}x + \mathbf{B}u, \tag{13.31}$$

$$y = \mathbf{C}x \tag{13.32}$$

with the following vectors and matrices

$$x = \begin{bmatrix} x_1 \\ x_2 \\ x_3 \end{bmatrix}, \quad u = [u_1], \quad y = [y_1],$$

$$\mathbf{A} = \begin{bmatrix} -1.0 \cdot 10^{-5} & 0.2 \cdot 10^{-5} & 0 \\ 0.1 & -3 & 0.3 \\ 0 & 0.5 & -4 \end{bmatrix}, \quad \mathbf{B} = \begin{bmatrix} 1 \\ 0 \\ 0 \end{bmatrix}, \quad \mathbf{C} = [1\ 0\ 0].$$

Construct its balanced realization for analytical model reduction.

Observe that although the system is stable because its state matrix is a conservation matrix, the eigenvalues of the system are different by fifth order of magnitudes. This calls for deleting the two "fast" states (x_2 and x_3).

For the example above we get from the procedure `balreal`:

$$\mathbf{AA} = \begin{bmatrix} 0.0000 & 0.0000 & 0.0000 \\ 0.0000 & -2.9137 & 0.2373 \\ 0.0000 & 0.2373 & -4.0863 \end{bmatrix},$$

$$\mathbf{BB} = \begin{bmatrix} 1.0000 \\ -0.0002 \\ 0.0000 \end{bmatrix}, \quad \mathbf{CC} = [1.0000 \; -0.0002 \; 0.0000]$$

with the gramian vector and transformation matrix

$$g = 1.0 \cdot 10^4 \begin{bmatrix} 5.0340 \\ 0.0000 \\ 0.0000 \end{bmatrix}, \quad \mathbf{T} = \begin{bmatrix} 1.0000 & -0.0002 & 0.0000 \\ 0.0338 & 221.5036 & -30.5964 \\ 0.0042 & 39.4998 & 285.9600 \end{bmatrix}. \tag{13.33}$$

■ ■ ■

Eliminating the Non-contributing States

The next step is to eliminate those transformed states which do not contribute to the joint controllability and observability gramian. States whose contribution is less than 10% of the largest contribution, that is

$$x_i: \quad g_i < \frac{g_1}{10}$$

can be omitted.

The elimination can be done using the MATLAB procedure

$$[\mathtt{AR}, \mathtt{BR}, \mathtt{CR}, \mathtt{DR}] = \mathtt{modred}(\mathtt{AA}, \mathtt{BB}, \mathtt{CC}, \mathtt{DD}, \mathtt{ELIM}),$$

where (**AA**, **BB**, **CC**, **DD**) are the balanced state space realization matrices *ELIM* is the index vector of the states to be eliminated and (**AR**, **BR**, **CR**, **DR**) are the reduced state space model matrices of appropriate dimension.

The reduction is performed in such a way that the static gains of the system are conserved. The time derivative of the states to be eliminated is set to zero (e.g. $\mathrm{d}x_j/\mathrm{d}t = 0$) and the resultant set of algebraic equations is substituted into the remaining differential equations.

This reduction step is also illustrated by continuing the previous example.

EXAMPLE 13.3.4 (Analytical model reduction of a simple LTI system (Example 13.3.3 continued)). From the obtained gramian vector (13.33) of the example we can safely conclude, that the transformed state variables $\bar{x}_2$ and $\bar{x}_3$ can be omitted. Therefore, we set the elimination vector `ELIM` to be

$$\mathtt{ELIM} = [2; 3].$$

Using the MATLAB procedure `modred` for the reduction we obtain the following reduced SSR:

$$\mathbf{AR} = [-9.9325 \cdot 10^{-6}], \quad \mathbf{BR} = [1.0000], \quad \mathbf{CR} = [1.0000],$$

$$\mathbf{DR} = [-7.8905 \cdot 10^{-9}]. \tag{13.34}$$

■ ■ ■

13.3.5. Relation between Model Reduction and Model Structure Simplification

It is easy to see from the above and from the description of the model structure simplification steps that *model reduction is the analytical version of a sequence of variable removal structure simplification steps* if one does not apply the transformation to get a balanced realization.

In order to show this, consider the reduction principle and formulate this in mathematical terms. First we partition the state variable vector x into x_1 to be kept and x_2 to be eliminated and partition all the SSR matrices accordingly:

$$\begin{bmatrix} \frac{dx_1}{dt} \\ \frac{dx_2}{dt} \end{bmatrix} = \begin{bmatrix} \mathbf{A}_{11} & \mathbf{A}_{12} \\ \mathbf{A}_{21} & \mathbf{A}_{22} \end{bmatrix} \begin{bmatrix} x_1 \\ x_2 \end{bmatrix} + \begin{bmatrix} \mathbf{B}_1 \\ \mathbf{B}_2 \end{bmatrix} u. \tag{13.35}$$

$$y = [\mathbf{C}_1 \quad \mathbf{C}_2]\, x + \mathbf{D}u. \tag{13.36}$$

Next we set the derivative of x_2 to zero and the resulting algebraic equation is solved for x_2 and substituted into the remaining ones to get:

$$\frac{dx_1}{dt} = \left[\mathbf{A}_{11} - \mathbf{A}_{12}\mathbf{A}_{22}^{-1}\mathbf{A}_{21}\right] x_1 + \left[\mathbf{B}_1 - \mathbf{A}_{12}\mathbf{A}_{22}^{-1}\mathbf{B}_2\right] u, \tag{13.37}$$

$$y = \left[\mathbf{C}_1 - \mathbf{C}_2\mathbf{A}_{22}^{-1}\mathbf{A}_{21}\right] x_1 + \left[\mathbf{D} - \mathbf{C}_2\mathbf{A}_{22}^{-1}\mathbf{B}_2\right] u. \tag{13.38}$$

Note that the MATLAB procedure `modred` can be used to compute the weights of the reduced structure graph model after variable removal simplification steps.

■

EXAMPLE 13.3.5 Analytical model reduction and model structure simplification by variable removal of a simple LTI system (Example 13.3.4 continued).

If we want to compute the state space model parameters after simplification by variable removal we set the elimination vector `ELIM` to be

$$\mathtt{ELIM} = [2;\, 3]$$

to eliminate states x_2 and x_3 but now from the original state space realization.

Using the MATLAB procedure `modred` for the reduction again we obtain the following reduced SSR:

$$\mathbf{AQ} = [-9.9325 \cdot 10^{-6}], \quad \mathbf{BQ} = [1], \quad \mathbf{CQ} = [1], \quad \mathbf{DQ} = [0]. \tag{13.39}$$

Comparing this reduced model with the original one in Eq. (13.34) using balanced realization we see that there is almost no difference in this case. However, the results are *structurally different* because in the simplified structure, we do not have a direct influence from the input to the output. In the reduced case, however, we have one but its magnitude is negligible compared to the other influences. ■ ■ ■

13.4. SUMMARY

In this chapter, we have investigated some key concepts applicable to models specifically developed for process control applications. In particular the concepts of observability and controllability are extremely important for control applications. The idea of stability was introduced and investigated in relation to linear time-invariant models. These constitute a very large class of problems in control applications and are amenable to analysis through well established techniques.

The second part of the chapter has been devoted to the analysis of structural dynamical properties which hold for a class of systems with the same structure. The concepts and analysis techniques of structural controllability, observability and stability are based on graph theory.

Finally, analytical and structural methods for model simplification and reduction have also been presented and compared.

13.5. FURTHER READING

The analysis of dynamic process models in itself is rarely the subject of any book, papers, conference or other information source, but it appears as an integral part of process design or process control studies. Therefore, there is a big overlap between the *information sources* of dynamic analysis and control or design of process systems. The most important sources are as follows.

- *Journals*
  - Journal of Process Control
  - Industrial and Engineering Chemistry Research
  - Computers and Chemical Engineering
  - AIChE Journal
- *Conferences*
  - AIChE Meetings
  - ESCAPE Conferences
  - IFAC Conferences "ADCHEM" and "DYCORD"
  - IFAC World Congresses
  - PSE Conferences
- *Web* Control Engineering Virtual Library, with homepage: `http://www-control.eng.cam.ac.uk/extras/Virtual_Library/Control_VL.html`.

There are a few "hot topics" in connection with the analysis of dynamic process modelling which deserve special attention.

- *Interaction between process design and control*
  The interaction between process design and control had been recognized a long time ago and the use of open-loop dynamic indicators has been suggested for assessing the controllability and operability of design alternatives [89].
  There are advanced spectral methods for the analysis and design of process dynamics which are reviewed and applied in a recent PhD Thesis [90].
  There are several recent research papers which perform integration of design and control through model analysis, such as [91].
- *Operability analysis of process plants*
  Operability is the property of a process plant characterizing its behaviour under changing operating conditions. Design alternatives are assessed with respect to their operability using structural and simple linearized operability measures [92].
  A simulation-based methodology has been reported in [93] for evaluating the stability of control structures and finding the best flowsheet structure from a controllability point of view.
- *Model simplification and reduction*
  Various techniques for model simplification and reduction have been proposed ranging from the simplest linear techniques to very complex nonlinear ones.
  A wavelet-based model reduction method has been proposed in [94] for distributed parameter systems.

13.6. REVIEW QUESTIONS

Q13.1. Give the problem statement of state controllability and observability. (Section 13.1).

Q13.2. Compare the notions of BIBO and asymptotic stability. Which one is the stronger? Give a process example which is BIBO stable but not asymptotically stable! (Section 13.1)

Q13.3. What is the relation between a dynamical property (say stability) and its structural counterpart (structural stability). Does the structural dynamical property imply that property for *every* dynamical system in the class? (Section 13.2)

Q13.4. What is the *physical condition* which causes the state structure matrix of a process system to almost always be of full structural rank? Give simple process examples. (Section 13.2)

Q13.5. Give conditions which ensure the structural controllability of a system model. In which case are they equivalent? (Section 13.2)

Q13.6. Give conditions which ensure the structural observability of a system model. In which case are they equivalent? (Section 13.2)

Q13.7. When do we call a matrix a conservation matrix? Are conservation matrices of full structural rank? Why? (Section 13.2)

Q13.8. Give the problem statement of model simplification. Derive its special case for linear state space models. (Section 13.3)

Q13.9. What are the two model simplification operations one applies for simplification of linear process models? Do they change the structural controllability and observability properties of the model? (Section 13.3)

Q13.10. Compare the analytical model reduction with model structure simplification. What are the similarities and the differences? Is it possible to use a combination of the two? How? (Section 13.3)

13.7. APPLICATION EXERCISES

A13.1. Compute the observability matrices of the LTI systems given in Examples 10.2.2 and 10.2.3 and check if the systems are observable. (Section 13.1)

A13.2. Compute the controllability matrices of the LTI systems given in Examples 10.2.2 and 13.1.3 and check if the systems are controllable. (Section 13.1)

A13.3. Derive the signed structure matrix form of the linearized state space model matrices of the batch water heater in Example 14.3.1 derived from its SDG model shown in Fig. 14.2. Compare them with that derived from the lumped model equations (14.36) and (14.37). (Section 13.2)

A13.4. Derive the signed structure graph model of the LTI state space model matrices in Example 10.2.1. Is this model structurally controllable and observable? Compare your findings with the analytical results given in Examples 13.1.2 and 13.1.5. Comment on the similarities and possible differences. (Section 13.2)

A13.5. Check if the state space matrices of the LTI system models of the Examples 10.2.1 and 10.2.3 are conservation matrices. Are the corresponding model structures structurally stable? Compare your findings with the analytical results given in Examples 13.1.7 and 13.1.8. Comment on the similarities and possible differences. (Section 13.2)

A13.6. Is the state space model of the batch water heater (Eqs. (14.36) and (14.37)) structurally observable and controllable? Check it by graph theoretic methods on its SDG model shown in Fig. 14.2. Compare the result with the one obtained by computing the rank of the observability and controllability matrices ($\mathcal{O}$ and $\mathcal{C}$ of the same model giving numerical value to the model parameters). (Section 13.2)

A13.7. Is the state space model of the batch water heater (Eqs. (14.36) and (14.37)) structurally stable? Check it by the method of conservation matrices using the signed structure matrix derived in Exercise A13.3. Compare the result with the one obtained by computing the eigenvalues of the state matrix $\mathbf{A}$ of the same model giving numerical value to the model parameters. (Section 13.2)

A13.8. A linear time invariant (LTI) system model leads to the following state space model equations:

$$\frac{\mathrm{d}x}{\mathrm{d}t} = \begin{bmatrix} -2.0 & 1.3 & 0 \\ 1.2 & -4.0 & 1.2 \\ 0 & 0 & -3.0 \end{bmatrix} \cdot x + \begin{bmatrix} 1.4 \\ 0 \\ 0 \end{bmatrix} \cdot u,$$

$$y = \begin{bmatrix} 4.2 & 0 & 0 \end{bmatrix} \cdot x.$$

(a) Give the SSR matrices and determine the dimension of the state, input and output vectors.
(b) Draw the structure graph of the system.
(c) Investigate the structural controllability and observability of this system and comment on your findings.

14 PROCESS MODELLING FOR CONTROL AND DIAGNOSTIC PURPOSES

Process control and diagnosis is the application area where dynamic process models are extensively used. This chapter contains the necessary material for a *process engineer* to understand and successfully use modern model-based process control and diagnostic methods. The presentation of the material is therefore much more application oriented than the case would be in a book designed for control engineers. The methods and procedures are described in terms of *Problem statements* and detailed derivation of the methods or procedures are usually left out. The conditions of applicability of the particular methods and the evaluation of the results from a process engineering perspective are emphasized.

Dynamic process models for control and diagnostic purposes are almost always lumped process models in the form of DAE systems with given initial conditions. Therefore, only lumped dynamic process models are considered throughout this chapter.

- In Section 14.1 the modelling requirements for process control are formulated and analysed together with the problem statements and conceptual solution of the most frequently used state space model-based control problems.
- The process models suitable for diagnosis and the most important approaches for process diagnosis are discussed in Section 14.2.
- Finally a brief survey of qualitative, logical and artificial intelligence process models and their use for control and diagnostic purposes is given in Section 14.3.

14.1. MODEL-BASED PROCESS CONTROL

Until now we have mainly dealt with *systems analysis*, when process systems have been investigated as they are, without any intention to influence their natural behaviour. *Control theory deals with the problems, concepts and techniques to influence the behaviour of systems to achieve some prescribed goal.*

Besides control in the above narrow sense, all the other subtasks which are either related to control or involve some decision concerning dynamic systems belong to the area of control theory, such as

- identification (model parameter and structure estimation),
- filtering, state filtering, prediction and smoothing,
- diagnosis (fault detection and isolation).

In this section, we briefly review the most important control related tasks relevant to process engineering.

14.1.1. Feedback

Feedback is a central notion in control theory. Most of the controllers, but not all of them, apply either state or output feedback to compute the control action.

It is intuitively clear that one should take into account the effect of the previously applied input values when computing the present one through the dependence of the input value on the observed outputs or the states – that is to apply feedback.

If the present control action, that is the applied input depends on the value of the state variables, then a state feedback is applied for control

$$u = F(x), \tag{14.1}$$

where F is a known, possibly nonlinear function.

Output feedback is defined similarly but then

$$u = F(y). \tag{14.2}$$

Feedbacks can be classified according to the function F in the feedback equation as follows:

1. *static feedback*—when F is a time invariant algebraic operator,
2. *full feedback*—when F depends on every element in the state or output vector respectively,
3. *linear feedback*—when F is a linear operator.

For LTI systems most of the controllers apply *linear static full state feedback* in the following form:

$$u = \mathbf{K}x, \tag{14.3}$$

where the constant matrix $\mathbf{K}$ with appropriate dimension is the *control gain*.

14.1.2. Pole-placement Control

The pole-placement controller design is a technique for setting the system response by adjusting the eigenvalues of the closed-loop system. It should use calibrated and validated continuous time LTI versions of process models. Therefore, linearization and model parameter estimation should be performed before we can apply this method. The problem statement and the solution of the pole-placement controller design is as follows.

POLE-PLACEMENT CONTROLLER DESIGN

Given:

- A continuous time LTI state space model of a SISO system in the form:

$$\frac{\mathrm{d}x}{\mathrm{d}t} = \mathbf{A}x + \mathbf{B}u, \quad y = \mathbf{C}x \tag{14.4}$$

  with its matrices $(\mathbf{A}, \mathbf{B}, \mathbf{C})$, where $\mathbf{A}$ determines the *poles of the open-loop system by its characteristic polynomial*

$$a(s) = \det(s\mathbf{I} - \mathbf{A}). \tag{14.5}$$

- *A desired closed-loop characteristic polynomial* $\alpha(s)$ with $\deg(a(s)) = \deg(\alpha(s)) = n$, i.e. they have the same degree.

Compute:
The *static feedback gain vector*

$$k = [k_1, k_2, \ldots, k_n] \tag{14.6}$$

of a full state feedback such that the closed-loop characteristic polynomial is $\alpha(s)$ given by

$$\alpha(s) = \det(s\mathbf{I} - \mathbf{A} + \mathbf{B}k), \tag{14.7}$$

Solution [70]:
Full state feedback can arbitrarily relocate the poles if and only if the system is state controllable and then

$$k = (\underline{a} - \underline{\alpha})\mathbf{T}^{-\mathrm{T}}\mathcal{C}, \tag{14.8}$$

where $\underline{a}$ and $\underline{\alpha}$ are vectors containing the coefficients of the open-loop and closed-loop characteristic polynomial respectively, $\mathbf{T}$ is a Toeplitz-matrix and $\mathcal{C}$ is the controllability matrix of the system.

There are several important remarks and observations related to pole-placement control of process systems including:

1. Pole-placement control is applied in process control systems quite rarely because it is difficult to specify the desired poles of the closed-loop system and it is only applicable for SISO systems.
2. It has a theoretical importance, however, because it states that the necessary and sufficient condition for stabilizing an LTI system is its state controllability.

3. *The controller design method does not require the solution of the model equations but only uses the model parameter values.* It means that as soon as the controller parameters are computed the controller does not perform heavy computations in real time.
4. The software package MATLAB has a simple procedure (**place**) to compute the feedback gain vector from the SSR matrices of the open-loop system and from the desired poles.

EXAMPLE 14.1.1 (Pole-placement controller design for a simple LTI system). Consider a simple SISO LTI system in the form of Eq. (14.4) with the matrices

$$\mathbf{A} = \begin{bmatrix} 3 & 4 \\ 5 & 6 \end{bmatrix}, \quad \mathbf{B} = \begin{bmatrix} 7 \\ 8 \end{bmatrix}, \quad \mathbf{C} = [1 \quad 0]. \tag{14.9}$$

Let the desired poles of the closed-loop system be -1 and -3. Then the pole-placement controller design procedure checks the full rank of the controllability matrix and gives the following feedback gain vector:

$$k = [0.5455 \quad 1.2727]. \tag{14.10}$$

14.1.3. Linear Quadratic Regulator (LQR)

Linear quadratic regulators (LQRs) are widely used in process control. Its popularity is explained by the fact that it is quite robust with respect to modelling uncertainties. Generalised versions are also available for classes of linear time-varying, special nonlinear and robust linear models.

The problem statement and the solution of the LQR design problem is as follows:

LINEAR QUADRATIC REGULATOR DESIGN

Given:

- A continuous time LTI state space model of a MIMO system in the form:

$$\frac{dx}{dt} = \mathbf{A}x + \mathbf{B}u, \quad x(0) = x_0, \tag{14.11}$$

$$y = \mathbf{C}x \tag{14.12}$$

  with its matrices $(\mathbf{A}, \mathbf{B}, \mathbf{C})$
- A loss functional $J(x, u)$

$$J(x, u) = \frac{1}{2} \int_0^T \left[x^T(t)\mathbf{Q}x(t) + u^T(t)\mathbf{R}u(t)\right] dt \tag{14.13}$$

  with a positive semidefinite state weighting matrix $\mathbf{Q}$ and with a positive definite input weighting matrix $\mathbf{R}$ where identically zero state reference value $x^{(ref)} = 0$ is assumed. (See Section A.4.1 in the appendix for the definition of positive definite and semidefinite matrices.)

Compute:
A control signal

$$\{ u(t) \mid 0 \leq t \leq T \}, \tag{14.14}$$

that minimizes the loss function (14.13) subject to model equations (14.11).

Solution [70]:
For the stationary case, when $T \to \infty$, the solution is a full state feedback controller which stabilizes the system and at the same time gives the minimal loss provided the system is jointly controllable and observable.

The solution of the mathematical problem in the stationary case leads to the solution of the so-called control algebraic Ricatti equation (CARE), which is a quadratic equation for a square matrix. In this case this equation has a unique solution from which the full state feedback matrix can be easily computed.

The software package MATLAB has a simple procedure (**lqr**) to compute the feedback gain matrix from the state-space representation matrices of the open-loop system and from the weighting matrices in the loss functional.

EXAMPLE 14.1.2 (LQR design for a simple LTI system introduced in Example 14.1.1). Consider a simple SISO LTI system in the form of Eq. (14.4) with the matrices in Eq. (14.9).

Let the weighting matrices in the loss function (14.13) be as follows:

$$\mathbf{Q} = \begin{bmatrix} 1 & 0 \\ 0 & 1 \end{bmatrix}, \quad \mathbf{R} = \begin{bmatrix} 0.5 \end{bmatrix}. \tag{14.15}$$

This means that we weight the error in the state or output much higher than the effort in changing the input.

Then the LQR design procedure in MATLAB gives the following feedback gain matrix which is now a row vector because we have a SISO system:

$$k = [0.9898 \quad 0.8803]. \tag{14.16}$$

The poles of the closed-loop system

$$[-15.2021 \quad -8.7690]$$

indeed have negative real parts which means that the closed-loop system is stable.

14.1.4. State Filtering

Both of the above controller design methods apply full state feedback to realize the controller. Therefore one should be able to compute the non-measurable state variables from the measured input and output variables using only the system model. In the case of an observable deterministic LTI system, there is a direct way how to do this making use of the observability matrix $\mathcal{O}$.

However, *in the presence of any random elements in the model, either measurement errors or system noise, we have to estimate the value of the state variables from the measured input and output variables using the system model. This procedure is called state filtering*. The problem statement and the conceptual steps of the solution are described below.

STATE FILTERING

Given:

- A *discrete time stochastic LTI state space model* represented by

$$x(\tau+1) = \mathbf{\Phi}x(\tau) + \mathbf{\Gamma}u(\tau) + v(\tau), \quad x(0) = x_0, \tag{14.17}$$

$$y(\tau) = \mathbf{C}x(\tau) + e(\tau), \quad \tau = 0, 1, 2, \ldots \tag{14.18}$$

  with its model matrices ($\mathbf{\Phi}$, $\mathbf{\Gamma}$, $\mathbf{C}$) *and with the properties of an additive system and measurement noise* $v(\tau)$ *and* $e(\tau)$ *being independent random variables at any time* τ.
- A *measurement record* being a set of data corrupted by measurement noise

$$\mathcal{D}[1, k] = \{(u(\tau), y(\tau)) \mid \tau = 1, \ldots, k\}. \tag{14.19}$$

Compute:
An estimate of the current state vector $x(k)$ based on the measurement record $\mathcal{D}[1, k]$ and the system parameters.

Solution [74]:
For the LTI case when both the state noise and measurement noise processes are Gaussian white noise processes, the famous Kalman-filter can be used to solve the problem. The Kalman filter uses a simple linear filter equation to be computed in real time:

$$\hat{x}(\tau+1) = \mathbf{\Phi}\hat{x}(\tau) + \mathbf{\Gamma}u(\tau) + \mathbf{K}(\tau)(y(\tau) - \mathbf{C}\hat{x}(\tau)), \quad x(0) = x_0, \tag{14.20}$$

where $\hat{x}(\tau)$ is the state estimate in time τ and $\mathbf{K}(\tau)$ is the Kalman gain. Note that this filter equation is very similar to the state equation (14.17) but the state noise is replaced by its estimate based on the output equation (14.18).

The sequence of Kalman gains $\mathbf{K}(\tau)$, $\tau = 0, 1, \ldots$ can be precomputed using the model parameters in a recursive way. This sequence is then used to compute the estimate using the filter equation (14.20).

The software package MATLAB has procedures to compute the sequence of Kalman gains as well as the sequence of the state estimate (Kalman).

14.1.5. Model-based Predictive Control

Model-based predictive control (MPC) is the most popular and widely used method for advanced process control. Its popularity is partially explained by the availability of reasonable good and reliable process models and implementations on commercially available control systems.

Moreover the basic idea behind model-based predictive control is simple, easy to understand and yet powerful. If one has a dynamic process model which is reasonably accurate and validated then this model can be used for predicting how the real process system would behave by applying a specific control signal, either discrete or continuous to its input. This predicted output behaviour is then compared to the desired or reference output using a suitably chosen distance between signals (measured by a signal norm). If the predicted behaviour is not "close enough" to the desired one then another control signal is tested. Otherwise, the control signal is applied to the real process system.

The problem statement of model-based predictive control and the conceptual steps of the solution are described below.

MODEL-BASED PREDICTIVE CONTROL

Given:

- *A parametrized sampled predictive dynamic system model* in the form of

$$y^{(M)}(k+1) = \mathcal{M}(\mathcal{D}[1,k]; p^{(M)}), \quad k = 1, 2, \ldots \tag{14.21}$$

  with the model parameters $p^{(M)} \in \mathcal{R}^{\nu}$ being known, a measured data record $\mathcal{D}[1,k] \in \mathcal{R}^{\delta}$ up to time k and a vector-valued output variable $y^{(M)}(k) \in \mathcal{R}^{m}$.
- A set of measured data which is called a *measurement record* to emphasize the time dependence of the measured data

$$\mathcal{D}[1,k] = \{(u(\tau), y(\tau)) \mid \tau = 1, \ldots, k\}. \tag{14.22}$$

- A *reference output signal*

$$\{y^{(r)}(k) \mid k = 1, 2, \ldots\}. \tag{14.23}$$

- A *prediction horizon* N
- A *loss functional* $J^{(N)}$

$$J^{(N)}(y^{(M)} - y^{(r)}, u) = \sum_{i=1}^{N} \left[r^{\mathrm{T}}(k+i)\mathbf{Q}r(k+i) + u^{\mathrm{T}}(k+i)\mathbf{R}u(k+i) \right], \tag{14.24}$$

  where r is the residual

$$r(k) = y^{(r)}(k) - y^{(M)}(k), \quad k = 1, 2, \ldots \tag{14.25}$$

  with a positive semidefinite output weighting matrix $\mathbf{Q}$ and with a positive definite input weighting matrix $\mathbf{R}$.

Compute:
A discrete time control signal

$$\{u(k+i) \mid i = 1, \ldots, N\}, \tag{14.26}$$

that minimizes the loss function (14.24) subject to the model equations (14.21).

Conceptual solution method:
In the general case we can solve the problem as an optimization problem minimizing the functional (14.24) as a function of the discrete time control signal (14.26). This breaks down the problem into an optimization and a simulation step in a similar way as for the parameter estimation of dynamic nonlinear models in Section 12.4.3.

14.2. MODEL-BASED PROCESS DIAGNOSIS

The basic *principle of model-based process diagnosis* is as follows. Let us assume we have a process model describing the normal non-faulty "healthy" behaviour of the system and possibly also other models describing the behaviour when a particular fault occurs. The model which corresponds to normal behaviour will be called the "normal model" and labelled by 0. The other ones are called fault model i and they are labelled by the fault mode. We need to have as many fault models as the number of faulty modes considered ($i = 1, \ldots, N_F$).

With these models and the set of measurements taken from the real process we can decide on the faulty model of the process by comparing

- the measured data with the data predicted by the models of different faulty modes. This is called *prediction-based diagnosis*;
- the estimated parameters from the data with the given values in different faulty modes. This is called *identification-based diagnosis*.

14.2.1. Prediction-based Diagnosis

The problem statement of prediction-based diagnosis resembles the one for MODEL STRUCTURE ESTIMATION described in Section 12.2.

PREDICTION-BASED DIAGNOSIS

Given:

- *The number of considered faulty modes* N_F with the normal non-faulty model labelled by 0.
- *A set of predictive discrete time parametrized dynamic system models describing the system in various faulty modes*:

$$y^{(Fi)}(k+1) = \mathcal{M}^{(Fi)}(\mathcal{D}[1,k];\ p^{(Fi)}), \quad k = 1, 2, \ldots, \quad i = 0, \ldots, N_F \tag{14.27}$$

  depending on the model parameters $p^{(Fi)}$ being known, on the measured data record $\mathcal{D}[1,k] \in \mathcal{R}^{\delta}$ up to time k and on the vector-valued output variable $y^{(Fi)}(k) \in \mathcal{R}^m$.
- A set of measured data in a *measurement record*

$$\mathcal{D}[1,k] = \{(u(\tau), y(\tau)) \mid \tau = 1, \ldots, k\} \tag{14.28}$$

- A *loss functional* $J^{(Fi)},\ i = 0, \ldots, N_F$

$$J^{(Fi)}(y - y^{(Fi)}, u) = \sum_{\tau=1}^{k} \left[r^{(i)T}(\tau) \mathbf{Q} r^{(i)}(\tau) \right], \tag{14.29}$$

  where $r^{(i)}$ is the residual

$$r^{(i)}(\tau) = y(\tau) - y^{(Fi)}(\tau), \quad \tau = 1, 2, \ldots \tag{14.30}$$

  with a positive definite weighting matrix $\mathbf{Q}$.

Compute:
The faulty mode of the system, which is the model index i that minimizes the loss function (14.29) subject to model equations (14.27).

Conceptual solution method:
The solution is computed in two steps.

1. Perform the prediction with each of the models $\mathcal{M}^{(\mathrm{F}i)}$, $i = 0, \ldots, N_\mathrm{F}$ and compute the loss (14.29).
2. Select the model index with the minimum loss.

14.2.2. Identification-based Diagnosis

Identification-based diagnosis uses the estimated set of model parameters in each of the fault models and then compares them to their "true" values as seen from the problem statement below.

IDENTIFICATION-BASED DIAGNOSIS

Given:

- *The number of considered faulty modes* N_F with the normal non-faulty model labelled by 0.
- *A set of parametrized discrete time dynamic system models describing the system in various faulty modes*:

$$y^{(\mathrm{F}i)}(k+1) = \mathcal{M}^{(\mathrm{F}i)}(\mathcal{D}[1,k]; p^{(\mathrm{F}i)}), \quad k = 1, 2, \ldots, \quad i = 0, \ldots, N_\mathrm{F} \tag{14.31}$$

  depending on the model parameters $p^{(\mathrm{F}i)}$ being known, on the measured data record $\mathcal{D}[1,k] \in \mathcal{R}^{\delta}$ up to time k and on the vector-valued output variable $y^{(\mathrm{F}i)}(k) \in \mathcal{R}^m$.
- A set of measured data in a *measurement record*

$$\mathcal{D}[1,k] = \{(u(\tau), y(\tau) \mid \tau = 1, \ldots, k\}. \tag{14.32}$$

- A *loss functional* $J^{(\mathrm{F}i)}$, $i = 0, \ldots, N_\mathrm{F}$ depending now on the parameters

$$J^{(\mathrm{F}i)}(p^{(\mathrm{estF}i)} - p^{(\mathrm{F}i)}) = \rho^{(i)T}\mathbf{Q}\rho^{(i)} \tag{14.33}$$

  where $\rho^{(i)}$ is the parameter residual

$$\rho^{(i)} = p^{(\mathrm{estF}i)} - p^{(\mathrm{F}i)} \tag{14.34}$$

  with a positive definite weighting matrix $\mathbf{Q}$ and $p^{(\mathrm{estF}i)}$ is the estimated parameter vector of the model with faulty mode i.

Compute:
The faulty mode of the system, which is identified by the model index i that minimizes the loss function (14.33) subject to the model equations (14.31).

Conceptual solution method:
The solution is computed in two steps.

1. Perform a MODEL PARAMETER ESTIMATION procedure for every model $\mathcal{M}^{(Fi)}$, $i = 0, \ldots, N_F$ assuming $p^{(Fi)}$ unknown and compute the loss (14.33).
2. Select the index with the minimum loss.

14.3. QUALITATIVE, LOGICAL AND AI MODELS

In comparison with "traditional" engineering models, qualitative, logical and artificial intelligence (AI) models have a special common property: *the range space of the variables and expressions in these models is interval valued.* This means that we specify an interval $[a_{\ell_t}, a_{u_t}]$ for a variable a at any given time within which the value of the variable lies. Every value of the variable within the specified interval is regarded to be the same because they all have the same qualitative value. Hence, the value of the variable a can be described by a finite set of non-intersecting intervals covering the whole range space of the variable.

In the most *general qualitative case* these intervals are real intervals with fixed or free endpoints. The so-called *universe* of the range space of the interval-valued variables $U_{\mathcal{I}}$ in this case is

$$U_{\mathcal{I}} = \{[a_\ell, a_u] \mid a_\ell, a_u \in \mathcal{R}, a_\ell \leq a_u\}.$$

There are models with *sign valued* variables. Here the variables may have the qualitative value "+" when their value is strictly positive, the qualitative values "−" or "0" when the real value is strictly negative or exactly zero. If the sign of the value of the variable is not known then we assign to it an "unknown" sign value denoted by "?". Note that the sign value "?" can be regarded as a union of the above three other sign values meaning that it may be either positive "+" or zero "0" or negative "−". The corresponding universe $U_{\mathcal{S}}$ for the sign valued variables is in the form

$$U_{\mathcal{S}} = \{+,\ -,\ 0;\ ?\}, \quad ? = + \cup 0 \cup -. \tag{14.35}$$

Finally, *logical models* operate on logical variables. Logical variables may have the value "**true**" and "**false**" according to traditional mathematical logic. If we consider time varying or measured logical variables then their value may also be "**unknown**". Again, note that the logical value "**unknown**" is the union of "**true**" and "**false**". The universe for logical variables is then:

$$U_{\mathcal{L}} = \{\textbf{true},\ \textbf{false},\ \textbf{unknown}\}.$$

There is no matured agreement in the literature on the common definitive characteristics of AI methods, however it is commonly believed that they

- solve difficult, complicated tasks,
- use techniques that contain heuristic elements,
- provide solutions that are "human-like".

The most widely used and common logical and AI models for process control applications are as follows:

1. *rules*, which may be static or dynamic,
2. *qualitative (dynamic) models* where one may have
   - Confluences of Qualitative Physics [95],
   - Constraint type qualitative differential equations of Qualitative Simulation [95],
   - Signed Directed Graph (SDG) models [71,96].
3. objects, frames, semantic nets,
4. low-level and coloured Petri nets,
5. fuzzy models.

Note that Petri net models, both low level and coloured are the subject of Chapter 15 in the context of discrete event dynamic models.

The aim of this section is to highlight the relation of some of the above AI models to the process engineering models. We shall investigate the construction and analysis of the following AI models here:

- SDG models
- Confluences
- Rules

The AI modelling methods will be analysed and compared using the following simple example.

EXAMPLE 14.3.1 (Batch Water Heater).

Process system

Consider a perfectly stirred tank with a water flow in and out. The in- and outflow is controlled by valves. Let us assume that the tank is adiabatic, and moreover it contains an electric heater. The electric heater is controlled by a switch. The flowsheet is shown in Fig. 14.1.

Modelling goal

Describe the dynamic behaviour of the batch water heater.

Assumptions[$\mathcal{A}$4.]

$\mathcal{A}$1. Perfect mixing.
$\mathcal{A}$2. Only liquid phase water is present.
$\mathcal{A}$3. The effect of evaporation is negligible.
$\mathcal{A}$4. Adiabatic operation.
$\mathcal{A}$5. Cylindrical vessel with constant cross section A.
$\mathcal{A}$6. Constant physico-chemical properties.
$\mathcal{A}$7. Binary (open/close) valves and switch.

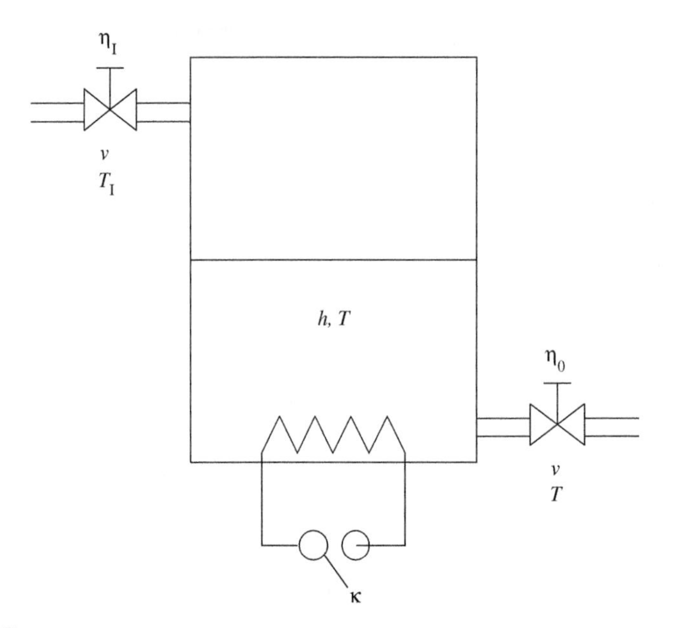

FIGURE 14.1 Batch water heater

Model equations

Differential (balance) equations

$$\frac{dh}{dt} = \frac{v}{A}\eta_I - \frac{v}{A}\eta_O, \tag{14.36}$$

$$\frac{dT}{dt} = \frac{v}{Ah}(T_I - T)\eta_I + \frac{H}{c_p \rho h}\kappa, \tag{14.37}$$

where the variables are

t	time [s]
h	level in the tank [m]
v	volumetric flowrate [m^3/s]
c_p	specific heat [J/kg K]
ρ	density [kg/m^3]
T	temperature in the tank [K]
T_I	inlet temperature [K]
H	heat provided by the heater [J/s]
A	cross section of the tank [m^2]
η_I	binary input valve [1/0]
η_O	binary output valve [1/0]
κ	binary switch [1/0]

Initial conditions

$$h(0) = h_0, \quad T(0) = T_0.$$

14.3.1. Logical and Sign Operations

Logical and sign operations have a lot in common. Sign values can be viewed as multivalued logical values. The definition of both logical and sign operations are usually given in the form of *operation tables* in which the value of the operation is enumerated for every possible value of the operands.

Logical Operations

The properties of the well-known logical operations are briefly summarized here in order to serve as a basis for their extension to the sign operations.

The operation tables of logical operations are also called *truth tables*. For example the following truth tables in Tables 14.1 and 14.2 define the logical **and** ($\wedge$) and the **implication** ($\rightarrow$) operations.

The logical operations ($\wedge$ = **and**, $\vee$ = **or**, $\neg$ = **neq**, $\rightarrow$= **impl**) have the following well-known algebraic properties:

1. *Commutativity*:

$$(a \wedge b) = (b \wedge a), \quad (a \vee b) = (b \vee a).$$

2. *Associativity*:

$$(a \wedge b) \wedge c = a \wedge (b \wedge c), \quad (a \vee b) \vee c = a \vee (b \vee c).$$

3. *Distributivity*:

$$a \wedge (b \vee c) = (a \wedge b) \vee (a \wedge c), \quad a \vee (b \wedge c) = (a \vee b) \wedge (a \vee c).$$

4. *De Morgan identities*:

$$\neg(a \wedge b) = \neg a \vee \neg b, \quad \neg(a \vee b) = \neg a \wedge \neg b.$$

With the logical identities above, every logical expression can be transformed into *canonical form*. There are two types of canonical forms:

- the *disjunctive* form with only the operations $\neg$ and $\vee$
- the *conjunctive* form with only $\neg$ and $\wedge$.

The traditional two-valued logic is usually extended for real world applications with a third value being **unknown** to reflect the fact that the value of a variable may

TABLE 14.1 Operation Table of the "and" Operation

$a \wedge b$	False	True
False	False	False
True	False	True

TABLE 14.2 Operation Table of the "Implication" Operation

$a \rightarrow b$	False	True
False	True	True
True	False	True

not be known. Note that **unknown** can be interpreted as "either **true** or **false**", i.e.

$$\textbf{unknown} = \textbf{true} \vee \textbf{false}.$$

The result of any logical operation with any of its operand being **unknown** is usually **unknown**. That is, an additional column and row is added to the operation tables with all the values in them being **unknown**. Table 14.3 shows the extended operation table for the logical **or** operation. It is seen from the second row and second column of the table that the logical value **true** in any of the operands will "improve" the uncertainty given in the **unknown** value of the other operand.

TABLE 14.3 Extended Operation Table of the "and" Operation

$a \vee b$	False	True	Unknown
False	False	True	Unknown
True	True	True	True
Unknown	Unknown	True	Unknown

Sign Algebra

The sign algebra is applied for variables and expressions with sign values, where their range space is the so-called *sign universe* defined in Eq. (14.35).

We can consider the sign universe as an extension of the logical values (**true**, **false**, **unknown**) and define the usual arithmetic operations on sign value variables with the help of operation tables. The operation table of the sign addition ($+_S$) and that of the sign multiplication ($*_S$) are given in Tables 14.4 and 14.5.

TABLE 14.4 Operation Table of Sign Addition

$a +_S b$	+	0	−	?
+	+	+	?	?
0	+	0	−	?
−	?	−	−	?
?	?	?	?	?

The operation table for the sign subtraction and division can be defined analogously. Note that the operation tables of operators, such as sin, exp, etc., can be generated from the Taylor expansion of the functions and the application of the operation tables of the elementary algebraic operations.

It is important to note that the sign algebra has the following important properties.

1. *growing uncertainty with additions* which is seen from the sign addition table as the result of "$+ \ +_S \ -$" being unknown "?",
2. the usual algebraic properties of addition and multiplication, i.e. commutativity, associativity and distributivity.

TABLE 14.5 Operation Table of Sign Multiplication

$a *_S b$	+	0	−	?
+	+	0	−	?
0	+	0	−	**0**
−	−	0	+	?
?	?	**0**	?	?

14.3.2. Signed Directed Graph Models

The structure of a linear or linearized state space model given by Eq. (13.2) can be described by an SDG model. There is a one-to-one correspondence between the graph and the structure of an SDG model. Moreover, the SDG model of a system is nothing but its structure graph weighted by signs as seen in Section 10.5.2.

A signed directed graph $G = (V, E; w)$ can be characterized by its set of vertices V, set of edges E and the weight function W. In the case of an SDG model associated with a state space model with SSR $(\mathbf{A}, \mathbf{B}, \mathbf{C})$ the elements of the graph are as follows:

- the set of vertices is partitioned into three disjoint sets:

$$V = \{x_1, \ldots, x_n; u_1, \ldots, u_r; y_1, \ldots, y_p\}$$

  i.e. a vertex is associated with each dynamic variable,
- the set of directed edges corresponds to the structure of the state space model equations:

$$(v_i, v_j) \in E \quad \text{if } v_i, v_j \in V$$

  and v_j is present on the right-hand side for v_i,
- the edge weights $w(v_i, v_j) \in U_S$ are equal to the sign of the corresponding state space model matrix elements.

The *occurrence matrix* of the graph is composed of the state space model matrices as follows:

$$\mathbf{O}_G = \begin{bmatrix} [\mathbf{A}]_S & [\mathbf{B}]_S & 0 \\ 0 & 0 & 0 \\ [\mathbf{C}]_S & 0 & 0 \end{bmatrix}. \tag{14.38}$$

EXAMPLE 14.3.2 (SDG model of the batch water heater of Example 14.3.1). In order to distinguish between the partitions of the SDG model vertices, circles will be applied for the state, double circles for the input and rectangles for the output variables. With this notation the SDG model of the batch water heater is shown in Fig. 14.2.

Application of SDG Models

SDG models can be applied in the analysis of structural dynamic properties of process systems in the following areas:

1. *System analysis*
   The application of SDG models for structural controllability and observability as well as for structural stability has been discussed before in Section 13.2.

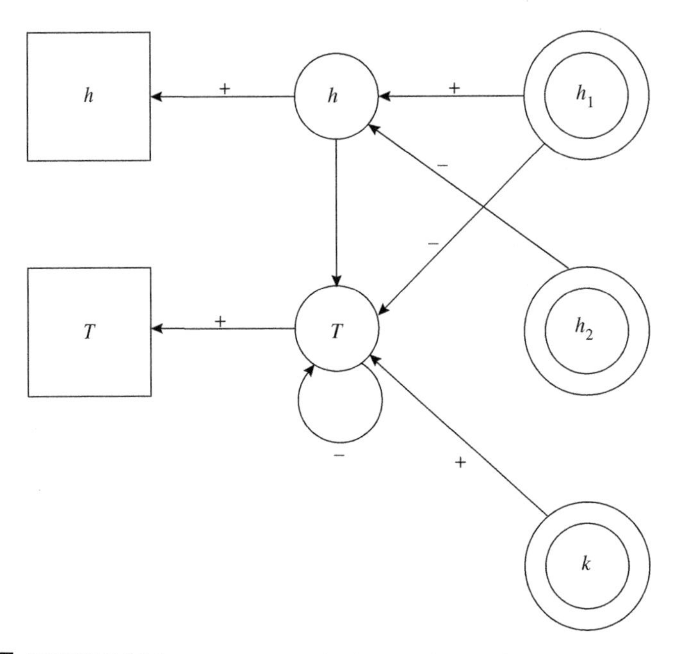

FIGURE 14.2 Structure digraph of the batch water heater.

2. *Qualitative prediction of the unit step response*
   Using sign algebraic manipulation it is possible to determine [97] the
   - *sign of the initial deviation of the step response* by computing the value of the shortest path in SDG. The sign value of a path in an SDG is the sign product of the signs as edge weights forming the directed path.
   - *sign of the steady state deviation* by computing the solution of a linear sign set of equations corresponding to the steady state algebraic model equations.

It is important to note that because of the properties of the sign addition we often have indeterminate (denoted by ?) values as qualitative prediction.

EXAMPLE 14.3.3 (Qualitative prediction of the unit step response of the batch water heater of Example 14.3.1). The SDG model of the batch water heater is shown in Fig. 14.2.

Let us compute the sign of the initial deviation of the unit step response of output T_0 (where T_0 is the output temperature variable while the state temperature variable is denoted by T) to the unit step in input κ. The shortest path between κ and T is (κ, T), (T, T_0). The sign on both of the edges involved is "+" therefore the sign value of the path is again "+". This means that the initial deviation of the temperature when putting a positive step change to the binary switch of the heater is positive.

14.3.3. Confluences

The notion of confluences as a kind of qualitative model has been introduced in AI by de Kleer and Brown [98] in their theory called "Qualitative Physics". *Confluences can be seen as sign versions of (lumped) balance equations from a process engineering perspective.*

They can be *formally derived* from lumped process model equations in the following steps:

1. *Define qualitative variables* $[q]$ *and* δq *to each of the model variables* $q(t)$ *as follows*:
$$q \sim [q] = \text{sign}(q), \quad dq/dt \sim \delta q = \text{sign}(dq/dt),$$
where sign $(\cdot)$ stands for the sign of the operand;
2. *operations are replaced by sign operations*, i.e.
$$+ \sim +_S, \; * \sim *_S \text{ etc.}$$
3. *parameters are replaced by* $+$ *or* $-$ *or* 0 forming *sign constants* in the confluence equations, i.e. they virtually disappear from the equations.

The *solution of a confluence* is computed by simple enumeration of all possible values of the qualitative variables and arranged in an *operation (truth) table of the confluence*. The operation (truth) table of a confluence contains all the possible values satisfying it and resembles the operation table for sign operations.

The above concepts are illustrated on the example of the batch water heater introduced in Example 14.3.1.

EXAMPLE 14.3.4 (Confluences of the batch water heater of Example 14.3.1). The confluences are derived from the model equations (14.36)–(14.37) in the following steps:

1. *Qualitative variables*
From their physical meaning the sign value of the variables is as follows:
$$[\eta_1] \in \{0, +\}, \quad [\eta_2] \in \{0, +\}, \quad [\kappa] \in \{0, +\}$$
2. *Sign constants*
The sign of all parameters is strictly positive, so that all sign constants are equal to "+".
3. *Confluences*
$$\delta h = [\eta_1] -_S [\eta_2], \tag{14.39}$$
$$\delta T = [T_I - T] *_S [\eta_1] +_S [\kappa]. \tag{14.40}$$

The truth table of the confluence (14.39) is shown in Table 14.6.

The truth table of the other confluence (14.40) has more columns on the right hand because we have more variables. Observe that a composite qualitative variable $[T_I - T]$ is present in the first right hand column of Table 14.7.

Application of Confluences

The solution of confluences, being various interpretations of the truth table, can be applied for process monitoring and diagnosis purposes. Some of the characteristic applications are as follows.

1. *Sensor validation* performed in the following logical steps:
   - measure the qualitative value of model variables $q_i(t)$, $\delta q_i(t)$ at any time t,
   - check the corresponding row in the table for sign consistency.

   The above procedure validates a group of sensors connected by a balance equation from which the confluence is derived. It is a simple and quick but not too sensitive method for sensor validation.

2. *Generating rule sets* from confluences
   A row in the truth table of a confluence can be interpreted as a rule by "reading the row from the right to the left". For example row number two in Table 14.6 can be read as follows:
   "**if** η_2 is open **and** η_1 is closed **then** h is decreasing"

TABLE 14.6 Truth Table of a Confluence on Height

δh	$[\eta_1]$	$[\eta_2]$
0	0	0
−	0	+
+	+	0
?	+	+

TABLE 14.7 Truth Table of a Confluence on Temperature

δT	$[T_I - T]$	$[\eta_1]$	$[\kappa]$
+	0	0	+
+	+	0	+
+	−	0	+
+	0	+	+
+	+	+	+
?	−	+	+
0	0	0	0
0	+	0	0
0	−	0	0
+	0	+	0
+	+	+	0
?	−	+	0

14.3.4. Rules

Rules are the most widespread form of knowledge representation in expert systems and other AI tools. Their popularity is explained by their simplicity and transparency from both theoretical and practical point of views. This implies that rule sets are relatively easy to handle and investigate.

The logical validation of a rule set, which checks for consistency and absence of contradiction is a hard problem from an algorithmic viewpoint. The problem is NP-hard. Rule sets mostly describe black-box type heuristic knowledge in process applications, as well as other application areas. Therefore rule sets are difficult to validate against other types of engineering knowledge, such as process models.

This section gives hints how one can relate process models and rules as well as giving a brief introduction into the syntax and semantics of rules, and reasoning with them.

Syntax and Semantics of Rules

A rule is nothing but a conditional statement: "**if** ... **then** ...". The syntax of a rule consists of the following elements.

1. *Predicates*
   Predicates are elementary logical terms. Their value can be from the set {**true**, **false**, **unknown**}. They usually contain arithmetic relations ($=$, $\neq$, $\leq$, $>$, $<$) and they may contain qualitative or symbolic constants (e.g. **low**, **high**, **very small**, **open** etc.). Simple examples are

   $$p_1 = (\kappa = \textbf{on}); \quad p_2 = (T < 300); \quad p_3 = (h = \textbf{low}),$$
   $$p_4 = (error = \text{"}tank\ overflow\text{"})$$

   where p_1, p_2, p_3 and p_4 are arithmetic predicates.
2. *Logical expressions*
   A logical expression contains:
   - logical variables, which can be either predicates or logical constants,
   - logical operations ($\neg$, $\wedge$, $\vee$, $\rightarrow$)

   and obeys the syntax rules of mathematical logic.
3. *Rules*
   A rule is in the following syntactical form:

   if *condition* **then** *consequence*;

   where *condition* and *consequence* are logical expressions. An equivalent syntactical form of the above rule is in the form of an implication:

   condition $\rightarrow$ *consequence*

   Note that a rule is a logical expression itself.

The *semantics of a rule*, or its meaning when we use it, depends on the goal of the reasoning. Normally the logical expression *condition* is first checked if it is **true** using the values of the predicates. If this is the case then the rule can be applied or executed (the rule "fires"). When *applying or executing a rule* its *consequence* is *made* **true** by changing the value of the corresponding predicates.

EXAMPLE 14.3.5 (A simple rule set). Consider a simple rule set defined on the following set of predicates:

$$P = \{p_1, p_2, p_3, p_4\}, \tag{14.41}$$

$$\textbf{if } (p_1 \textbf{ and } p_2) \textbf{ then } p_3; \tag{14.42}$$

$$\textbf{if } p_3 \textbf{ then } p_4; \tag{14.43}$$

The equivalent implication form of the rule set above is

$$(p_1 \wedge p_2) \rightarrow p_3;$$
$$p_3 \rightarrow p_4;$$

■ ■ ■

Rules Derived from Confluences

One of the application area for confluences outlined in Section 14.3.3 is to derive rules from them. The truth table such as Table 14.6 can serve as a basis for rule generation. At least one rule is generated from each row of the table in the following steps.

1. Take the qualitative value of the variables appearing on the right-hand side of the confluence listed in the right part of the particular row and form predicates from them.
2. Form the *condition* part of the rule by making the conjunction ($\wedge$) of the predicates.
3. The *consequence* part of the rule is just the predicate made of the qualitative value of the variable appearing on the left-hand side of the confluence listed in the left part of the particular row.

If the qualitative value of the *consequence* part of a rule derived from a particular row is unique for the whole table occurring only once on the left-hand side of the table then that particular rule is an **if and only if** rule. Its "inverse" statement formed by changing the role of the *condition* and *consequence* parts can also be formed.

The above procedure is illustrated on the example of the batch water heater.

EXAMPLE 14.3.6 (Rules derived from a confluence of the batch water heater of Example 14.3.1). We derive a set of rules from the truth table of the confluence (14.39) shown in Table 14.6.

The rules in implication form derived from the 1st row are as follows:

$$(\delta h = 0) \rightarrow ((\eta_1 = 0) \wedge (\eta_2 = 0)),$$
$$((\eta_1 = 0) \wedge (\eta_2 = 0)) \rightarrow (\delta h = 0).$$

Note that two rules can be generated as the qualitative value 0 for δh is unique in the table.

From the 2nd row, only a single rule can be generated (cf. 4th row):

$$((\eta_1 = 0) \wedge (\eta_2 = +)) \rightarrow (\delta h = -).$$

■ ■ ■

Note that a *set of rules* can be generated from the operation (truth) table of a confluence. The generated set of rules is always

- *complete* (i.e. "nothing is missing")
- *consistent* (i.e. "no contradictions")

ensured by the consistency and completeness of the process model equations from which the confluence is generated and by the construction of the truth table which enumerates all possible qualitative values for which the confluence is valid.

Reasoning with Rules

Rules are used for *reasoning* in intelligent systems, mainly in *expert systems*. The data base of an expert system is called the *knowledge base* and it consists of two main parts:

- a static part (*relations*) containing the rules in encoded form,
- a dynamic part (*facts*) where the changing values of the predicates are stored.

The value of all predicates forms the *state of the knowledge base*. This state is described by a point in the so called *predicate space* spanned by all predicates:

$$\mathcal{P} = \{[p_1, p_2, \ldots, p_P]^T \mid p_i \in \{\textbf{true}, \textbf{false}\}\},$$

where P is the number of all predicates.

The reasoning is performed by executing the rules which are applicable, i.e. those which "fire". Formally, for a rule in the form of

$$condition \quad \rightarrow \quad consequence,$$

the execution performs the following manipulation:

$$\textbf{if } (condition = \textbf{true}) \textbf{ then } (consequence := \textbf{true}).$$

Note that one moves in the predicate space by executing rules as the value of the predicates appearing in the *consequence* are changed by executing the rule. The part of the expert system performing the reasoning is called the *inference engine*.

There can be more than one applicable rule in a situation. This is called a *conflict situation* and various heuristic techniques can be applied to select the rule to be executed from them for *conflict resolution*. It is an important property of a given inference engine as to which conflict resolution techniques are applied and how these techniques can be tuned according to the need of the particular application.

There are two different methods for reasoning with rules driven by the reasoning goal:

1. *Forward chaining* used for prediction
   Here the task is *to find all consequences of the facts (i.e. set of predicate values) by executing the rules in their normal direction*, reasoning from *conditions* to *consequences*.
2. *Backward chaining* used for diagnosis
   Here one wants *to find all causes of the facts (i.e. set of predicate values) by using the rules opposite to the normal direction.*

The following simple example shows the principles of forward and backward chaining.

EXAMPLE 14.3.7 (Reasoning with the simple rule set of Example 14.3.5). Consider a simple rule set given in Eqs. (14.42) and (14.43) based on the predicate set (14.41). First we classify the predicates in the set P according their place in the rules: the "root" predicates are the ones which only occur on the *condition* part of the rules. In our case they are:

$$P_{\text{root}} = \{p_1, p_2\}.$$

The root predicates should have values determined by the outside world, that is from measurements before we start reasoning with forward chaining.

Forward chaining
Let us assume we have the following values for the root predicates:

$$p_1 = \textbf{true}, \quad p_2 = \textbf{true}$$

and all the other predicates have **unknown** values. Then rule (14.42) fires and we get $p_3 =$ **true** by executing it. Now rule (14.43) fires and gives $p_4 =$ **true** after execution.

Backward chaining
Now we assume no value for the root predicates:

$$p_1 = \textbf{unknown}, \quad p_2 = \textbf{unknown},$$

but a definite value for the predicate p_4

$$p_4 = \textbf{true}, \quad p_3 = \textbf{unknown}.$$

First, we use rule (14.43) where we have the *consequence* part defined and we get $p_3 =$ **true** by backward executing it. Now rule (14.42) is applicable for backward reasoning and gives $p_1 =$ **true** and $p_2 =$ **true** after execution. ■ ■ ■

14.4. SUMMARY

The question of model applications in a range of control scenarios is investigated in this chapter covering such techniques as LQR and model predictive control. The form and use of the model is illustrated in these applications.

The second part of this chapter looks at typical models applied to the problem of process diagnosis. This is an important aspect of model development and process engineering.

Finally, dynamic process models with interval-valued variables are discussed which are the form of process models applicable to intelligent control and diagnosis. These models take various forms ranging from qualitative models through to rule based representations.

14.5. FURTHER READING

The area of chemical process control is a broad and rapidly developing one with books, conferences and journals entirely devoted to it. Here, dynamic process modelling has been widely recognized as one of the key areas. Therefore, we do not attempt to make a literature review on this broad topic from the process modelling point of view but direct the readers' attention to the most important information sources. Note that the information sources for process control are overlapping with that of model analysis, therefore we also refer to the sources given in Section 13.5. Further information sources are:

- *Web*
  Control Engineering Virtual Library, with homepage:
  `http://www-control.eng.cam.ac.uk/extras/Virtual_Library/Control_VL.html`
- *Textbooks*
  There are a few special textbooks available on process control for those who have process engineering background such as [99–101].

14.6. REVIEW QUESTIONS

Q14.1. Give the problem statement of pole-placement control. What are the conditions for its application? What are the properties of the controller? (Section 14.1)

Q14.2. Give the problem statement of the LQR. What are the conditions for its application? What are the properties of the controller? Why is state filtering important in process control? (Section 14.1)

Q14.3. Give the problem statement of model-based predictive control. What are the conditions for its application? What are the steps of its conceptual solution? (Section 14.1)

Q14.4. Give the problem statement of prediction-based diagnosis. What are the steps in its conceptual solution? (Section 14.2)

Q14.5. Give the problem statement of identification-based diagnosis. What are the steps in its conceptual solution? (Section 14.2)

Q14.6. Compare the logical and sign operations! What are the similarities and differences? What are the strength and weaknesses of logical and AI methods which are based on them? (Section 14.3)

Q14.7. Compare the SDG model, the confluences and the rule-set derived from the confluences of the same lumped process model. What is the relation between them? What are the similarities and differences? Is there any one-to-one relation among them? (Section 14.3)

14.7. APPLICATION EXERCISES

A14.1. Is the pole placement controller applicable for stabilizing the state space model (Eqs. (14.36) and (14.37)) of the batch water heater? Why? Formulate the special case of the LQR controller design to stabilize the model. (Section 14.1)

A14.2. Derive the confluences from the nonlinear state space model of a CSTR derived in Example 10.3.1 in Section 10.4. Give the truth table of the confluences. (Section 14.3)

A14.3. Derive the complete set of rules for both of the confluences obtained in the previous Exercise (A14.2). Which are the rows in the truth tables giving rise to two rules (in both direction)? (Section 14.3)

A14.4. Derive the complete set of rules for the truth Table 14.6. What are the properties of that rule set and why? (Section 14.3)

A14.5. Derive the complete set of rules for the truth Table 14.7. Is this set complete? How would you show this? (Section 14.3)

A14.6. Apply forward chaining reasoning with the rule set of Example 14.3.5 assuming the following values of the root predicates

$$p_1 = \textbf{true}, \quad p_2 = \textbf{false}$$

(Section 14.3)

A14.7. Apply backward chaining reasoning with the rule set of Example 14.3.5 assuming the following values of the predicates

$$p_1 = \textbf{false}, \quad p_2 = \textbf{unknown}$$
$$p_3 = \textbf{unknown}, \quad p_4 = \textbf{false}$$

Is your result unique? Why? (Section 14.3)

15 MODELLING DISCRETE EVENT SYSTEMS

This chapter discusses the modelling and model analysis issues related to discrete event dynamic systems applied in process systems engineering. The term *discrete event system* [102] stands for dynamic systems with discrete valued variables: state, input and output variables with a finite number of possible values. Time is also considered to be discrete in discrete event systems but the time instances are not necessarily equidistant. An "event" occurs when any of the discrete valued variables change their value, and this generates a discrete time instance.

The theory of discrete event systems has been developed within systems and control theory and computer science where most of the applications are found.

The material presented in this chapter is arranged as follows:

- The characteristic properties and the application areas of discrete event systems in process systems engineering are briefly outlined in Section 15.1.
- The most important approaches to model representation are introduced and compared in Section 15.2.
- The solution of discrete event dynamic models performed via discrete event simulation is the subject of Section 15.3.
- Finally, the most important properties of discrete event systems, and tools for analysing these properties, are described in Section 15.4.

It is important to note here that process systems are most often composed of *both* discrete and continuous subsystems. Chapter 16 deals with these so-called hybrid systems.

15.1. CHARACTERISTICS AND ISSUES

Process systems are often influenced by external or internal events which occur instantaneously and often have a discrete range space. In other words, they may take only a finite number of possible values. Binary on/off or open/close valves, switches or operator actions are the simplest examples of external discrete events. Rapid phase changes, such as sudden evaporation or condensation are examples of internal events. Some of the more complex nonlinear behaviour like weeping or flooding in distillation columns may also be viewed as internal discrete events.

If one is only interested in describing the behaviour of a process system to cover events that influence it and the subsequent event sequences which may arise then some kind of discrete event process system model is used. Such a discrete event model focuses only on the sudden discrete changes and approximates the continuous behaviour of the continuous system elements by event sequences. From a system theoretic point of view all variables including the state, input and output variables of the process system are then described by time-dependent variables with discrete range space:

$$x(t) \in \mathbf{X} = \{x_0, x_1, \ldots, x_n\}, \quad u(t) \in \mathbf{U} = \{u_0, u_1, \ldots, u_n\},$$
$$y(t) \in \mathbf{Y} = \{y_0, y_1, \ldots, y_n\}. \tag{15.1}$$

Here, the elements in the range space or the possible discrete values of the variables can be either numerical values or disjunct intervals or even symbolic values. In this sense there is a close relationship between discrete event and qualitative dynamic models.

A change in the discrete value of any of the variables is called an event. With the operation of a discrete event system its variables change their values in time causing sequences of events. The set of all discrete time of these events forms the discrete time domain of the system.

Discrete event process system models naturally arise in the following application areas:

1. Discrete event process system models are traditionally and effectively used in *design, verification and analysis of operating procedures* which can be regarded as sequences of external events caused by operator interventions.
2. The *scheduling and plant-wide control of batch process plants* is another important, popular and rapidly developing field.

15.2. APPROACHES TO MODEL REPRESENTATION

There are various related approaches to describe a discrete event dynamic system with discrete time and discrete valued variables.

Most of the approaches representing such systems use combinatorial or finite techniques. This means that the value of the variables including state, input and output variables, are described by finite sets and the cause-consequence relationships between these discrete states are represented by directed graphs of various kinds.

Examples of these approaches outlined in this section are,

1. *finite automata* models in Section 15.2.1,
2. extended *Petri nets* in Section 15.2.2.

Together with the description of these approaches their relationship will also be pointed out. The approaches are compared using the following simple example which will be used later in the chapter.

EXAMPLE 15.2.1 (A reactor–filter system).

Process system

Consider a perfectly stirred tank reactor to which raw materials and solvent are fed. A chemical reaction takes place in the reactor for a given time interval to produce a solution of the product which crystallizes. Having finished the reaction, the mixture is fed to one of the filters. The filtering takes some time and then the solid product in crystallized form is taken out and the solvent is fed back to the reactor when it is available.

The flowsheet is shown in Fig. 15.1.

Modelling goal

Describe the dynamic behaviour of the reactor–filter system.

Assumptions

$\mathcal{A}1$. Perfect mixing in the reactor.
$\mathcal{A}2$. Full conversion of the raw material to the product in the reactor.
$\mathcal{A}3$. Total crystallization and ideal filtering in the filters.
$\mathcal{A}4$. Isothermic operation in all process units.
$\mathcal{A}5$. Constant physico-chemical properties.
$\mathcal{A}6$. Binary ideal (open/close) valves.
$\mathcal{A}7$. Fixed and known processing times for every unit.
$\mathcal{A}8$. No transportation delay for any of the material in the system.
$\mathcal{A}9$. No malfunctions or failures in the units.

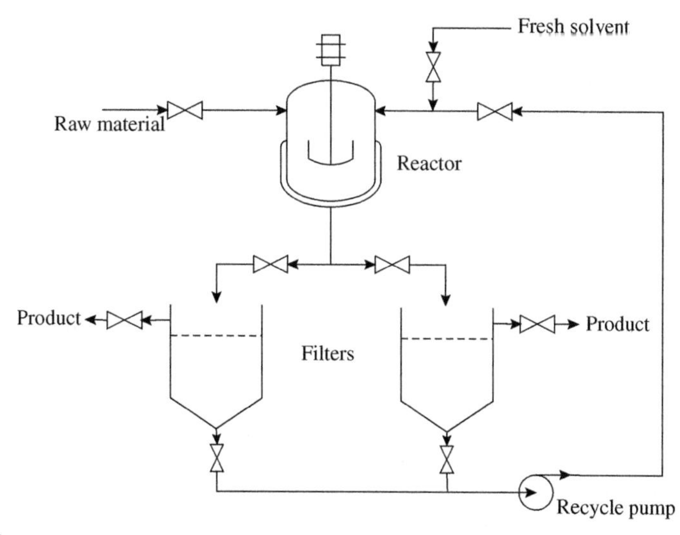

FIGURE 15.1 Reactor–filter system.

15.2.1. Finite Automata

The notion of finite automata is a basic and central one in computer science for analyzing algorithmic (computational) complexity [69]. A famous example is the so-called Turing machine of which various types exists.

A finite automata is an abstract discrete machine with at least one input tape containing symbols from a predefined finite set called an alphabet. The automata operates in discrete steps and it has a predefined set of states. When performing a step the automata changes its state depending on its current state and the subsequent symbol, read from the input tape. Thereafter, it may or may not move the reading head of the input tape one step ahead. This instruction to move or not to move the head is encoded in the state–symbol pair before the step.

The above abstract operation is formally described as follows. The finite automata is a triplet

$$\mathbf{A} = (Q, \Sigma, \delta), \tag{15.2}$$

where Q is the finite set of discrete states in the automata. The state of the reading head (whether it is to be moved or not to be moved) is also encoded in the state of the automata. Σ is a finite set of input alphabet. This set may contain a special symbol "#" denoting the end of the tape. The input alphabet set is used in the automata to describe external events, as will be explained in more detail later in Section 15.2.3 and δ is the state transition function of the finite automata, which determines the next state, after a symbol from the input tape has been read. Formally,

$$\delta : Q \times \Sigma \rightarrow Q. \tag{15.3}$$

This is normally a partial function, defined only for some of the state–symbol pairs.

Note that bounded Petri nets (in Section 15.2.2) can also be described by this set of states, considering the nodes in the related reachability graph as states.

The state transition function is visualized using the so called state transition graph, which is a weighted directed graph $G_A = (V_A, E_A; w_A)$, where V_A is the set of vertices, E_A is the set of edges with edge weights w_A. The vertices of the state transition graph are the discrete states, that is $V_A = Q$. There is a directed edge between two vertices if the state transition function is defined for the starting state vertex of the edge resulting in the ending state vertex of the edge for some symbol on the input tape. The symbol is put onto the edge as its edge weight.

In order to describe the operation of the automata, we need to define special subsets of discrete states: Q_{I} the set of admissible initial states $Q_{\mathrm{I}} \subseteq Q$ and Q_{F} the set of final states $Q_{\mathrm{F}} \subseteq Q$.

The automata can be started from an initial state with the given "content" of the input tape. It works according to the state transition function δ until it reaches a final state from set Q_{F}, and then it stops. The automata may also stop if the state transition function δ is not defined for a particular (Q_j, s_i) pair or when it reaches the end of the input tape (denoted by the end of tape symbol "#"). Finally, the automata may go into an "infinite cycle", and never stop.

Note that in a finite automata model, there are only discrete time steps without any duration. It means that only precedence or cause–consequence relations are modelled.

The ideas are best understood if one considers a simple everyday example: a hot/cold drink automata.

EXAMPLE 15.2.2 (Hot/cold drink automata). *Imagine that we have an automata which only accepts $1 coins and gives a glass of hot or cold drink. The type of drink is selected on the operation panel of the automata by pushing the knob of the desired drink. Describe the normal (that is fault-free) operation of the automata.*

In developing the finite automata model, we may proceed using the following steps:

1. *Verbal description of the operation process*
   If everything goes smoothly we find the automata in its "ready" state. We either insert a $1 coin, then press the button of the desired drink or do these in a reverse order. Then we will get our drink.

2. *Construction of the model elements*
   The possible states of the drink automata are as follows:

$$Q = \{\text{“}ready\text{”}, \text{“}operate\text{”}, \text{“}coin\_inserted\text{”}, \text{“}button\_pushed\text{”}\}$$

   The input alphabet contains the symbols needed to code the user input:

$$\Sigma = \{\text{“\$1”}, \text{“}n\text{”}, \text{“\#”}\}$$

   where n is the code of the drink. With these symbols a possible (acceptable) content of the input tape may be:

$$\text{“\$1”}, \text{“3”}, \text{“\#”}$$

3. *Construction of the state transition diagram*
   This is done by drawing all states first and then figure out from the operation process how the state changes occur. One step corresponds to one edge labelled by the symbol at the reading head on the input tape causing the state to change from the state at the origin of the edge to the state at the destination of the edge.

The resulting state transition diagram is seen in Fig. 15.2.

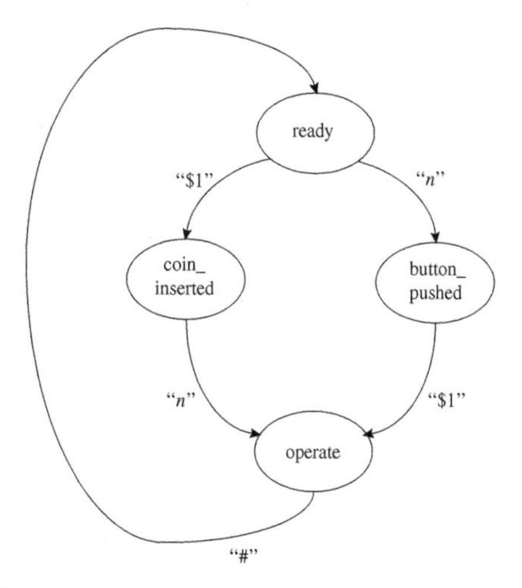

FIGURE 15.2 State transition diagram for the hot/cold drink automata.

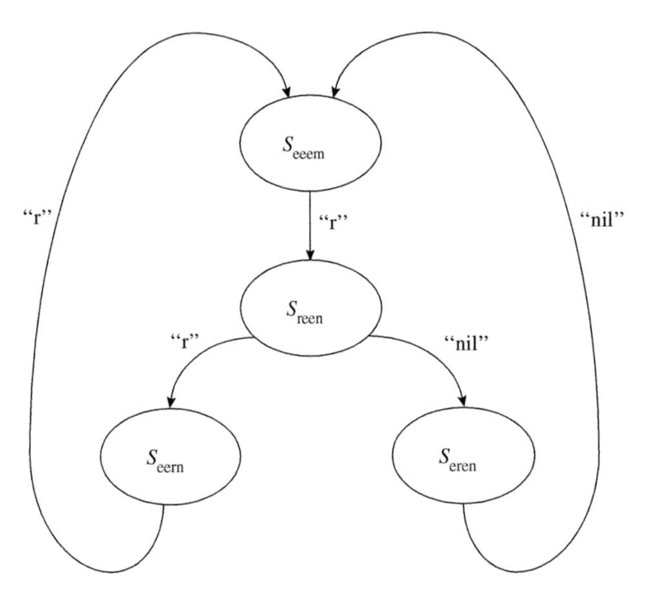

FIGURE 15.3 State transition diagram for the reactor–filter system.

The finite automata approach is further illustrated on the process example of the reactor–filter system in Example 15.2.3 below.

The state transition diagram of the finite automata model of the reactor–filter system is shown in Fig. 15.3, when *filter2* is given preference.

EXAMPLE 15.2.3 (A finite automata model of the Reactor-Filter System introduced in Example 15.2.1). In order to construct a finite automata model of the system, consider an initial state for it, when all the equipment is empty and ready for operation and a batch of raw material and solvent is available. The following formal elements are needed.

Q_{RF} is the finite set of discrete states as follows:

$$Q_{RF} = \{S_{eeem}, S_{reen}, S_{eren}, S_{eern}\}, \tag{15.4}$$

where S_{xyzw} describes states of the reactor, the first and second filter and the movement of the reading head. The letters in the subscript indicate whether a particular element is empty and ready to operate ("e"), finished operation ("r"), the reading head of the input tape is to be moved forward ("m") or not to be moved ("n").
Σ_{RF} is a finite set of input alphabet

$$\sum_{RF} = \{\mathbf{r}, \mathbf{nil}\}, \tag{15.5}$$

where **r** denotes the presence of a batch of the raw material and **nil** is the null or "empty" symbol,

δ_{RF} is the state transition function of the finite automata

$$\delta_{RF}(S_{eeem}, \mathbf{r}) = S_{reen}, \tag{15.6}$$

$$\delta_{RF}(S_{reen}, \mathbf{r}) = S_{eern}, \tag{15.7}$$

$$\delta_{RF}(S_{reen}, \mathbf{nil}) = S_{eren}, \tag{15.8}$$

$$\delta_{RF}(S_{eern}, \mathbf{r}) = S_{eeem}, \tag{15.9}$$

$$\delta_{RF}(S_{eren}, \mathbf{nil}) = S_{eeem}. \tag{15.10}$$

Q_I is the finite set of initial states

$$Q_I = \{S_{eeem}\}$$

Q_F is the finite set of final states

$$Q_F = \{S_{eeem}\}.$$

■ ■ ■

15.2.2. Petri Nets

Petri nets were originally developed for modelling sequences of finite automata and have been found to be useful for simulation and analysis of resource allocation problems of discrete event dynamic systems with both sequential and parallel elements. There are a number of extensions to the original concept of Petri nets in the literature including timed, hierarchical, multi-valued logical as well as stochastic elements [103,104].

The main application area in process engineering is the modelling and analysis of batch process systems and that of specifying/describing operating procedures [105,106].

Low Level Petri Nets: Structure and Dynamics

A Petri net model of a discrete event dynamic system represents its possible (discrete) states and the state changing events that relate to them. An event can only occur if its preconditions described by a particular discrete value of the system state holds. The occurrence of an event changes the state of the system into the post-conditions or consequences of the event. In a Petri net structure, the states or their values as conditions are generalized to the notion of *places*, and events are described by so called *transitions*.

The structure of a low level Petri net is formally described by the following four-tuplet:

$$\mathbf{C} = (\mathbf{P}, \mathbf{T}, \mathbf{I}, \mathbf{O}), \tag{15.11}$$

where $\mathbf{P} = \{p_1, \ldots, p_n\}$ is the non-empty finite set of places, $\mathbf{T} = \{t_1, \ldots, t_m\}$ is the non-empty finite set of transitions, $\mathbf{I} : \mathbf{T} \to \mathbf{P}^{\infty}$ is the input function relating places to a transition forming its pre-condition and $\mathbf{O} : \mathbf{T} \to \mathbf{P}^{\infty}$ is the output function relating places to a transition forming its post-condition or consequence.

Note that the input and output functions give the logical relationships between places and transitions. In the general case, a place may appear in the pre-conditions or post-conditions of a transition several times (in multiplication) that is denoted by the symbol $\mathbf{P}^{\infty}$.

The structure of a Petri net is usually described in a graphical way in the form of a bipartite directed graph $G_P = (V_P, E_P)$ because the set of places and that of transitions are disjoint, that is

$$P \cap T = \emptyset.$$

The vertex set V_P of the graph is composed of the set of places and that of transitions:

$$V_P = P \cup T.$$

Places are denoted by circles and transitions by bold intervals on the graph.

The input and output functions correspond to the directed edges in the graph. Because of the bipartite nature of the graph, edges join place-transition pairs only. A directed edge (p_i, t_j) or (t_j, p_o) exists if p_i is present in the pre-conditions or p_o is present in the post-conditions of the transition t_j, formally

$$(p_i, t_j) \in E_P \quad \text{if } p_i \in \mathbf{I}(t_j),$$
$$(t_j, p_o) \in E_P \quad \text{if } p_O \in \mathbf{O}(t_j).$$

Note that the formal and graphical description is fully equivalent.

The formal and graphical description of a simple Petri net structure is given in the following example.

EXAMPLE 15.2.4 (Formal description of a simple Petri net structure). The graphical description of the Petri net structure is shown in both of the sub-figures of Fig. 15.4.

The equivalent formal description is as follows:

$$\mathbf{C}_S = (\mathbf{P}_S, \mathbf{T}_S, \mathbf{I}_S, \mathbf{O}_S), \tag{15.12}$$

where

$$\mathbf{P}_S = \{p_{\mathrm{rr}}, p_{\mathrm{sr}}, p_{\mathrm{Rr}}, p_{\mathrm{pr}}\}$$
$$\mathbf{T}_S = \{t_{\mathrm{r}}, t_{\mathrm{R}}\}$$
$$\mathbf{I}_S(t_{\mathrm{r}}) = \{p_{\mathrm{rr}}, p_{\mathrm{sr}}\}$$
$$\mathbf{I}_S(t_{\mathrm{R}}) = \{p_{\mathrm{Rr}}, p_{\mathrm{Rr}}\}$$
$$\mathbf{O}_S(t_{\mathrm{r}}) = \{p_{\mathrm{Rr}}, p_{\mathrm{Rr}}\}$$
$$\mathbf{O}_S(t_{\mathrm{R}}) = \{p_{\mathrm{pr}}, p_{\mathrm{sr}}\}$$

Observe that the graph in Fig. 15.4 contains multiple edges directed to and from the place p_{Rr}. This means that the condition corresponding to place p_{Rr} appears twice in the pre-condition of t_{R} and in the post-condition of t_{r}.

The *dynamics or operation of a Petri net* is described using the notion of *markings*. *A discrete state of the process system described by a Petri net is represented by the state of the places, i.e. by the state of conditions.* A place may be active with its associated condition fulfilled or passive when it is not. A *token* is put in a place

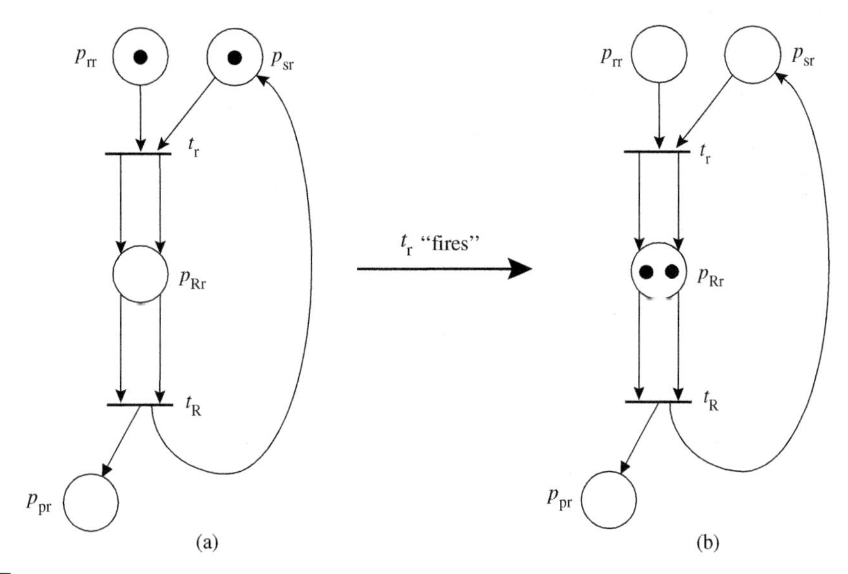

FIGURE 15.4 Simple Petri net

when it is active. *A marking is then a complete set of the states of the places, or the distribution of the tokens on places in the Petri net.*

The graphical representation of the above example shown in Fig. 15.4 illustrates the notion of markings by giving two examples. The marking on the left-hand sub-figure corresponds to the case when the places p_{rr} and p_{sr} are active.

The engineering interpretation of a marking is based on the meaning of a token put on a place which can be given in two alternative ways:

1. *Workpiece-based interpretation*
   We can imagine a token as a workpiece put on a plate (the place) waiting for further processing performed by a place processing equipment (a transition). This is the usual interpretation in the manufacturing industries.
2. *Value-based interpretation*
   We can also associate a logical variable with a place. The presence of a token indicates its logical "**true**" value. The place then represents an event described by a logical expression, for example the event of *temperature* $> 25°$C.

The formal description of a marking is performed by constructing an *integer-valued marking function over the set of places* as follows:

$$\mu : \mathbf{P} \rightarrow \mathcal{N}, \quad \mu(p_i) = \mu_i \geq 0, \tag{15.13}$$

$$\underline{\mu}^T = [\mu_1, \mu_2, \ldots, \mu_n], \quad n = |\mathbf{P}|, \tag{15.14}$$

where $\mu(p_i)$ is the number of tokens on the place p_i.

The graphical representation of a marking is shown on the directed bipartite graph of the Petri net. The presence of a token on a place is indicated by drawing a bold dot inside the circle representing the place.

The execution of a Petri net with a given marking represents one step in the dynamic state change of the discrete event dynamic system. The state of the system is represented by a specific marking. When an event occurs then we say that the corresponding *transition fires*. If the preconditions of a transition t_j hold in a marking $\underline{\mu}^{(i)}$ then the event (or transition) t_j takes place (that is t_j fires) to give rise to a new marking $\underline{\mu}^{(i+1)}$. The preconditions of the transition become **false** and its post-conditions become **true**. This means that the places connected to the firing transition as preconditions may become inactive and the places connected to the transition as post-conditions will be active. In notation, we write

$$\underline{\mu}^{(i)}[t_j > \underline{\mu}^{(i+1)} \tag{15.15}$$

for the event of firing the transition t_j.

The graphical representation of the execution is shown in the directed bipartite graph representation of the Petri net by moving the corresponding tokens around the firing transition. One token from each of the precondition places (those with an arrow directed towards the transition) is taken away and one token is inserted in each of the post-condition places (those with an arrow directed from the transition).

The formal and graphical description of the execution of a simple Petri net introduced in Example 15.2.4 is as follows.

EXAMPLE 15.2.5 (Execution of a simple Petri net of Example 15.2.4). The graphical description of firing transition t_r is shown in the sub-figures of Fig. 15.4. The marking before the firing is drawn on sub-figure (a), while the marking after the firing is on sub-figure (b).

The situation depicted in sub-figure (a) can be interpreted as follows: we have the solvent ready (token on the place p_{sr}) and the reagent(s) ready (token on the place p_{rr}). After finishing the filling (described by the transition t_{f}) we have both components in our equipment ready for further processing (two tokens on place p_{Rr}) shown in sub-figure (b).

The equivalent formal description is as follows:

$$\underline{\mu}_S^{\mathrm{T}} = [\mu_{\mathrm{rr}}, \mu_{\mathrm{sr}}, \mu_{\mathrm{Rr}}, \mu_{\mathrm{pr}}], \tag{15.16}$$

$$\underline{\mu}_S^{(0)\mathrm{T}} = [1, 1, 0, 0], \tag{15.17}$$

$$\underline{\mu}_S^{(1)\mathrm{T}} = [0, 0, 2, 0], \tag{15.18}$$

$$\underline{\mu}_S^{(0)}[t_{\mathrm{r}} > \underline{\mu}_S^{(1)}. \tag{15.19}$$

A *finite or infinite sequence of operations* is a sequence of the steps of execution connected by joint markings, i.e.

$$\underline{\mu}^{(0)}[t_{j_0} > \underline{\mu}^{(1)}[t_{j_1} > \cdots [t_{j_k} > \underline{\mu}^{(k+1)}. \tag{15.20}$$

The marking or the state of the Petri net $\underline{\mu}^{(k+1)}$ is called *reachable from* marking $\underline{\mu}^{(0)}$ if there is a sequence of transitions forming a finite sequence of operations above moving the net from the state $\underline{\mu}^{(0)}$ to $\underline{\mu}^{(k+1)}$.

It is important to note that there can be more than one transition able to fire for a given marking or state of the Petri net. This marking may have more than one subsequent marking. The fireable transitions can be executed in parallel, when they do not have any joint place in their preconditions, or they are *in conflict*. There are various extensions to the original Petri net concept to handle transitions in conflict and to make the net conflict-free. We can either

- *introduce priorities* between the transitions in conflict positions, or
- *extend the original net by* so called *inhibitor edges*. If a precondition place p_i is connected to a transition t_j by an inhibitor edge then a token on place p_i will inhibit the firing of t_j. That is, t_j can only fire if there is no token on place p_i.

The set of markings reachable from a given marking $\underline{\mu}$ is called the *reachability set* and is denoted by $\mathbf{R}(\underline{\mu})$. *The structure of the reachability set* is depicted using the so called *reachability graph of the Petri net*. The vertices of the reachability graph are the set of markings. There is a directed edge between two vertices if there is a transition the firing of which moves the net from the starting vertex to the destination vertex. The construction of the reachability graph of a Petri net is done by generating all possible sequences of operations starting from the given original marking. *The reachability graph can be regarded as the solution of the Petri net and it is the basis of analyzing its properties*. It is important to note that the construction of the reachability graph is a computationally hard problem in the general case.

The operation of the reactor–filter system introduced in Example 15.2.1 is discussed in the following example using its Petri net model.

EXAMPLE 15.2.6 (The operation of the reactor–filter system of Example 15.2.1 analyzed by a Petri net model). The graphical description of the Petri net of the reactor–filter system is shown in Fig. 15.5. The operation of any of the units such as the reactor and the two filters is described by a single transition t_{GR}, t_{Gf1} and t_{Gf2} respectively. The surroundings of these transitions are much the same: they can start to operate if they are empty (places $p_{\mathrm{Rempt}}, p_{\mathrm{f1empt}}, p_{\mathrm{f2empt}}$) and they have the necessary material to process (places p_{raw} and p_{solv} for the reactor). After finishing operation they produce their product (places p_{semi} and p_{product}) and become empty. Filters also produce the solvent to be recirculated (place p_{solv}).

A marking corresponding to a possible initial state of the system with all three operating units empty and solvent and raw material available is also shown in Fig. 15.5.

The equivalent formal description of the marking is as follows

$$\underline{\mu}^{\mathrm{T}} = [\mu_{\mathrm{raw}}, \mu_{\mathrm{solv}}, \mu_{\mathrm{Rempt}}, \mu_{\mathrm{semi}}, \mu_{\mathrm{f1empt}}, \mu_{\mathrm{f2empt}}, \mu_{\mathrm{product}}], \quad (15.21)$$

$$\underline{\mu}^{(0)\mathrm{T}} = [1, 1, 1, 0, 1, 1, 0]. \quad (15.22)$$

In this initial state transition, t_{GR} can only fire to produce marking $\underline{\mu}_S^{(1)}$ when the semi-product is available and the reactor is empty again, but there is no more raw material and solvent available.

$$\underline{\mu}^{(1)\mathrm{T}} = [0, 0, 1, 1, 1, 1, 0], \quad (15.23)$$

$$\underline{\mu}^{(0)}[t_{\mathrm{GR}} > \underline{\mu}_{\mathrm{S}}^{(1)}. \quad (15.24)$$

The situation described by marking $\underline{\mu}_{\mathrm{S}}^{(1)}$ *is clearly a conflict situation*: both filters can process the semi-product because both of them have their operation conditions fulfilled, both are empty and have the semi-product available. Here, the modeller has to design how to solve the conflict: by, for example, setting priorities for the filters to cause *filter1* to work if both are available.

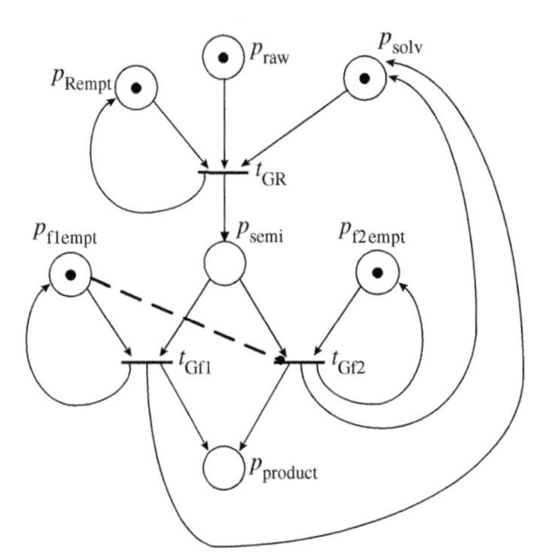

FIGURE 15.5 Petri net of the reactor–filter system

The same effect is achieved if one introduces an inhibitor edge from p_{f1emp} to t_{Gf2} as shown in Fig. 15.5. It will cause the second filter to be inhibited if the first is available for processing.

If we use the second filter when the next batch of raw material is available and the first otherwise, then the reachability graph of the system starting from its initial state is exactly the same as the state transition graph of the finite automata model of the same system seen in Fig. 15.3.

Extended Petri Nets: Hierarchical, Coloured and Timed Petri Nets

The original discrete Petri net concept is a transparent way of describing the state transition graph of finite bounded automata (cf. with Section 15.2.1). It can be shown that the union of all reachability graphs of a Petri net for every possible initial marking is the state transition graph of the finite automata model of the same underlying discrete event system [107].

A Petri net describes sequences of events without time, the precedence relations between them and with only discrete two-valued elements. Extensions of Petri nets aim at removing some or all of the above constraints.

The following summarizes the main ideas of the most important extensions. More details can be found in Jensen and Rosenberg [107].

1. *Timed Petri nets*

The firing of a transition is regarded as instantaneous in a Petri net and the places take and provide their tokens instantaneously. *In timed Petri nets a firing time (t^F) and a waiting time (t^V) can be allocated to every transition and place respectively. There is also the possibility of introducing clocks and timers as special places to the net.*

A clock as a special place can be imagined to produce a token in every given interval, say in every second. The given interval, which is the frequency rate for producing a token, is the parameter of the clock.

Part of the system dynamics of a process can be described by a timed Petri net model if one adds a firing time to every transition equal to the processing time of the corresponding process operation. This solution can also solve some of the conflict

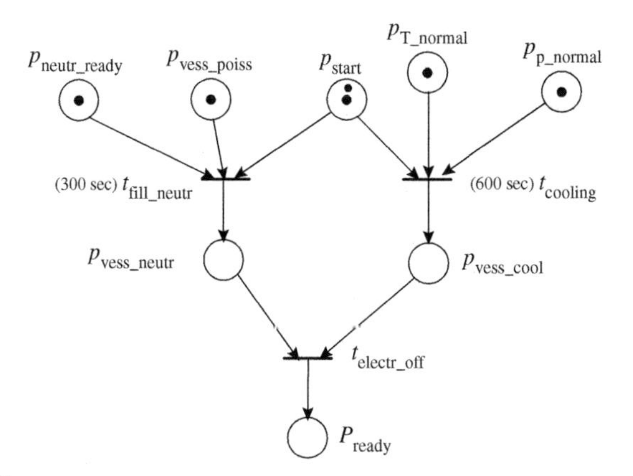

FIGURE 15.6 Petri net of maintenance shutdown procedure.

situations present in the original, non-timed Petri net model as can be illustrated on the reactor–filter example of Example 15.2.6. If we allocate different processing times for the two filters and the reactor, say $t^{\text{F}}(t_{\text{Gf1}})$, $t^{\text{F}}(t_{\text{Gf2}})$ and t^{F}_{Gr} such that

$$t^{\text{F}}(t_{\text{Gf1}}) > t^{\text{F}}(t_{\text{Gf2}}), \quad t^{\text{F}}(t_{\text{Gf1}}) > t^{\text{F}}_{\text{Gr}}, \tag{15.25}$$

then we have implicitly prescribed a definite event sequence for the case when two subsequent batches are present for processing.

A simple example of a maintenance shutdown procedure illustrates the use of timed Petri nets for describing operating procedures with sequential and parallel elements.

EXAMPLE 15.2.7 (A simple maintenance shutdown procedure). The usual verbal description of the maintenance shutdown procedure of a high pressure heated vessel storing poisonous gas intermediate is as follows:

1. Start the shutdown procedure in the normal operating mode.
2. Start cooling the vessel; it takes approximately 600 s to cool down.
3. Start filling the vessel with inert gas bleeding the content of the vessel to another storage. It takes approximately 300 s.
4. When the vessel has cooled down and it is filled with inert gas then switch off electric power.

The corresponding timed Petri net is shown in Fig. 15.6.

2. *Hierarchical Petri nets*

Experience with Petri net models shows that they can get very large quickly even for simple systems with complex operation modes. Moreover, they often contain repeated, similar or identical subgraphs of the original bipartite directed Petri net structure. (See for example the subgraphs around the filter operation transitions t_{Gf1} and t_{Gf2} in Example 15.2.6). Hierarchical Petri nets provide a tool for structuring the Petri net model in a hierarchical way: *one can substitute a complete Petri net called a sub-net into a transition in the higher level net called a super-net.* This is the Petri net equivalent for procedure calling: one calls the sub-net if the transition it belongs to is to be fired in the super-net.

Note that a sub-net in a hierarchical Petri net can also have its own sub-nets which are "sub-sub-nets" of the overall super-net. This way a *model hierarchy of Petri net models* can be gradually defined as one refines a Petri net model. It is also important to note that hierarchical Petri nets can be constructed for every type of Petri net model including timed and coloured Petri nets.

The concepts in developing a hierarchical Petri net and its use are again illustrated on the example of the reactor–filter system of Example 15.2.6.

EXAMPLE 15.2.8 (A simple hierarchical timed Petri net model of the reactor-filter system). Let us assume that we have started the modelling process by constructing the Petri net model of the reactor–filter system shown in Fig. 15.5 and extended it by the firing times given in Eq. (15.25) to have a timed Petri net. *This model can be regarded as the super-net of the hierarchical model.*

We may discover that we need to have a finer model of the operation of the reactor itself and build a Petri net model shown in Fig. 15.7. The new sub-model is to be inserted into the place of transition t_{GR} of the original model, that is the part of the Petri net model shown in Fig. 15.7 in between the transitions t_r and t_R. The rest of the net is to indicate the environment of the sub-net.

Here we have described the discrete dynamics of the heating operation within the reactor by the following additional model elements:

p_{heater}	heater is ready,
p_{Rr}	reactor is ready (filled),
p_{rr}	reactor operation is completed,
t_h	heating.

The corresponding timed Petri net in Fig. 15.7 is regarded as a sub-net to be placed into the transition t_{GR} of the original super-net. The transitions t_r and t_R together represent t_{GR} such that t_r corresponds to the "entrance" part and t_R to the "exit" part of it towards the environment of t_{GR} in the super-net.

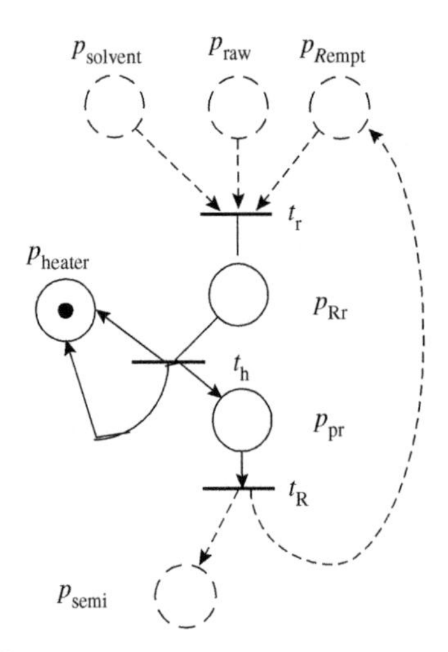

FIGURE 15.7 sub-net of reactor operation.

3. *Coloured Petri nets (CPNs)* [108]
In coloured Petri nets all the purely logical elements, for example the value of events or the firing preconditions and post-conditions, are extended to handle multivalued items. Tokens in a CPN are distinguishable as they may have a discrete value taken from a predefined set. For example tokens may carry the values **very high, high, low, very low** marked by colours instead of just carrying **true** for nonzero and **false** for zero. The set of possible values of the tokens on a place p_i is denoted by $\mathbf{col}(p_i)$ and the value of any one token on this place (denoted by $\mathbf{val}(p_i)$) can be an element from this set, so that

$$\mathbf{val}(p_i) \in \mathbf{col}(p_i).$$

In order to handle the processing of these multiple discrete valued edges and transitions, these elements are extended to have so called prescriptions, which are discrete functions to describe the operation of the corresponding element. Incoming edges to a transition t_j have a Boolean value prescription function associated with them with independent variables being the value of the token(s) on the place they start at. They are imagined to transfer a Boolean value to the transition t_j. The transition fires if all of its incoming edges transfers the logical value **true**.

Outgoing edges from a transition t_j determine the colour of the token(s) to be put on the end place as a consequence of firing the transition. Therefore, the outgoing edge from a transition t_j to the place p_o has a prescription function which takes its values from the colour set of place p_o.

Note that colours and prescriptions around a transition describe a dynamic state equation in its qualitative form (compare with Section 14.3).

We illustrate the use of coloured Petri nets on part of the reactor-filter system, on the operation of the reactor in the Example 15.2.9.

EXAMPLE 15.2.9 (A simple coloured Petri net model of part of the reactor-filter system). In order to have a finer description of the operation of the reactor in the reactor-filter system we develop a coloured extension of the original Petri net model given in Example 15.2.6. The extension describes the operation of the reactor when the storage tanks for the raw material and the solvent can have material for a maximum of 3 batches.

The structure of the coloured Petri net model is shown in Fig. 15.8. Observe that we have added "reverse" edges to the places p_{solv} and p_{raw} in order to describe the decrease of the material in the raw material and solvent storage tank as compared to the original model shown in Fig. 15.5. The following additional model elements (prescriptions) are needed to construct a coloured Petri net model:

Colour sets for the multi-valued discrete variables

$$\mathbf{col}(p_{\text{raw}}) = \mathbf{col}(p_{\text{solv}}) = \{\text{“}e\text{”}, \text{“}1b\text{”}, \text{“}2b\text{”}, \text{“}3b\text{”}\}$$
$$\mathbf{col}(p_{\text{Rempty}}) = \mathbf{col}(p_{\text{semi}}) = \{\text{“}true\text{”}, \text{“}false\text{”}\}$$

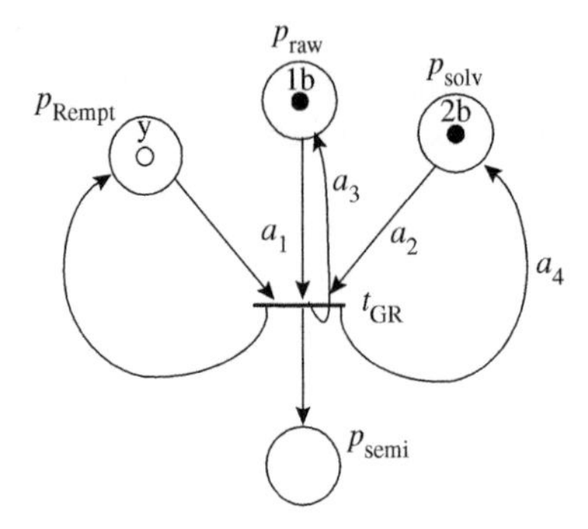

FIGURE 15.8 Coloured Petri net model of reactor operation.

Arc prescriptions

a_1: **if** (**val**(p_{raw})) $\in$ {"1*b*", "2*b*", "3*b*"} **then** "*true*"
otherwise "*false*"

a_2: **if** (**val**(p_{solv})) $\in$ {"1*b*", "2*b*", "3*b*"} **then** "*true*"
otherwise "*false*"

a_3: **val**(p_{raw}) := **case val**(p_{raw}) = "1*b*" : "*e*"
case val(p_{raw}) = "2*b*" : "1*b*"
case val(p_{raw}) = "3*b*" : "2*b*"

a_4: **val**(p_{solv}) := **case val**(p_{solv}) = "1*b*" : "*e*"
case val(p_{solv}) = "2*b*" : "1*b*"
case val(p_{solv}) = "3*b*" : "2*b*"

■ ■ ■

15.2.3. Discrete Event System Models as Dynamic System Models

Discrete event systems are dynamic systems with variables of discrete finite range space and with discrete time. Therefore their usual models, in the form of finite automata or Petri nets, should be special cases of the general nonlinear discrete time state space model of lumped systems in the form of

$$x(k+1) = f(x(k), u(k)), \quad x(0) = x^0, \tag{15.26}$$

$$y(k) = g(x(k), u(k)), \quad k = 1, 2, \ldots, \tag{15.27}$$

where x, u and y are the state, input and output variables respectively, k is the discrete time and f and g are nonlinear functions. Note that this equation is just the discrete time counterpart of the continuous time equation (13.1).

Finite Automata Models

There is a straightforward one-to-one explicit correspondence between the model elements of the finite automata model (Eq. (15.2)) and the discrete time dynamic state space model (15.26) with discrete range spaces (15.1).

State space model	Automata model	Remark
$x(k) \in \mathbf{X}$	$x(k) \in Q$	Set of possible states
$u(k) \in \mathbf{U}$	$u(k) \in \Sigma$	Set of possible inputs
$f(x(k), u(k))$	$\delta : Q \times \Sigma \rightarrow Q$	State transition function

Petri Net Models

In the case of Petri nets the correspondence between the model elements of the Petri net model (Eq. (15.11)) and the discrete time dynamic state space model (15.26) with discrete range spaces as in Eq. (15.1) is not so simple and even not explicit as it was the case for automata models.

It is important to realize that *we need to use coloured Petri nets* to make a one-to-one correspondence between a discrete time dynamic state space model (15.26) with discrete range spaces as in Eq. (15.1) and a Petri net model to be able to describe the value of the variables in the model. The following table summarizes the relation between the model elements.

State space model	Peri net model	Remark
$\mathbf{X}, \mathbf{U}$	$C = \{\mathbf{col}(p_i) \mid i = 1, \ldots, n\}$	Range spaces coded by colours Places are for both states and inputs
$x(k) \in \mathbf{X}$	$\underline{\mu}^{(k)}$	Set of possible states and inputs
$u(k) \in \mathbf{U}$	$\mu_i^{(k)} \in \mathbf{col}(p_i)$	
$f(x(k), u(k))$	$G_P(V_P, E_P)$	Encoded in the structure graph (digraph and prescriptions)

15.2.4. Exception Handling in Discrete Event System Models

One of the most difficult problems in developing a discrete event model of a process system is to ensure that the system model covers every possible scenario. This usually involves extending the model to include events for all possible faulty modes. It should encounter all exceptions which may occur during "normal" operation.

The theoretical problem behind exception handling is the *completeness* of a discrete model. A discrete event model is called complete if it is defined for every possible state-input pair. Remember that finite automata models in Section 15.2.1 are *not* necessarily complete because the state transition function, δ, in Eq. (15.3) is normally a partial function only, defined for some of the state–input symbol pairs. From this understanding it is clear, at least theoretically, how one can make a discrete event system model complete in order to handle all exceptions correctly: the state transition function should be extended to make it complete. That is it should be defined for every state–input symbol pair. In practice, however, extension of the set of states as well as that of input symbols is usually necessary.

The following example shows how the consideration of every possible exception (even in the simplest way!) makes the discrete event system model extremely complicated.

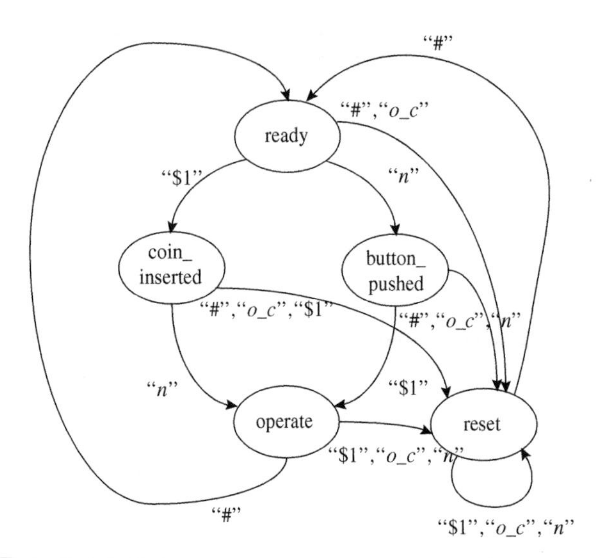

FIGURE 15.9 Hot/cold drink automata model with exceptions.

EXAMPLE 15.2.10 (Hot/cold drink automata model with exception handling). In order to obtain the simplest possible way of handling exceptions we introduce a new state of the automata model described in example 15.2.2 corresponding to the "reset" mode of operation, that is

$$Q_E = \{\text{"}ready\text{"}, \text{"}operate\text{"}, \text{"}coin\_inserted\text{"}, \text{"}button\_pushed\text{"}, \text{"}reset\text{"}\}$$

The input alphabet is again extended to have a symbol for "wrong coin" user input:

$$\Sigma = \{\text{"\$1"}, \text{"}n\text{"}, \text{"\#"}, \text{"}o\_c\text{"}\}$$

Moreover, we assume that the input tape contains several "#" symbols at the end of user inputs (we artificially put them there to indicate the end of the tape).

With these extensions we can imagine the exception handling as follows. If the hot/cold drink automata does not get what it expects then it goes to "reset" mode when it gives the user back the inserted coins, displays an error message and goes to the initial "ready" mode. This can be formally described (without the concrete actions taken in "reset" mode) by the extended state transition diagram shown in Fig. 15.9. The multiple labels on the figure abbreviate multiple edges with one label each from the set.

Observe that the complete model with proper exception handling is a *complete digraph*, such that every pair of vertices is connected by an edge.

15.3. SOLUTION OF DISCRETE EVENT DYNAMIC SYSTEM MODELS

Before we turn to the solution techniques of discrete event dynamic models the notion of "solution" of such models should be clarified. Remember that the solution of continuous or discrete time state space models with real vector valued variables is a set of time dependent functions. These describe the time evolution of the state and output variables over a given time interval assuming the time evolution of the input variables and suitable initial conditions for the state and output variables are given.

In the case of discrete event system models, we need to specify the following items for a well posed mathematical problem of determining its solution:

1. *A discrete event dynamic mathematical model*—for example, a finite automata or a Petri net model,
2. *A time interval* $[t_0, t_M]$—with a given initial time t_0 and final time t_M over which the solution will be computed,
3. *Initial conditions for the discrete valued state variables*—in the initial time instance t_0,
4. *A record for the input variables*—containing a sequence of its discrete values over the whole interval $[t_0, t_M]$.

With these ingredients the solution of the discrete event model will be the time evolution of the state and output variables, similar to the case of the state space models with real vector valued variables. But now the time evolution will be given in terms of a set of possible sequences of the discrete values for the state and output variables in the form of

$$S[v](t_0, t_M) = \{(v(t_0), v(t_1) \ldots, v(t_M)) \mid v(t_0) = v_0, \quad v(t_i) \in \mathbf{V}, \;\; i = 1, 2, \ldots, M\} \tag{15.28}$$

for the variable v, where v_0 is its initial condition and $\mathbf{V}$ is its discrete range space as in Eq. (15.1).

The solution of discrete event dynamic models is usually performed via discrete event simulation. Here, the solution method is given in terms of a discrete algorithm with executable steps instead of a numerical procedure or method. The combinatorial nature of the solution of discrete event models will be seen on the examples of finite automata and Petri net models described in the following two subsections.

15.3.1. Solution of Finite Automata Models

If the discrete event system model of a process system is given in the form of a finite automata according to Eq. (15.2) then *the ingredients needed for the solution* become the following items:

1. *A discrete event dynamic model*—in the form of Eq. (15.2). It is important to note that the set of discrete states of the automata Q is now constructed from vectors $S^{(i)}$, where the jth entry $S_j^{(i)}$ corresponds to the state variable x_j of the system with range space $\mathbf{X_j}$

$$Q = \{S^{(i)} \mid i = 0, 1, \ldots, N, \; S_j^{(i)} \in \mathbf{X_j}\}.$$

2. *A time interval* $[t_0, t_M]$—with a given initial time t_0 and final time t_M over which the solution will be computed. In case of non-timed models only the required number of steps, M, is specified.
3. *Initial conditions for the discrete valued state variables*—a concrete $Q_0 \in Q_I$.
4. *A record for the input variables*—which is now the content of the input tape. In case of vector valued input variable the elements of the finite alphabet Σ are

again vectors $\Sigma^{(i)}$, where the jth entry $\Sigma_j^{(i)}$ corresponds to the input variable u_j of the system with range space $\boldsymbol{U}_j$

$$\Sigma = \{\Sigma^{(i)} \mid i = 0, 1, \ldots, L, \Sigma_j^{(i)} \in \boldsymbol{U}_j\}.$$

Thereafter, the solution is generated in the following *conceptual algorithmic steps*:

A1 Take the given initial state Q_0 as the current state of the automata and set the reading head of the input tape to the start of the tape. Put the initial state at the head of the state sequence (or list) forming the solution of the model.
A2 Test if there is any unfinished solution sequence. If there is not then STOP.
A3 Apply the state transition function δ to the current $(S^{(j)}, \Sigma^{(j)})$ pair to get the next state $S^{(j+1)}$ and increase the time. Check if the final time t_M is not over. If so, then STOP.
- If this is unique then append the new state to the already generated state sequence.
- If $S^{(j+1)}$ is not unique then form two identical state sequences for the set of solutions to be continued.
- If a final state is reached $S^{(j+1)} \in Q_F$ then close the solution sequence by appending $S^{(j+1)}$ to it and GO TO step A2.
- If the state transition function is not applicable to the current $(S^{(j)}, \Sigma^{(j)})$ pair then close the solution sequence and GO TO step A2.

A4 Set the state $S^{(j+1)}$ as the current state, move the reading head accordingly and GO TO step A3.

The algorithm above is illustrated on the example of the reactor–filter system introduced in Example 15.2.1 using its finite automata model given in 15.2.3.

EXAMPLE 15.3.1 (Solution of the finite automata model of the reactor–filter system developed in Example 15.2.3). *Compute the solution of the model until an infinitely large final time t_M with the following initial condition:*

$$Q_0 = S_{\text{eeem}}$$

that is, all equipment is ready and with the content of the input tape:

$$(\text{“}\mathbf{r}\text{”}, \text{“}\mathbf{nil}\text{”}, \#), \tag{15.29}$$

which corresponds to the case when a single batch is to be processed.

The state transition diagram shown in Fig. 15.3 will be the representation of the model, we use for the solution.

Following the algorithm of solving finite automata models, we can easily compute the following unique solution sequence:

$$(S_{\text{eeem}}, S_{\text{reen}}, S_{\text{eren}}, S_{\text{eeem}}), \tag{15.30}$$

15.3.2. Solution of Petri Net Models

In the case of discrete event process system models in the form of a Petri net *the ingredients needed for the solution* are as follows:

1. *a discrete event dynamic model*—in the form of a Petri net structure. Here the current state of the system is described by a marking with the places corresponding to the entries in the state vector of the discrete event system.
2. *a time interval* $[t_0, t_M]$—with a given initial time t_0 and a final time t_M over which the solution will be computed. In case of untimed models only the required number of steps, M, is specified.
3. *initial conditions for the discrete valued state variables*—which is now a concrete initial marking.
4. *a record for the input variables*—the input variables correspond to places in the Petri net model with no inward directed edges. In order to imitate the occurrence of the change in input signal sequence we associate a new place acting as an external source. These source places put an appropriate token in case of an input event. The external sources of tokens may be synchronized by a special clock or by a timer, therefore we need a timed Petri net model in case of multiple inputs for this purpose.

Thereafter the solution is generated by executing the Petri net model in the way it was described in Section 15.2.2. The solution is obtained as set of finite or infinite sequences of operations in the form of Eq. (15.20).

The solution of extended Petri net models such as timed, hierarchical and coloured ones, proceeds much the same as that of the low level one, taking into account the specialities of the models. The following changes should be made when computing the solution of an extended Petri net model:

1. *Timed Petri net models*
   Intermediate states (markings) should be introduced to describe the state of the net during the firing times of a transient t_j. Such an intermediate state can be described by a marking $\underline{\mu}^{(t_j)}$ where the places of the pre-condition of the transition loose their tokens, but places in the post-condition do not receive any token.
2. *Coloured Petri net models*
   The firing pre-condition of any of the transitions t_j should be computed from the prescriptions of the incoming edges (they are all Boolean-valued expressions). Similarly, the post-condition, that is the colour of the tokens put onto the places adjacent to the outgoing edges, is computed from the prescriptions of the outgoing edges.

The solution of a low level Petri net model is illustrated on the example of the reactor–filter system introduced in Example 15.2.1 with its Petri net model of Example 15.2.6.

EXAMPLE 15.3.2 (Solution of the Petri net model of the Reactor-Filter System developed in Example 15.2.6). *Compute the solution of the model when all equipment is ready for processing and two batches are present to be processed.*

The digraph of the Petri net model is shown in Figure 15.5. Recall, that the marking vector has the following entries:

$$\underline{\mu}^T = [\mu_{\text{raw}}, \mu_{\text{solv}}, \mu_{\text{Rempt}}, \mu_{\text{semi}}, \mu_{\text{f1empt}}, \mu_{\text{f2empt}}, \mu_{\text{product}}].$$

With this notation the initial marking is as follows:

$$\underline{\mu}^{(0)} = [2, 1, 1, 0, 1, 1, 0]^{\mathrm{T}}. \tag{15.31}$$

If one performs the execution of the Petri net model two different solution sequences will be obtained which are as follows:

$$S^{(1)}(\underline{\mu}) = ([2, 1, 1, 0, 1, 1, 0]^{\mathrm{T}},\ [1, 1, 1, 1, 1, 1, 0]^{\mathrm{T}},\ [1, 2, 1, 0, 1, 1, 1]^{\mathrm{T}},\ [0, 1, 1, 1, 1, 1, 1]^{\mathrm{T}},\ [0, 2, 1, 0, 1, 1, 2]^{\mathrm{T}}) \tag{15.32}$$

s and

$$S^{(2)}(\underline{\mu}) = ([2, 1, 1, 0, 1, 1, 0]^{\mathrm{T}}, [1, 1, 1, 1, 1, 1, 0]^{\mathrm{T}},\ [0, 0, 1, 2, 1, 1, 0]^{\mathrm{T}},\ [0, 1, 1, 1, 1, 1, 1]^{\mathrm{T}},\ [0, 2, 1, 0, 1, 1, 2]^{\mathrm{T}}). \tag{15.33}$$

■ ■ ■

15.4. ANALYSIS OF DISCRETE EVENT SYSTEMS

The analysis methods for discrete event dynamic systems are significantly different from that of continuous systems even for similar properties. The properties of a discrete event system fall into two categories:

- *behavioural properties*—such as reachability, deadlocks or conflict situations, which are initial state dependent,
- *invariance properties*—characterizing cyclic behaviour, which are properties of the underlying bipartite directed structure graph of the discrete event model.

As behavioural properties are much more important than structural ones from the viewpoint of applications we shall concentrate on behavioural properties in the following.

In the case of behavioural properties, the analysis requires the solution of the model equations, in other words it requires dynamic simulation. Therefore the analysis of discrete event systems is usually computationally expensive or even computationally hard.

The analysis of discrete event systems deals with the following most important dynamic properties.

1. *Reachability or controllability*
   The notion of state controllability described in Section 13.1 for continuous time systems can be extended to discrete event systems as well, where it is termed reachability.
2. *Deadlocks*
   A deadlock is a non-final state, from which we cannot move the system. In other words, this is an unwanted stoppage of the discrete event system.

3. *Finiteness or cyclic behaviour*
   In some discrete event systems there can be cyclic processes without any natural stop. This makes the system non-finite.
4. *Conflict situations*
   Conflict situations are usually discrete states for which the next state is non-unique. The potential future states are in conflict and the choice has to be resolved. The presence of conflict situations indicates ambiguity in the discrete event model.

The first step in the analysis is the generation of the state transition diagram introduced in Section 15.2.1 for the discrete event system modelled by finite automata. Equivalently, the reachability graph of the discrete event system modelled by Petri nets is needed for the analysis.

The state transition diagram of the discrete event system contains the states, events and the state transition function. This explains the need for discrete event simulation when constructing the state transition diagram of discrete event systems.

15.4.1. Controllability or Reachability

A discrete event system is called *(state) controllable or reachable* if we can always find an appropriate input function to move the system from its given initial state to a specified final state in finite time. This applies to every given initial state final state pair.

Given the state transition diagram of the discrete event system one can easily check if the system is reachable. If the state transition diagram seen as a directed graph is in itself a strongly connected component then the underlying discrete event system is reachable. Note that a directed graph is a strongly connected component in itself if there is at least one directed path from any vertex to any other vertex.

This means that every vertex pair in a strongly connected component is connected by at least one directed path. The labels on the path form the input sequence moving the system from the initial state vertex to the final one.

15.4.2. Deadlocks and Finiteness

Deadlock is a state of the discrete event system from which any event, either external or internal, cannot move the system out of it.

Cyclic behaviour is a set of states which occurs spontaneously and from which any external event cannot move the system out of it. The presence of cyclic behaviour causes the system to be non-finite in the sense that it "never stops".

Both deadlock and cyclic behaviour present in the discrete event system can be analyzed based on its state transition diagram. If we view the state transition diagram as a weighted or labelled directed graph, then

- deadlock states can be detected as vertices with no outward directed edge,
- cyclic behaviour can be detected by finding a set of strongly connected state vertices with no outward directed edge labelled by an external input event.

15.4.3. Conflict Situations

If the next discrete state of a given state with a given event is not unique then a conflict situation arises: the next possible states are in conflict with each other. In Example 15.2.6 a simple conflict situation is shown. The ambiguity of the next state is caused by the improper definition of the state transition function when it has multiple values for a given (*state*, *input_symbol*) pair.

Conflict situations are easily detected on the state transition diagram seen as a labelled directed graph. If a state vertex is found which has two or more outward directed edges with the same label leading to different vertices then these state vertices are in conflict with each other. This is a conflict situation in the discrete event system model.

The analysis methods are illustrated on the example of the reactor–filter system.

EXAMPLE 15.4.1 (The analysis of the reactor–filter system of Example 15.2.1). The state transition diagram of the discrete event model of the discrete reactor–filter system is shown in Fig. 15.3.

The reactor–filter system is *reachable* because the state transition graph in Fig. 15.3 is a strongly connected component in itself. The input sequence "(**r**, **nil**)" moves the system from state S_{eeem} back to itself.

There is no deadlock or cyclic behaviour present in the reactor–filter system because there is at least one edge labelled by an external input event starting from every state vertex.

The state transition diagram in Fig. 15.3 depicts the situation when the natural conflict situation present in the reactor–filter system with two filters working in parallel has been resolved using priorities: if the next batch of the raw material is present then choose *filter1* else *filter2*. This is seen in the labels of the outward directed edges of state S_{reen}. With these labels the discrete event system does not contain any conflicts. ■ ■ ■

15.5. SUMMARY

This chapter deals with discrete event systems from the viewpoint of process modelling and model analysis. This type of model is used when external or internal events have a major impact on the behaviour of the process system and modelling focuses on the sequential and/or parallel events taking place.

Based on the characteristic issues and main application areas of discrete event process models the most important model representation methods, namely finite automata and Petri net models are discussed. Following this, the solution of discrete event process models via simulation and the methods for analysis of their behavioural properties, such as reachability, finiteness, deadlocks and conflict situation have been described.

15.6. FURTHER READING

The *sources* of the literature on the field of discrete event systems and their application in process systems engineering are to be found in books and papers and also on the

Web. The following list contains the most important information sources relevant to the field:

- *Journals*
  - Computers and Chemical Engineering
  - AIChE Journal
- *Conferences*
  - ESCAPE Conferences
  - IFAC Symposia "On-line fault detection and supervision in the chemical process industries"
  - International Conferences "Application and Theory of Petri Nets", where the Proceedings are available as volumes in *Lecture Notes in Computer Science* by Springer Verlag.
- *Web*
  - Linkøping University Division of Automatic Control homepage: `http://www.control.isy.liu.se`

Because of the novelty of the field, the results on modelling discrete event process systems can mainly be found in journal and conference papers, as well as research reports. Some review papers can also be found from the field of systems and control theory, such as [109,110].

The relevant papers can be most easily recognized based on the characteristic application areas: batch process scheduling and process fault detection and isolation. Here we mention only some of them classified according to the technique they use.

- *Finite automata techniques*
  A fault detection and isolation method based on the observation of events applying finite automata for describing discrete event systems is presented in [111].

  A framework is presented for synthesizing logic feedback controllers for event-driven operations, such as startup and shutdown operations, emergency procedures and alarm handling in [112].
- *Discrete event simulation techniques*
  An autonomous decentralized scheduling system based on discrete-event simulation is found in [113].

  Short-term batch plant scheduling has been performed using a combination of discrete-event simulation and genetic algorithms in [114].
- *Grafcet description*
  A planning framework has been developed and applied for automating the synthesis of operating procedures for batch process plants using *Grafcet*, a discrete event modelling concept in [115] and [116].
- *Petri net approaches*
  A flexible, efficient and user-friendly system for HAZOP analysis of batch processing plant has been reported in [117] and [118] where the process model is constructed as a combination of high-level Petri nets and cause and effect digraphs.

  An intelligent diagnostic system based on Petri nets is reported in [119].

  Petri nets have been used for modelling a supervisory control system of a multipurpose batch plant [120].

15.7. REVIEW QUESTIONS

Q15.1. What are the key components of a process system which contribute to the discrete event behaviour? (Section 15.1)

Q15.2. What are the key application areas in discrete event system modelling and simulation? (Section 15.1)

Q15.3. What approaches are available in the literature to model a discrete event dynamic system? (Section 15.2)

Q15.4. What are the elements of the formal description of a finite automata? How do these elements appear in the state transition diagram of the automata? (Section 15.2.1)

Q15.5. What are the elements of the formal description of a low level discrete Petri net model? How are these elements visualized in the graphical description of the Petri net? How do we describe the operation or dynamics of a Petri net in a formal and in a graphical way? (Section 15.2.2)

Q15.6. Compare the finite automata and the Petri net description of a discrete event dynamic system. What is the relationship between the state transition diagram of the automata and the reachability graph of the Petri net? (Sections 15.2.1 and 15.2.2)

Q15.7. What kind of extended Petri nets are available for modelling? How can we use extended Petri nets for modelling discrete event systems? (Section 15.2.2)

Q15.8. Which dynamic properties of discrete event systems are the subject of dynamic analysis? What is the definition and practical significance of these properties? (Section 15.4)

Q15.9. What are the ingredients needed to perform the solution of a finite automata model? What do you obtain as a result? How do we describe the solution method of an automata model? (Section 15.3)

Q15.10. What are the ingredients needed to perform the solution of a Petri net model? What do you obtain as a result? Describe the algorithm of solving a Petri net model. (Section 15.3)

Q15.11. Which is the discrete event system model used for the analysis of dynamic properties? What is the first step of the analysis? Is the analysis of dynamic properties of a discrete event system model computationally hard? Justify your answer. (Section 15.4)

15.8. APPLICATION EXERCISES

A15.1. Draw the state transition diagram of the reactor–filter system defined in Example 15.2.1. Compare it to the state transition diagram shown in Fig. 15.3.

A15.2. Develop a Petri net model for the problem of the hot/cold drink automata described in Example 15.2.2. Extend it with processing times to get a timed Petri net model. Show that the reachability graph of the model is the same as the finite automata model shown in Fig. 15.2.

A15.3. Develop an extended version of the Petri net model of the emergency shutdown procedure of example 15.2.7 to handle the following exceptions:

1. there is no inert gas available,

2. there is no cooling available,
3. the switch for turning off electricity from the equipment is out of service.

Draw the extended Petri net model starting from Fig. 15.6. How would you proceed if you had to check the completeness of your model?

A15.4. Perform the simulation of the timed Petri net model of the emergency shutdown procedure of Example 15.2.7 shown in Fig. 15.6 starting with the initial conditions (initial marking) shown on the figure up until a very large final time. Draw the reachability graph of the model.

A15.5. Perform the simulation of the timed Petri net model of the emergency shutdown procedure extended with exception handling in Application Exercise A15.3 starting with initial conditions (initial marking) which enable the firing of both $t_{cooling}$ and $t_{fill\text{-}neutr}$. Assume that final time for the simulation is very large. Draw the reachability graph of the model.

A15.6. Repeat the simulation of the Petri net model of the reactor–filter system described in example 15.3.2 by adding finite firing (processing) times to the transitions t_{GR}, t_{Gf1} and t_{Gf2} such that

$$t_{GR}^{F} < t_{Gf1}^{F} < t_{Gf2}^{F}. \qquad (15.34)$$

Is the solution unique this time? Compare the obtained solution(s) with the ones given in Example 15.3.2.

A15.7. Unite the super-net and sub-net of the hierarchical Petri net of the reactor–filter system described by Example 15.2.8, that is insert the sub-net shown in Fig. 15.7 into the super-net shown in Fig. 15.5. Give a suitable initial marking of the united Petri net. Perform several steps in executing the united net and observe the operation of the sub-net part.

A15.8. Perform the simulation of the finite automata model of the hot/cold drink automata extended with exception handling in Example 15.2.10 shown in Fig. 15.9. Let the initial state be Q_0 = "ready" and the input tape contents be:

(a) : ("$1", "$1", "2", "#")

(b) : ("2", "3", "#")

(c) : ("$1", "#", "2", "#")

A15.9. Let us imagine a batch processing plant composed of two reactors of equal size, a large tank for cleaning solvent, another large tank for paint solvent, three small vessels for pigments and three storage tanks for the prepared paints. The paint of different colours is made in a single processing step in one of the reactors and then fed to the appropriate paint storage tank. The reactors need to be cleaned after processing if the next batch is a different colour. The processing and cleaning times are given such that the cleaning takes much shorter time. The time needed for the filling operation is negligible.

(a) Construct a discrete event model of the plant.
(b) Perform simulations for different paint orders and under different conditions.
(c) Extend your model for the most common exceptions you may imagine.

A15.10. Analyse the behavioural properties of the timed Petri net model of a maintenance shutdown procedure of Example 15.2.7 shown in Fig. 15.6. Try different initial conditions (initial marking) including the one shown on the figure.

Can there be any conflict situations in the model? When?

How would you modify the model to avoid conflict situations? What would this change mean in practice?

16 MODELLING HYBRID SYSTEMS

This chapter will discuss the modelling issues surrounding process systems which are classified as either "discrete–continuous" or "hybrid" in nature. Some in the literature also refers to these systems as "event driven". They are composed of elements which contain discrete events at particular times or conditions combined with elements which are basically continuous in behaviour. These represent an important class of process problems.

The tools and techniques available for modelling and analysing discrete event dynamic process systems were treated in Chapter 15. This chapter will cover the basic ideas behind the modelling methodologies employed for hybrid systems with both discrete and continuous parts, their analysis as well as the particular challenges for their numerical solution, whether small or large-scale problems.

16.1. HYBRID SYSTEMS BASICS

16.1.1. Characteristics and Issues

It can be convincingly argued that all process engineering systems are hybrid in nature. We need to investigate what is meant by this statement. There are several reasons why systems consist of discrete behaviour either driven by events occurring in the system or by external influences, whether planned or uncontrolled disturbances.

Figure 16.1 shows a typical process system emphasizing some of the key components which contribute towards hybrid system behaviour.

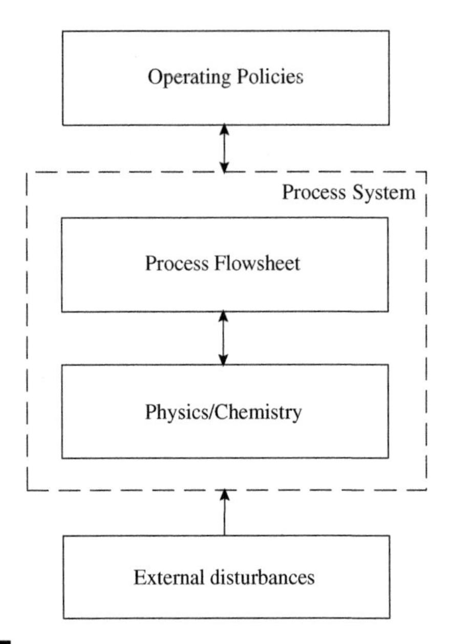

FIGURE 16.1 Hybrid system characteristics.

The various components can be considered to see where discrete event behaviour might occur.

Process Flowsheet

The process flowsheet defines the processing units of the system and how they are interconnected. Here we include the physical equipment and control system. It is obvious that the equipment has various geometric characteristics which lead to discrete behaviour.

These include:

- fixed capacity of tanks and vessels which can be filled, emptied or overflowed,
- decanter systems which have specified baffle levels for overflow,
- compressor surge behaviour,
- changed use of flow paths depending on operating conditions, which may switch feeds or energy streams,
- limiting conditions on operating or equipment parameters such as maximum pressure before a safety valve lifts or bursting disc operates,
- operating limits in the control elements such as control valves fully shut or fully open or saturation of transmitter and controller outputs,
- "soft" constraints imposed by safety considerations whereby process conditions are limited,
- operation of safety systems which limit process behaviour, switch protective systems on or off and activate controlled or emergency shutdown systems.

All these, and more, impose discrete events on the underlying continuous system by virtue of the flowsheet design.

Process Chemistry and Physics

This aspect, shown in Fig. 16.1 represents the underlying chemistry and physics of the process. Here, there are naturally occurring phenomena which contribute to discrete behaviour. These characteristics can include:

- flow transitions between laminar, transitional and turbulent flow as represented by flow models,
- chemical reactions which accelerate rapidly or are "switched on" by operating conditions of temperature, pressure or composition,
- operational regimes such as boiling or non-boiling in steam–water or other vapour–liquid systems,
- particle systems which are characterized by discrete items,
- appearance and disappearance of vapour, liquid or solid phases due to chemical activity and phase stability.

In many cases, such as metallurgical systems the phase behaviour is extremely complex and leads to many discrete events which have to be handled within the modelling of the system. Other processing operations such as those dealing with particle enlargement or comminution reflect the basic discrete physical behaviour of the physics.

Operational Policies

The operational policies imposed on the process can contribute to specific discrete event behaviour of a system. In many cases, processes are run continuously or in batch mode due in part to the underlying chemistry and the required production rate. Some processes include continuous elements combined with batch stages. The principal operational issues contributing to hybrid system behaviour include the following:

- Production campaigns for multiple chemical products from a single continuous process, where product switching is done. Acetate ester production can be an example of this operational policy.
- Regular switching of raw material feeds to the process, an example being a variety of crude oil feeds in a refinery.
- Regular product slate switches such as glass or fibreglass insulation thickness changes to meet market demands.
- Changes in energy supply streams to processes during start-up and energy utilization on an integrated processing site.

These externally imposed changes or operating policies drive the underlying continuous system through a series of discrete events which clearly reflect the imposed operating policy.

External Disturbances

One major area giving rise to discrete events is the effect of uncontrolled disturbances on the process. The following illustrate typical examples of such events:

- thunder storms, rain, snow or atmospheric events,
- discrete changes in steam supply pressures from boiler systems,

- power spikes in electrical supply systems,
- contamination of feeds to the process or specific feed characteristics such as quality,
- market or seasonal demands for products.

The four fundamental sources of discrete event behaviour are usually represented in some form in every process. Whether these characteristics are captured within the modelling will depend on the modelling goal, the time frame considered and the degree of fidelity required.

In this next section, we consider some typical application areas, where hybrid models find use in describing the system.

16.1.2. Applications

Key application areas in hybrid system modelling and simulation include the following:

- batch processes for speciality chemicals, pharmaceuticals and paints,
- continuous processes during start-up, shut-down and product changeover,
- manufacturing systems where discrete items are processed on discrete machines,
- design of systems with units described by discrete variables such as staged operations and binary variables (0,1) where they are present (1) or not present (0).

A typical example of modelling for hybrid systems is given by the following process which represents a forced circulation evaporator system which could be used for concentrating sugar liquor or a fruit juice. Here, we are interested in the start-up of the system. Figure 16.2 shows the flowsheet of the plant with a number of controls for pressure, product concentration and level.

In considering the start-up of such a unit the steps of the procedure are shown in Table 16.1.

This start-up sequence involves a number of discrete changes to flows, controller settings and the start of pumps and other plant devices. The underlying process is continuous but in this case the operational procedure imposes a series of discrete events on the system.

Figure 16.3 shows the evaporator mass holdup during start-up, which clearly shows the transient nature of the process as external changes take place.

The following section discusses the options available for representing hybrid system behaviour. The approaches can be used for process systems and manufacturing applications, for start-up, design and scheduling. The best representation depends on the final application of the modelling and the basic characteristics of the system being modelled.

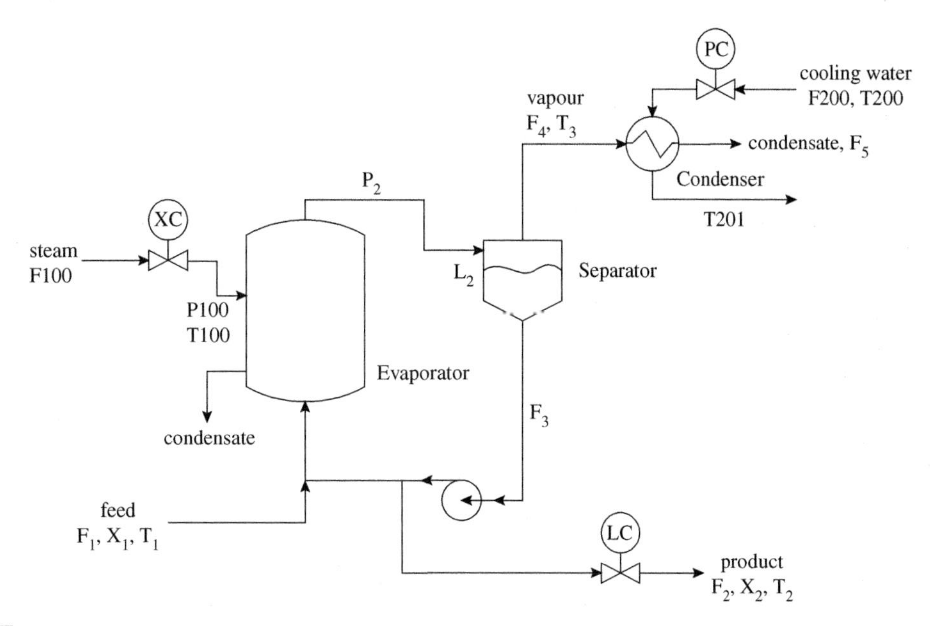

FIGURE 16.2 Forced circulation evaporator system.

TABLE 16.1 Start-up procedure for forced circulation evaporator

Time (min)	Status or operation to be done
Time < 5	Feed flow = 0 Recirculation flow = 0 All controllers in manual with 0 output
5 ≤ Time ≤ 7	Set feed flow to fill shell in 2 min All controllers in manual with 0 output Recirculation flow = 0
7 ≤ Time ≤ 10	Feed flow = 0 Recirculation flow = 0 All controllers in manual with 0 output
10 ≤ Time ≤ 20	Feed flow ramped to 5 kg/min Recirculation flow ramped to 50 kg/min Level controller set to automatic, setpoint = 1 m Pressure controller set to automatic, setpoint = 50.5 kPa Composition controller set to automatic, setpoint = 25%
20 ≤ Time ≤ 80	Feed flow constant at 5 kg/min Recirculation flow constant at 50 kg/min
Time > 80	Feed flow increased to 10 kg/min
Time > 90	Feed flow increased to 15 kg/min
Time > 100	Feed flow increased to 20 kg/min
Time > 110	Feed flow increased to 25 kg/min
Time > 120	Feed flow increased to 30 kg/min

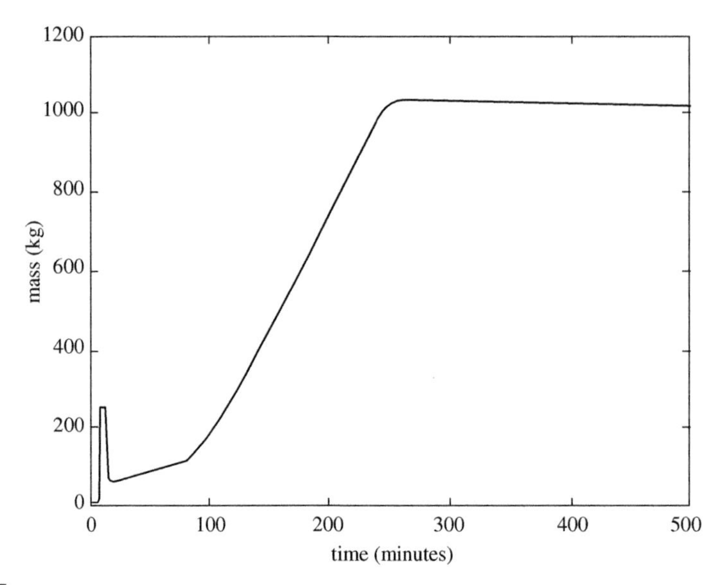

FIGURE 16.3 Evaporator mass changes during start-up.

16.2. APPROACHES TO MODEL REPRESENTATION

There are various but related approaches to describe a hybrid dynamic system with both continuous and discrete elements. From a system theoretic point of view a hybrid system consists of three components:

- a discrete event dynamic system with discrete time and discrete valued variables,
- a continuous system with continuous variables,
- an interface between them.

Most approaches use combinatorial or finite techniques such as graphs or automata extended by the capability to handle continuous elements. In this way, the continuous elements are *imbedded* into a discrete dynamic structure.

Examples of these approaches outlined in this section are

1. extended finite automata models (*hybrid automata*) in Section 16.2.1,
2. *hybrid Petri nets* in Section 16.2.2,
3. differential-algebraic equations in Section 16.2.3.

The approaches are compared on the simple example of a reactor–filter system described in Chapter 15 in Example 15.2.1. This process system is regarded as a hybrid system with all the equipment assumed to operate continuously forming the continuous part of the hybrid system. The steps of the operating procedures, such as feeding, switching on/off equipments, etc. are discrete events acting on the system.

16.2.1. Hybrid Automata

In the hybrid automata approach

1. the discrete event dynamic system is described by a finite automata according to Section 15.2.1;
2. the continuous system is represented by its lumped nonlinear state space model; and
3. the interface describes how the events generated by the finite automata part are transformed to act on the inputs and disturbances of the continuous part and how the continuous part generates events to be fed to the finite automata.

Motivated by the above general parts and their functionality the general hybrid automata model $\mathcal{HA}$ of hybrid systems consists of three components:

- a discrete event dynamic system (DEDS) described by finite automata D_{HA},
- a continuous nonlinear system model described by a nonlinear state space model C_{HA},
- an interface between them described by special functions I_{HA}.

Thus, the formal description of a hybrid automata model is a triple [121]:

$$\mathcal{HA} = (D_{\text{HA}}, C_{\text{HA}}, I_{\text{HA}}). \tag{16.1}$$

The components of the above general hybrid automata model contain special elements and properties to ensure their consistency and cooperation.

1. *Discrete event dynamic system*
The DEDS component of a hybrid automata model is the special case of the finite automata model (15.2) in the form of

$$D_{\text{HA}} = (Q_{\text{HA}}, \Sigma_{\text{HA}}, \delta_{\text{HA}}).$$

Q_{HA} : The elements or coordinates in any state vector $q_{\text{HA}} \in Q_{\text{HA}}$ can be divided into three different parts. First, there are elements that correspond to actuator or input variables. Second, there are elements related to different operating regions due to geometric of physical structure of the continuous system. Finally, there are elements corresponding to ordinary discrete states which are not related to the continuous part.

Σ_{HA} : The elements of the input alphabet describe *events* in the DEDS model. The set Σ_{HA} is partitioned into three parts:

$$\Sigma_{\text{HA}} = \Sigma_{\text{HA}}^{(\text{p})} \cup \Sigma_{\text{HA}}^{(\text{v})} \cup \Sigma_{\text{HA}}^{(\text{cr})}. \tag{16.2}$$

It consists of autonomous plant events $\sigma_{\text{p}} \in \Sigma_{\text{HA}}^{(\text{p})}$ generated from the continuous process by the event generator $\gamma_{\text{HA}}^{(\text{p})}$ in the interface. It also contains the events $\Sigma_{\text{HA}}^{(\text{v})}$, which occur spontaneously without the influence of any modelled dynamics. Finally, there is the set of controllable input events $\Sigma_{\text{HA}}^{(\text{cr})}$.

δ_{HA} : is the usual state transition function.

2. *Continuous system*
This part describes the continuous subsystem of the hybrid system using a special version of the general nonlinear state space model equations (13.2):

$$\dot{x}_p(t) = f_p(x_p(t), q_p(t), u(t), v(t)), \tag{16.3}$$

$$y(t) = g_p(x_p(t), q_p(t), u(t), v(t)), \tag{16.4}$$

$$z_p(t) = h_p(x_p(t), u(t), v(t)), \tag{16.5}$$

where $x_p(t) \in \mathcal{R}^n$ is the state vector of the continuous system, $q_p(t) \in Q_{\mathrm{HA}}$ is the piecewise constant function from the DEDS, $u(t) \in \mathcal{R}^m$ is an external constant control vector, $v(t) \in \mathcal{R}^k$ is a continuous disturbance vector affecting the continuous states and $y(t) \in \mathcal{R}^s$ are the observable outputs.

The function h_p gives the signal $z_p(t) \in Z_p \subseteq \mathcal{R}^d$ which is the input to γ_p the event generator in the interface.

Finally, to express jumps in the continuous trajectory, a transition function ϕ_p is invoked

$$x_p(t^+) = \phi_p(x_p(t), q_p(t), \sigma_p), \tag{16.6}$$

which determines the starting values of the continuous state after an event $\sigma_p \in \Sigma_p$.

3. *Interface*
This part determines the generation of a plant event $\sigma_p \in \Sigma_p$ from the continuous system to the DEDS by the event generator

$$\gamma_p : Z_p \to \Sigma_p. \tag{16.7}$$

Events may occur for two reasons:

- if the system enters a region with a different behaviour and therefore is described by a different model equation set from that previously used, or
- if there is a continuous state jump modelled by an internal event.

The above hybrid automata approach is illustrated on the example of the reactor–filter system introduced already in Example 15.2.1.

EXAMPLE 16.2.1 (A hybrid automata model of the Hybrid Reactor–Filter System introduced in Example 15.2.1)

The hybrid automata model of the reactor–filter system is constructed from the finite automata model described in Example 15.2.3.

Discrete event dynamic system
First, observe that a discrete state vector $q \in Q_{\mathrm{RF}}$ consists of four elements:

$$q = [s_{\mathrm{r}}\ s_{1\mathrm{f}}\ s_{2\mathrm{f}}\ s_{\mathrm{inp}}]^{\mathrm{T}}, \tag{16.8}$$

$$s_{\mathrm{r}},\ s_{1\mathrm{f}},\ s_{2\mathrm{f}} \in \{r, e\}, \quad s_{\mathrm{inp}} \in \{n, m\}, \tag{16.9}$$

where s_{r} is the state of the reactor, $s_{1\mathrm{f}}$ and $s_{2\mathrm{f}}$ are that of *filter1* and *filter2*, and s_{inp} is that of the reading head of the automata.

The input alphabet or the events to the system are extended compared to the set (15.5) to follow the decomposition in (16.2) in the following way:

$$\Sigma_p = \{r_{\rm r}, r_{\rm 1f}, r_{\rm 2f}\}, \tag{16.10}$$

$$\Sigma_v = \{e_{\rm r}, e_{\rm 1f}, e_{\rm 2f}, r_{\rm solv}\}, \tag{16.11}$$

$$\Sigma_i = \{\mathbf{r}, \mathbf{nil}\}, \tag{16.12}$$

where $r_{\rm w}$, $w = $ r, 1f, 2f and $e_{\rm w}$, $w = $ r, 1f, 2f represent the occurrence of the ready and empty status of the equipment, $r_{\rm solv}$ is the availability of the solvent and $\mathbf{r}$ is that of the raw material.

The state transition function is again an extension of Eqs. (15.6)–(15.10):

$$\delta_{\rm RF}(S_{\rm eeem}, (\mathbf{r}, r_{\rm solv})) = S_{\rm reen}, \tag{16.13}$$

$$\delta_{\rm RF}(S_{\rm reen}, r_{\rm r}) = S_{\rm eern}, \tag{16.14}$$

$$\delta_{\rm RF}(S_{\rm reen}, r_{\rm r}) = S_{\rm eren}, \tag{16.15}$$

$$\delta_{\rm RF}(S_{\rm eern}, r_{\rm 1f}) = S_{\rm eeem}, \tag{16.16}$$

$$\delta_{\rm RF}(S_{\rm eren}, r_{\rm 2f}) = S_{\rm eeem}, \tag{16.17}$$

Continuous system

The continuous system contains all three units: the reactor and the two filters. Therefore, the continuous state vector x_p contains three parts:

$$x_p = [x_{\rm r}\ x_{\rm 1f}\ x_{\rm 2f}]^{\rm T} \tag{16.18}$$

where $x_{\rm r}$, $x_{\rm 1f}$ and $x_{\rm 2f}$ are the state vectors of the reactor and the two filters respectively. The function f_p in Eq. (16.3) is composed of the corresponding state functions of the reactor and the two filters acting on the corresponding state vector elements in (16.18). These equations are active only if the equipment is working.

The working state of the equipment is represented by the following discrete state vectors q_p present in the argument list of ϕ_p and f_p in Eqs. (16.6) and (16.3):

$$\text{reactor: } q_p^{(\rm r)} = [\rm e\, e\, e\, m]^{\rm T}, \tag{16.19}$$

$$\text{filter1: } q_p^{(\rm 1f)} = [\rm r\, e\, e\, n]^{\rm T}, \tag{16.20}$$

$$\text{filter2: } q_p^{(\rm 1f)} = [\rm r\, e\, e\, n]^{\rm T}. \tag{16.21}$$

The signal function h_p in Eq. (16.5) can be defined on the continuous state vector in such a way that the event signal $z_p(t)$ takes four different values $z_p^{(\rm r)}$, $z_p^{(\rm 1f)}$ and $z_p^{(\rm 2f)}$ if the *reactor*, *filter1* and *filter2* is ready, and $z_p^{(0)}$ if any of the processing steps are taking place. For example,

$$z_p(t) = z_p^{(\rm r)} \quad \mathbf{if}\ x_{\rm r} \geq x_{\rm r}^{\rm ready},\ x_{\rm 1f} = x_{\rm 1f}^{\rm empty},\ x_{\rm 2f} = x_{\rm 2f}^{\rm empty}. \tag{16.22}$$

Finally, the jump function ϕ_p in Eq. (16.6) adjusts the continuous state vector to its new initial state when an event in the discrete event dynamic system has taken place.

For example the following jump is needed when the operation of the reactor starts:

$$x_p(t^+) = \left[x_{\mathrm{r}}^{\mathrm{empty}}\ x_{\mathrm{1f}}^{\mathrm{empty}}\ x_{\mathrm{2f}}^{\mathrm{empty}}\right]^T$$
$$= \phi_p(x_p(t), q_p^{(\mathrm{r})}, (\mathbf{r}, r_{\mathrm{solv}})). \tag{16.23}$$

Interface

Events in the continuous reactor–filter system may only occur if the system enters a region with a different behaviour. The event is detected by the event signal $z_p(t)$ in the continuous time system model. The event generator function γ_p maps the value of the event signal into a plant event. The event generator function for the hybrid reactor–filter system is as follows:

$$r_{\mathrm{r}} = \gamma\left(z_p^{(\mathrm{r})}\right), \tag{16.24}$$

$$r_{\mathrm{1f}} = \gamma\left(z_p^{(\mathrm{1f})}\right), \tag{16.25}$$

$$r_{\mathrm{2f}} = \gamma\left(z_p^{(\mathrm{2f})}\right). \tag{16.26}$$

■ ■ ■

16.2.2. Hybrid Systems Described by Petri Nets

Petri nets (see Section 15.2.2) are one of the most powerful and popular tools for modelling discrete event systems. In order to be able to handle the continuous part of a hybrid system the original concept of Petri nets should be largely extended. Hybrid systems can be described by Petri net in at least two different ways:

1. *Timed Coloured Petri nets* [105]

The main idea of the approach is to convert the continuous part of the hybrid system to a coloured Petri net by discretizing the time and the range space of the variables. In this way the original nonlinear state space model of a lumped continuous time system is converted to a set of constraint type qualitative differential equations having an equivalent timed coloured Petri net representation. The hybrid model is then converted to a fully discrete model of an approximate discrete event system to the original hybrid system.

2. *Hybrid Petri nets* [122]

Hybrid Petri nets use ordinary Petri nets for describing the discrete event dynamic system.

The continuous system is modelled as a continuous Petri net. In this model, the following extensions to the ordinary Petri net structure are defined:

- Infinitely many tokens can occur at all places. Real values of the Petri net state described by the marking vector elements are approximated in such a way with arbitrary precision.
- Instead of firing the transitions at certain time instants with zero duration there is a continuous firing with a flow that may be externally generated by an input signal and may also depend on the continuous marking vector.

The continuous places and transitions are denoted by double circles and double lines. In this way, a nonlinear state space model can be described using the continuous Petri net.

The interface is modelled by special interface structure, such as switches, state jumps and boundary generators connecting discrete and continuous Petri net elements.

16.2.3. Differential-Algebraic Systems

A natural representation of hybrid systems arises from the basic process engineering modelling approach used in this book, namely in terms of conservation equations and constitutive relations. However, in the case of hybrid systems there exists extra relations which help to fully describe the system behaviour. These relations are necessary to define the events which occur in the system and the conditions that apply after the switching. As such, the overall description in terms of equations has an equivalent representation in the form of a state transition graph. This concept was introduced in Section 15.2.1.

A description in mathematical terms for a hybrid system consists of:

- The state equations which can be in various forms including:
  - ordinary differential equations (ODE),
  - differential-algebraic equations (DAE) from lumped parameter models,
  - partial differential and ordinary differential-algebraic equations (PODAE) from distributed system modelling,
  - partial and integro differential-algebraic equations (PIODAE), arising from the use of population balances within the model, such as those for granulation dynamics,
- the conditions under which the state switching occurs. These are typically in the form of logical event statements containing boolean operators. They are often encoded as IF... THEN... ELSE... statements in the programming language being used.
- The state of the system after the transition or switch has been made. This allows the system solution to proceed.

The following sections set out in more precise terms these modelling elements for hybrid systems. Following this, an example is given which illustrates the use of these elements.

State Equations

The state equations are needed for each of the possible states in the state transition graph for the process. If we assume that there are n discrete states $\left(\sum_{[i]},\ i = 1, \ldots, n\right)$ in which the system exists, then we can write the dynamic state equations for each state i as

$$f_i(\dot{x}_i, x_i, u, d, t) = 0, \quad \forall\, i = 1, \ldots, n, \tag{16.27}$$

where x_i denotes the vector of states applicable to $\sum_{[i]}$ with u and d denoting the system inputs and disturbances. These equations can also include the nonlinear algebraic constraints associated with the DAE system.

If the system state equations are either in the form of PODAEs or POIDAEs then the equations become:

$$f_i(\dot{x}_i, x_i, x_{i_r}, x_{i_{rr}}, u, d, r, t) = 0, \quad \forall\, i = 1, \ldots, n, \tag{16.28}$$

where x_{i_r}, $x_{i_{rr}}$ represent the first and second derivatives of the states with respect to the spatial variable r.

Even though each state is represented by one of the n equations, it is clear that in most process systems only a small set of equations change across the transition. Hence, the actual total number of equations is often not greatly enlarged .

Event Equations

The event equations describe the conditions under which a transition is made between the discrete state $\sum_{[i]}$ and $\sum_{[j]}$. These conditions take the general form of:

$$E^{(i,j)} = E\left(x_i, x_{i_r}, x_{i_{rr}}, u, z, t\right), \tag{16.29}$$

and can be logical or functional statements which involve time, the state vectors and/or their derivatives at the discrete state $\sum_{[i]}$. Hence, the event equations can be

- simple time events, where a transition occurs at a specified time, are given by
  IF *time* $= t_1$
  THEN
  action needed
  ENDIF
- state events where time is not known in advance but the switching occurs as a function of the states, as given by
  IF *state condition function*
  THEN
  action needed
  ENDIF

The actions which occur in the system can be categorized into the following types:

- *Vector field switching* where the equation set changes in terms of equations and variables. Typical examples are in the area of phase transitions like non-boiling to boiling regimes.
- *State jumps* where the underlying model does not change but the values of some of the new states x_j in $\sum_{[j]}$ do. Situations such as an instantaneous or almost instantaneous addition of mass or energy to the system or implementation of gain scheduling control can produce this form of switching.
- *Task descriptions*, where processing can take various forms such as
  - serial tasks: *task*1 $\rightarrow$ *task*2 $\rightarrow \cdots \rightarrow$ *taskN* $\rightarrow$ *end tasks*
  - parallel tasks: $\rightarrow$ *task*1 $\rightarrow$ *task*2 $\rightarrow \cdots \rightarrow$ *taskN* $\rightarrow$ *end tasks*
    $\rightarrow$ *task*1*P* $\rightarrow$ *task*2*P* $\rightarrow \cdots \rightarrow$ *taskMP* $\rightarrow$ *end tasks*
  - iterative tasks:$\rightarrow$ *task*1 $\rightarrow$ *task*1 $\rightarrow \cdots \rightarrow$ *task*1 $\rightarrow$ *end tasks*

Transition Functions

These functions help relate the past states to the new states thus providing a unique starting point for the solution to continue. They are typically of the form:

$$T^{(i,j)}(x_i, x_j, r, u, t) = 0. \tag{16.30}$$

Hence, after a jump transition affecting a mass holdup $x_i^{[1]}$ in which an amount K of component [1] is added to the system, the transition function could be:

$$x_j^{[1]} = x_i^{[1]} + K. \tag{16.31}$$

The procedure of representing the system in these fundamental forms can also be applied to each of the subsystems within an overall system. Hence, large problems can be decomposed into coupled subsystems and similar descriptions applied at the lower levels. The overall behaviour is a summed total of all the individual contributions.

DAE Index Issues

Another important aspect of hybrid systems expressed as DAEs is the concept of changing index of the equation set. The index can change significantly because of the imposition of new constraints on the system differential (continuous) variables imposed by the event equations and the transition functions. DAE systems that were originally index-1 can take on higher index properties as switching occurs and the reverse can occur. The example problem has this behaviour.

EXAMPLE 16.2.2 (The water cistern system). Consider the following simple problem as an illustration of the key hybrid concepts.

The system in Fig. 16.4 represents a typical overflow design for a water cistern. Water enters the outer tank at a rate m_i and drains out at rate m_1. If the outlet flow from tank 1 is less than the inlet flow then the water will overflow into the second internal tank at rate m_t. At this point the outer tank has a fixed liquid height of h. If the inner tank level continues to climb, then it too will reach a liquid height determined by the inner tank wall. Liquid height continues to rise above height h, until the tank overflows at height H. Hence, it can be seen that this system has a number of states and transitions which determine the state transition graph. Other transitions occur. Figure 16.5 shows the four principal states of the system ($\sum_{[i]}, i = 1, \ldots, 4$).

The following mathematical description gives the state equations, the event equations and transitions when the event condition is violated.

$$A_1 \frac{dh_1}{dt} = m_i - m_1 - m_t, \tag{16.32}$$

$$A_2 \frac{dh_2}{dt} = m_t - m_2 \tag{16.33}$$

with the conditions that define the states being:

State 1:

$$E^{(1,2)}, \quad h_1 < h,$$

$$E^{(1,3)}, \quad h_2 < h,$$

$$m_t = 0.$$

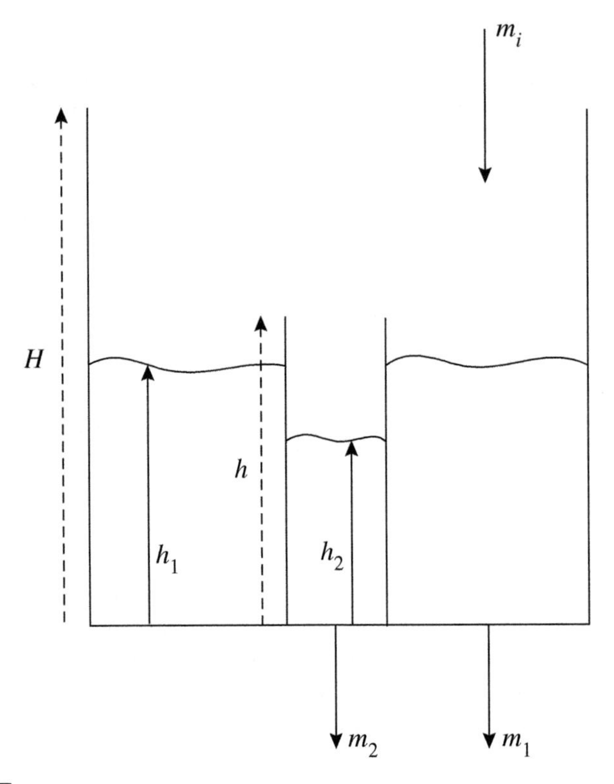

FIGURE 16.4 Water cistern schematic.

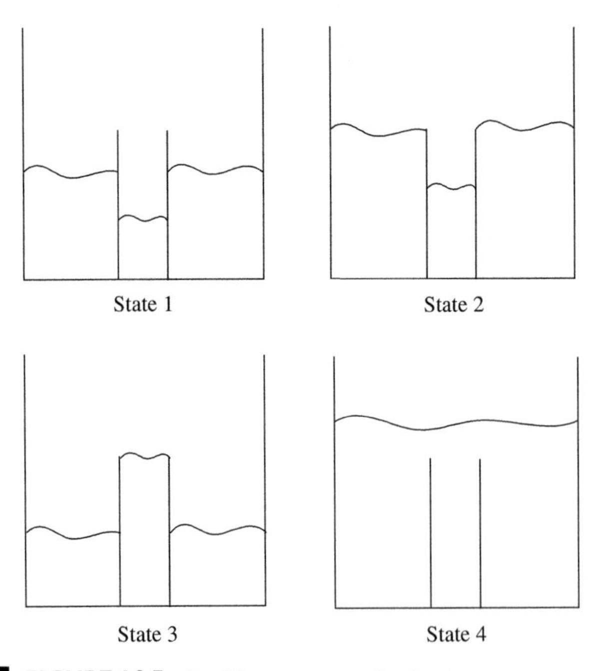

FIGURE 16.5 Possible system states for the cistern.

State 2:

$$E^{(2,1)}, \quad m_t > 0,$$
$$E^{(2,4)}, \quad h_2 < h,$$
$$h_1 = h.$$

State 3:

$$E^{(3,4)}, \quad h_1 < h,$$
$$E^{(3,1)}, \quad m_t < 0,$$
$$h_2 = h.$$

State 4:

$$E^{(4,3)}, \quad h_1 > h,$$
$$E^{(4,2)}, \quad h_2 > h,$$
$$h_1 = h_2.$$

The state transition graph is given in Fig. 16.6. If any of the conditions in a particular state are violated, then the particular event equation switches the system to a new state. For example, if in state 3 ($\sum_{[3]}$) the transfer flow m_t becomes zero, then the event equation $E^{(3,1)}$ activates and the system moves to state 1 ($\sum_{[1]}$). It is clear that apart from state 1, all other states lead to DAEs of index-2 because of the imposed algebraic conditions on the continuous (differential) variables.

The solution of the equation systems requires special solvers that can handle the index-2 problem or further transformation to reduce the system to index-1 via differentiation of the constraint equations due to the transitions. ■ ■ ■

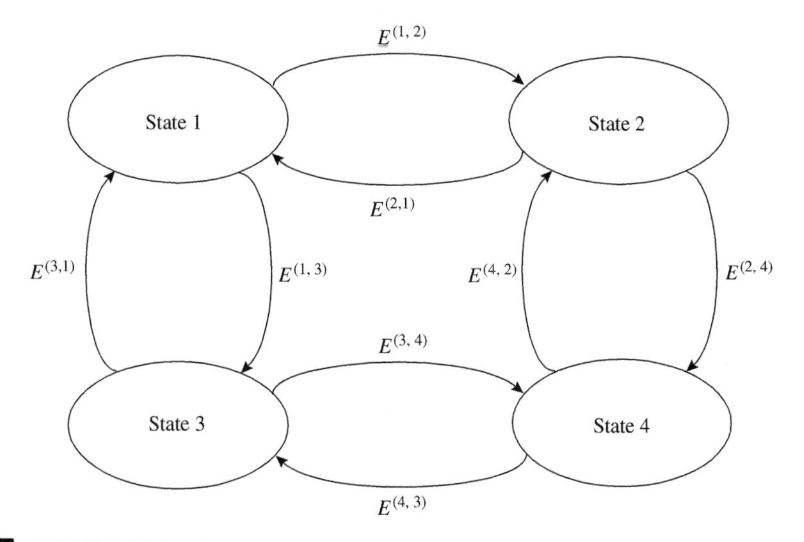

FIGURE 16.6 State transition graph for water cistern model.

16.3. ANALYSIS OF HYBRID SYSTEMS

The analysis techniques for hybrid systems use a combination of methods developed for the discrete event and continuous dynamic part. The properties which can be investigated are either defined similarly for the two parts or they only reflect the properties of the discrete event part.

Hybrid systems are highly nonlinear because of the presence of the discrete event part therefore *the dynamic properties of the hybrid system are strongly dependent on the initial state of both parts*. For the same reasons the analysis usually requires solution of the model equations, in other words it requires dynamic simulation. Therefore, the analysis of hybrid system is usually computationally expensive or even hard.

Similar to the analysis of discrete event systems described in Section 15.4 in the previous chapter the analysis of hybrid systems deals with the following most important dynamic properties.

1. *Reachability or controllability*
   As the reachability of the discrete event system is defined analogously to the usual reachability notion of continuous systems, the definition of reachability for hybrid systems is the natural combination of the two. Reachability is the joint property of the continuous and discrete event parts.

2. *Deadlocks*
   A deadlock is an unwanted stop of the hybrid system therefore it is an important but unwanted property. Deadlocks can be discovered by jointly analyzing the discrete event and continuous part. The notion of deadlocks for discrete event systems and the way to discover them has been described in Section 15.4.2 before.

3. *Finiteness or cyclic behaviour*
   In some of the hybrid systems there can be cyclic processes without any natural stop, making the system non-finite. The non-finiteness of the continuous part can also make the hybrid system non-finite.

4. *Conflict situations*
   Conflict situations are usually discrete states for which the next state is non-unique. The presence of conflict situations is a property of the discrete event part. The way to analyse conflict situations are therefore described in the previous chapter dealing with discrete event systems in Section 15.4.3.

Analysis of hybrid systems for properties such as reachability is difficult. The influence of a continuous state space on discrete behaviour implies that, even though a state may be reachable in a discrete sense, the continuous dynamics may be such that a discrete transition is never triggered.

Clearly, the only appropriate solution is to solve both discrete and continuous dynamics, but this is not feasible for most systems. A significant amount of work has been done on systems whose continuous behaviour is linear. Current methods of analysis therefore simplify the continuous system by partitioning its state space and then performing analysis on this simplified model (see [123]).

One significant exception has been work by Dimitriadis *et al.* [124] where verification of a safety system has been cast as an optimization problem. While computationally hard for all but the simplest of problems, this method fully accounts for the nonlinear dynamics.

16.4. SOLUTION OF HYBRID SYSTEM MODELS

16.4.1. Differential-Algebraic Equation Systems

These systems are the most typical of dynamic problems in hybrid modelling. As such the numerical solution methods need to specifically address the characteristics of hybrid problems. Several approaches are available to address the problems, however most numerical solution software for these problems remains tied to specific simulation packages with little in the public domain that addresses the challenges.

Key Challenges

The key challenges in solving hybrid DAE systems are as follows:

- Efficient solution of small and large problems, which might exploit structural aspects like parallelism and sparsity in the equation system.
- Robustness of solution such that a reliable solution is generated in reasonable time.
- The ability to handle time events, where switching times are known in advance.
- The ability to handle state variable events, where the time of the event is not known.
- The ability to detect state variable event occurrences.
- The ability to locate the time at which state variable events occur.
- The ability to restart efficiently at the new state.
- The ability to handle situations where the number of variables/equations changes for various states.
- The ability to handle higher index problems (typically index-2).

Differential Systems

The solution of pure differential systems has already been dealt with in Chapter 6. There are numerous codes available to solve these problems. However, there are few codes which routinely handle the ODE problem, where discontinuous behaviour is present. This is even less so with DAE systems. Specialized numerical codes are needed for such tasks and are normally part of a larger simulation environment such as *Speedup*, *gPROMS* [125] or *Daesim Studio* [126]. Most, if not all, only deal with index-1 DAE problems. There are several codes for higher index problems, which are based on Runge–Kutta formulae including those by Hairer and Wanner [127], Williams *et al.* [50]. Only the latter addresses large sparse hybrid DAE problems of index-1 or -2.

Little exists for solution of hybrid partial-ordinary differential and algebraic equations or partial integro-differential and algebraic equation systems, which require a choice of discretization method for the partial differential equations using such techniques as finite difference, orthogonal collocation or finite elements. Some of these techniques are given in Chapter 8. In the case of partial integro-differential equations, decisions have to be made about handling the integral terms. For particulate systems, Hounslow's technique [128] has found widespread application.

Event Detection

A key requirement for hybrid system solvers is the ability to detect when an event has occurred which moves the process to a new node on the state transition graph. The easiest event detection is the time event which is typically prespecified by the user in an event list. The numerical method can be made to step directly to the specified event time within the machine accuracy, activate the state transition, then resume solution.

The much greater challenge is efficient state variable event detection where the transition is determined by the event equations $E^{(i,j)} = E^{ij}\left(x_i, x_{i_r}, x_{i_{rr}}, u, z, t\right)$. Here the time of event is not known but is implicit in the event equations. These equations can be checked at each step of the integration routine to see whether a change of sign has occurred. If so, then a switching point has been passed in the current step of the solver. This is seen in Fig. 16.7 which shows the behaviour of a number of event equations over a solution step (t_n, t_{n+1}).

In this case, there are two event functions which change sign. $E^{(j,k)}$ changes $(+) \rightarrow (-)$ and $E^{(k,l)}$ changes $(-) \rightarrow (+)$, with $E^{(i,j)}$ remaining $(+)$. However the time t_{s1} at which $E^{(j,k)}$ changes sign occurs before event $E^{(k,l)}$ at time t_{s2}. Hence, this change must be handled first. In all cases of event detection over a solution step, there must be a precedence ordering of all the event function sign changes with the first located and then activated. Of course, this activation of the first event can then change the pattern displayed by subsequent event functions. Thus, the detection algorithm must identify and sort all active events over the step.

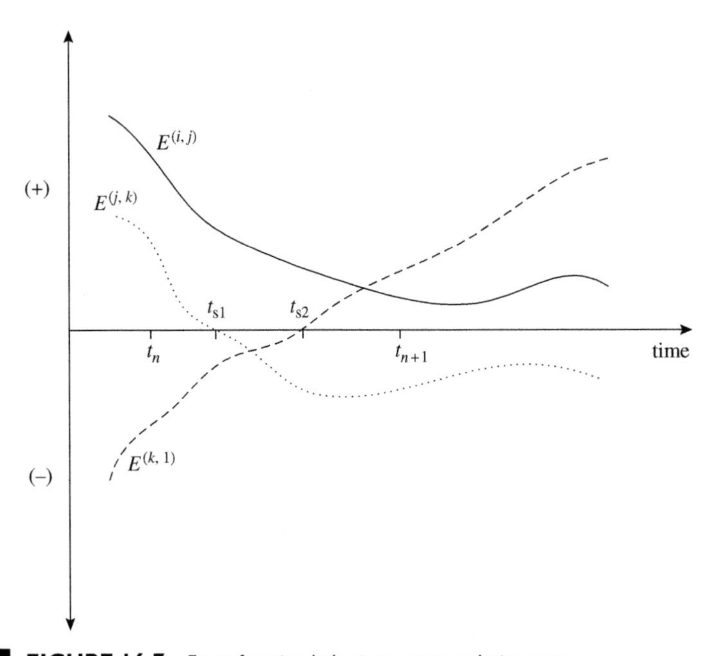

FIGURE 16.7 Event function behaviour over a solution step.

Even though in Fig. 16.7, the event functions are shown as continuous functions, it is normally not the case that much information is known about the event function behaviour between discrete solution outputs. Simply what is known in most cases, such as the use of linear multi-step methods, is a change of sign from t_n to t_{n+1}. In some cases, the internal stage calculations can actually help detect the event function change before the end of the current step.

Alternatively, information can be kept about the event functions $E^{(i,j)}$ which allows prediction of possible sign changes before the current step is taken. This represents a "look ahead" strategy for event times. This can be useful in improving the efficiency of the event detection.

Event Location

Having identified that an event, or several events may have occurred in the last step, it is necessary to locate within a prescribed tolerance the time of the first event. This normally requires an interpolation routine applied iteratively until the event time is located to a user set tolerance. In many cases only a few iterations are needed to locate the event time, depending on the quality of the interpolant.

Because event functions can consist of either state and/or algebraic variables, the location algorithm must handle both cases.

After location of the event, it is necessary to ensure that the new state is activated and the transition conditions given by the function $T^{(i,j)}$ imposed. This can be difficult at times, since a number of issues could be important, including:

- change of index. From index-1 to index-2 or higher as well as the reverse situation,
- change of numbers of equations and variables from one state to another in the state transition graph.

Restart of Solution

There are several issues which must be addressed when restarting the solution after a state transition. This depends directly on the type of solver being used.

Linear multi-step methods In the case of the LMMs, whether explicit or implicit formulae, these techniques rely on past solution points to generate a high order representation of the solution. Unfortunately, once a state transition takes place, the past information is invalid from the perspective of the polynomial representing the solution. Hence, LMMs typically revert to order 1 methods which are either Euler or backward Euler methods depending on whether they are explicit or implicit. The LMMs then gradually build up the order of the representation via the standard step/order change mechanisms.

In order to handle highly discontinuous problems, it is possible to restart the solution with a higher order Runge–Kutta formula which generates sufficient information to then switch back to the LMM after a few steps. No generally available production codes are available for this purpose.

Runge–Kutta methods The advantages of Runge–Kutta methods for highly discontinuous problems is obvious from their use in LMMs for restart. The key advantage is the maintenance of high order across a state transition. Being single step methods they use the last solution point only as the starting point for the numerical solution over the current step. Hence, the main issue is related not to maintenance of order but to a rapid step increase mechanism which quickly brings the steplength to the appropriate level to maintain local error tolerances.

High-Index Solvers

As was mentioned in Section 16.2.3, state transitions can often mean changes not only in the size of the equation set but also its differential index. The example of the water cistern system in Section 16.2.3 shows how index can change from 1 to 2 and back again. In these cases, high-index solvers can be used. However, there are few high-index solvers available. Most reliable index-2 (or higher) solvers are based on Runge–Kutta formulae. Programs such as RADAU5 [129] and ESDIRK provide efficient solution of index-2 problems [50].

Linear multistep methods such as the BDF can be made to solve index-2 problems but usually at a significant cost in efficiency.

Transformations for DAE Systems of High Index

There are a number of modelling transformations which can be used to alleviate high index DAE solution problems. In the DAEPACK by Barton [47] this problem is handled by the introduction of dummy derivatives. Other techniques requiring the differentiation of subsets of the algebraic system can also be used [40].

16.5. SUMMARY

This chapter has developed the background theory and applications of hybrid system models. It can be said that all real processes are essentially hybrid in nature, only that in many situations we pretend that they are continuous for reasons of simplicity and solvability.

The chapter has investigated the current major representation of these systems highlighting the characteristics of hybrid automata, Petri nets, and differential-algebraic systems.

These systems continue to be a major challenge for researchers and application software experts.

There is much to be done in this area from a systems perspective, in order to provide robust and transparent modelling systems for the process industries.

16.6. FURTHER READING

The literature on modelling of hybrid process systems is mainly centered around the particular tools and techniques applied for solving a particular problem. Unfortunately little is known and almost nothing has been published on the comparison of the different available approaches.

Because of the scattered and rapidly developing nature of the field further reading is recommended mainly for recent journal and conference papers. There are, however, some review papers mainly from the field of systems and control theory, such as

[109,110]. We only cite some of the most characteristic application papers from the field of process systems engineering arranged in groups driven by the technique they use.

- The *discrete event system modelling approach* applied to process systems with hybrid nature has been presented by Preisig [130].
- An *optimization problem*, the derivation of an optimal periodic schedule for general multipurpose plants comprising batch, semi-batch and continuous operations resulted in a mixed integer nonlinear programming (MINLP) problem in [131]. The basis of the hybrid system model is the general resource-task network (RTN) coupled with a continuous representation of time.

  A similar problem, on-line scheduling of a multiproduct polymer batch plant has been solved in [132]. Instead of solving the resultant large, non-convex MINLP problem by conventional optimization packages, they applied genetic algorithms to generate the schedules.
- The *simulation approach* is used in [133] for complex operability analysis of dynamic process systems under the influence of exceptional events, such as failures and sudden changes.

16.7. REVIEW QUESTIONS

Q16.1. What are the key components of a process system which contribute to the hybrid behaviour? (Section 16.1)

Q16.2. What are the key application areas in hybrid system modelling and simulation? (Section 16.1)

Q16.3. What are the three components in a hybrid system? What approaches are available in the literature to model the discrete event component? (Section 16.2)

Q16.4. What are the elements of the formal description of a hybrid finite automata? How do these elements appear in the state transition diagram of the automata? (Section 16.2.1)

Q16.5. Give the formal description of the three components of a hybrid automata model: the discrete event dynamic system, the continuous system and the interface. What are the extensions needed in the discrete event component and continuous component to fit them into the hybrid automata framework? (Section 16.2.1)

Q16.6. What kind of extended Petri nets are available for modelling hybrid systems? How can we use Petri nets for modelling hybrid systems? (Section 16.2.2)

Q16.7. Which dynamic properties of hybrid systems are subject of dynamic analysis? What is the definition and practical significance of these properties? (Section 16.3)

Q16.8. Which is the hybrid system model used for the analysis of dynamic properties? What is the first step of the analysis? Is the analysis of dynamic properties of a hybrid model computationally hard? Justify your answer. (Section 16.3)

Q16.9. What are the key issues in numerical solution of hybrid DAE models? (Section 16.4)

16.8. APPLICATION EXERCISES

A16.1. Draw the extended state transition diagram of the reactor–filter system defined in Example 16.2.1 as the discrete event part of the hybrid model. Compare it to the state transition diagram shown in Fig. 15.3.

A16.2. Consider the state transition graph of the water cistern model shown in Fig. 16.6. Is the underlying hybrid system reachable? Has it got any deadlocks or conflict situation?

A16.3. Construct the hybrid automata model of the water cistern system in Section 16.2.3.

A16.4. Construct the hybrid Petri net model of the water cistern system in Section 16.2.3.

17 MODELLING APPLICATIONS IN PROCESS SYSTEMS

In this chapter, we outline some modelling applications on a variety of process systems. The purpose of these applications is to show the general methodologies discussed in earlier chapters. Other important issues to do with plant data acquisition and the application of numerical techniques for solution are highlighted. Also the need to revise the assumptions underlying the model and hence the mathematical model itself is an important issue, where there are significant discrepancies between the model predictions and the plant behaviour.

The examples have been chosen to illustrate a wide variety of process application areas and the intended use of the models in an industrial or research environment. The models make no claim to being the most accurate or comprehensive, but simply illustrate various approaches to the problems being addressed.

The modelling examples addressed in this chapter include:

- high temperature copper converter dynamics (minerals processing),
- phenol destruction in waste water using a photochemical reaction process (environmental engineering),
- dynamics of drum granulators and granulation circuits (particulate processing),
- prefermentation in wastewater treatment (wastewater engineering),
- dynamics of an industrial depropanizer (petrochemical processing).

17.1. COPPER CONVERTER DYNAMICS

This application concerns the modelling for dynamic analysis and optimal control of a 320 tonne reactor for liquid copper processing. This is a fed batch reactor with both continuous and discrete addition and extraction of materials.

17.1.1. Process Description

In the industrial process for making copper metal from chalcopyrite ores, copper concentrate being about 5–10% copper with the rest being sulphides, iron and other minerals is reacted in a high intensity smelter with oxygen. This produces a matte copper of around 50% copper with substantial amounts of sulphur and iron. These impurities must be removed so that liquid copper can be cast into anode plates for electro-refining. To do this, a high temperature Pierce–Smith converter is used to give 95.5% copper product. Figure 17.1 shows a typical industrial converter of around 350 tonnes capacity. The converter is 4.5 m diameter and 15 m in length.

The operation of such a converter requires the discrete addition of matte copper from 25 ton ladles as well as many discrete additions of silica flux, solid re-work copper, anode skim and other materials over the time of the complete cycle. Constant sparging of air through injection tuyeres located 300 mm below the bath surface promotes the oxidation of sulphur compounds, the formation of slag phases and finally liquid copper. Slag, which contains unwanted iron compounds is periodically drained off by "rolling out" the reactor. The operation is typically divided into two distinct periods, known as the "slag making phase" and the "copper making phase".

FIGURE 17.1 Depiction of rolled-out converter pouring to a pot.

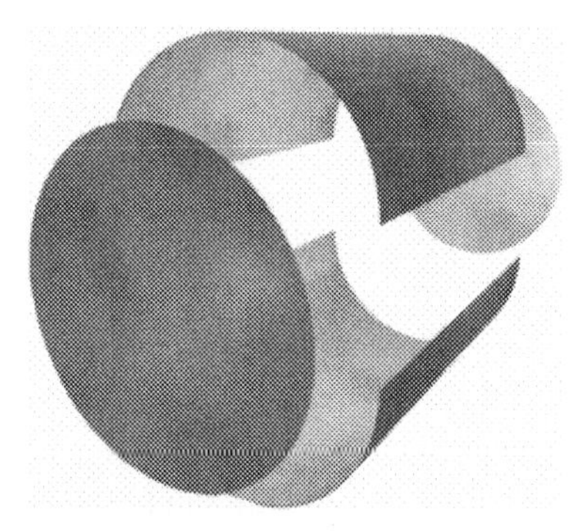

FIGURE 17.2 Scenario 1.

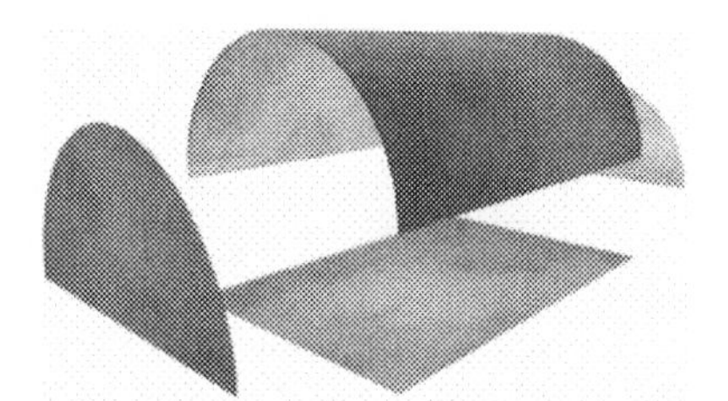

FIGURE 17.3 Scenario 2.

The process is characterised by intense exothermic chemical reactions, high radiative heat transfer and complex phase behaviour with up to nine phases present.

There are three key operational periods for the converter:

1. The idle period, which is between batches, where cooling and possible solidification of residual contents occurs. This could be 45 min in duration.
2. The slag making period, where sulphur is oxidized and released as SO_2 and iron is oxidized into a silica saturated slag. This period lasts about 90 min with the bath being mainly Cu_2S. This is termed "white metal".
3. The copper making period, where liquid copper begins to form as Cu_2S is oxidised by air injection leaving a "blister" copper of around 1.5–1.9 wt % sulphur and a significant amount of oxygen. This period is about 150 min.

The graphics in Figs. 17.2–17.4 show the three principal scenarios for radiation calculations. Scenario 1 relates to an empty converter during the idle period. Scenario 2 is when the converter is charged and "rolled in", whilst Scenario 3 covers the case where the converter is charged but "rolled out" and where losses to the hood are important.

Modelling Goal

To predict the dynamic behaviour of the system as a result of a specified event sequence.

FIGURE 17.4 Scenario 3.

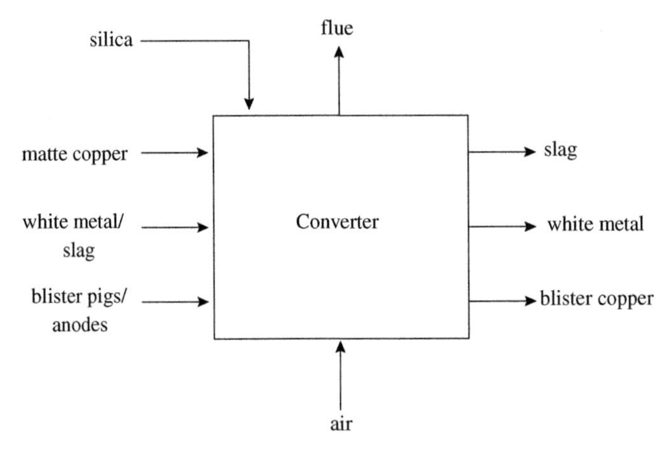

FIGURE 17.5 Overall converter mass flows.

17.1.2. Controlling Factors

The principal controlling factors in this system are

- complex exothermic reactions of oxygen and sulphur compounds,
- formation of slag phases consisting of complex iron compounds (FeO, Fe_3O_4),
- radiation heat transfer from bath surfaces and converter internals,
- bubble formation at tuyeres and liquid phase mass transfer of oxygen to the bath,
- reaction and temperature control using discrete material additions.

Figure 17.5 shows the principal material flows associated with the converter.

17.1.3. Assumptions

The principal assumptions used include:

$\mathcal{A}1$. Reactions are very fast and the bath contents are quickly in equilibrium.
$\mathcal{A}2$. Radiation plays a major role in the energy balance.
$\mathcal{A}3$. Penetration of metal into the walls contributes significantly to energy dynamics.
$\mathcal{A}4$. External heat transfer from the converter surface is important.

$\mathcal{A}5$. Residual bath contents can solidify between cycles and must be modelled.
$\mathcal{A}6$. During the reaction phases the bath is assumed to be well mixed.
$\mathcal{A}7$. The reaction stages exhibit distinct condensed phases which must be predicted.
$\mathcal{A}8$. No physical losses in additions and extractions during operations.
$\mathcal{A}9$. Constant heat capacities.

17.1.4. Model Development

Model development consisted of several key component and liquid phase energy balances. This included detailed submodels describing the following phenomena:

- Radiation effects under various converter configurations during the principal cycles as well as the filling, operating and "roll-out" positions.
- A distributed parameter submodel of the metal penetration, solidification and subsequent melting within the refractory lined walls.
- A sequence of operational regions exists where certain chemical species appear based on their thermodynamic feasibility.

The full model is described by Waterson and Cameron [134]. What follows are some of the key aspects of the modelling.

Mass Balances

Mole balances are performed for each individual species in the bath. They have the general form:

$$\frac{\mathrm{d}N_i}{\mathrm{d}t} = N_{i,\mathrm{in}} - N_{i,\mathrm{out}} + N_{i,\mathrm{gen}}, \tag{17.1}$$

where $N_{i,\mathrm{in}}, N_{i,\mathrm{out}}$ are molar flows of species via discrete additions or deletions from the system and $N_{i,\mathrm{gen}}$ refers to generation or consumption of a species.

In particular, conservation balances are performed on elemental species so that Eq. (17.1) becomes

$$\frac{\mathrm{d}N_i}{\mathrm{d}t} = N_{i,\mathrm{in}} - N_{i,\mathrm{out}}, \quad i = \mathrm{Cu, Fe, S, O, N, SiO_2}. \tag{17.2}$$

Energy Balance

The energy balance over the bath contents, after simplification and writing in intensive terms gives

$$\sum_{i=1}^{c} n_i C_{p_i} \frac{\mathrm{d}T_B}{\mathrm{d}t} = \dot{E}_B, \tag{17.3}$$

where n_i is the moles of species i, C_{p_i} is the heat capacity of species i and $\dot{E}_B$ is the rate of energy change in the bath. This is a complex function of addition and deletion rates of discrete materials, their temperature and bath operational temperature which could drop below the liquidus–solidus boundary. The right-hand term also includes all conductive and radiative heat losses.

Radiative Energy Losses

Radiation heat transfer was computed using detailed estimates from a Monte–Carlo ray-tracing algorithm to evaluate radiative heat exchange amongst the vessel interior and the converter mouth. The model incorporates gas radiation from SO_2 fractions within the enclosures, and also considers mirror-like reflections from the bath surface. The radiation enclosures were selected based on the geometric scenarios previously mentioned.

The model accounts for external radiation from the shell, the ends and the converter mouth.

Conductive Energy Losses

Unsteady-state heat conduction is used to describe heat losses from the internal walls of the converter:

End enclosure model

$$\frac{\partial T}{\partial t} = \alpha \frac{\partial^2 T}{\partial x^2}$$

Main body

$$\frac{\partial T}{\partial t} = \alpha \frac{\partial^2 T}{\partial r^2} + \frac{\alpha}{r}\frac{\partial T}{\partial r}. \tag{17.4}$$

Appropriate boundary conditions are applied to solve these equations.

Particle Population Balance

When considering the addition of silica particles a population balance was used to help describe the dissolution process. This is described by

$$\frac{\partial f}{\partial t} = D\frac{\partial f}{\partial L} + \frac{\partial D}{\partial L} f + \sum_i n_i x_i \dot{m}_i, \tag{17.5}$$

where f is the number population distribution, L the characteristic particle size, D the linear dissolution rate, n_i the population density for the ith silica stream, x_i the mass fraction of silica in the ith particulate stream and $\dot{m}_i$ the mass flowrate of the ith particle stream.

17.1.5. Reaction Equilibria

In this model, it was assumed that due to the operating conditions, equilibrium would be maintained in the reactor throughout the slag making phase. For the ith reaction, of the form:

$$\alpha A + \beta B + \cdots \longleftrightarrow \phi M + \theta N + \cdots, \tag{17.6}$$

where $K_i = e^{-\Delta G_i/RT}$ is the equilibrium coefficient, the equilibrium relations are written as:

$$K_i = \frac{a_M^{\phi} \cdot a_N^{\theta} \cdots}{a_A^{\alpha} \cdot a_B^{\beta} \cdots},$$

where $a_j = \gamma_j x_j$ or the activity of species j, and x_j is the mole fraction of j.

These equations are combined with the elemental and phase mole balances to give the set of equations describing the compositional changes with time. Correlations for the activity coefficients were taken from Goto [135].

For the copper making phase, when Cu_2S is being oxidized, a number of complex reaction sequences take place from the final elimination of FeS to the formation of oxidized copper metal, Cu_2O. In the model, five distinct substages were defined and the appropriate model selected based on the bath compositions.

17.1.6. Model Solution and Simulation

Initially, the model equations were programmed into a standalone series of Fortran routines, linked together by an executive program. Event files for the individual discrete additions of material were generated, these used to drive the overall simulation. Individual subprograms were independently verified to ensure correct coding and check predictions.

17.1.7. Process Experimentation and Model Performance

A series of plant measurements were taken over several operating shifts to obtain concentration and temperature profiles of the key species. This involved live tapping through the injection tuyeres in order to release molten copper at 1200–1300°C for collection and subsequent laboratory analysis for sulphur and oxygen content. Temperatures were obtained from optical pyrometers mounted to view the bath contents. Figure 17.6 shows the model and plant behaviour for FeS during the slag making period while Fig. 17.7 gives the profiles of total sulphur during copper making. Both model predictions are in reasonable agreement with the gathered data, given the great difficulty and uncertainty in the data acquisition process. Figures 17.8 and 17.9 give model predictions and measured data for the temperature profiles throughout the two key operational periods. There is some over and underprediction of the profiles but the result is good considering no fitting of the model to plant data has taken place.

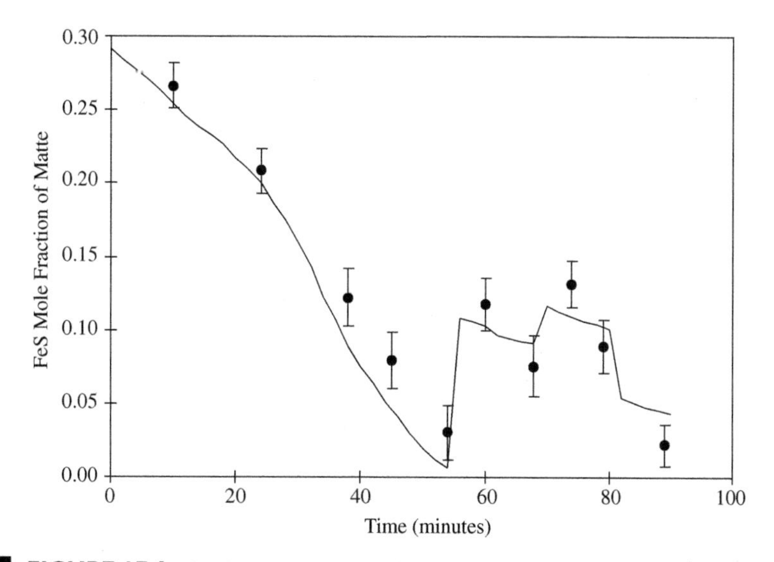

FIGURE 17.6 Predicted and measured FeS mole fraction during slag making stage.

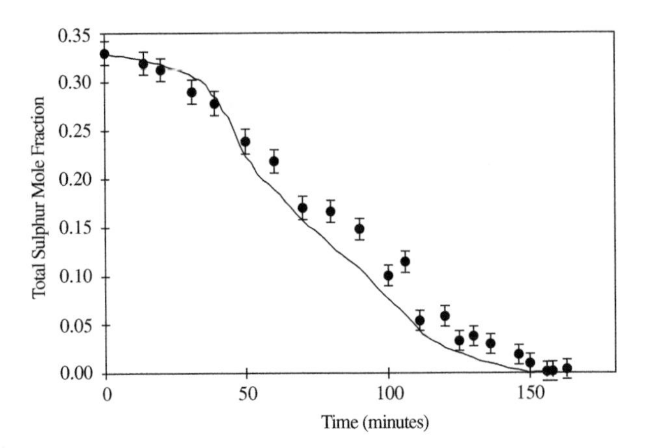

FIGURE 17.7 Predicted and measured sulphur mole fraction during copper making stage.

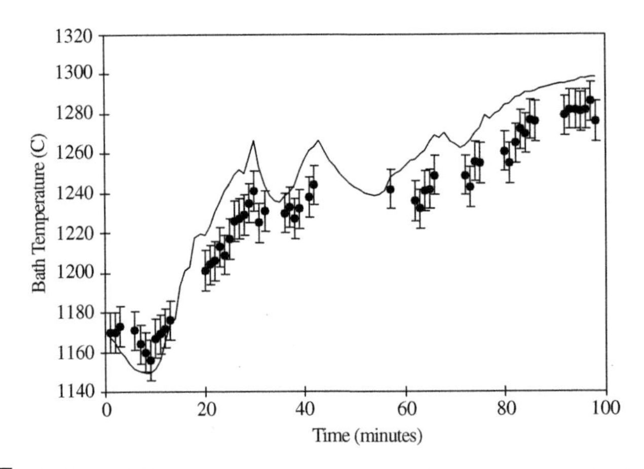

FIGURE 17.8 Predicted and measured bath temperature during slag making stage.

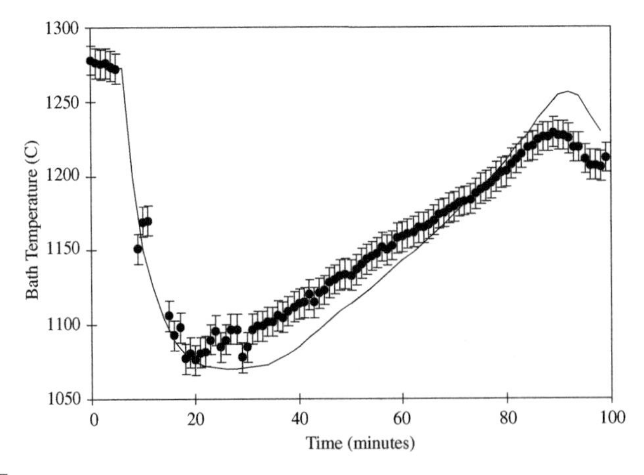

FIGURE 17.9 Predicted and measured bath temperature during copper making stage.

17.2. DESTRUCTION OF PHENOL IN WASTEWATER BY PHOTOCHEMICAL REACTION

This application looks at the time-dependent destruction of phenol in wastewater in a recirculating batch system involving a ultra-violet (UV) reactor. The dynamics of the process are described and the model is validated against plant data [136].

17.2.1. Process Description

A pilot scale rig is shown in Fig. 17.10. A pump recirculates an industrial waste stream from a holding tank through a UV light reactor. The waste stream contains approximately 10 ppm of phenol. The flowrate of wastewater through the reactor is regulated. Phenol is broken down in two main ways. Photolysis occurs in the presence of UV light and photo-oxidation occurs in the presence of both UV light and an oxidant, hydrogen peroxide.

The two reactions involving phenol (P) and oxygen (O) which are believed to occur are

$$\mathrm{P} \overset{\mathrm{UV}}{\rightarrow} \text{products.} \tag{17.7}$$

$$\nu_{\mathrm{P}} P + \nu_{\mathrm{O}} \mathrm{O} \rightarrow \text{products.} \tag{17.8}$$

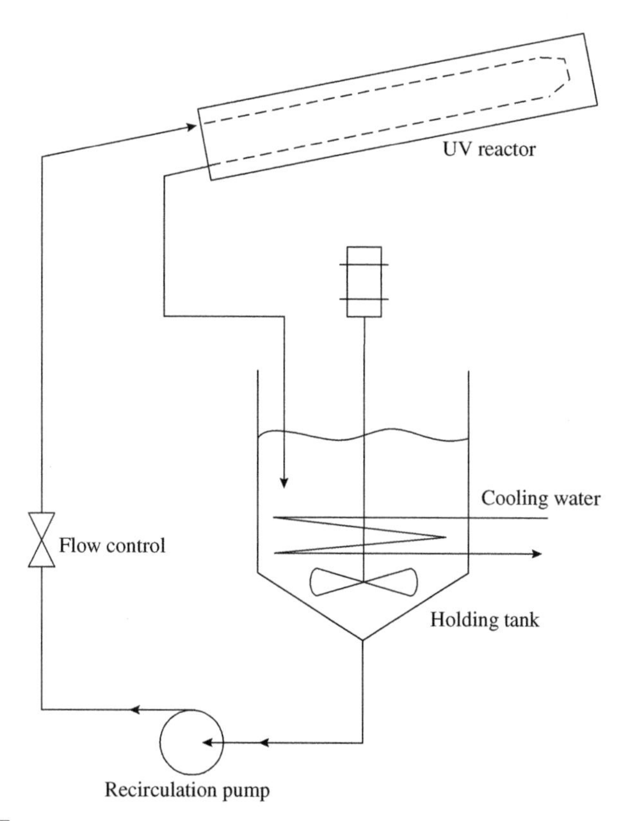

FIGURE 17.10 Schematic of UV reactor system.

Modelling Goal

The model should describe the time varying phenol degradation as a function of recirculation flowrate and initial concentration of hydrogen peroxide.

17.2.2. Controlling Factors

The controlling factors for this system are

- chemical reaction due to the UV light,
- radiative heat transfer in the UV light reactor,
- forced convection heat transfer in the recirculation tank.

17.2.3. Assumptions

The following assumptions were applied in developing the model:

$\mathcal{A}1$. Recirculation tank is well mixed.
$\mathcal{A}2$. No heat losses from tank walls and piping.
$\mathcal{A}3$. Agitator speed is constant and shaft work is negligible.
$\mathcal{A}4$. Cooling coil contents are well mixed.
$\mathcal{A}5$. Liquid holdup in pipelines and reactor is negligible compared with the tank.
$\mathcal{A}6$. Kinetic and potential energy effects are negligible.
$\mathcal{A}7$. Reaction rate constants are independent of temperature over the operating range.
$\mathcal{A}8$. There is no physical mass loss or gain in the system.
$\mathcal{A}9$. Physical properties of the solution are those of water.
$\mathcal{A}10$. No side reactions occur and no UV reaction occurs outside the reactor.
$\mathcal{A}11$. There is no heat loss from the UV reactor.
$\mathcal{A}12$. The UV light provides a constant heat source evenly distributed along the reactor.
$\mathcal{A}13$. For each mole of phenol, half a mole of hydrogen peroxide is consumed.
$\mathcal{A}14$. Heat transfer coefficients are constant.
$\mathcal{A}15$. Coiling coil mass holdup is constant.

17.2.4. Model Development

Three principal balance volumes were identified for the system. These include:

- volume of the UV reactor,
- volume of the recirculation tank,
- volume of the cooling coil within the recirculation tank.

Recirculation Tank Balances

Component balances The component mole balances for phenol (P) and oxidant (O) are given by

$$\frac{dP_T}{dt} = \frac{1}{V_T}(P_R q_R - P_T q_T), \tag{17.9}$$

$$\frac{dO_T}{dt} = \frac{1}{V_T}(O_R q_R - O_T q_T), \tag{17.10}$$

where P_T, O_T are concentrations in [mol/l] and q_R and q_T are volumetric flowrates in [l/s].

Energy balance The energy balance in simplified intensive form is given by

$$\frac{dT_T}{dt} = \frac{1}{V_T}(q_R T_R - q_T T_T) - \frac{Q_C}{V_T \rho c_p} \tag{17.11}$$

Cooling Coil Balances

The mass balance is not required as the holdup is assumed constant ($\mathcal{A}$15).

Energy balance The energy balance for the cooling coil is given by

$$\frac{dT_C}{dt} = \frac{1}{V_C}(q_C T_{C,in} - q_C T_{C,out}) - \frac{Q_C}{V_T \rho c_p}, \tag{17.12}$$

where $T_{C,in}$, $T_{C,out}$ are the cooling water temperatures in and out of the coil ([K]), q_C is the cooling water volumetric flow ([l/s]).

Lumped Parameter UV Reactor Balances

The first approach uses a lumped parameter model of the UV reactor.

Component balances The component balances are given by

$$\frac{dP_R}{dt} = \frac{1}{V_R}(P_T q_T - P_R q_R) - r_1 - \nu_P r_2, \tag{17.13}$$

$$\frac{dO_R}{dt} = \frac{1}{V_R}(O_T q_T - O_R q_R) - \nu_O r_2. \tag{17.14}$$

Energy balance The energy balance in simplified intensive variable form is given by

$$\frac{dT_R}{dt} = \frac{1}{V_R}(q_T T_T - q_R T_R) + \frac{1}{\rho c_P}(r_1 \Delta H_{r_1} + r_2 \Delta H_{r_2}) - \frac{Q_L}{V_R \rho c_P}, \tag{17.15}$$

where r_1, r_2 are reaction rates for the two key reactions [mol/l · s], ΔH_{r_1} and ΔH_{r_2} are heats of reaction [J/mol] and Q_L is the rate of heat transfer from the UV light [J/s].

The previous development assumed that the reactor could be regarded as a well mixed system. An alternative approach considers the UV reactor to be a one dimensional distributed parameter system. Using this approach the following equations can be developed.

One-dimensional Distributed Parameter UV Reactor Balances

This model assumes no radial variation in concentration or temperature ($\mathcal{A}16$).

Component balances The two component balances are

$$\frac{\partial C_P}{\partial t} = \mathcal{D}\frac{\partial^2 C_P}{\partial x^2} - \frac{q_R}{A}\frac{\partial C_P}{\partial x} - r_1 - \nu_P r_2, \tag{17.16}$$

$$\frac{\partial C_O}{\partial t} = \mathcal{D}\frac{\partial^2 C_O}{\partial x^2} - \frac{q_R}{A}\frac{\partial C_O}{\partial x} - r_1 - \nu_O r_2, \tag{17.17}$$

where C_P, C_O are the key concentrations along the reactor, $\mathcal{D}$ is a diffusion coefficient.

Energy balance The energy balance for the reactor is given by

$$\frac{\partial T_R}{\partial t} = \frac{\kappa}{\rho c_P}\frac{\partial^2 T_R}{\partial x^2} - \frac{q_R}{A}\frac{\partial T_R}{\partial x} + \frac{Q_R}{\rho c_P V_R} + \frac{r_1 \Delta H_{r_1}}{\rho c_P} + \frac{r_2 \Delta H_{r_2}}{\rho c_P} \tag{17.18}$$

where κ is a thermal diffusivity term $[\mathrm{J/l \cdot s.K.m^2}]$.

The accompanying boundary conditions at the inlet and outlet of the UV reactor are

$$C_{P,R}(t,0) = C_{P,T}, \quad \frac{\partial C_{P,R}(t,l)}{\partial x} = 0, \tag{17.19}$$

$$C_{O,R}(t,0) = C_{O,T}, \quad \frac{\partial C_{O,R}(t,l)}{\partial x} = 0, \tag{17.20}$$

$$T_R(t,0) = T_T, \quad \frac{\partial T_R(t,l)}{\partial x} = 0. \tag{17.21}$$

This indicates that the inlet conditions are simply the conditions in the recirculation tank. The outlet conditions imply that reaction ceases once the waste leaves the UV reactor.

Constitutive Relations

Reaction rates These are given by

$$r_1 = k_1 P_R^a, \tag{17.22}$$

$$r_2 = k_2 P_R^b O_R^c. \tag{17.23}$$

Heat transfer

$$Q_C = UA(T_T - T_C). \tag{17.24}$$

Heat transfer in the UV reactor is fixed as stated in assumption $\mathcal{A}12$.

17.2.5. Model Solution and Verification

The lumped parameter model was coded in MATLAB and solved using the stiff ODE solver ODE15S. For the distributed parameter model, orthogonal collocation was used to discretize the UV reactor model using three internal collocation points. The resulting equation set was solved in MATLAB.

17.2.6. Process Experimentation and Model Performance

A number of experiments were performed to assess the assumptions and establish the unknown parameters. These included:

1. Assess the significance of the heating effects of the UV lamp and cooling effects of the coil.
2. Establish kinetic parameters and heat of reaction for degradation of phenol by photolysis.
3. Establish kinetic parameters and heat of reaction for degradation of phenol by photolysis and photo-oxidation.
4. Validation data for the models, reducing hydrogen peroxide concentration.
5. Validation data for the models, reducing circulation flowrate of wastewater.

Sampling was done at the reactor outlet as well as from the holding tank and subsequently analysed using a spectrophotometer. Experiments were carried out over a period of 45–60 min.

Model Fitting with Photolysis only

Experiment 1 allowed the estimation of the kinetics parameters for the photolysis reaction to be made. Following this it was then possible to estimate the photo-oxidation parameters.

Model Fitting with Photolysis and Photo-oxidation

The introduction of hydrogen peroxide at a concentration of 0.15 vol.% leads to an increased rate of phenol degradation compared with peroxide free solutions. Figures 17.11 and 17.12 show the measured data and the fitted models for both lumped and distributed parameter models. The more complicated distributed parameter model gave similar results. At this stage all relevant parameters have been estimated. It is now worthwhile to test the model on a run which changed the initial peroxide concentration.

Model Validation and Improvement

The reactor was run using only half the concentration of peroxide compared with the runs to estimate parameters. Initial model predictions, although correct in trend, did not at all represent the data well.

In order to improve the model's ability to represent the independent data, the assumption of a 0.5 molar ratio of peroxide to phenol was removed and the data points

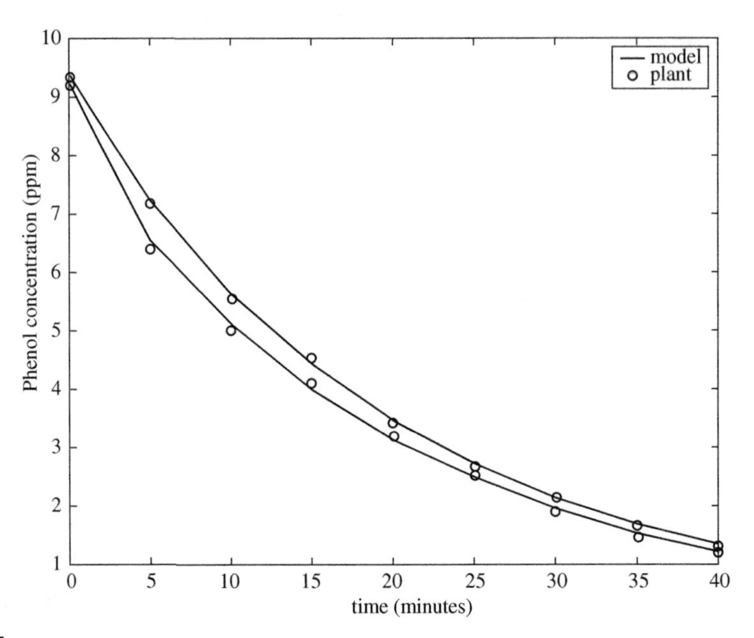

FIGURE 17.11 Lumped model, reaction parameter estimation.

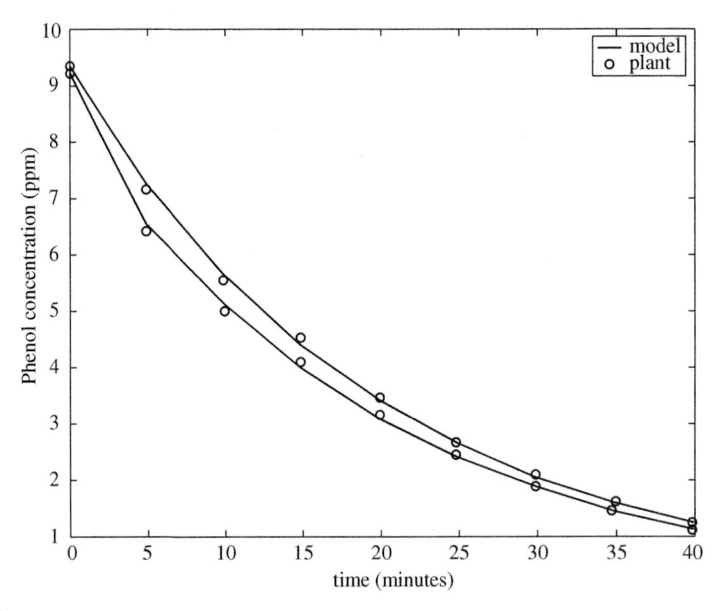

FIGURE 17.12 Distributed parameter model, reaction parameter estimation.

were adjusted to allow for the time delay in sampling. Parameters were recalculated using the original data and then the model tested against the validation data. Figure 17.13 shows the performance of the revised model, which indicates a significant improvement.

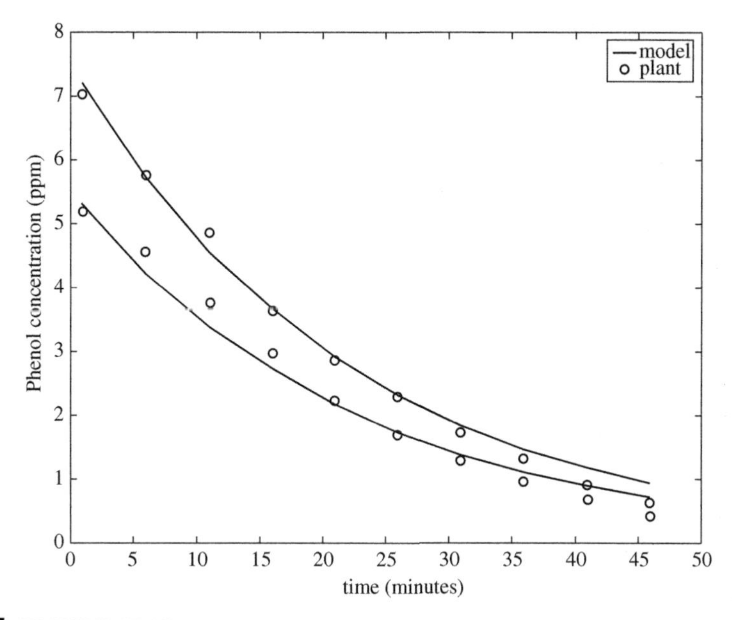

FIGURE 17.13 Full lumped parameter model preformance.

17.3. PREFERMENTER SYSTEM FOR WASTEWATER TREATMENT

This example considers the dynamics of a prefermenter unit used in a wastewater treatment facility.

17.3.1. Process Description

Prefermentation is a relatively new operation in biological nutrient removal (BNR) of wastewater treatment systems. The prefermentation role is to produce volatile fatty acids (VFAs) which assist in the subsequent removal of nitrogen and phosphorus from the wastewater. The prefermenter relies on complex biological reactions taking place producing VFAs via an anaerobic degradation process. The dynamic modelling of these systems has recently been completed by von Münch [137]. A schematic of the overall process is seen in Fig. 17.14.

Modelling Goal

The goal is to predict the dynamic behaviour of VFA production, pH and soluble chemical oxygen demand (SCOD) under the influence of changing wastewater feed composition or flow.

17.3.2. Controlling Factors

Key factors in the system include:

- two major processes of hydrolysis and acidogenesis take place;
- hydrolysis is catalysed via hydrolytic enzymes excreted by bacteria;

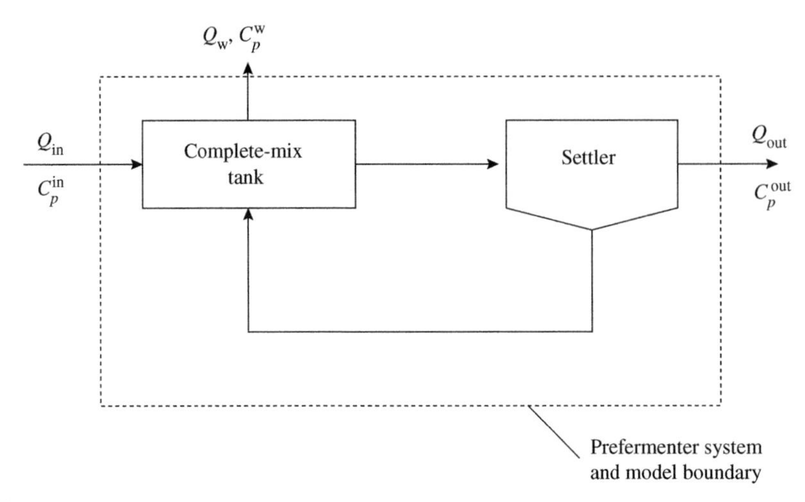

FIGURE 17.14 Over schematic for prefermenter.

- three (3) key organic substrate groups are broken down, including carbohydrates, proteins and lipids;
- acidogenesis process uses sugars, amino acids and long chain fatty acids as a carbon and energy source;
- bacteria produce VFAs, lactic acid, ethanol and hydrogen;
- hydraulic residence time (HRT) and solids residence time (SRT) are important in determining overall dynamics.

17.3.3. Assumptions

The principal assumptions used in constructing the model include:

$\mathcal{A}1$. Reaction is considered as a well-mixed continuous reactor.
$\mathcal{A}2$. Reactor volume is considered constant.
$\mathcal{A}3$. Inert insoluble or soluble substrate is not considered in the model.
$\mathcal{A}4$. Growth rates of acidogenic and methanogenic bacteria are described by Monod kinetics.
$\mathcal{A}5$. Only two main groups of bacteria are in the prefermenter, acidogens and methanogens.
$\mathcal{A}6$. Methane generation from CO_2 and H_2 is not considered.
$\mathcal{A}7$. Methanogenesis can occur from all VFAs.
$\mathcal{A}8$. Hydrolytic enzyme production is limited to the hydrolysis rate via a yield coefficient.
$\mathcal{A}9$. Hydrolytic enzymes undergo denaturation via first order kinetics.
$\mathcal{A}10$. Ammonia is taken up by bacteria for cell synthesis and liberated on cell death.
$\mathcal{A}11$. Organic nitrogen in feed is assumed to be particular proteins.
$\mathcal{A}12$. System is regarded as iso-thermal.

17.3.4. Model Development

In developing the model, balances in the system of Fig. 17.14 relate to

- solids (insoluble substrate, biomass, proteins),
- liquid components (soluble substrate, VFA, enzymes).

Mass Balances The solid mass balances become

$$V_R \frac{dC_{p,i}}{dt} = Q_{in} C_{p,i}^{in} - Q_w C_{p,i}^{w} - Q_{out} C_{p,i}^{out} + V_R \sum_j \upsilon_{ji} \mathcal{R}_j, \tag{17.25}$$

where V_R is the prefermenter reactor volume [litres], $C_{p,i}$ the particulate component i concentration [mg/l], Q the flowrate of stream [m^3/h], υ_{ji} the stoichiometric coefficient for component i in process j, $\mathcal{R}_j$ is the rate equation for process j.

Liquid Component Balances

$$V_R \frac{dC_{L,i}}{dt} = Q_{in} C_{L,i}^{in} - Q_w C_{L,i}^{w} - Q_{out} C_{L,i}^{out} + V_R \sum_j \upsilon_{ji} \mathcal{R}_i. \tag{17.26}$$

The structure of the prefermenter model is shown in Fig. 17.15 where the individual species definitions are given in Table 17.1.

Solid lines show COD mass balances and dashed lines represent nitrogen mass balances. Y_k represents yields for various reactions k, $Y_{N/X}$ are stoichiometric ammonia-N requirements for biomass synthesis and r represent the various reaction rates.

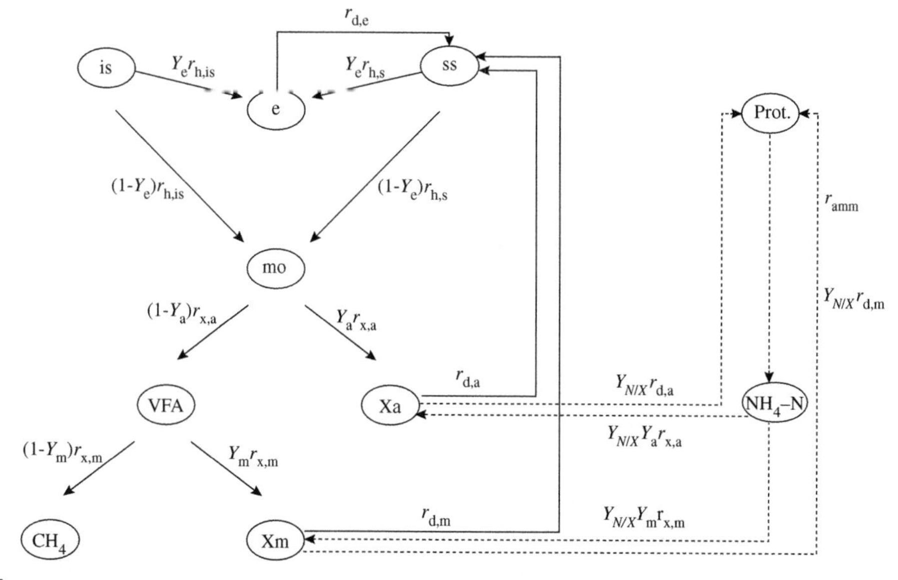

FIGURE 17.15 Prefermenter model structure.

TABLE 17.1 Model Nomenclature

State	Description	Rate	Description
C_{is}	Insoluble substrate	$r_{h,is}$	Hydrolysis of insoluble substrate
C_{ss}	Soluble substrate	$r_{h,s}$	Hydrolysis of soluble substrate
C_{mo}	Monomer species	r_{amm}	Ammonification of proteins
C_{vfa}	Volatile fatty acids	$r_{x,a}$	Growth of acidogens
C_e	Hydrolytic enzymes	$r_{x,m}$	Growth of methanogens
C_{Xa}	Acidogenic bacteria	$r_{d,a}$	Decay of acidogens
C_{Xm}	Methanogenic bacteria	$r_{d,m}$	Decay of methanogens
C_{prot}	Particulate proteins	$r_{d,e}$	Decay of hydrolytic enzymes
C_{NH_4}	Ammonium nitrogen		

The above mass balances can be rewritten to include the HRT and SRT directly giving

$$\frac{dC_{p,i}}{dt} = \frac{1}{\mathrm{HRT}} C_{p,i}^{\mathrm{in}} - \frac{1}{\mathrm{SRT}} C_{p,i} + \sum_j \upsilon_{ji} \mathcal{R}_j, \tag{17.27}$$

$$\frac{dC_{L,i}}{dt} = \frac{1}{\mathrm{HRT}} \left(C_{L,i}^{\mathrm{in}} - C_{L,i} \right) + \sum_j \upsilon_{ji} \mathcal{R}_j, \tag{17.28}$$

with

$$\mathrm{HRT} = V_{\mathrm{R}}/Q_{\mathrm{in}}$$

$$\mathrm{SRT} = V_{\mathrm{R}} Cp/(C_p^{\mathrm{out}} Q_{\mathrm{out}} + C_p^{\mathrm{w}} Q_{\mathrm{w}})$$

Model Parameters

The previously developed model leads to a system which includes some 27 parameters related to the kinetics of the biological reactions as well as parameters associated with the pH function and system retention times. Most of the biological parameters are available in the literature covering anaerobic digestion, others were obtained from parameter estimation. In all, three parameters were fitted using pilot plant data.

17.3.5. Model Solution and Simulation

The model was developed within the Daesim Studio modelling and simulation environment [126] with the resultant DAE model being solved by a sparse DIRK method.

17.3.6. Experimentation, Calibration and Validation

Experimentation was carried out on a pilot scale prefermenter system fed by a sidestream of wastewater going into a major wastewater treatment plant. A glucose solution was added into the sewage holding tank which fed the prefermenter system. Influent flowrate to the prefermenter was approximately 70 l/h with the HRT being 11 h.

The system was then analysed over the next 60 h and the model predictions compared with the plant data. Figures 17.16–17.18 show a comparison of model predictions with the plant data indicating a good prediction of the system dynamics for VFA, SCOD and Ph.

The model provides a reasonable basis for further detailed control and design studies.

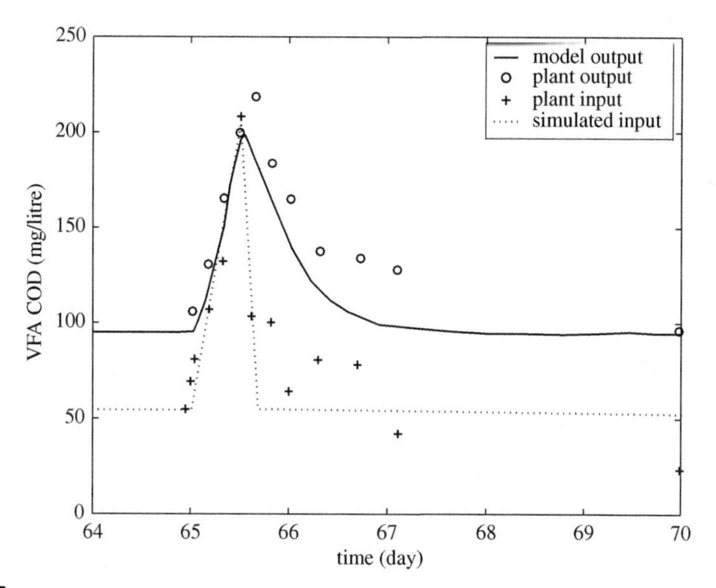

FIGURE 17.16 VFA prediction versus plant data.

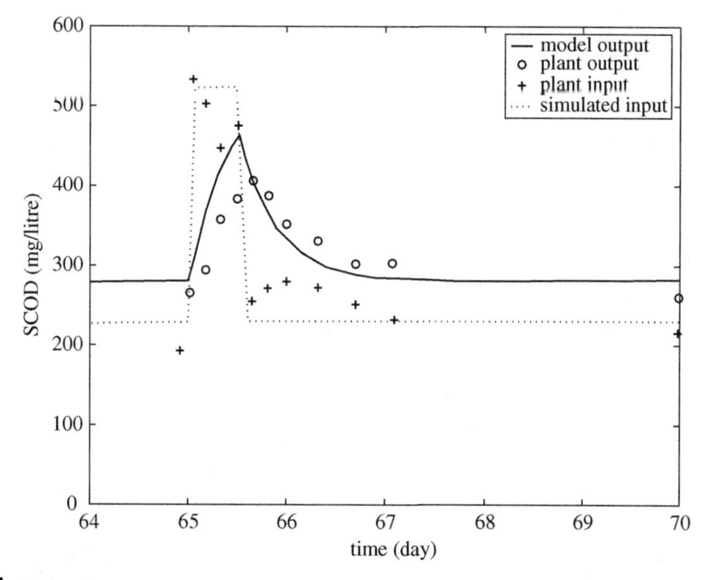

FIGURE 17.17 SCOD predictions versus plant data.

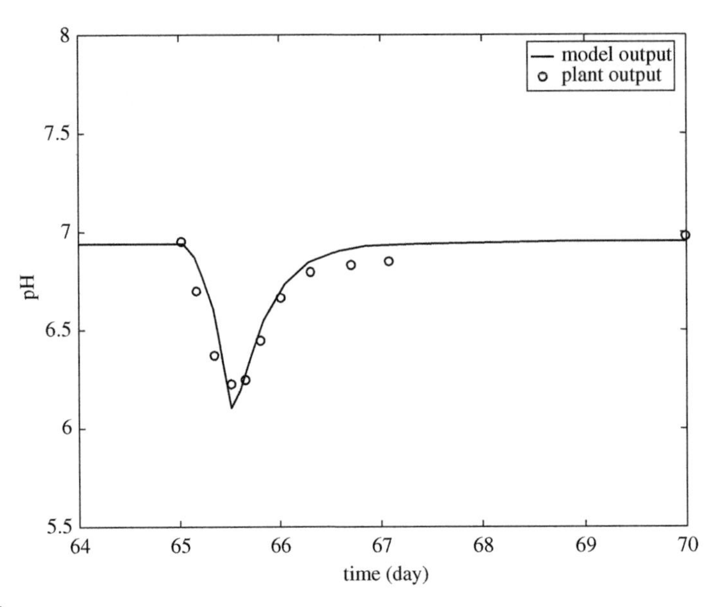

FIGURE 17.18 pH prediction versus plant data.

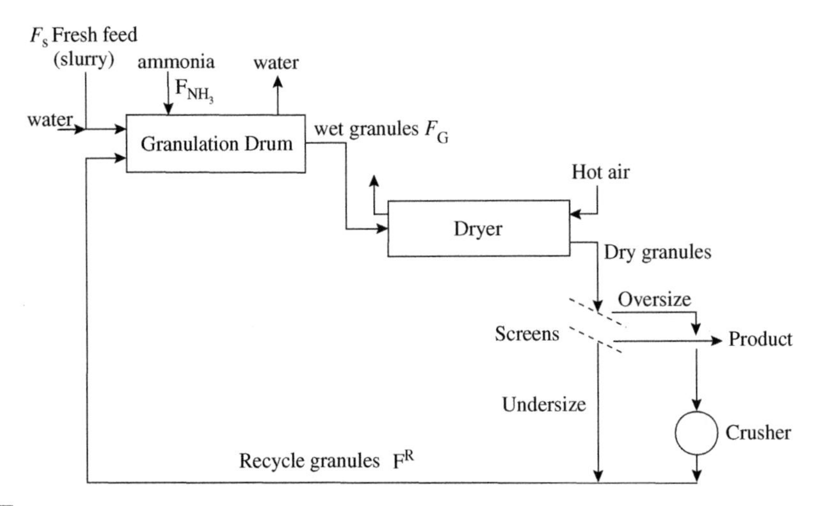

FIGURE 17.19 Typical fertilizer granulation circuit.

17.4. GRANULATION CIRCUIT MODELLING

This application considers the dynamic performance of a fertilizer granulation circuit for the industrial production of diammonium phosphate (DAP) fertilizers.

17.4.1. Process Description

The process represents a typical commercial fertilizer circuit design. This is shown by the flowsheet in Fig. 17.19. Feed slurry enters the granulation drum, where it mixes with recycled fertilizer granules which are normally in the undersize range. This recycle consists of either true undersize material or crushed oversize from the screening operation. Product from the granulator is fed to a high temperature dryer for moisture reduction and consolidation before passing through a series of screens which separate out the product size (3–5 mm diameter). Undersize goes directly into the recycle stream and oversize passes through a crusher then into the recycle stream for return to the granulator.

Operation of commercial granulation circuits is prone to problems of large variability and poor stability. Development of advanced control strategies is important for addressing productivity issues.

Modelling Goal

To predict the circuit dynamics in the face of disturbances in feed rate and moisture in order to test control schemes for improved circuit performance.

17.4.2. Controlling Factors

In this application, there are several controlling factors:

- granule growth within the granulator drum, which is both moisture and size dependent;
- particulate drying within the rotary dryer;
- screening of particles in vibratory screen system;
- hammer mill crusher performance for particle reduction;
- transportation delays in granulator, dryer and recycle systems.

17.4.3. Assumptions

The principal assumptions are:

$\mathcal{A}0$. No nucleation occurs in the slurry.
$\mathcal{A}1$. All remaining MAP reacts with ammonia fed to the granulator.
$\mathcal{A}2$. Particles only undergo coalescence due to surface liquid being present.
$\mathcal{A}3$. Layering occurs first in the drum then coalescence.
$\mathcal{A}4$. No particle breakage is considered.
$\mathcal{A}5$. Operating temperature along the drum is constant.

17.4.4. Model Development

The key unit in the flowsheet is the granulator. Here a number of mechanisms are important such as

- reaction
- layering
- coalescence

The reaction is exothermic with ammonia reacting with mono-ammonium phosphate (MAP) to form DAP and liberating water.

Layering occurs which increases particle mass but not particle number, whilst coalescence affects particle number.

Overall Mass Balance

Here slurry, recycle and fresh ammonia combine to give wet granules and water

$$F^{\mathrm{S}} + F^{\mathrm{R}} + \mathrm{F}^{\mathrm{f}}_{\mathrm{NH_3}} = F^{\mathrm{G}} + F^{\mathrm{evap}}_{\mathrm{water}}. \tag{17.29}$$

Granule Mass Balance

Fresh DAP includes DAP in the slurry plus the formation from MAP.

$$F^{\mathrm{N}}_{\mathrm{DAP}} = F^{\mathrm{S}}_{\mathrm{DAP}} + (F^{\mathrm{S}}_{\mathrm{MAP}} + F^{\mathrm{f}}_{\mathrm{NH_3}}). \tag{17.30}$$

The total DAP includes both fresh and recycled

$$F^{\mathrm{G}}_{\mathrm{DAP}} = F^{\mathrm{N}}_{\mathrm{DAP}} + F^{\mathrm{R}}_{\mathrm{DAP}} \tag{17.31}$$

Particle Population Balance

Using volume of particle as the internal co-ordinate it is possible to discretize the continuous volume domain into a geometric series as described by Hounslow *et al.* [128].

The particle population balance is given by

$$\frac{\partial N(v_i, t)}{\partial t} = -v\frac{\partial N(v_i, t)}{\partial z} + B_i^{(1)} + B_i^{(2)} - D_i^{(1)} - D_i^{(2)}, \tag{17.32}$$

where birth and death terms are

$$B_i^{(1)} = \sum_{j=1}^{i-2} 2^{j-i+1}\beta_{i-1,j}N_{i-1}N_j \tag{17.33}$$

birth in *i*th interval from coalescence of particle in $(i-1)$th interval with particles in first to $(i-2)$th interval

$$B_i^{(2)} = \frac{1}{2}\beta_{i-1,i-1}N_{i-1}N_{i-1} \tag{17.34}$$

birth in *ith* interval from two particles in $(i-1)$th interval

$$D_i^{(1)} = \sum_{j=1}^{i-1} 2^{j-1}\beta_{i,j}N_iN_j \tag{17.35}$$

death of particle in *ith* interval due to coalescence with particle in first to $(i-1)$*th* interval

$$D_i^{(2)} = \sum_{j=1}^{\infty} \beta_{i,j}N_iN_j \tag{17.36}$$

death in *ith* interval due to coalescence between particle in interval with a larger particle.

The coalescence kernel developed by Adetayo [138] was used where,

$$\beta_{i,j} = \begin{cases} a_1 S_{\mathrm{sat}} & t \le t_1, \\ a_2(v_i + v_j) & t > t_1 \end{cases} \tag{17.37}$$

Here a_1, a_2 are parameters from experimental data and t_1 is a switching time based on granulation mechanisms and S_{sat} is the fractional saturation of the granules.

Besides the granulator, other key equipment items in the circuit included a dryer, screens, crusher and transport conveyor.

Rotary Drier

In this case, a distributed parameter model of a drier was used based on the work of Wang *et al.* [139]. The assumptions included:

$\mathcal{A}6$. Air velocity is assumed constant along the drier length.
$\mathcal{A}7$. No particle breakage or granulation occurs in the drier.

Solid phase, air phase and moisture mass balances were developed, along with the solid phase energy balance and air energy balance. This was coupled with a drying rate model based on work by Brasil [140]. Heat transfer coefficients were computed using the modified work of Friedman and Mashall [141] as done by Brasil [140]. The solids retention time model, which takes into account drum angle, solid mean size, gas velocity and drum cross-section was used from Brasil [140].

The resulting equations were then solved by using orthogonal collocation to give a set of DAEs.

Mechanical Equipments (Screens, Crusher and Conveyor)

The screen model used in this work was based on Adetayo's [20] validated model as developed by Whiten [142]. This involves the use of a probability function which describes the probability of a particular particle size not passing through the screen. Hence, the oversize material in the ith size interval is given by

$$O_i = F_i Y_{d_i}, \tag{17.38}$$

where F_i is the mass flowrate of particles in *ith* size interval entering screen and Y_{d_i} is the probability that particles in *ith* size interval will not pass through the screen.

This value is a complex function of particle size, screen aperture size and mesh diameter.

The crusher model was based on a validated model of Lynch [143]. The assumptions include

$\mathcal{A}9$. Single fracture breakage occurs.
$\mathcal{A}10$. The breakage function is constant for a size interval.
$\mathcal{A}11$. The system is perfectly mixed.
$\mathcal{A}12$. Efficiency does not change with flowrate.

The product fraction P_i from the crusher is defined as

$$P_i = (1 - S_i)F_i + \sum_{j=1}^{i-1} B_{i,j} S_j F_j, \tag{17.39}$$

where $B_{i,j}$ is the breakage function, being the probability of a particle in the ith size interval being broken into particles less than the top size in the *ith* size interval.

The selection function S_i represents the probability that a particle in the ith size interval is selected for breakage. Below a certain size, S_{low} there is no selection. Above S_{upp} all particles are selected for breakage.

The final model is that for a conveyor system. In this case, the mass balance gives

$$\frac{\partial M}{\partial t} = -V_b \frac{\partial M}{\partial z}, \tag{17.40}$$

where M is the particle mass holdup per unit belt length, and V_{b} is the belt velocity. This model can be discretized using finite differences or collocation methods for solution.

17.4.5. Model Solution

The individual models were built into the library structure of the dynamic simulation and modelling system Daesim Studio [126]. The overall flowsheet included all processing units plus control equipment such as sensors, controllers and actuators.

The large, sparse DAE system of equations was then solved using BDF and DIRK methods.

17.4.6. Model Experimentation

The model was built with the purpose of examining the overall circuit dynamics, effects of drier and crusher characteristics and to examine possible standard or advanced control applications. Several case studies were carried out to assess control performance of various strategies.

The case studies included:

- Feedback only control using the water content of the feed slurry as the manipulative variable whilst using the fraction of oversize and undersize particles in the recycle for control purposes.
- Feedforward only control based on a simplified model of the granulator moisture as a function of the key stream properties.
- A combined feedforward–feedback controller which used a simplified process model for the feedforward part and feedback control to compensate for inaccuracies in the process model and unmeasured disturbances.

In the first case study, slurry flowrate was increased by 10%, whilst the $N:P$ ratio was decreased by 5%.

Figure 17.20 shows the predicted responses of the circuit to such changes and the performance of the control strategies. The increase in slurry flow, with water input initially at the steady operating conditions leads to a significant increase in undersize material in the recycle. The predictions show the relative performance of the three control strategies with superior performance demonstrated by the feedforward–feedback strategy. Significant time delays in the granulator, drier and recycle system contribute to the difficulty of circuit control.

The second study considered a simulated partial blockage in recycle from a time of 10 min to 60 min leading to a reduction in recycle flow of 10%. Figure 17.21 shows the responses of the under-size mass fraction in the recycle for various control

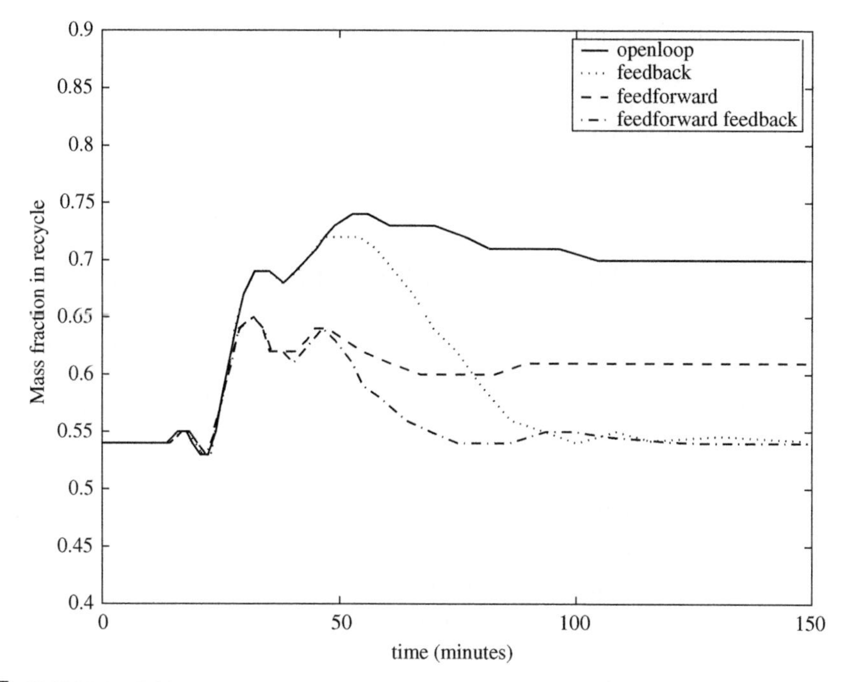

FIGURE 17.20 Dynamic transition of undersize fraction in recycle.

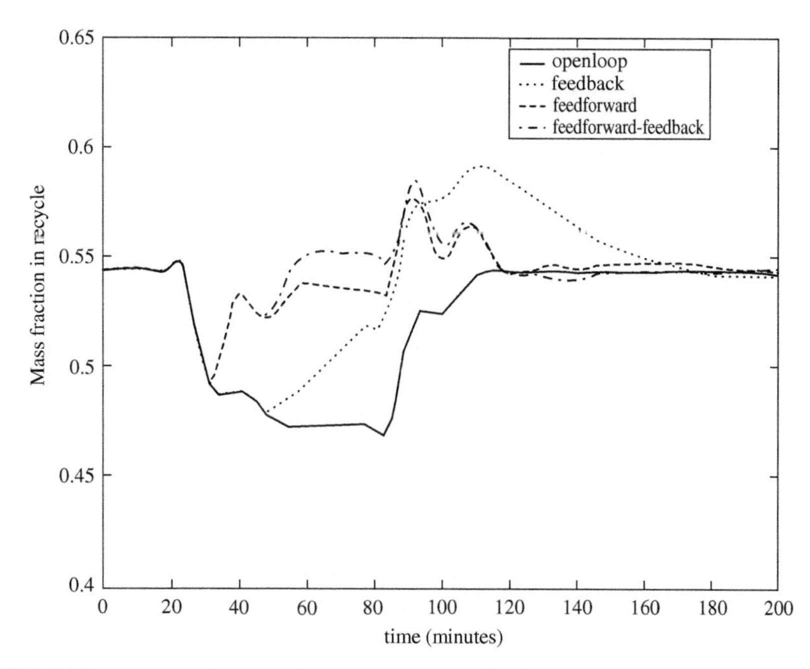

FIGURE 17.21 Dynamic transition of undersize fraction in recycle with simulated partial blockage, in recycle from 10 to 60 min.

strategies. Again, the feedforward-feedback strategy performs better than the simpler control approaches such as feedback or purely feedforward.

The study was able to identify a number of promising strategies for circuit control which considered a range of potential control and manipulated variables. The model has also been used to study the application of model predictive control (MPC) to the circuit.

17.5. INDUSTRIAL DEPROPANIZER USING STRUCTURAL PACKING

This example discusses the modelling of an industrial depropanizer as part of an oil refinery. In this case, the tower uses a structured packing and the modelling leads to a distributed parameter system. The model uses a rate based approach in developing the dynamic description of the process [144,145].

17.5.1. Process Description

Depropanizers are often used in refinery processing. In this case, the column is an integral part of an alkylation unit. Here, the modelling is required to help derive improved control under the influence of large disturbances on the system.

17.5.2. Controlling Factors

Key factors in this system are

- the use of structured packing (Sulzer Mellapak 250/300) in top and bottom sections of the column;
- mass and heat transfer in packed column operations;
- importance of major holdups in the system including the reboiler and reflux drum;
- the number of individual components to be considered in the model versus plant.

17.5.3. Assumptions

The main assumptions include:

$\mathcal{A}1$. The vapour phase holdup is regarded as negligible.
$\mathcal{A}2$. No significant heat losses from the column or associated equipment.
$\mathcal{A}3$. Simplified Stefan–Maxwell equations are applicable and represented by Wilke's equation.
$\mathcal{A}4$. Mass and heat transfer represented by correlations.
$\mathcal{A}5$. Pressure profile is considered constant.
$\mathcal{A}6$. Thermal capacity of the column and packing are negligible.
$\mathcal{A}7$. No chemical reactions occur.
$\mathcal{A}8$. No variation in the radial direction.
$\mathcal{A}9$. Assume simple relationship between liquid flow and holdup.
$\mathcal{A}10$. Assume liquid–vapour interface is at equilibrium for the two-film model.
$\mathcal{A}11$. No significant heat of mixing in the system.
$\mathcal{A}12$. Vapour phase controls the mass transfer rate.

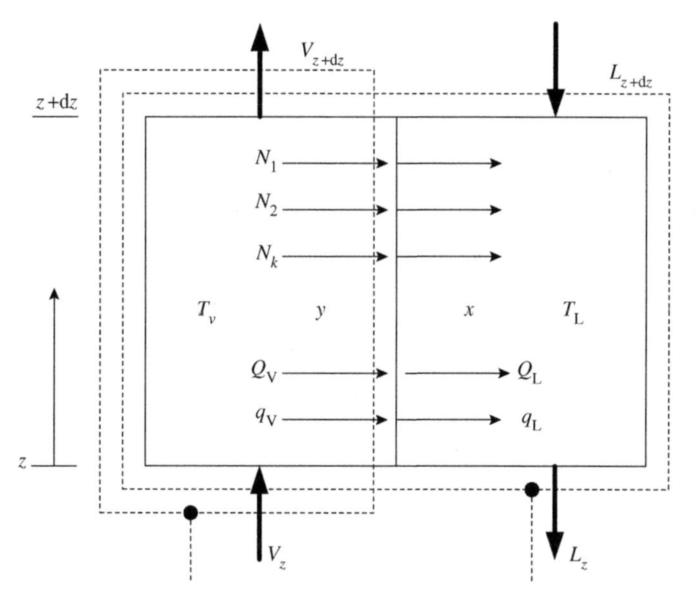

FIGURE 17.22 Two film model of a packed column.

17.5.4. Model Development

The model development included the following key items:

- a model for a packed column section,
- a model for the feed section,
- a model for the reflux drum,
- a model for the reboiler.

Mass and energy conservation equations were written for each key item. In the case of the packed column section, this involved a distributed parameter model for both mass and energy. In the other units, the model was regarded as a lumped parameter system.

Packed Column Sections

Overall mass balances A typical two film model, shown in Fig. 17.22 was used as the basis for the conservation balances. An overall mass balance, around balance volume 1 was performed as well as a vapour phase balance over volume 2 for molar holdups of H_V and H_L.

This leads to

$$\frac{\partial H_V}{\partial t} + \frac{\partial H_L}{\partial t} = \frac{\partial L}{\partial z} - \frac{\partial V}{\partial z} \tag{17.41}$$

or

$$\frac{\partial H_L}{\partial t} = \frac{\partial L}{\partial z} - \frac{\partial V}{\partial z}, \tag{17.42}$$

where the vapour phase holdup (H_V) is ignored.

For the vapour phase, we obtain

$$\frac{\partial H_V}{\partial t} = -\frac{\partial V}{\partial z} - \sum_{k=1}^{c} N_k \tag{17.43}$$

or

$$0 = -\frac{\partial V}{\partial z} - \sum_{k=1}^{c} N_k, \tag{17.44}$$

when the vapour holdup is negligible. Here, N_k is the molar flux of component k between vapour and liquid phases, A is the interfacial area of the packing and S is the column.

Component balances Component balances for $C-1$ components were also needed. These were written for both liquid and vapour phases as

$$\frac{\partial}{\partial t}(H_L x_k) = \frac{\partial}{\partial z}(L x_k) - \frac{\partial}{\partial z}(V y_k) \tag{17.45}$$

and

$$0 = -\frac{\partial}{\partial z}(V y_k) - N_k \tag{17.46}$$

which neglects vapour holdup.

Energy balance The overall energy balance is obtained from balance volume 1 as

$$\frac{\partial}{\partial t}(H_L h + H_V H) = \frac{\partial}{\partial z}(Lh) - \frac{\partial}{\partial z}(VH) \tag{17.47}$$

or

$$\frac{\partial}{\partial t}(H_L h) = \frac{\partial}{\partial z}(Lh) - \frac{\partial}{\partial z}(VH) \tag{17.48}$$

by neglecting vapour holdup.

The vapour phase energy balance over volume 2 gives

$$0 = -\frac{\partial}{\partial z}(VH) - Q_V - q_V$$

which ignores the vapour holdup and also accounts for the convective heat transfer to the liquid phase (Q_V) and the conductive heat transfer (q_V).

Noting that there is no accumulation of energy at the interface we can write

$$Q_L - q_L = Q_V - q_V \tag{17.49}$$

so that the final energy balance is

$$\sum_{i=1}^{C} N_i H_i + q_V = \sum_{i=1}^{C} N_i h_i + q_L \tag{17.50}$$

Constitutive Equations

There are many constitutive equations required in the model which relate to rate expressions, property definitions and equilibrium relations. The key constitutive equations include:

Mass transfer This can be defined as

$$N_k = (K_{OG})_k A \cdot S \cdot (y_k - y_k^*), \tag{17.51}$$

where K_{OG} is the overall gas phase mass transfer coefficient which can be defined using the Onda [146] correlations and the Bravo, Rocha and Fair correlations [147] or from Speigel and Meier [148] with

$$\frac{1}{(K_{OG})_i} = \frac{1}{k_{i,\mathrm{eff}}^{\mathrm{V}}} + \frac{m_i}{k_{i,\mathrm{eff}}^{\mathrm{L}}} \tag{17.52}$$

and the effective liquid and vapour phase coefficients being given using the previous correlations.

Heat transfer Heat transfer coefficients were estimated using the Chilton–Colburn analogy, such that:

$$h^{\mathrm{V}} = k_{\mathrm{av}}^{\mathrm{V}} C_p^{\mathrm{V}} (Le^{\mathrm{V}})^{2/3}, \tag{17.53}$$

$$h^{\mathrm{L}} = k_{\mathrm{av}}^{\mathrm{L}} C_p^{\mathrm{L}} (Le^{\mathrm{L}})^{2/3}, \tag{17.54}$$

where

$$k_{\mathrm{av}}^{\mathrm{V}} = \sum_{i=1}^{\mathrm{C}} y_i k_{i,\mathrm{eff}}^{\mathrm{V}} \tag{17.55}$$

$$k_{\mathrm{av}}^{\mathrm{L}} = \sum_{i=1}^{\mathrm{C}} x_i k_{i,\mathrm{eff}}^{\mathrm{L}} \tag{17.56}$$

The convective and conductive heat transfer terms are given by

$$Q_{\mathrm{V}} = \sum_{k=1}^{\mathrm{C}} N_k H_{\mathrm{k}}, \tag{17.57}$$

$$Q_{\mathrm{L}} = \sum_{k=1}^{\mathrm{C}} N_k h_{\mathrm{k}}, \tag{17.58}$$

$$q_{\mathrm{V}} = h^{\mathrm{V}} AS(T_{\mathrm{V}} - T_{\mathrm{i}}), \tag{17.59}$$

$$q_{\mathrm{L}} = h^{\mathrm{L}} AS(T_{\mathrm{i}} - T_{\mathrm{L}}) \tag{17.60}$$

Phase equilibrium To compute the interfacial equilibrium compositions y_k^*, we use

$$y_k^* = \frac{\gamma_k \mathcal{F}_k P_k^{\mathrm{vap}}}{P} x_k, \tag{17.61}$$

where γ_k is the activity coefficient, $\mathcal{F}_k$ the fugacity, P_k^{vap} the vapour pressure and P is the system pressure.

Property definitions Simplified enthalpy expressions were used for liquid and vapour phases. The component enthalpies were simple polynomials in temperature

$$h_k = a_k T^2 + b_k T + c_k, \tag{17.62}$$

$$H_k = d_k T^3 + e_k T^2 + f_k T + g_k \tag{17.63}$$

with the overall enthalpies given by

$$h = \sum_{k=1}^{C} x_k h_k, \tag{17.64}$$

$$H = \sum_{k=1}^{C} y_k H_k. \tag{17.65}$$

Liquid holdup A liquid holdup relation was used to relate H_L to packing parameters (p), liquid flowrate and vapour properties such that

$$H_L = f(L, p, \text{ vapour properties}). \tag{17.66}$$

This was based on work by Billet [149] and Bravo [150].

Other thermo-physical properties Many other property predictions for phase viscosity, heat capacity, diffusivities, surface tension and thermal conductivities are also needed. Exact expressions are given in Sadeghi [145].

17.5.5. Model Solution

The model was coded as several submodels within the Daesim Studio modelling and simulation system. This consisted of:

- overhead condenser
- reflux drum
- rectification and stripping sections
- feed section
- reboiler
- control loop elements

In order to solve the PDE models representing the structured packing sections, orthogonal collocation on finite elements was used. This converted the PDEs to a set of ODEs with appropriate boundary conditions at each end of the sections.

The collocation approach allowed the user to select the number of elements and the order of the orthogonal polynomials in each section.

Individual models were then assembled graphically into a final flowsheet and the large sparse DAE set was solved using BDF and DIRK numerical methods.

17.5.6. Process Experimentation and Model Performance

An industrial depropanizer was used as a case study for the modelling. Key details for the column are given in Table 17.2.

The steady-state predictions of the model were compared against the plant data given in Tables 17.2 and 17.3 with the results given in Table 17.4.

These results show good agreement for propylene compositions at top and base of column but some overprediction of iso-butane at the top and under prediction at the base of the column.

The intention of the model was to examine the system dynamics and so a number of simulation tests were done to assess the dynamic performance of the model against qualitative data from the plant. Quantitative time series of major compositions for feed, overhead and bottoms products were not available from the process.

Figures 17.23 and 17.24 show the dynamic behaviour of the column under a series of reflux rate changes indicating the propylene and iso-butane compositions in the bottom of the column as a function of time.

Of importance here is the inverse response of the system due principally to differing underlying responses of the fluid dynamics and the composition dynamics.

The model was used to study optimal control policies for such a system.

TABLE 17.2 Parameters of the depropanizer column

Packing material	Mellapak type	Mellapak 250 Y
	Specific area	$250\,m^{-1}$
	Void fraction	0.95
	Corrugation angle (θ)	45°
	Crimp height (h)	0.0119 m
	Corrugation side (S)	0.0171 m
	Corrugation base (B)	0.0241 m
	Operating pressure	16.1 atm
Rectifying section	Diameter	0.91 m
	Height	4.08 m
Stripping section	Diameter	0.91 m
	Height	10.35 m
Condenser	Surface area per shell	$250.1\,m^2$
	No. of shells	2
	Nominal heat duty	4.23×10^6 kJ/h
	Heat transfer coefficient	3.02×10^3 kJ/h m^2K
	O.D. of tube	19.05 mm
	Length of tube	6.10 m
	No. of tube/shell	697
Reflux drum	I.D.	1.07 m
	Length	4.72 m
	Distillate rate	108.9 kg mol/h
Reboiler	Nominal heat duty	36.94×10^6 kJ/h
	Nominal liquid holdup	28.3 kg mol

TABLE 17.3 Steady-state data of the depropanizer column

Item	Property	Value
Feed condition	Vapour flow rate (V_f)	25.4 kg mol/h
	Liquid flow rate (L_f)	8.555 kg mol/h
	Temperature ($T_{vf} = T_{lf}$)	344.2 K
Liquid composition (x_{lf})	Ethane (C_2H_6)	0.005
	Propane (C_3H_8)	0.075
	Propylene (C_3H_6)	0.190
	Iso-butane (C_4H_{10})	0.730
Vapour composition (y_{lf})	Ethane (C_2H_6)	0.026
	Propane (C_3H_8)	0.119
	Propylene (C_3H_6)	0.354
	Iso-butane (C_4H_{10})	0.501

TABLE 17.4 Steady-state composition results

Location	Simulation		Plant	
	Propylene	iso-Butane	Propylene	iso-Butane
Base reflux				
Top	0.693	0.038	0.698	0.010
Base	0.103	0.851	0.095	0.870
+20% reflux				
Top	0.711	0.018	0.708	0.002
Base	0.087	0.865	0.088	0.876

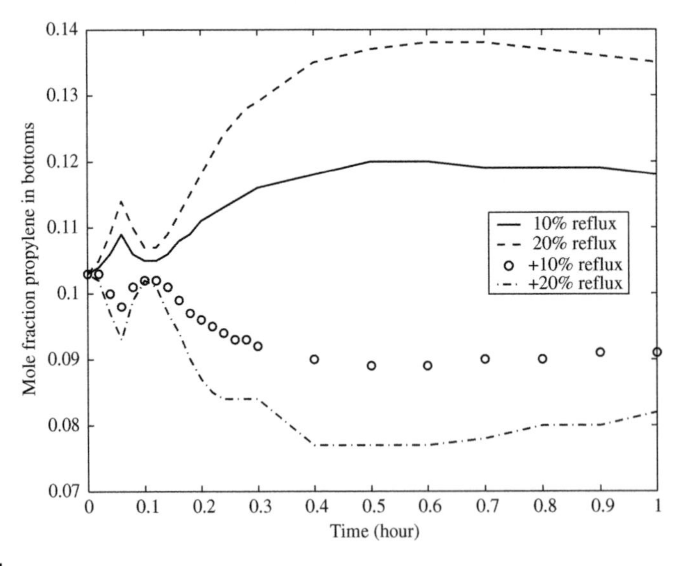

FIGURE 17.23 Propylene composition in column bottoms.

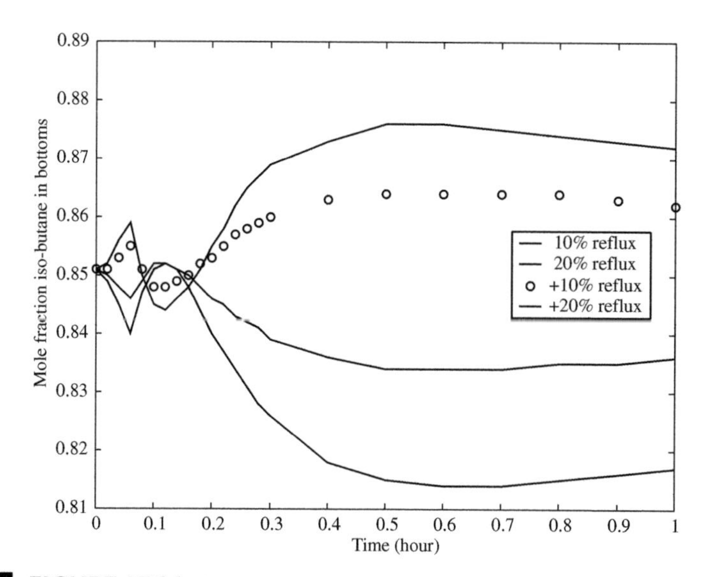

FIGURE 17.24 Iso-butane compositions in column bottoms.

17.6. SUMMARY

This chapter has presented some examples of industrial modelling practice. The models displayed in the chapter cover a range of process applications and also model characteristics. What is evident in these examples is the often difficult problems in obtaining good plant data for calibration and validation purposes as well as the need to make simplifying assumptions in order to make the modelling process tractable. Inevitably in the modelling of real world problems for a range of applications, the required model fidelity needs to be assessed so that there is a clear endpoint to the modelling effort.

18 COMPUTER AIDED PROCESS MODELLING

This chapter discusses the general principles behind the development of computer based model building environments—or computer aided process modelling (CAPM) systems. Recognition is given to the current range of process flowsheeting tools but the main emphasis is on established or recently developed tools specifically aimed at effective process model development. We review a number of approaches proposed by various workers in this field as well as assess the scope and applicability of the current generation of CAPM tools. This has been an important research and development area for almost two decades with earlier seminal ideas dating back almost 40 years [53]. Even after such a long gestation period, only a few CAPM systems are in regular industrial use.

The chapter is arranged as follows:

- An overview of the importance and scope of computer aided modelling is given in Section 18.1.
- In Section 18.2, we highlight recent industrial input into the future development of CAPM tools and also extend those issues through discussion of other practical, pertinent and necessary developments to make the tools useful in the widest industrial modelling context.
- In Section 18.3, we discuss the basic ideas underlying any computer aided modelling environment pointing out the importance of these for modellers carrying out the modelling of new or non-traditional systems.
- In Section 18.4, we review a number of representative systems which provide computer aided facilities for modelling both general systems and specifically process systems.

18.1. INTRODUCTION

In the earlier chapters in the book it has been pointed out that process modelling is a fundamental activity underlying the effective commercialization of process ideas and the ongoing production of goods and services. It has found great utility in the design, control and optimization of process systems, and more recently, in the general area of risk assessment. In carrying out the necessary simulation studies, most process engineers make use of canned models offered in commercial simulation packages such as *Aspen Plus*, *Aspen Dynamics* [30,151] or *Hysys* [152] to generate solutions to process engineering problems. Sometimes, customization of existing models is needed and in some cases completely new models must be developed. Tools such as *Aspen Custom Modeler* provide extensions to commercial flowsheet simulators or the ability to link to existing models written in C or Fortran. In most cases the commercial flowsheeting tools do not adequately support the user who wishes to build, test and document new models based on a fundamental understanding of the chemistry and physics of the system. Some professional process modellers most often write stand-alone applications in C/C++, Fortran, Java or make use of other tools such as MATLAB and Simulink [153] to develop process applications.

The improvement and increasing importance of computational fluid dynamics (CFD) packages has provided a computer based modelling approach at the most complex end of the modelling spectrum, whilst general process model building tools still remain largely an academic curiosity. The CFD tools are extremely important for resolving complex fluid flow, heat and mass transfer problems in process vessels or in large-scale environmental systems. Their utility in these areas will inevitably be coupled with standard flowsheeting tools to span the complete spectrum of process modelling.

In this chapter, we view CAPM as an approach which provides an environment that aids in model building, model analysis and model documentation. It guides the modeller in developing the model, based on an understanding of the chemistry and physics of the system, rather than simply forcing the user to write and code equations into C functions, Java classes or Fortran subroutines. This naturally limits the scope of the tools discussed in this chapter. Commercial flowsheeting packages still remain the backbone of process engineering simulation but effective modelling tools are essential to expand the range and depth of the applications into new and advanced areas.

In what follows, we consider the basic ideas behind CAPM systems, the key approaches to structuring the modelling problem and specific instances of these approaches. The chapter does not give a complete review of all available systems but covers a number of the currently available environments and uses them for illustrative purposes. For a more complete review, readers are referred to the work of Marquardt [154], Jensen [155], and Foss *et al.* [156].

18.2. INDUSTRIAL DEMANDS ON COMPUTER AIDED MODELLING TOOLS

Ultimately, the key users of process modelling tools are the process industries. When we consider modelling in the process industries, there are a small number of very mature areas and a huge number of potential users. The petroleum and petrochemical industries have dominated the development of flowsheeting and modelling tools. However, significant advances in providing the benefits of process modelling have been made in the minerals processing sector [157] and the environmental engineering area. The uptake of these tools and their widespread use has often been hampered by the difficulty of using the canned models in particular applications. The lack of tools to tackle process modelling at a level higher than the equation systems, where enormous amounts of time, effort and money are expended, has often been an excuse for not taking up process model based applications.

In a useful review, Foss *et al.* [156] carried out an industrial study of modelling practice within some major European process companies. From a practitioner's viewpoint, the key requirements for computer aided model building systems were identified as:

1. Models should not be regarded as simply equation sets but incorporate such facets as the underlying modelling assumptions, the limitations of the models and a record of all decision making leading to the model.
2. The need to support the development of model families and the documentation which goes with the models.
3. A need to perform the modelling at a phenomenological level rather than at the equation level.
4. A need to have modelling "know-how" for use in guiding the model development.
5. Support for the reuse and subsequent modification of existing models.
6. A library of predefined model building blocks of fine granularity to be available for model building.
7. Model building automation for certain parts of the model building process including such areas as equation generation, documentation and report generation.

These attributes were regarded as vital to a fully functional system to aid in model development and generation. These attributes link logically into the overall modelling methodology outlined in Chapter 2. There are further issues which also need attention in the development of CAPM tools, including:

1. model calibration and validation, including a plant data repository along with the developed models;
2. the structured development and documentation of hybrid discrete–continuous models;
3. the ability to develop, analyse and solve large-scale models which integrate distributed, lumped and population balance systems;
4. model sharing and standardization of model, properties and solution interfaces in line with *CAPE-OPEN* initiatives [158];
5. business-to-business (B2B) CAPM environments utilizing the advantages of internet delivery.

In the following sections, we consider these issues, pointing out the necessary components of a full-fledged CAPM environment and assessing how the principal tools have tackled the issues.

18.3. BASIC ISSUES IN CAPM TOOLS

The previous section has briefly outlined some of the key attributes needed to be addressed in the development of effective CAPM tools. What then are the more basic, underlying issues which need to be considered? In this section, an outline is given of the key concepts underlying CAPM tools in terms of the modelling, model analysis, modelling workflow and documentation.

18.3.1. Modelling Structures

In any model building tool, there must be a clear definition of the underlying building blocks to be used. Any application problem must be decomposed into basic or canonical modelling objects and the definition of those objects should be unambiguous. For any modelling exercise the holdup of mass, energy and momentum are the key conserved extensive quantities in the identified balance volumes. There is also the recognition that convective flows or diffusive fluxes must be part of the system description. These flows transfer extensive quantities under the influence of a driving potential. They do not store extensive quantities but simply transfer from one balance volume to another. Hence, there are essentially two basic entities represented by the balance volumes and system flows.

Figure 18.1 shows the conceptualization of a continuous stirred heated tank in terms of three balance volumes (Σ), three diffusive fluxes (Q) and four convective flows. Other possibilities can also be considered.

There are a number of CAPM environments which tackle the modelling problem at the fundamental level by defining a clear set of structured objects as the basis for any model development. Different terms are used in this systems for equivalent modelling objects. In the *Modeller* system [159], for example, these objects are termed SYSTEMS and CONNECTIONS, *ModKit* [160], uses the terms DEVICES and CONNECTIONS, whereas *ModDev* [155] refers to these objects as SHELLS and CONNECTIONS. These constitute the canonical modelling objects for the particular system.

One of the earliest comprehensive representations of modelling objects was done by Stephanopoulos and co-workers as the basis for the *Model.la* environment [161].

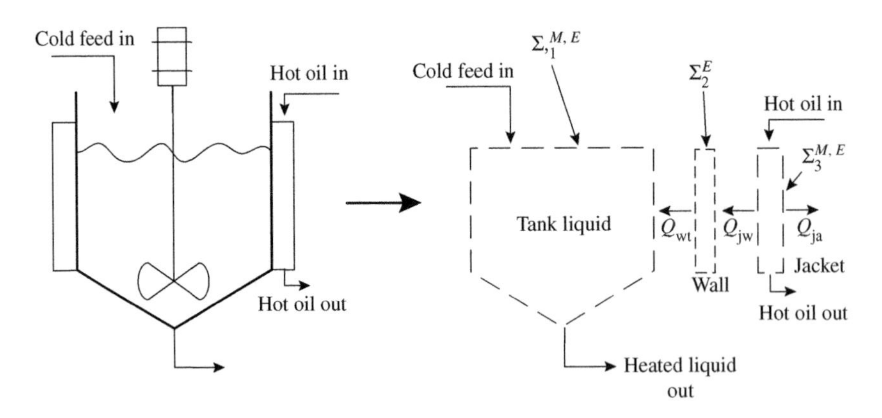

FIGURE 18.1 Heated tank and key modelling objects.

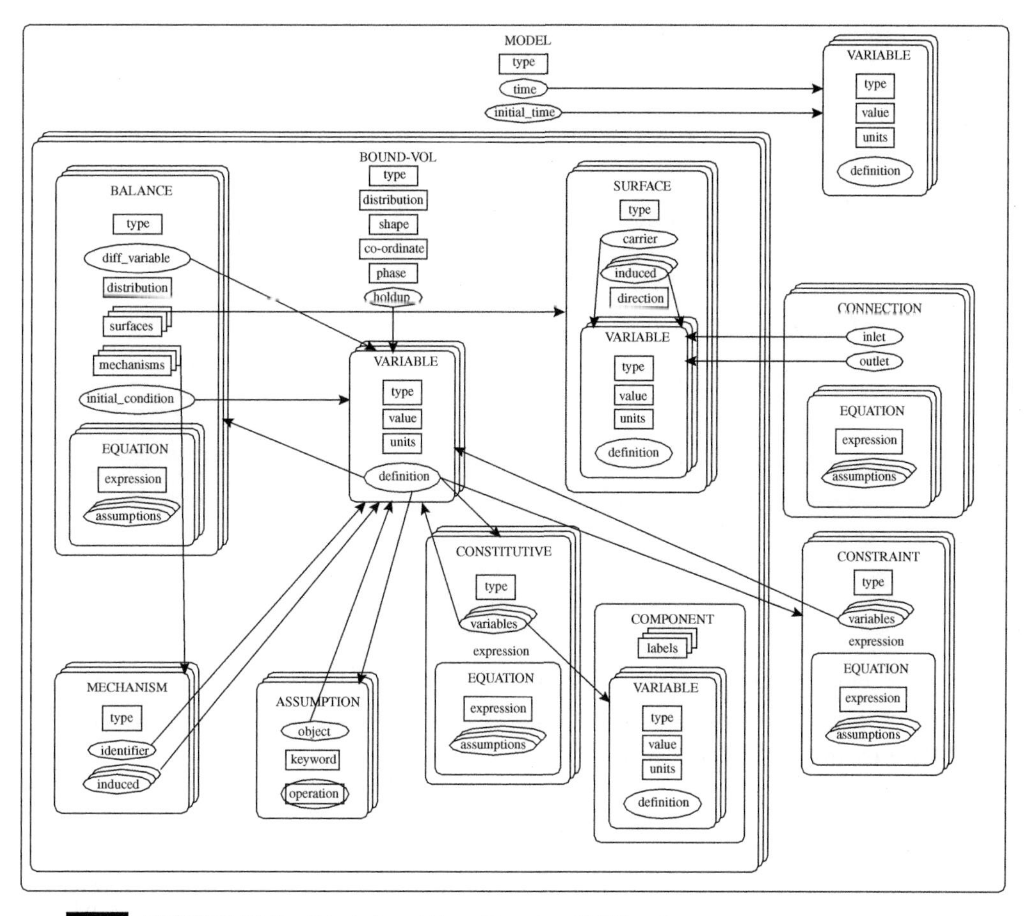

FIGURE 18.2 Principal modelling objects in a CAPM environment.

Other work has been done by Marquardt [160] as the basis for the VEDA and VDDL language development within *ModKit*. A particular representation of the key objects in process modelling is seen in Fig. 18.2 as developed by Williams *et al.* [162]. This shows the key modelling objects, their interrelationships and their attributes as the basis for an object oriented model building environment implemented in Java.

18.3.2. Structural Characterization

It is not only necessary for CAPM tools to define principal modelling objects but those objects must also be characterized by giving their specific properties or attributes. It is important that the properties of balance volumes, flows and fluxes are specified. CAPM tools must provide facilities for the user to define such properties to produce a specific instance of a general class of modelling object. These class specific attributes must be defined clearly within the modelling system.

Figure 18.3 shows a dialogue box within the *ModDev* system which helps characterize the behaviour of a balance volume ("shell") used to represent a liquid phase

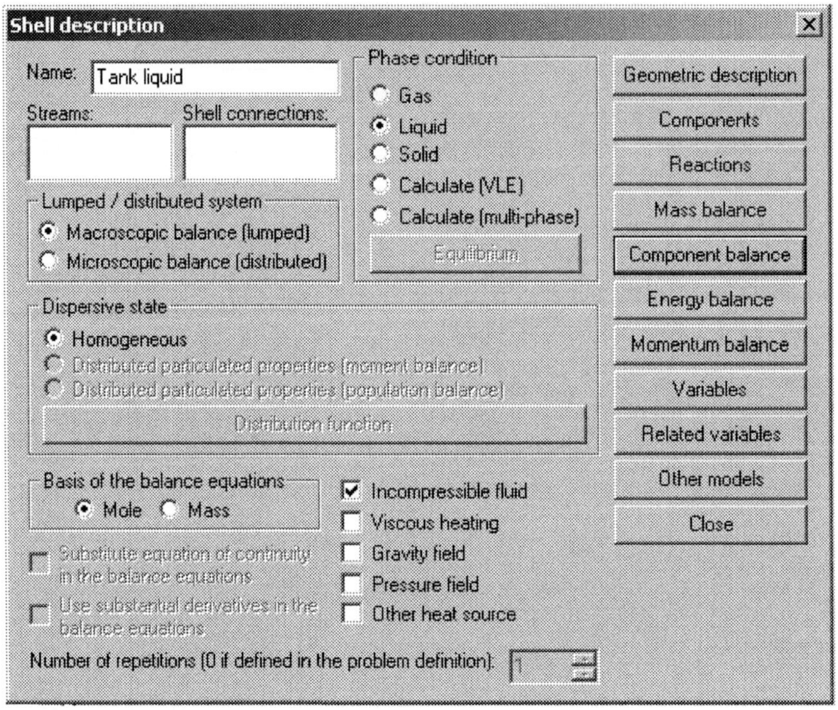

FIGURE 18.3 Setting attributes of a balance volume ("shell") object.

within a tank. The dialogue facility provides opportunity to customize the particular instance of a balance volume by addressing such issues as:

- the spatial variation within the system as either being distributed or lumped,
- the geometric description of the system balance volume,
- the dispersive state of the balance volume as being homogeneous or having dispersed particulate character,
- the possible phase condition,
- the type of reactions taking place,
- the presence of total, component mass balances as well as energy or momentum balances.

Hence, the model builder can tailor the generic building block to reflect the perceived characteristic of the balance volume.

In an alternative approach taken by Williams *et al.* [162] the balance volume attributes are developed in text form based on a hierarchical definition of process modelling assumptions in line with the key modelling objects and attributes shown in Fig. 18.2. This is typical of the facilities available in a wider range of CAPM tools.

The importance of the specific instance of a modelling object is that it can then be used as a basis for automatic generation of the governing conservation equations and constitutive relations.

18.3.3. Multi-level Modelling Hierarchies

One of the important aspects of computer aided modelling is the need for model representation on various levels of granularity. This could be at the overall flowsheet level, unit level or equipment level. The general idea is seen in Fig. 18.4, which shows a particular hierarchy of model representation. In this figure, we see system decomposition from the flowsheet level down to the reactor bed level, and aggregation in the opposite direction. The decomposition could continue down to the individual catalyst pellet level, where the fundamental behaviour could be described in terms of a distributed parameter model. These low level building blocks provide the basis for the modeller to aggregate the lower level systems up to the point of analysis needed for the application. The ability to reuse the fundamental building blocks in a variety of ways provides powerful facilities for process modelling.

Of course, this aspect of modelling large-scale systems has been available in commercial flowsheeting packages for decades. However, in the area of CAPM this concept allows the user to develop building blocks at much lower levels and then aggregate them into higher level structures as required. A common petrochemical example might be the model of a distillation column tray which can be re-used and aggregated into a model of a distillation column. Many systems incorporate such multi-level modelling hierarchies. A particular example illustrating this concept is

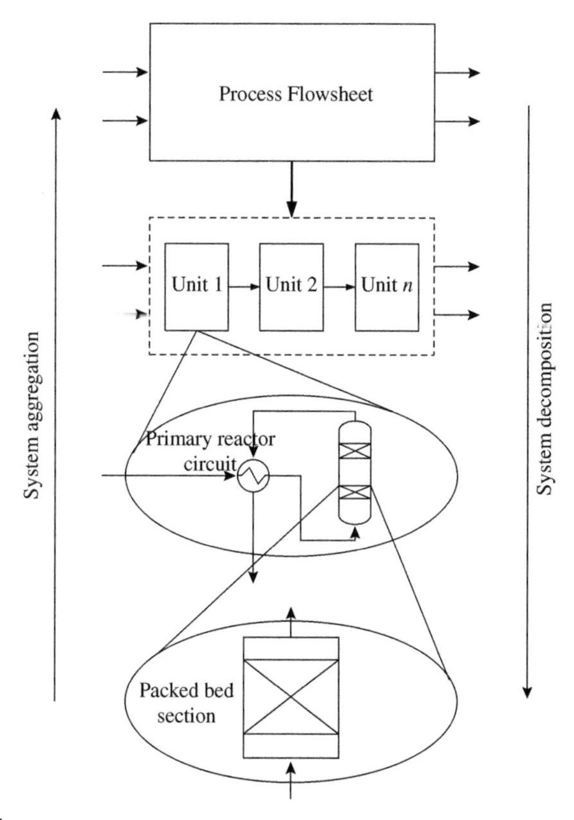

FIGURE 18.4 Multi-level modelling hierarchy.

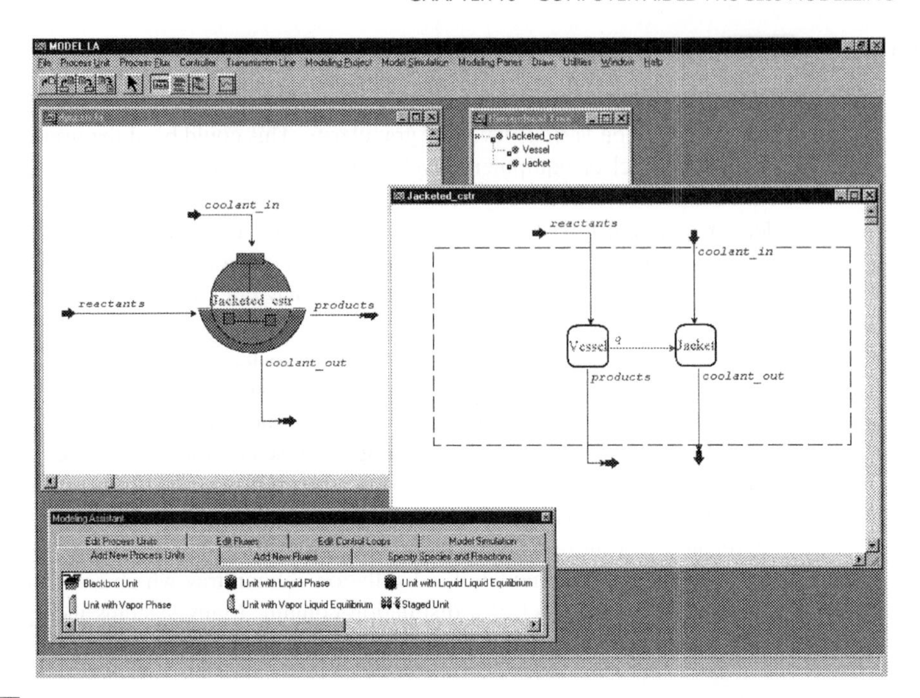

FIGURE 18.5 Hierarchical model development in *Model.la.*

seen in Fig. 18.5, which shows the hierarchical decomposition of a CSTR within the *Model.la* environment. In this case, the left pane shows the overall CSTR model being built, showing the incident convective flows into and out of the reactor and the cooling jacket. The right-hand pane shows the extended view of the reactor vessel, clearly made up of two balance volumes—one for the vessel contents and the other for the cooling jacket. At this level the conductive energy flux between the two principal balance volumes can be seen and described. Further hierarchical decomposition of the system is possible.

The aggregation of such fundamental building blocks from the finest granularity to the upper levels of the hierarchy can be achieved either through textual means as done in *gPROMS* or graphically as can be seen in *Model.la, ModDev* or *ModKit.* Figure 18.6 shows the hierarchy tree describing the aggregation of the fundamental buildings blocks of `vessel` and `Jacket` along with the jacket properties to describe the CSTR model within *Model.la* as already given in Fig. 18.5.

18.3.4. Multi-scale Modelling Hierarchies

In contrast to the multi-level aspects of computer aided modelling, multi-scale modelling refers to the ability to capture within the model varying levels of behaviour or characteristics. The ability to move seamlessly up and down a multi-scale model hierarchy is an important area for model development. This ability to develop models on multi-scales often reflects the stage of a project or the intended application of the

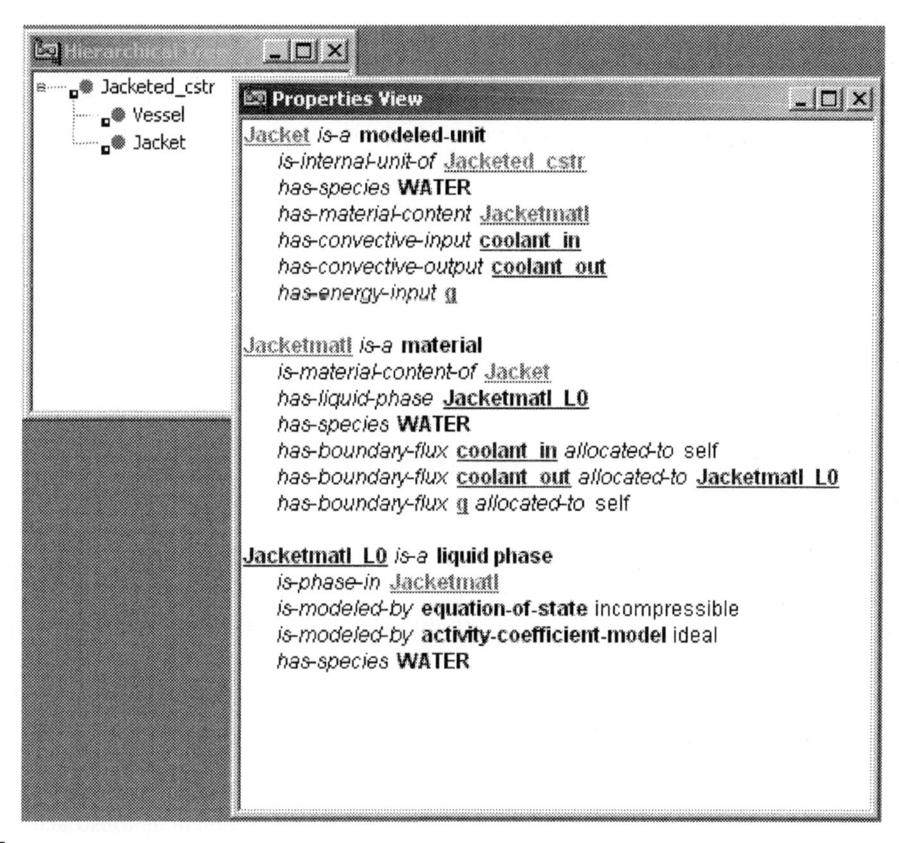

FIGURE 18.6 Model hierarchy tree as used in *Model.la*.

model. It still remains a major challenge for CAPM tools. The hierarchy of multi-scale modelling as seen in Chapter 9 can incorporate a number of dimensions that include:

- timescales, where different characteristic times are incorporated into a hierarchy of models;
- dimension scales, where the model hierarchy captures phenomena at the nano, micro, meso or macroscale;
- complexity scale, where the variation between models in the hierarchy reflect the incorporated complexity in the model in terms of the underlying physics and chemistry.

We can view this issue as a 3D multi-scale modelling space for a model whose spatial attributes (distributed or lumped) have already been determined. Hence, a whole family of models then exists determined by the multi-scale assumptions associated with that model.

In most circumstances, there do not exist environments which help model building in a general multi-scale environment apart from those where choices relating to the importance of spatial distribution are incorporated in the modelling object attributes.

In these circumstances, the spatial distribution can be chosen as well as the underlying geometry to generate models related through their degree of spatial distribution, from 3D to fully lumped systems. Other attributes such as time scales, complexity and dimension scales must normally be introduced manually by the user.

18.3.5. Model Analysis and Solution Tools

The generation of process equations from the fundamental understanding of the chemistry and physics is just one step in a much larger set of modelling tasks. It is often not the most difficult task to undertake. There are, however, key issues that must be addressed that are related to posing the problem in a final form which is solvable. These issues include:

- selection of the DOF for the problem,
- the assessment of the index of the resulting differential-algebraic system and tools to treat high index problems,
- the presence of discontinuous behaviour, either by way of time or state events,
- the treatment of distributed parameter models and their numerical solution,
- the treatment of integro-differential systems which may arise in population balance modelling,
- the provision of solution diagnostic tools for problems encountered during the numerical solution.

A lot of effort has been expended in this area to address these issues. Many CAPM tools incorporate interactive analysis of the DOF problem as illustrated in Fig. 18.7 where the model test-bed (*MoT*) from the Integrated Computer Aided System (*ICAS*) can be seen. Here the variables are classified and DOF satisfied through choices which are analysed for index issues.

These tools provide a visual, interactive means of analysing the equation-variable patterns in order to choose appropriate variables to satisfy the DOF or to investigate equation ordering to enhance solution methods. Structural analysis methods can be found in various forms in *ModKit*, *Model.la*, *gPROMS*, *ASCEND* and other CAPM environments.

In most cases, the diagnostic capabilities of most simulation and CAPM systems are very poor. Expert knowledge is often needed to debug intransigent solution problems and little effective help is available to pinpoint the main problem areas. This is an area that can be extremely important when problems arise—and they inevitably will!

18.3.6. Model Documentation and Reporting Systems

Model documentation is one of the key requirements for effective CAPM environments and probably one of the least developed. The importance is seen in the following issues:

- the systematic recording of all the underlying assumptions which are made during the course of model development;

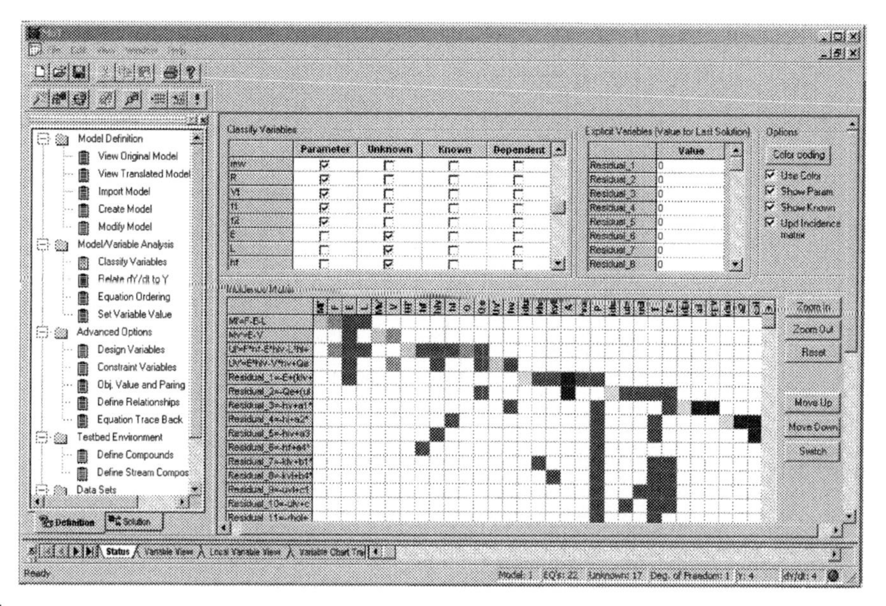

FIGURE 18.7 Interactive model equation-variable selection system from the *ICAS* model test-bed.

- the documentation of the decisions or reasoning made in arriving at a particular model type;
- the generation of readable model descriptions and final reports for communication and archival purposes.

Inadequacies in the area of documentation can reduce communication of the modelling effort within organizations, lead to major problems in model re-use or further development at a later time and create problems of maintainability. A CAPM system should have self-documenting capabilities that are strongly determined by the underlying structure of the objects within the system as well as the tools available to the user to help document model development. Bogusch *et al.* [160] have suggested the use of hypertext systems for users to note down decisions made as the modelling proceeds. This can be complemented by the use of an issue based information system (IBIS) [163] which has been used successfully by other workers in engineering design environments such as Bañares-Alcantará [164,165]. The IBIS structure consists of several elements including:

- *Issues*: a statement relating to a modelling problem being faced. This could be whether a particular energy balance volume such as a reactor vessel wall should be included or excluded in a model application.
- *Positions*: a statement setting out the possible positions to be taken and the preferred position on the issue. In this case it could be to exclude the wall energy holdup or to include it in the model.
- *Arguments*: statements which are used to justify the position taken on resolving the issue. In the case of excluding the wall energy holdup, this might include such justifications as the wall is thin; that the heat content is small

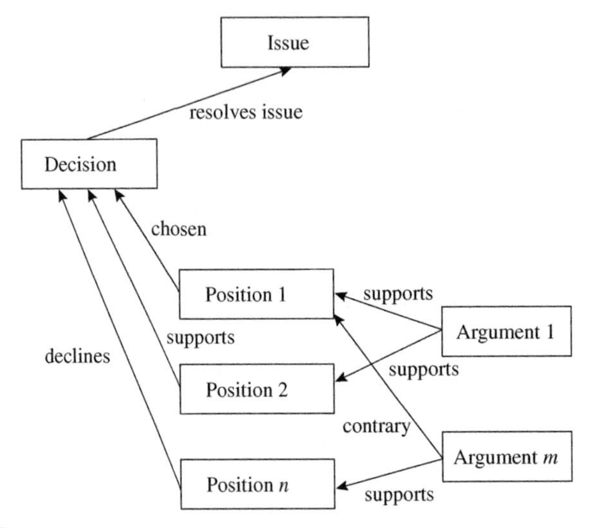

FIGURE 18.8 Typical IBIS structure for decision documentation.

compared with the reactor contents or that, in the time frame considered, the wall temperature can be regarded as constant.

Key IBIS concepts and their interrelation are shown in Fig. 18.8. The elements of the decision making can be stored compactly and retrieved easily by either textual or graphical means.

18.3.7. Model Simulation Environments

In many cases, the modelling environment is integrated with a simulation environment such as in *Model.la*, which contains its own numerical solution routines. In other cases, the CAPM tools export final equation systems in a form which can be directly read into powerful equation solving packages. In the case of *ModKit*, options allow for the export of models into code readable by *gPROMS* or *SpeedUp*. In these cases, the modelling environment is limited to the model building and generation phases, numerical solution being done outside the CAPM system.

What is essential is the availability of reliable simulation engines that can handle a range of underlying model characteristics. In most cases, current systems are limited to the solution of index-1 DAE systems. Some systems such as *gPROMS* and *Model.la* provide a wider range of solution capabilities such as handling the solution of PDEs.

18.3.8. Modelling Decision Support Systems

Decision support systems (DSSs) are commonly used in a range of management applications. However, little if any use is found in process modelling environments apart from typical help systems for commercial simulators and some tools in the area of solvability analysis. DSSs provide timely advice in the most appropriate form on which the user can take action. The DSS in no way makes the decision for the user. DSSs within CAPM tools would help in the following areas:

- guidance on the modelling process and what is required at key steps of the process;
- advice on the likely implications of making certain modelling assumptions on the complexity, solvability and model characteristics;
- guidance on what is required in the process modelling in order to meet the functional specifications for the model;
- advice on selection of DOF and the implications choices have on solvability;
- advice on addressing difficult numerical solution problems;
- advice on model calibration and subsequent model validation procedures.

In many cases, DSS tools rely on assessment of the current state of the modelling through analysis techniques and the interpretation of the results. The design of DSSs is extremely challenging as they must support both novice and advanced users. This implies analysis and presentation at varying depths of detail.

18.3.9. The Modelling Goal

All modelling is goal oriented. Models are built for a particular end-use and not to fill in idle time of the modeller! In Chapter 1, we discussed the wide ranging application of models in PSE. It is certain that one model does not fit all uses, be it for control, design or safety studies. It is also imperative that as far as possible the model builder has a clear modelling specification to work with in developing the model. In this case, it is certain that general specifications can be set down at the outset of modelling with the specification assuming more detail as the modelling and testing proceeds. The challenge in this issue is to provide guidance to model builders as to the appropriate choices required throughout the modelling tasks with regard to model characteristics, form and fidelity.

The area of clearly stating modelling goals and using these in terms of model building is sometimes implicitly assumed by expert modellers and completely forgotten by novices. No existing CAPM tools appear to address this issue in any structured way and certainly not in any depth. It remains a major challenge for the next generation of CAPM tools ... or the generation after that!

18.4. APPROACHES TO CAPM TOOL DEVELOPMENT

In this section, we discuss more fully the types of tools which have been developed or are under development to tackle computer aided process modelling. Some systems have already been used to illustrate particular aspects of CAPM requirements. Current CAPM tools fall into a number of categories although some cross several categories simultaneously. Various general categorizations have been done by Marquardt [154] and Jensen [155].

In this section, we consider the following categories, give some details on specific tools and comment on the key features:

1. flowsheet modelling environments (FME),
2. general modelling language systems (GML),
3. process knowledge based environments (PKBE).

Here, we discuss the basic features of the systems and some of the specific implementations.

18.4.1. Flowsheet Modelling Environments

These systems are dominated by the large commercial flowsheeting packages which have dominated the petroleum and petrochemical industries for the last four decades. These tools are the most widely used environments for carrying out large scale studies on process performance and dynamic behaviour. The level of decomposition in these systems stops at the unit block which can represent a reactor, mixer or distillation column. Table 18.1 gives a representative list of such systems currently available and gives an indication of their applicability to both steady state and dynamic modelling applications.

These environments are not specifically designed for the development of new models but are designed for building integrated flowsheets from a library of predefined models covering a range of unit operations and unit processes. Some facilities are available in certain systems to add new models such as the *Aspen Custom Modeler* system for *Aspen*.

TABLE 18.1 Flowsheet Modelling Environments

Flowsheet environment	Features		Reference
	Steady state	Dynamic	
Aspen Plus	✓		Aspen Technology Inc. [151]
Aspen Dynamics		✓	Aspen Technology Inc. [151]
Hysys	✓	✓	AEA Hyprotech [152]
PRO II	✓		Simulation Sciences Inc. [166]
CHEMCAD	✓		Chempute Software [167]
SuperPro Designer	✓		Intelligen Inc. [168]

18.4.2. General Modelling Languages (GML)

These can be seen as extensions to the class of equation oriented simulation languages, where users write models in a structured language. The language being used is not necessarily process oriented. Many of these systems are strongly object oriented in design, which helps in model reuse and model modification. Such systems include *ASCEND* [169,170], *gPROMS* [125] and *Modelica* [171]. *ASCEND* is described as "a type definition language that uses and extends object oriented concepts, including refinement hierarchies, generalized arrays, part/whole modeling, partial and complete

TABLE 18.2 General Modelling Languages

GML system	Features				Reference
	Steady state	Dynamic	Distributed	Hybrid	
gPROMS	✓	✓	✓	✓	PSE Ltd [125]
ASCEND	✓	✓			Carnegie Mellon [170]
Modelica	✓	✓			Modelica [171]

merging, deferred binding and universal types" [169]. Table 18.2 gives details of some of the available systems and their capabilities.

The concepts within these languages reinforce the non-specific domain of the modelling system. The modelling tools which utilize GML are applicable to a wide range of application areas. They simply provide a mechanism to develop individual models, link them where necessary and obtain simulated solutions. Process engineering concepts are developed within these systems from the use of more generic objects.

Underlying these GML systems are the specific building blocks which are used to develop the user application. These are not necessarily in the form of modelling objects as discussed in Section 18.3.1 but are language objects suitable for writing equations. In the case of *ASCEND*, the fundamental building blocks consist of ATOMS and MODELS, whereas *gPROMS* utilizes VARIABLES and MODELS. Figure 18.9 shows the generic building blocks for both *ASCEND* and *gPROMS*. All ATOMS are inherited from other ATOMS or from base types such as integer, boolean, real or string. MODELS are defined in terms of keywords, model variables and equation definitions.

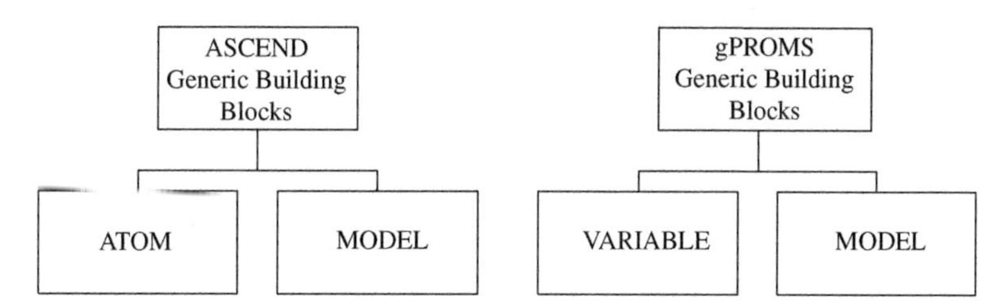

FIGURE 18.9 Generic building blocks in GMLs.

The application of the generic modelling objects is seen in the following example using the language constructs found in *ASCEND*.

EXAMPLE 18.4.1 (Use of these building blocks within ASCEND). This is seen in the following ATOM:

```
ATOM VariableType REFINES BaseClassName
   DIMENSION measure unit dimension
   DEFAULT value {measure unit}
   lower_bound := value {measure unit}
   nominal := value {measure unit}
   upper_bound := value {measure}
END VariableType
```

and in the following MODEL:

```
MODEL ModelType REFINES BaseModelName
   VariableName1 IS-A VariableType1
   ModelName1 IS-A ModelType1
   . . .
   Equations declared
   . . .
END
```

Similar to *ASCEND*, *gPROMS* code uses basic objects to build applications. A partial example of a *gPROMS* model is given in Example 18.4.2 below.

EXAMPLE 18.4.2 (Use of generic objects within a *gPROMS* model code).

```
MODEL BufferTank
    PARAMETER
        Rho AS REAL
        .
        .
    EQUATION
        # Mass holdup
        $Holdup = FlowIn - FlowOut ;
        # Calculation of liquid level from holdup
        Holdup = CrossSectionalArea * Height * Rho ;
        .
        .
    ASSIGN    # Degrees of freedom
        T101.FlowIn = 20 ;
    INITIAL    # Initial conditions
        T101.Height = 2.1 ;
    SCHEDULE    # Operating procedure
        CONTINUE FOR 1800
END
```

The *gPROMS* system provides a range of tools for the efficient solution of large-scale equation systems covering both lumped and distributed parameter systems as well as hybrid models requiring the detection of both time and state discontinuities. It provides state-of-the-art solution tools for nonlinear optimization and has been widely used in a number of multinational organizations.

The *Modelica* system represents a major effort in providing a physical modelling system built firmly on object oriented principles [171]. It was aimed at developing a uniform modelling language with the hope that it would become a *de facto* standard in modelling languages [172]. It has three main features:

- non-causal modelling based on DAEs,
- multi-domain modelling capabilities,
- a general type system that unifies object orientation, multiple inheritance and templates within a single class construct.

Modelica allows multi-domain modelling to be done which can simultaneously combine electrical, mechanical and hydraulic systems in the one model. Through the use of a Graphical User Interface (GUI), the system allows compositional modelling to be done using high level descriptions of the objects through the use of building block icons. The facilities of *Modelica*, are extended by the design environment *MathModelica* which provides a Mathematica-like notebook capability providing documentation of:

- parts of models,
- text or documentation,
- dialogue for for specification and modeification of input data,
- results tables,
- graphical result representation,
- diagrams to aid in the visualization of the model structure.

Modelica models generated by the system can be solved in the simulation system *Dymola* or can be generated as C++ code for direct execution.

General modelling languages provide powerful features to address general problems which are not restricted solely to process systems. They can be equally applicable for developing large-scale national economic models as they are for developing process specific models. They rely heavily on the user developing and coding directly all equations and variables associated with the application. These GML systems have a long established history within PSE and are widely used by the engineering community. New systems such as *ABACUSS II* also address the general modelling area with features for hybrid modelling [173].

18.4.3. Process Knowledge Based Environments (PKBE)

In contrast to GML systems, process modelling languages (PMLs) are specifically designed for the application area of process engineering. Fisher *et al.* [174] point out that in order to improve utility and enhance usefulness, a modelling tool should not add complexity to the task but simplify it. Providing a user with a language directly related to the problem domain should make the process of modelling more transparent. Most of these PMLs are embedded within a larger modelling environment which typically uses a GUI to build the model from the key building blocks. These systems we can denote as process knowledge based environments. In these cases, the modelling language is used to describe the model in process engineering terms and as such is required to be flexible enough to encompass all possible process modelling situations. In most cases the user sees the model description in the language but does not necessarily write directly in that language. Tools such as *Model.la* [161,175], *ModDev* [155] and *Modeller* [159] belong to this class. Facilities are often available to post-edit the equation system which is generated in the chosen PKBE.

The use of domain specific concepts also makes the transition from a text based interface to a graphical interface much simpler as icons can replace meta-words. This can be seen for example, in the user interfaces of *Model.la* and also *ModKit* shown in Figs. 18.11 and 18.12.

Model.la Environment

In the case of *Model.la* the interaction (Fig. 18.11) takes place through a GUI which provides access to all the construction and analysis tools. Icons within the *Modeling*

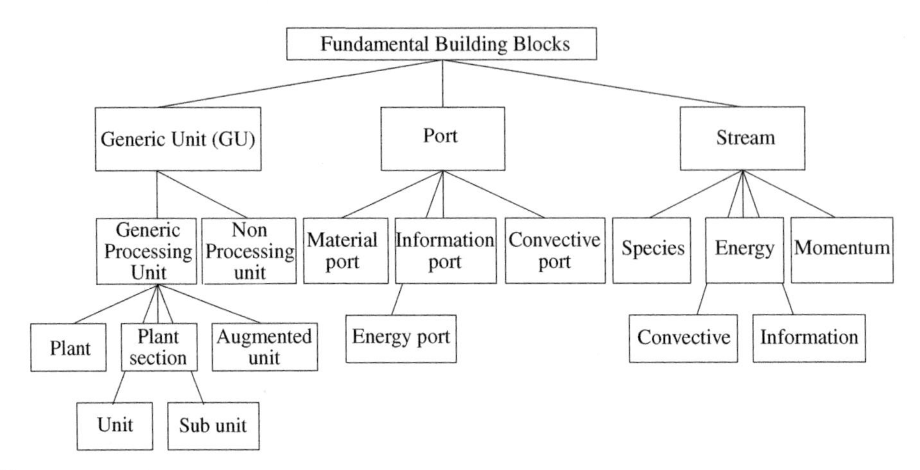

FIGURE 18.10 Fundamental building blocks for *Model.la*.

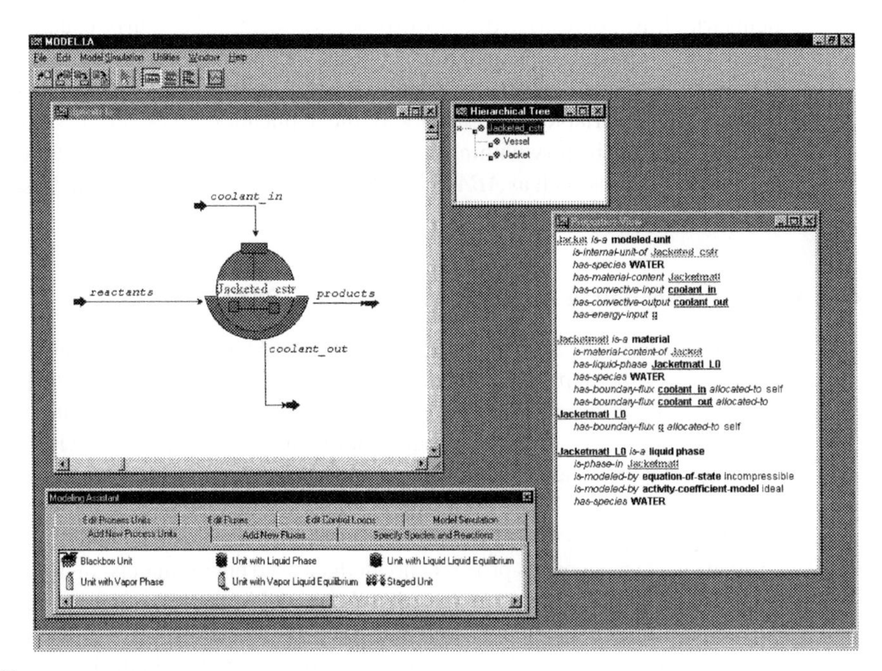

FIGURE 18.11 *Model.la* user interface.

Assistant help the user to develop the model from the underlying modelling objects seen in Fig. 18.10. Balance volumes, convective and diffusive flows can be used and instantiated for the specific application. There is a suggested order in the building of an application which closely mimics the modelling methodology advocated in Chapter 2. The resultant model description in the *Properties View* pane is easily readable by someone with a process engineering background. Facilities also exist for analysing DOF, checking the equation system index and running simulations. Options for distributed parameter modelling are also available and a variety of numerical solution techniques are given.

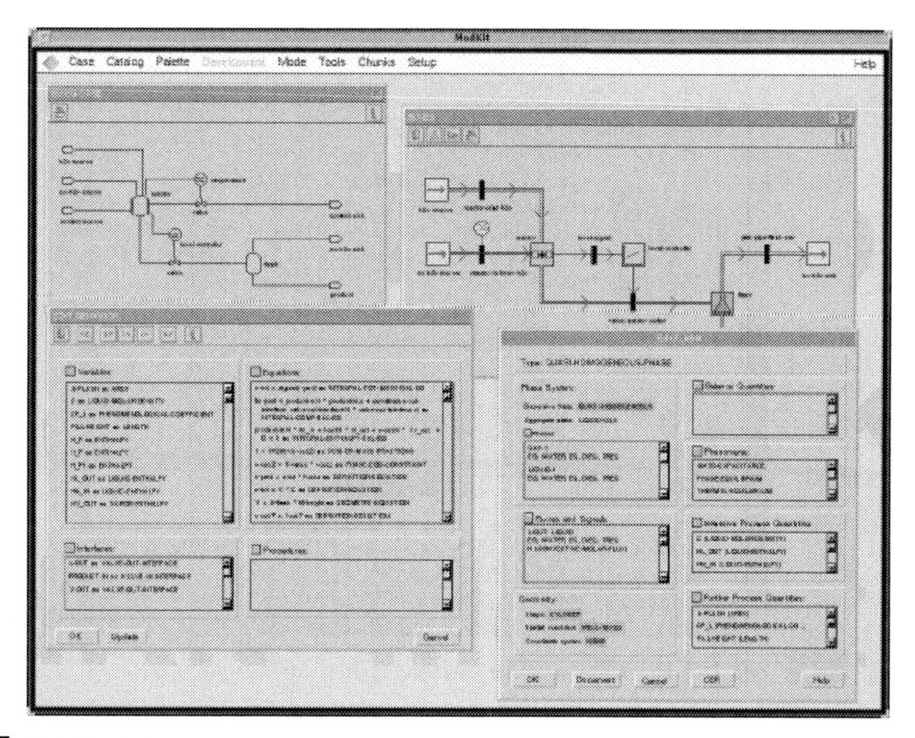

FIGURE 18.12 *ModKit* user interface.

ModKit Environment

The *ModKit* system is an advanced process modelling tool built upon a well-defined object oriented data model VEDA which utilizes a formal process engineering representation language VDDL (VEDA data definition language). It is a development of the process systems group at the Lehrstuhl für Prozesstechnik at Aachen University of Technology. Figure 18.12 shows a typical GUI for the system which is built upon the *G2* expert system software from Gensym Corporation [176].

The GUI displays some of the key aspects of the interactivity available in the system including the model building pane at the top right where the fundamental building blocks are assembled. As well, the attributes of a specific modelling object can be selected and set through interactive windows. In this case the lower right pane shows the attributes associated with a flash unit which is part of the overall flowsheet. Equations and variables can also be accessed for particular sections or models of the flowsheet under construction.

ModKit provides features for model documentation through its IBIS system supplemented by hypernotes which are attached to the IBIS structure. Other facilities include the ability to analyse the equations for DOF and index, and generate the code suitable for simulation within *SpeedUp* or *gPROMS.*

The system embodies most of the attributes of a fully functional process modelling tool and continues to undergo refinement and expansion.

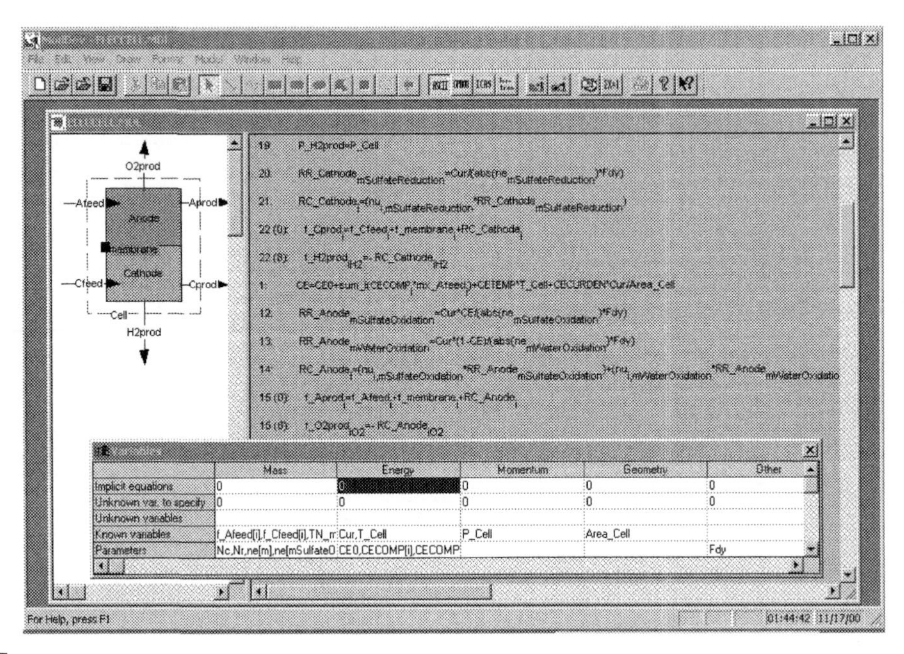

FIGURE 18.13 *ModDev* user interface.

ModDev Modelling System

ModDev is a modelling environment which is part of a much larger ICAS developed at the Danish Technical University. It specifically aims at developing process models from first principles using an advanced object oriented framework driven through a GUI. A typical interface for *ModDev* is seen in Fig. 18.13, where a model for an electrolytic cell is being developed. In this case the basic modelling objects are assembled graphically in the left pane and then the specific attributes of the balance volumes and flows are set through dialogue windows which are activated by double clicking the modelling object. As the modelling proceeds and information on the description is completed, the conservation and constitutive equations relating to the model appear in the equation pane. These can subsequently be modified or analysed within the *ModDev* framework or via the *MoT* as another module of *ICAS*.

ModDev uses the simulation facilities within the *ICAS* with a variety of DAE or structured solution approaches being available. The system provides a database of constitutive equations which aids the user in selecting the appropriate form for the model. The system allows model output in a number of formats which include *gPROMS* input code, Fortran or *ICAS* compliant input. The tools provided with the system allow advanced analysis of the equation systems and include:

- dimensional analysis and selection of degrees of freedom,
- determination of equation structure with subsequent ordering and partitioning,
- index analysis,
- analysis of stiffness for numerical solution.

The system provides facilities for modelling lumped parameter and distributed parameter systems and leads the modeller in a systematic way in order to build the final model.

Daesim Studio Environment

Daesim Studio [126] is an integrated modelling and simulation system which uses a process modelling language to develop models in a wide range of application areas including power generation, food, minerals processing and wastewater treatment. The underlying language is a structured text language based on an International Electrotechnical Commission (IEC) standard. The language provides significant flexibility to the modeller to define variable types, flow types and structures within the model. Multi-dimensional arrays are easily defined within the system, so that systems incorporating population balances with multiple internal co-ordinates can be described. The system is written as a Java application and provides access to advanced simulation tools for large scale DAE systems which can be hybrid in nature. Hybrid models can be described directly in the structured language and sequencing control for operational modes is generated through a GUI which implements the *Grafcet* standard. The system is based on object oriented principles.

Figure 18.14 shows the GUI for *Daesim Studio* where a model of a switched heated tank was built and simulated. The left-hand pane shows the individual components making up the application whilst the right-hand pane shows the *Grafcet*

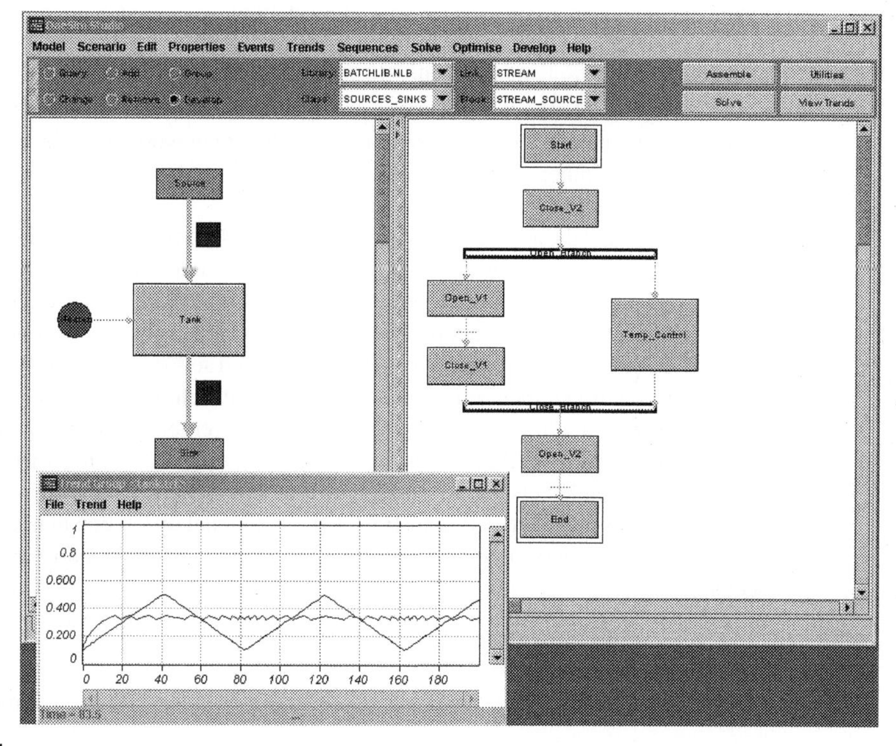

FIGURE 18.14 User interface for *Daesim Studio.*

diagram for the sequenced control of the heater. Simulated behaviour is shown in the results pane.

Daesim Studio provides facilities for managing model libraries which are maintained in classes and members. The structured text language provides advanced logic within the model description allowing the development of hybrid models. Simulation facilities are supported by large scale DAE solvers which handle both index-1 and index-2 systems that can be conventional or hybrid in character. Model libraries of modelling primitives can be developed and then used to assemble complex models consisting of these building blocks.

These previous sketches of CAPM tools have given some idea of the types of systems now being developed which are capable of tackling process modelling applications. Most of them share common ideas in their construction and there is a growing number of the tools available. Some like *ModKit* or *ModDev* are designed to generate output compatible with well known simulation systems. Others such as *Model.la* and *Daesim Studio* provide both modelling and simulation environments in the one environment.

It is clear that these tools will continue to grow in importance as process systems engineers demand more advanced tools to carry out modelling of new and non-traditional processes not covered by commercial flowsheet simulators.

18.5. SUMMARY

CAPM is an extremely important area of modelling practice. Conventional simulation tools provide some ability to develop new models or modify existing models. However, CAPM tools that specifically address the issue of model building are vital for better uptake of modelling in the process industries. It has been seen that several approaches are possible in developing these tools. Some work at the level of writing equations whilst others work at a much higher conceptual level, where most users would naturally work.

It is likely that the approaches which work at the physical description level dealing with balance volumes, flows and fluxes will become increasingly important. This is where most engineers and scientist naturally think about the systems they wish to model. New developments in structured assumption hierarchies [177] could play an increasingly important role in model development and self-documentation of models. Other work on model hierarchies and the ability to develop families of models with different complexity, time or dimension scales for different applications will play an important future role in CAPM tools. Finally, developments to try and guide modellers based on the intended end-use of models will be important in encouraging uptake of these techniques in the process industries. Efficient model development is an incentive for the increased use of process models in a wide range of applications and industry sectors.

19 EMPIRICAL MODEL BUILDING

19.1. INTRODUCTION

Process modelling is an iterative and incremental engineering activity. Theoretically, well-grounded elements based on first engineering principles and empirical parts based on measured data, operational experience and engineering heuristics or any combination of these are present in most models. In a realistic process modelling activity there is almost always a need for empirical model building. This is the subject of this chapter.

The purely empirical model elements are called *black-box* elements in contrast to the white-box elements based on first engineering principles. It is important to recognize, however, that process models are always grey-box models in nature. That is, they contain both black and white parts. Therefore, empirical or black-box modelling is treated in the context of grey-box modelling.

The material in the chapter is presented in the following sections:

- First the SEVEN STEP MODELLING PROCEDURE introduced in Section 2.3 is revisited in Section 19.2 to examine the need for empirical modelling and the problems which may arise during this procedure calling for an empirical or black-box model element.
- Thereafter, the most popular and most widely used black-box model structures are presented in Section 19.3 together with the relationship of black-box modelling to model calibration and identification.
- Finally, the most common traps and pitfalls in empirical model building are described in Section 19.4 along with methods to detect and deal with them.

19.2. THE MODELLING PROCEDURE REVISITED

In Section 2.3 of Chapter 2 a systematic SEVEN STEP MODELLING PROCEDURE is described which gives a solid framework for the model building process. The iterative nature of the modelling is also emphasized. It means that we may need to step back to an earlier modelling step in the procedure if the results of a later step are not satisfactory.

19.2.1. Grey-box Models: Models with Black-box Elements

The notion of grey-box models has been introduced as early as Chapter 2 these being models with both white (purely mechanistic) and black (purely empirical) model elements. Thereafter, in Section 12.1 of Chapter 12 on *Grey-Box Models and Model Calibration*, grey-box models were defined as process models developed using first engineering principles but with part of their model parameters and/or structure unknown. From this understanding and description we can infer what kind of elements in a process model may be black-box types.

If one follows the systematic SEVEN STEP MODELLING PROCEDURE then the ingredients of a mathematical model developed using first engineering principles will always be as follows:

- balance volumes,
- balance equations being (ordinary or partial) differential equations,
- constitutive equations which are algebraic equations,
- initial and boundary conditions.

The classification and properties of the terms in balance equations, as well as the main categories of constitutive equations and their properties fixes some additional mathematical properties, like the possible type of the PDEs derived as balance equations.

It follows from the above that the black-box model elements in a process model should also fall into the categories above and respect the properties of the category they belong to. It is also clear that black-box elements may have very different complexity ranging from simple parameters up to balance equations being PDEs with partially unknown structure. This view on grey-box models is much broader than that used in Section 12.1 where the unknown parameter case was only considered in most of the cases.

19.2.2. Steps Calling for Black-box Elements

The mathematical model of a process system is developed in steps 1–4 of the SEVEN STEP MODELLING PROCEDURE where we return from later steps if we want to refine the model. In carrying out these steps, various circumstances may indicate the need for empirical or black-box model elements as follows:

2. *Identify control mechanisms*
   When recognizing controlling mechanisms important for the modelling problem, we may find rather unusual or poorly understood mechanisms, such as biomass degradation, for example, together with relatively well-described

ones, like evaporation, heat and mass transfer, etc. Such a poorly understood or less investigated controlling mechanism will call for a black-box model of its own if any measured data is available.

3. *Evaluate the data for the problem*
   When encountering the data needed for the model we may find out that some of the model parameters, for example reaction kinetic constants, are not available in the literature. Then we have to find alternative ways to obtain or estimate them, most often by using measured data. For this purpose, we may use either the whole original model or just a special part of it or some other model built for the estimation purpose.
4. *Develop the set of model equations*
   In formulating the model in step 4 it may turn out that we cannot get data for the model form we arrived at. This will result in a grey-box model where either the parameters or some of the model elements are unknown. We can either use model calibration to get the missing parameters using measured data or look for additional models which relate available data for the unknown model element.

The black-box model elements that we need to make a process model complete may be one of the following three main types arranged in their decreasing order of complexity. The types are defined according to the mathematical form of the black-box model element and its role in the overall model.

m: *model-type*

The most complex case is when it is a lack of a description of a mechanism based on first engineering principles. This means that there is an unknown (or partially known) mechanism, for example fermentation, present in our model. Therefore, the corresponding constitutive equations, terms in balance equations or complete balances are missing from the overall model.

The mathematical form of the black-box model element of this type is usually an explicit nonlinear algebraic equation or a differential equation which is first order in the time derivative.

The following simple example shows a grey-box model containing a black-box element of this type.

EXAMPLE 19.2.1 (A simple continuous stirred fermentor).

Problem statement

A perfectly stirred continuous fermenter produces alcohol from sugar. A solution of sugar in water is fed to the fermenter, where the biomass is already present. The outlet does not contain anything from the biomass but contains water, produced alcohol and remaining sugar. Describe the dynamic behaviour of the fermenter for control purposes.

Modelling assumptions

$\mathcal{A}1$. Perfect mixing.
$\mathcal{A}2$. Constant liquid volume.
$\mathcal{A}3$. Isothermal conditions.
$\mathcal{A}4$. Inlet contains only water and sugar.
$\mathcal{A}5$. The fermentation produces alcohol from sugar.

$\mathcal{A}6$. Constant physico-chemical properties.
$\mathcal{A}7$. Open-air tank with constant pressure.

Balance equations
Sugar mass balance

$$V\frac{dx_S}{dt} = vx_S^{(in)} - vx_S - \text{rate}_S(x_S, x_B, x_A), \tag{19.1}$$

Alcohol mass balance

$$V\frac{dx_A}{dt} = -vx_A + \text{rate}_S(x_S, x_B, x_A), \tag{19.2}$$

Overall biomass mass balance

$$V\frac{dx_B}{dt} = \text{rate}_B(x_S, x_B, x_A), \tag{19.3}$$

where V is the volume of the liquid in the fermenter, x_S, x_B and x_A is the concentration of the sugar, biomass and alcohol respectively, and v is the volumetric flowrate.

The fermentation mechanism is largely unknown which affects the component mass balances and the reaction rate expressions. In the equations above, we can identify the following black-box elements calling for empirical modelling.

1. The reaction rate expressions $\text{rate}_S(x_S, x_B, x_A)$ and $\text{rate}_A(x_S, x_B, x_A)$ and their relationship. This is a function type black-box element.
2. The overall biomass mass balance equation (19.3). If the required accuracy dictated by our modelling goal requires modelling the fermentation kinetics in more detail, then we need to construct more balances for the various kinds of bacteria in the biomass. This would call for a set of additional balance equations and then this is a model-type black-box element of our model.

■ ■ ■

f: *function-type*
If there is no structure and/or parameters available for a constitutive equation (functional relationship), then a function-type black-box model is to be constructed. This black-box model element is an algebraic equation in its mathematical form which may be implicit and is usually nonlinear.

$$0 = f_{y_i}(x_1, \ldots, x_n; y_1, \ldots, y_m), \tag{19.4}$$

where x_j is a differential variable, y_i is an algebraic variable and f_{y_i} is an unknown but deterministic function.

Examples of this type of black-box model element are usually rate equations, such as reaction or transfer rate equations. This is the case in Example 19.2.1 above where the fermentation rate expressions $\text{rate}_S(x_S, x_B, x_A)$ and $\text{rate}_A(x_S, x_B, x_A)$ and their relationship was unknown.

p: *parameter-type*
A parameter-type black-box element arises when one or more of the parameters in a process model are unknown. This is the case we can in fact handle: we use measured data and model calibration to get an estimate of the missing parameter(s).

As we shall see later in this chapter, parameter-type uncertainty is the one to which we transform both function-type and model-type uncertainties.

19.3. BLACK-BOX MODELLING

Black-box elements always appear in connection with white elements.

The *structure of a process model is white* which fixes:

1. the number and terms in the balance equations,
2. the functional form of some of the constitutive equations.

This implies that black-box elements are to be imbedded in a white structure. Therefore neither the overall model nor its elements are ever completely black, at least some of their structural properties are known *a priori*.

19.3.1. Model Calibration and Black-box Modelling

In Section 12.1 in Chapter 12 on *Model calibration*, we have seen that model calibration is essentially estimation of unknown model parameters of a given grey-box model with only parameters unknown using measured data and a measure of fit.

If not only parameters but functions or complete mechanisms described by balance and/or constitutive equations are missing in a grey-box model, then empirical or black-box modelling should be applied "to fill the gap". The two cases, however, are strongly related from model calibration point of view.

As we have already seen in Chapter 12 the MODEL STRUCTURE ESTIMATION is solved by constructing a set of candidate model structures with unknown parameters and performing a MODEL PARAMETER ESTIMATION to get a model with best fit. The same technique is applied when we build an empirical model for a model-type or function-type black-box model element. As we shall see in the following, we parametrize the problem by selecting a suitably rich nonlinear black-box model structure and then estimate its parameters using measured data and a measure of fit.

19.3.2. Sensitivity of Models with respect to their Parameters and Variables

If an empirical model or function type black-box model is to be constructed from measured data the first thing to find out is the set of independent variables and parameters of the model. Later on, when candidate model structures are constructed, we may want to verify their set of independent variables and parameters. Sensitivity calculations are suitable for both purposes.

Sensitivities with respect to model parameters are also used in transferring uncertainty information about a black-box model when imbedded into its white-box environment.

In order to discuss the most general case for sensitivity calculations the following general model structure is used:

$$\frac{dx}{dt} = f(x; p), \quad x(t_0) = x_0, \tag{19.5}$$

$$y = g(x; p), \tag{19.6}$$

where $x(t) \in \mathcal{R}^n$, $y(t) \in \mathcal{R}^r$ and $p \in \mathcal{R}^m$. Note that the above model is in the form of a DAE containing both model-type uncertainty in Eq. (19.5) and function-type uncertainty in Eq. (19.6). Note that the solution $[x(t), y(t)]$, $t_0 \leq t$ is a function of p being the vector of the parameters, that is $[x(t; p), y(t; p)]$, $t_0 \leq t$.

Parametric Sensitivity

The parametric sensitivity of a scalar variable $v = v(q)$ with respect to a scalar parameter q is characterized by the derivative

$$\frac{\mathrm{d}v}{\mathrm{d}q}$$

taken at any reference point $q = q^0$. In the process modelling context, static and dynamic models relate variables and their sensitivities with respect to the same parameter.

In the case of dynamic models in the form of (19.5), the parametric sensitivity $\varphi_{x,p}$ of the dynamic vector variable x with respect to the vector parameter p is a time-dependent matrix with the partial sensitivities as its elements:

$$[\varphi_{x,p}(t)]_{ij} = \frac{\mathrm{d}x_i}{\mathrm{d}p_j}(t), \quad i = 1, \ldots, n, \; j = 1, \ldots, m \tag{19.7}$$

taken at a reference point $p = p^0$.

If we take $\mathrm{d}p_j$ as a parameter and consider that the total derivative of $f_i(x(p), p)$ with respect to $\mathrm{d}p_j$ is

$$\frac{\mathrm{d}f_i}{\mathrm{d}p_j} = \frac{\partial f_i}{\partial p_j} + \sum_{\ell}^{n} \frac{\partial f_i}{\partial x_\ell} \frac{\mathrm{d}x_i}{\mathrm{d}p_j}, \tag{19.8}$$

we can obtain the total derivative of both sides of Eqs. (19.5) and (19.6) in the following form:

$$\frac{\mathrm{d}[\varphi_{x,p}(t)]_{ij}}{\mathrm{d}t} = \sum_{\ell}^{n} [\mathbf{J}_{f,x}]_{i\ell} [\varphi_{x,p}(t)]_{\ell j} + \frac{\partial f_i}{\partial p_j}, \quad [\varphi_{x,p}]_{ij}(t_0) = 0, \tag{19.9}$$

$$[\varphi_{y,p}(t)]_{ij} = \sum_{\ell}^{m} [\mathbf{J}_{g,x}]_{i\ell} [\varphi_{y,p}(t)]_{\ell j} \tag{19.10}$$

with $\mathbf{J}_{f,x}$ and $\mathbf{J}_{g,x}$ being the Jacobian matrices of the right-hand sides of the equations with respect to the variable x respectively. Note that all the Jacobian and sensitivity matrices should be taken at a reference point $p = p^0$. Finally, the above equations can be written in the following compact matrix-vector form:

$$\frac{\mathrm{d}\varphi_{x,p}(t)}{\mathrm{d}t} = \mathbf{J}_{f,x} \varphi_{x,p}(t) + \mathbf{J}_{f,p}, \quad \varphi_{x,p}(t_0) = 0, \tag{19.11}$$

$$\varphi_{y,p} = \mathbf{J}_{g,x} \varphi_{x,p}, \tag{19.12}$$

where $\mathbf{J}_{f,p}$ is the Jacobian matrix of the right-hand side of Eq. (19.5) with respect to the parameter p taken at a reference point $p = p^0$.

Note that the sensitivities $[\varphi_{x,p}(t)]_{ij}$ at any point t in time can be calculated by solving the differential equation (19.9). This can be conveniently performed simultaneously with the original differential equation (19.5) because the Jacobian matrices needed are available at any point t in time.

Variable Sensitivity

The sensitivity of a solution $[x(t), y(t)]$, $t_0 \leq t$ of Eqs. (19.5) and (19.6) at any point t with respect to the variables is characterized by the derivatives

$$\frac{dx}{dx_0}(t), \quad \frac{dy}{dy_0}(t),$$

where $y_0 = g(x_0)$ with the other parameters fixed. This case can be regarded as a special case of parametric sensitivity when the right-hand side functions f and g do not depend explicitly on the parameters $p = x_0$ or $p = y_0$ respectively.

Therefore, the sensitivity equations (19.11) and (19.12) should be solved in order to calculate the variable sensitivities with $\mathbf{J}_{f,p} = 0$. It is important to notice that the sensitivity equations (19.11) and (19.12) contain the Jacobian matrices with respect to the variable x. In this case, the dynamic sensitivity equation (19.11) is the same as the locally linearized version of Eq. (19.5) taken at a reference point $x = x^0$. This gives the *connection between variable sensitivities with model linearization* which was described in Section 10.4.

19.3.3. Black-box Model Structures

The core of a black-box model is an unknown function to be determined from measured data. This function can be present on the right-hand side of a conservation balance or in a constitutive equation. The estimation of a completely unknown function is an ill-posed problem from a mathematical point of view as we have already seen in the case of MODEL STRUCTURE ESTIMATION problem statement in Section 12.4. The way to make the problem mathematically feasible is to construct the *set of possible model structures* for the structure estimation problem and determine the best-fit model from this set by performing a set of model parameter estimation.

The same technique is applied here for black-box modelling. We usually select a black-box model structure set, which fits to the modelling problem and easy-to-handle from the estimation point of view at the same time. The elements in a black box model structure set have similar structure, and an element from the set is identified by its structural indices. There are various well known and popular black-box model structures which will be briefly described in this section.

The following requirements hold for a good black-box model structure:

- *simplicity*—in its mathematical form resulting in an easy computation, such as a power series,
- *generality (descriptive power)*—to ensure that the members of the structure set can describe all possible black-box models in question,
- *flexibility*—to have a wide descriptive range of the structure set.

The above requirements call for an approximation which is similar to an infinite functional series expansion.

How to select a suitable black-box model class which fits the modelling problem? The application area of the class should match the key characteristics of the problem. The following key characteristics can be taken into account.

1. *Type of the black-box modelling problem*
   A function-type black-box modelling problem calls for a static model structure while a model-type problem may require dynamic model structures.
2. *Other properties of the problem*
   The following properties—if known in advance—may be taken into account:
   - linearity–nonlinearity,
   - symmetry,
   - time-invariance (constant parameters).

The black-box model classes or model structures are described using the following items:

- *Mathematical description*
  The mathematical form of the general element from the set of black-box model structures are given. This determines the descriptive power and the flexibility of the structure set.
- *Structural parameters*
  The model parameters used to select one of the members from the set are described here together with how to estimate them from measured data.
- *Mathematical properties*
  The special mathematical properties of the general form as well as the related parameter and structure estimation are given here. These are in close relationship with the mathematical form of the structure set.
- *Parameter estimation*
  The parameter estimation methods available for the model class are described briefly for the parameter set including structural parameters.

Linear Static Model Structures

Mathematical description
This is the simplest case when we need to develop a black-box model for an unknown linear function in the overall model in the form of

$$y = \sum_{i=1}^{N} \chi_i p_i = \chi^{\mathrm{T}} p. \tag{19.13}$$

Structural parameters
N and the set of variables present in the vector χ.

Parameter estimation
The parameter estimation of linear static model structures is described in Section 12.3.1. It is important that the problem statement of the parameter estimation requires a loss function which serves as a measure of fit for the estimation.

The structural parameters can be estimated along the lines shown by the general MODEL STRUCTURE ESTIMATION problem statement in Section 12.2. We construct the *set of possible model structures* including every possible (N, χ) pair and determine the best-fit model from this set by performing a set of model parameter estimations.

The following example shows the use of linear static model structures to construct function-type black-box models.

EXAMPLE 19.3.1 (Reaction rate expression). Let us assume we have a partially known reaction kinetic expression

$$r = k_0 e^{-E/(RT)} C_A^{q_A} C_B^{q_B} \tag{19.14}$$

in a grey-box model with the reaction orders q_A and q_B possibly equal to zero, that is r may or may not depend on any of the concentrations.

We need to estimate the

- *parameters*: k_0, E,
- *structural parameters*: q_A, q_B.

First, we transform the reaction rate equation to linear in parameter form by taking the natural logarithm of both sides:

$$\ln r = \ln k_0 - \frac{E}{RT} + q_A \ln C_A + q_B \ln C_B. \tag{19.15}$$

Thereafter, we can notice that the structural parameters become ordinary parameters in this case and the following correspondence can be made with Eq. (19.13):

$$y = \ln r, \quad \chi = [1, \ -\frac{1}{RT}, \ \ln C_A, \ \ln C_B]^{\mathrm{T}}, \quad p = [\ln k_0, \ E, \ q_A, \ q_B]^{\mathrm{T}}.$$

Nonlinear Static Model Structures and Artificial Neural Networks

Nonlinear static model structures are used to approximate the general case of static functional relationship in Eq. (19.4).

Functional Series Approximation

Mathematical description

The simplest one is the Taylor series expansion:

$$y_i = c_0 + \sum_{i=1}^{n+m} c_i \chi_i + \sum_{i_1=1, i_2=1}^{n+m} c_{i_1 i_2} \chi_{i_1} \cdot \chi_{i_2} + \cdots + \sum_{i_1=1,\ldots,i_N=1}^{n+m} c_{i_1 \ldots i_N} \chi_{i_1} \cdots \chi_{i_N} + \ldots, \tag{19.16}$$

where the vector χ contains all the variables including both the differential and algebraic ones:

$$\chi = [x_1 \ldots x_n \quad y_1 \ldots y_m]^{\mathrm{T}}.$$

It is important to note that the Taylor series expansion is *linear in its parameters*

$$p = [(c_i, i = 1, \ldots, m+n) \quad (c_{i_1 i_2}, i_1, i_2 = 1, \ldots, m+n), \ldots, (c_{i_1 \ldots i_N}, i_1, \ldots, i_N = 1, \ldots, m+n) \ ]^{\mathrm{T}}.$$

Structural parameters
The order of the power series N and the set of variables present in the vector χ.

Parameter estimation
Because the Taylor series expansion is linear in its parameters and is static, we can apply the parameter estimation of linear static model structures as before.

The structural parameters can be estimated the same way as in the general MODEL STRUCTURE ESTIMATION problem statement.

EXAMPLE 19.3.2 (Empirical heat transfer in a heat exchanger). Let us assume that we have a function-type black-box model for a heat transfer rate in a heat exchanger and its dependence on the operational parameters (flowrates (v), pressures (P) and temperatures of the two phases (T_1 and T_2)).

$$r_H = f(v_1, v_2, P_1, P_2, T_1, T_2). \tag{19.17}$$

In the case of liquid phases with moderate temperature difference between the phases, we can use the simplest linear approximation as a special case of Eq. (19.16)

$$r_H = K_1(T_1 - T_2),$$

where $N = 1$, $y = r_H$ and $\chi = [T_1 - T_2]$.

Alternatively, we can consider an expression which accounts more closely for the effect of fluid velocity on the heat transfer coefficient giving

$$r_H = (K_2 + K_3 u^{1/2})(T_1 - T_2),$$

where u is the fluid velocity and is a function of v and $N = 2$, $y = r_H$ and $\chi = [(T_1 - T_2)u^{1/2}]$.

Orthogonal Function Series Expansion

The Taylor series expansion uses polynomials of various order to construct an approximation of an unknown nonlinear function. Instead of polynomials other functions can also be used.

Mathematical description
If the polynomial terms

$$P_{i_1 \ldots i_N}(\chi_{i_1}, \ldots, \chi_{i_N}) = P_i(\chi) = \chi_{i_1} \cdots \chi_{i_N}$$

in the expansion (19.16) are replaced by terms $w_{i_1 \ldots i_N}(\chi_{i_1}, \ldots, \chi_{i_N})$ where $w_i(\chi)$ are called basis functions which are "orthogonal" in the sense that

$$\int (w_i(\chi) \cdot w_j(\chi)) d\chi = 0,$$

then we have an orthogonal function series approximation. We should require that the set of basis function should be "complete" in the sense that they really form a basis in the space of nonlinear functions of interest. That is, they should span that infinite dimensional function space.

It is important to note that there are *orthogonal polynomial approximations* when the basis functions form a special infinite set of orthogonal polynomials, like the Legendre polynomials. We have already used orthogonal polynomials to approximate the solution of nonlinear partial differential equation models in Chapter 8.

There are other well-known basis function sets, we may use if the special properties of the black-box modelling problem call for them.

- *Trigonometrical basis functions: the Fourier series expansion*
  This type of approximation is traditional in mechanical and electronic engineering for linear dynamic models. It fits well to periodic and symmetric problems.
- *Wavelet basis functions*
  Wavelet basis functions use functions which are non-zero only on a finite interval. These intervals are disjoint for the members of the basis and therefore the elements in the basis are orthogonal. This type of approximation fits well to multi-scale or transient problems.

Structural parameters
The basis functions $w_i(\chi)$, the "order" of the approximation N and the set of variables present in the vector χ.

Neural Networks

Neural networks were originally invented as a novel computing paradigm working the same way as neurons in living organisms. Therefore, the elementary processing unit is called a *neuron* and a neural network is a set of connected neurons.

Mathematical description [178]
The elementary processing unit, called neuron, performs a static linear computation from its *synapsis vector* x weighted by a *weight vector* w to produce its scalar output z.

$$z = f(w^{\mathrm{T}}x), \tag{19.18}$$

where the nonlinear function $f(\cdot)$ is the *transition function* of the neuron.

A *neural network* is then a structure

$$\mathrm{NH} = (N, S; w), \tag{19.19}$$

where $N = \{\nu_1, \ldots, \nu_k\}$ is the set of neurons, $S \subseteq N \times N$ is the set of synapses, that is neuron connections consisting of ordered pairs of neurons and w is a weight associated with the synapses. Observe, that the structure (19.19) is a weighted directed graph. The inputs of a neuron ν_i are in the set of

$$\nu^{(\mathrm{in})} = \{\mu \in N \mid (\mu, \nu) \in S\}. \tag{19.20}$$

Neuron ν_j is connected to ν_k if $\nu_j \in \nu_k^{(\mathrm{in})}$. Note that if a neuron is connected to several other neurons then it transmits the same output z to every one of them.

In applications, the neurons are arranged into *layer*s which are pairwise disjoint subsets of the set of neurons N. A layer in a neural net is called *feedforward* if its neurons are connected only to neurons of a specific given other layer. In a *feedback* *layer* any neuron is connected to every other neuron from the same layer. The layers in a neural net are homogeneous with respect to the transition function, hence the same transition function is present in each neuron.

We may consider the transient operation of a neural net the same way as for the natural neurons but in process modelling and control applications only the static (that is steady state) operation mode is considered. The synchronized operation of a neural net, that is the neural computation when every neuron computes its output from the inputs at exactly the same time, is performed in the following steps:

1. set the synapsis weights w;
2. compute the output of every neuron

$$z_j(k+1) = f\left(\sum_{v_i \in v_j^{(in)}} w_{ji} z_i(k) + I_j(k)\right), \tag{19.21}$$

   where $I_J(k)$ is the external signal to the neuron v_j at the time instance k;
3. repeat the computation in step 2 until a steady state is reached.

Structural parameters
There are two different classes of structural parameters in a neural net.

1. *Connection properties*
   Connection properties determine the number and the connection mode of the neurons in the network. Formally speaking, they are specified by the sets N and S in the structure (19.19). Some of the connection properties, such as the number of neurons in the input and output layer of the net are determined by the approximation problem. The number of neurons in the input layer is equal to the number of independent variables and that of the output layer is equal to the number of nonlinear functions with the same set of independent variables to be estimated.
2. *Computation properties*
   With the inputs to the neural net given, the result of the neural computation is determined by
   - synapsis weighs w_{ij} and
   - transition functions of the neurons f.

   We usually regard the transition functions given and belonging to the basic structure of the neural net and adjust the synapsis weights to produce the desired output of the network.

Therefore, *the synapsis weights are the parameters of a neural net model* in the usual sense, while all the other parameters above are regarded as structural parameters.

The transition function is usually a *sigmoid function* $\sigma(x)$ which is

$$\sigma(x) = \begin{cases} 1 & \text{if } x \to \infty, \\ 0\ (-1) & \text{if } x \to -\infty. \end{cases}$$

The commonly used sigmoid functions are as follows.

- Step functions, for example

$$\text{step}_0(x) = \begin{cases} 0 & \text{if } x < 0, \\ 1 & \text{if } 0 \le x. \end{cases} \tag{19.22}$$

- Threshold function

$$\text{thresh}(x) = \begin{cases} -1 & \text{if } x < -1, \\ x & \text{if } -1 \le x \le 1, \\ 1 & \text{if } x > 1. \end{cases} \tag{19.23}$$

Mathematical properties
If we apply neural networks to approximate a nonlinear function, then it can be shown (see the famous Cybenko theorem [179]) that even a three-layer feedforward neural net will be enough to approximate the smooth nonlinear function with any given precision ε. This means that neural networks can be regarded as general approximation tools for multivariate nonlinear functions even when there is a simple neural network structure with just an input, a hidden and an output layer (a three-layer feedforward neural net).

Parameter estimation
The parameter estimation consists of estimating the weights w_{ij} using a set of measured data called the set of training data. The backpropagation algorithm [14] and its variants [180] are used for this purpose. These algorithms are based on the gradient method for optimizing the prediction error as a function of the unknown weights.

Part of the structural parameters (the number of input and output neurons) is fixed by the problem statement. The neural network structure and the transition function is usually selected according to the application area: most often feedforward neural network with a single hidden layer and with the threshold function as transition function is chosen. Therefore, the only "free" structural parameter is the number of neurons in the hidden layer which is most often selected on a trial-and-error basis.

More about the traps and pitfalls in applying a neural network model is found later in Section 19.4.

EXAMPLE 19.3.3 (A simple feedforward neural network example). A simple three layer feedforward neural network structure is shown in Fig. 19.1. The input layer consists of 4, the hidden layer 8 and the output layer 3 neurons. Nonlinear (usually sigmoid) transition functions are on both the input and hidden layer, while the neurons on the output layer have the identity function as their transition function, that is their output is computed as

$$z = w^{\mathrm{T}}x.$$

■ ■ ■

Linear Dynamic Model Structures: ARMAX Models

It is known that the ARMAX (autoregressive moving average with an exogeneous signal) models are a general form of the input-output models for discrete time linear time-invariant systems. Note that time-invariance implies that the model parameters

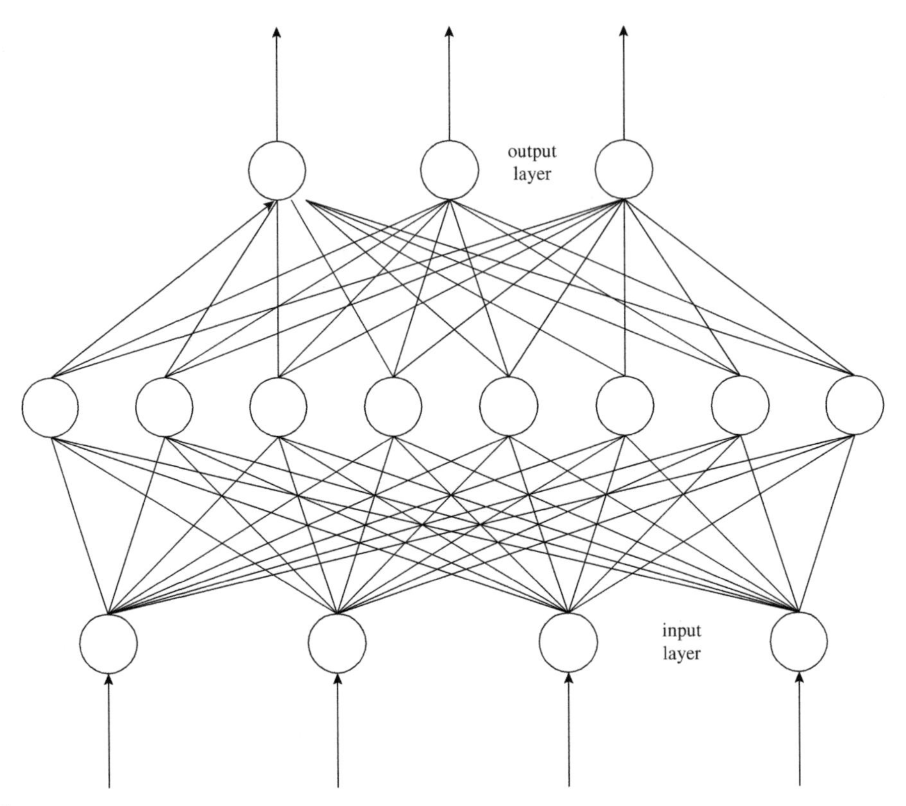

FIGURE 19.1 A simple neural net.

are constant, hence time-independent. Observe that we have to sample the continuous time models to transform them into their discrete time equivalents (see Section 11.1).

Mathematical description
For SISO systems an ARMAX model is in the form of a stochastic difference equation:

$$y(k) + a_1 y(k-1) + \cdots + a_p y(k-p)$$
$$= b_0 u(k) + \cdots + b_r u(k-r) + e(k) + c_1 e(k-1) + \cdots + c_p e(k-p), \tag{19.24}$$

where $y(k), k = 0, 1, 2, \ldots$ is the output, $u(k), k = 0, 1, 2, \ldots$ is the input signal (discrete time sequence), and $e(k), k = 0, 1, 2, \ldots$ is a discrete white noise stochastic process with identically (usually normally) distributed zero mean sequence of independent random variables.

For MISO systems the SISO equation (19.24) is generalized to have column vectors for $(u(k-j),\ j = 0, \ldots, r)$ and row vectors of the same dimension for $(b_j,\ j = 0, \ldots, r)$.

For the general MIMO case, we have column vectors for $((y(k-j), e(k-j)); j = 0, \ldots, p)$, square matrices for all of the coefficients $((a_i, c_i),\ i = 1, \ldots, p)$ and rectangular matrices for $(b_j,\ j = 0, \ldots, r)$.

Structural parameters
The structural parameters of the model (19.24) are as follows:

- the order of the approximation (p, r),
- the set of variables present in the vectors $y(k)$ and $u(k)$ and their dimension, that is $|y| = n$ and $|u| = m$,
- the probability distribution of the noise term $e(k)$.

Parameter estimation
There are reliable and well established methods to perform both the parameter and the structure estimation of ARMAX models [14]. These methods are implemented in the "System Identification Toolbox" of MATLAB.

The following simple example shows how we can construct the mathematical form and the parameters of various ARMAX models.

EXAMPLE 19.3.4 (A simple ARMAX model example). Let us start with the simplest SISO example in the form of

$$y(k) + a_1 y(k-1) = b_0 u(k) + b_1 u(k-1) + e(k),$$

where $p = 1$, $r = 1$, $c_1 = 0$ and $n = m = 1$. Here the parameters are the scalars

$$[a_1, \ b_0, \ b_1]^{\mathrm{T}},$$

Now let us extend the model with the same order ($p = r = 1$) to the MIMO case with $n = 2$ and $m = 2$. The algebraic form of the model remains the same, but now

$$y(k) = [y_1(k), \ y_2(k)]^{\mathrm{T}}, \quad u(k) = [u_1(k), \ u_2(k)]^{\mathrm{T}}, \quad e(k) = [e_1(k), \ e_2(k)]^{\mathrm{T}},$$

and the parameters are 2×2 matrices

$$a_1 = \begin{bmatrix} a_{11}^{(1)} & a_{12}^{(1)} \\ a_{21}^{(1)} & a_{22}^{(1)} \end{bmatrix}, \quad b_0 = \begin{bmatrix} b_{11}^{(0)} & b_{12}^{(0)} \\ b_{21}^{(0)} & b_{22}^{(0)} \end{bmatrix}, \quad b_1 = \begin{bmatrix} b_{11}^{(1)} & b_{12}^{(1)} \\ b_{21}^{(1)} & b_{22}^{(1)} \end{bmatrix}.$$

Nonlinear Dynamic Model Structures: Volterra and Hammerstein Models

The general structure of nonlinear dynamic systems is far from being simple even in the case of finite dimensional systems. Therefore, we shall restrict ourselves to the simplest finite dimensional discrete time deterministic input–output models. Their general form can be written in a so called *predictive form* as follows:

$$y(k) = F(U^{(k)}, Y^{(k-1)}), \tag{19.25}$$

where F is a nonlinear static function, $Y^{(k-1)}$ and $U^{(k)}$ are the records of past outputs and inputs:

$$U^{(k)} = \{u(j) \mid j = 0, 1, \ldots, k\}, \quad Y^{(k-1)} = \{y(j) \mid j = 0, 1, \ldots, k-1\}.$$

The nonlinear dynamic black-box model structures approximate the nonlinear function F by function series approach similar to the static nonlinear case.

In order to keep notation low the mathematical description is only given for the SISO case. For a more detailed treatment of the subject we refer to the recent book by Haber and Keviczky [181].

Mathematical description
The parametrized version of the most popular and widely used nonlinear dynamic black-box model structures can be regarded as second order polynomial type extensions of the linear time-invariant ARMAX model in Eq. (19.24).

The more general of these approximations is the so-called *parametric Volterra model* which is in the following general form:

$$y(k+d) = c_0 + \frac{B_1^L(q^{(-1)})}{A_1^L(q^{(-1)})}u(k) + \frac{B_1^H(q^{(-1)})}{A_1^H(q^{(-1)})}u^2(k) + \sum_{j=0}^{N^V} \frac{B_{2j}^V(q^{(-1)})}{A_{2j}^V(q^{(-1)})}u(k)u(k-j), \tag{19.26}$$

where the discrete linear filters

$$H^w(q^{(-1)}) = \frac{B^w(q^{(-1)})}{A^w(q^{(-1)})}, \quad w = H, L, V$$

are understood as

$$A^w(q^{(-1)})y(k) = B^w(q^{(-1)})u(k),$$

$$\sum_{j=0}^{n_a} a_j^w y(k-j) = \sum_{j=0}^{n_b} b_j^w u(k-j).$$

We see that the variable $q^{(-1)}$ in the polynomials A^w and B^w above is the so called backward shift operator which acts on a signal z as $q^{(-1)}z(k) = z(k-1)$. From the above filter equations it follows that we have infinite sums in the model structure (19.26) in the form of

$$H^w(q^{(-1)})u(k) = \sum_{j=0}^{\infty} h_j^w u(k-j).$$

It is then up to the user to find appropriate large integers N_1^{L}, N_1^{H} and N_{2j}^V, $j = 0, \ldots, N_V$ to truncate the above infinite sums.

There are three types of terms in the right-hand side of the parametrized Volterra model (19.26):

- the first term is purely linear with the upper index ^{L},
- the second is a quadratic with upper index ^{H}, and
- the third is the general bilinear (2nd order) term with upper index ^{V}.

If only the first two terms are in the Volterra model, that is $N^V = 0$ or

$$\frac{B_{2j}^V(q^{(-1)})}{A_{2j}^V(q^{(-1)})} = 0, \quad j = 0, \ldots, N^V,$$

then it is called a *generalized Hammerstein model.*

Structural parameters
The structural parameters fall into two categories:

- the order of the approximation that is the integers

$$N_1^{\mathrm{L}},\ N_1^{\mathrm{H}},\ N^V,\quad (N_{2j}^V,\ j=0,\ldots,N^V)$$

- the set of variables present in the vectors $y(k)$ and $u(k)$ and their dimension, that is

$$|y|=n,\qquad |u|=m.$$

Parameter estimation
Because the nonlinear second order expansion is linear in its parameters we can apply the parameter estimation method of the linear ARMAX model structures as before.

The structural parameters can be estimated the same way as in the general MODEL STRUCTURE ESTIMATION problem statement.

The following simple example shows how we can construct the mathematical form and the parameters of various nonlinear models.

EXAMPLE 19.3.5 (Simple dynamic nonlinear model examples). Let us start with a simple SISO Hammerstein model in the form of

$$y(k)=h_0^{\mathrm{L}}u(k)+h_1^{\mathrm{L}}u(k-1)+h_0^{\mathrm{H}}u^2(k)+h_1^{\mathrm{H}}u^2(k-1), \tag{19.27}$$

where $N_1^{\mathrm{L}}=1$, $N_1^H=1$, $N^V=0$. Here the parameters are the scalars

$$\left[h_0^{\mathrm{L}},\ h_1^{\mathrm{L}},\ h_0^{\mathrm{H}},\ h_1^{\mathrm{H}}\right]^{\mathrm{T}},$$

where h_0^{L} and h_1^{L} are the parameters of the linear submodel.

If we extend the above model to obtain a parametrized Volterra model with $N^V=1$, we get

$$\begin{aligned} y(k)=&\ h_0^{\mathrm{L}}u(k)+h_1^{\mathrm{L}}u(k-1)+h_0^{H}u^2(k)+h_1^{H}u^2(k-1)\\ &+h_{10}^{V}u(k)u(k-1)+h_{11}^{V}u(k-1)u(k-2). \end{aligned} \tag{19.28}$$

Here we have two new structural parameters $N_{20}^V=0$, $N_{21}^V=1$. The complete set of parameters is then

$$\left[h_0^{\mathrm{L}},\ h_1^{\mathrm{L}},\ h_0^{H},\ h_1^{H},\ h_{10}^{V},\ h_{11}^{V}\right]^{\mathrm{T}}$$

where h_0^{L}, h_1^{L}, h_0^H and h_1^H are the parameters of the Hammerstein sub-model. ■ ■ ■

19.3.4. Identification for Black-box Modelling

Identification as a tool for model calibration, that is model parameter and structure estimation for dynamic models is described in Section 12.4. The main emphasis there has been placed on parameter estimation.

Black-box modelling mainly needs structure estimation over special general and flexible model structures, called black-box model structures, which have simple structure indices. The subsection *Parameter estimation* given for the black-box model structures above gives guidelines on how to perform the model structure estimation relevant to the structure in question.

It is important to note that the result of identification is always in terms of

- estimate of the structural parameters,
- estimate of parameters of the model with the "best structure",
- uncertainty associated with the estimate(s).

19.3.5. The Information Transfer between the Black-box Model Element and the Overall Model

Having validated a black-box model based on measured data, the next step is to put together the overall grey-box model. This section is about how to incorporate black-box elements into a white-box structure.

Black-box Functional Relationship

The result of model validation is not only the structure and best parameters of the validated black-box submodel but also uncertainty ranges, usually confidence intervals associated with each of the parameters in the model. Then it is important to transfer the information contained in the static model with parametric uncertainty into the white-box structure.

If a black-box nonlinear function-type relationship $y = f(x, p)$ has been validated to obtain an estimate of the scalar parameters in the form of $p - \Delta_P \leq p \leq p - \Delta_P$, then the nonlinear function may damp or amplify the effect of the uncertainty in the parameters $2\Delta_P$. We can use the same principle as in the case of parametric sensitivity for static models, and derive the following estimate of the uncertainty in the variable y

$$\Delta_y = \max_{x \in Dom\, x} \left. \frac{\partial f}{\partial p}(x) \right|_{p=p^0} \Delta_P. \tag{19.29}$$

The above estimate can be easily generalized to the case of vector variables p:

$$\Delta_y = \max_{x \in Dom\, x} \left. \mathbf{J}_{f,x}(x) \right|_{p=p^0} \Delta_P, \tag{19.30}$$

where $\mathbf{J}_{f,x}$ is the Jacobian matrix of the function f with respect to the variables x.

Black-box Submodel

If a black-box submodel is to be imbedded into a white overall model structure, then we essentially have the same case as the relationship with models on different hierarchy levels in hierarchical modelling (see in Chapter 9). In the case of an unknown or partially known mechanism, we may use the methodology of multi-scale modelling, given by a model hierarchy driven by characteristic sizes (see in Section 9.2 in Chapter 9) to get an empirical model of the missing part of a grey-box model. Having performed the lower level modelling and the validation of this model, the information transfer requires the computation of volumetric integrals (averaging) for both the independent and dependent model variables.

19.4. TRAPS AND PITFALLS IN EMPIRICAL MODEL BUILDING

The various tools and techniques available for empirical model building indicate that it is usually not easy to obtain a reliable and easy-to-use black-box model which fits our purpose. We may fall into traps or come up with poor quality models. This section is about how to detect that we are approaching or have already fallen into some traps and then how to recover from them. In most cases these problems can be avoided if proper prevention steps are carried out.

The traps and pitfalls are described in the following items:

- *Causes and consequences*—This item gives a detailed explanation of the problem one may have with its possible causes and consequences.
- *Detection*—The possible ways of detecting the problem are briefly described here.
- *Prevention*—The possible precautions and steps to avoid the problem are also given.

19.4.1. Interpolation and Extrapolation: the Validity Range

Causes and consequences
The majority of empirical models are only valid within the closed (and usually convex) domain spanned by the measurement points. Therefore one should ensure that the measurement (or collocation) points cover the entire region: otherwise an ill-conditioned or poorly trained model is obtained. For further details on ill-conditioned models see section 19.4.3.

Detection and prevention
For nonlinear models the validity range is very narrow: one should not use them outside the region spanned by the measurement (or collocation) points.

Static Models

The main difference between linear and nonlinear static models from the viewpoint of their validity range is that nonlinear models are often very poor outside of their validity range. Therefore, their specification should contain an explicit description of the required validity range and the measurement points available for model validation should span the entire validity range.

The following example illustrates the fit of a polynomial to a reaction kinetic expression.

EXAMPLE 19.4.1 (The fit of a polynomial to a nonlinear reaction kinetic expression). Consider the following reaction kinetic expression

$$r_A = \frac{C_A^{1.5}}{1 - C_A} \tag{19.31}$$

over the validity range $0.2 \leq C_A \leq 0.8$.

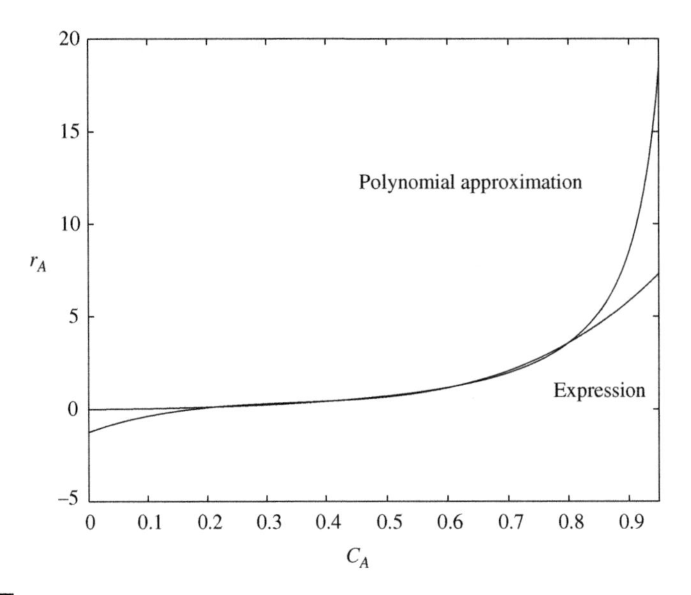

FIGURE 19.2 Extrapolation error with polynomial approximation.

Figure 19.2 shows the graph of the original functional relationship (19.31) together with its third-order polynomial approximation over the interval [0, 0.9]. It is seen that the approximation error increases dramatically when one leaves the validity range.

Dynamic Models

In the case of dynamic models one meets additional difficulties if the model is used for extrapolation. Besides the validity range in the space of the independent variables mentioned in section *Static models*, we have the validity range along the time variable as well.

Therefore, additional difficulties may arise which are as follows:

1. *The presence of slowly changing time-varying parameters*—which needs parameter tracking or adaptive (online) methods for model validation.
2. *The effect of growing error with prediction horizon*—which is a natural extension of leaving the validity range but now problems arise not only for nonlinear but also for stochastic models.

The following example illustrates the growth of uncertainty with prediction horizon, that is with advancing in time in the simplest case of a linear stochastic model.

EXAMPLE 19.4.2 (Prediction using autoregressive (AR) models). Consider the following simple single dimension AR model

$$y(k+1) = y(k) + e(k+1), \quad y(0) = e(0) \tag{19.32}$$

with $(e(i),\ i = 1, 2, \dots)$ being a sequence of independent random variables with zero mean and unit variance and $y(i)$ being the system output at discrete time instance i.

If we perform prediction with the model above for $k = 1, 2, \ldots$, the following explicit formulae is obtained for the predicted output:

$$y(k) = \sum_{i=0}^{k} e(k), \quad k = 0, 1, \ldots . \tag{19.33}$$

As $(e(i),\ i = 1, 2, \ldots)$ is a sequence of independent random variables with zero mean and unit variance, the variance of $y(k)$ is simply the sum of the variances of the variables on the right-hand side of Eq. (19.33):

$$\sigma^2_{y(k)} = k + 1, \tag{19.34}$$

which clearly grows in time in a monotonic way.

19.4.2. Non-informative Models

Causes and consequences
A model is not informative with respect to one or more of its variables and/or parameters if its dependent variable(s) do not change by changing them. It means that non-informative models are not sensitive enough with respect to their independent variable(s) and/or parameters.

Detection and prevention
The detection is performed by calculating the sensitivities of the validated model by all of its parameters and independent variables. The calculation can be performed analytically as it has been described in Section 19.3.2 and then check if the particular sensitivities differ significantly from zero.

Note that it is not always unfortunate if one figures out that there is a non-informative part in the model. The non-informative "unnecessary" part of the model should be simply left out from the model by neglecting the corresponding terms or considering the corresponding factor to be constant. This manipulation is then seen as a kind of model simplification.

The following simple example shows how one can visualize the notion of a non-informative functional dependence.

EXAMPLE 19.4.3 (Non-informative functional dependence). Figure 19.3 shows the graph of a two-variable nonlinear functional relationship. It is seen that there is a wide "flat" area of the function. This indicates the region where it does not depend on its variables, that is where it is non-informative.

19.4.3. Ill-conditioned or not Properly Trained Models

Causes and consequences
If the parameter and/or independent variable space is not covered by sufficient data then we end up with ill-conditioned or poorly trained models. Such models lead to uncertain, non-decisive or in some cases even non-correct decisions if one uses the model for control and/or diagnostic purposes or tries to use the model output for any kind of further processing.

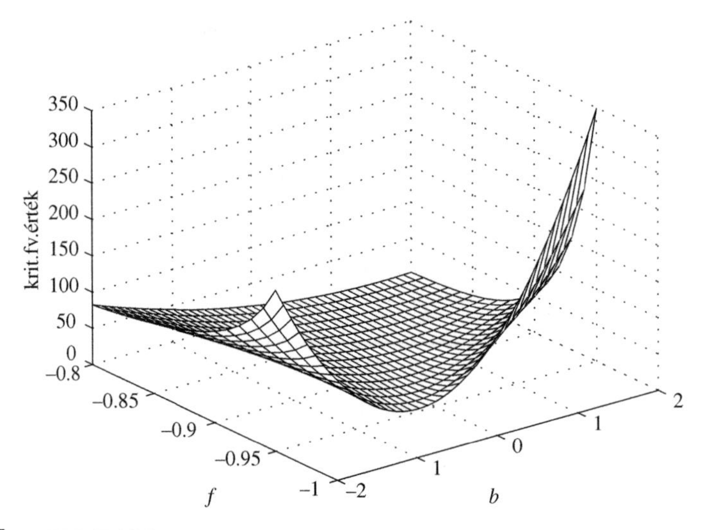

FIGURE 19.3 A non-informative functional relationship.

Detection and prevention

The method of detection depends on the type of black-box model which is used:

1. *Static and dynamic models linear in parameters*
   In this case, we can easily and safely detect ill-conditioned models via the covariance matrix of the parameters through the correlation coefficient ρ of the estimated parameters which is computed according to the formulae:

$$\rho_{ij} = \frac{[\mathbf{COV}\{\hat{p}\}]_{ij}}{s_{p_i} s_{p_j}}, \tag{19.35}$$

   where $[\mathbf{COV}\{\hat{p}\}]_{ij}$ is the off-diagonal element of the covariance matrix and s_{p_i}, s_{p_j} are the corresponding diagonal elements. Note that $1 \geq \rho \geq -1$ is a measure of the dependence between the estimated values: if it is small compared to 1 then we have a good, uncorrelated estimate for the parameters, otherwise the model is ill-conditioned or not properly trained.

   We have already seen an example of this phenomena in Section 12.5.3 where the estimated coefficients were highly correlated. Note that this poor quality estimate could have been avoided by experiment design, by properly selecting the measurement point in advance (see Sections 12.3 and 12.4).
2. *Neural net models*
   There are theoretical investigations on how many training samples are needed to properly train a neural net of a given architecture, say a feedforward three-layer neural net of a given number of neurons in the input, hidden and output layers [182]. This number can be rather high in practical cases compared with the number of available samples. Therefore, it is often the case that one uses a poorly trained neural net in real applications. Example 19.4.4 contains a concrete example on the number of necessary training samples.

   Poorly trained neural nets for classification may even misclassify unknown samples.

The usual way of testing whether the neural net is properly trained is to divide the training samples into two subsets. The first one is used for training and then the second is to test the computation error of the net.

EXAMPLE 19.4.4 (Poorly trained neural network models for fault classification [183]). Let us consider a three-layer feedforward neural net for fault classification with six neurons in the output (that is six fault types are considered) and three neurons in the hidden layer with all together 45 weights. If we want the net to perform the classification with less that 10% error, then we need to have 300 samples and train the network with them with a precision of 5%. This sample size grows to 2300 if we need less than 1% classification error.

More about neural networks in chemical engineering can be found in [184].

19.5. SUMMARY

Process models usually contain black-box model elements imbedded in a white-box model structure, therefore the methods and tools of empirical model building are of primary practical importance. The revised and extended SEVEN STEP MODELLING PROCEDURE provides a common framework for such grey-box models, incorporating the steps needed for empirical model building.

This chapter discusses the most commonly used black-box model structures in process modelling: linear and nonlinear static model structures including neural networks, as well as linear dynamic ARMAX and nonlinear dynamic Volterra and Hammerstein models. Together with the mathematical description, the structural parameters and the proposed identification method for each of the black-box model types are given.

Special emphasis is put into unusual and sensitive points related to empirical model building such as the sensitivity of the model output with respect to its parameters and variables, the information transfer between the black-box submodel and its white-box environment and to the traps and pitfalls in empirical model building.

19.6. FURTHER READING

There is a vast amount of literature on empirical model building typically in fields of high complexity and practical importance such as

- *Modelling of complex reaction kinetic expressions*
  An empirical kinetic model has been developed for the extraction of black cumin seed oil with supercritical carbon dioxide as solvent using neural net model in [185].

  A laminar falling film slurry photocatalytic reactor used for wastewater treatment is modelled in [186] using a simplified experimental model with only five dimensionless parameters. The parameters can be readily estimated from process data. A study of the sensitivity of the model to variations of the models parameter and validation of the model with experimental data are also presented.

- *Modelling of unusual complex processes*
  A grey-box model is constructed to describe the turbulent burning between vertical parallel walls with a fire-induced flow in [187]. Transport equations for mass, momentum, chemical components and enthalpy are combined with an empirical model for estimating the flame radiation energy to the burning wall.

A general systematic procedure is presented to support the development and statistical verification of empirical dynamic process model in [188]. Methods are presented to address structural identifiability and distinguishability testing, as well as optimal design of dynamic experiments for both model discrimination and improving parameter precision.

19.7. REVIEW QUESTIONS

Q19.1. In the context of the systematic SEVEN STEP MODELLING PROCEDURE outlined in Section 2.3 of Chapter 2 describe how black-box model elements of various types naturally arise. (Section 19.2)

Q19.2. What are the three main types of black-box model elements? Describe their general mathematical form together with their white-box environment. (section 19.2)

Q19.3. Explain the importance of parametric sensitivity in empirical modelling. What is the relationship between variable sensitivity and model linearization? (Section 19.3)

Q19.4. Describe the linear and nonlinear static model structures suitable for empirical model building. Give their mathematical description and their structural parameters. (Section 19.3)

Q19.5. Describe the linear and nonlinear dynamic model structures suitable for empirical model building. Give their mathematical description and their structural parameters. (Section 19.3)

Q19.6. How would you characterize the information transfer between the black-box submodel and the white overall model structure? What is the role of the sensitivities in this transfer? (Section 19.3)

Q19.7. What are the most common traps and pitfalls in empirical model building? Describe the problem, the detection and the prevention method(s). (Section 19.4)

19.8. APPLICATION EXERCISES

A19.1. Develop a grey-box model of a plug-flow tubular isotherm catalytic reactor where the heterogeneous reaction kinetic expression is unknown. Assume a function-type black-box model. (Sections 19.2 and 19.3)

A19.2. Develop a grey-box model of a plug-flow tubular isotherm catalytic reactor where the heterogeneous reaction kinetic expression is unknown. Use a model-type black-box model. Compare it with a hierarchical model driven by the characteristic sizes of the same system described in Section 9.2. Compare it also with the model developed in the previous exercise. (Sections 19.2 and 19.3)

APPENDIX: BASIC MATHEMATICAL TOOLS

A.1. RANDOM VARIABLES AND THEIR PROPERTIES

A.1.1. Random Variables with Gaussian (Normal) Distribution

Scalar Case

Let us take the one dimensional *1D scalar case* first, that is a real-valued random variable

$$\xi\colon\ \xi(\omega), \quad \omega \in \Omega, \ \ \xi(\omega) \in \mathcal{R}. \tag{A.1}$$

The random variable is *normally distributed* or *has a Gaussian (Normal) Distribution*

$$\xi \sim \mathcal{N}(m, \sigma^2), \tag{A.2}$$

if its probability density function f_ξ is in the following form:

$$f_\xi(x) = \frac{1}{\sqrt{2\pi}\,\sigma} \mathrm{e}^{-(x-m)^2/2\sigma^2}, \tag{A.3}$$

where m is the *mean value* and σ^2 is the *variance* of the random variable. Both m and σ^2 are real numbers. The shape of the 1D Gaussian probability density function is shown in Fig. A.1.

The mean value of the random variable ξ with probability density function f_ξ is computed as

$$\mathbf{E}\{\xi\} = \int x f_\xi(x)\, \mathrm{d}x. \tag{A.4}$$

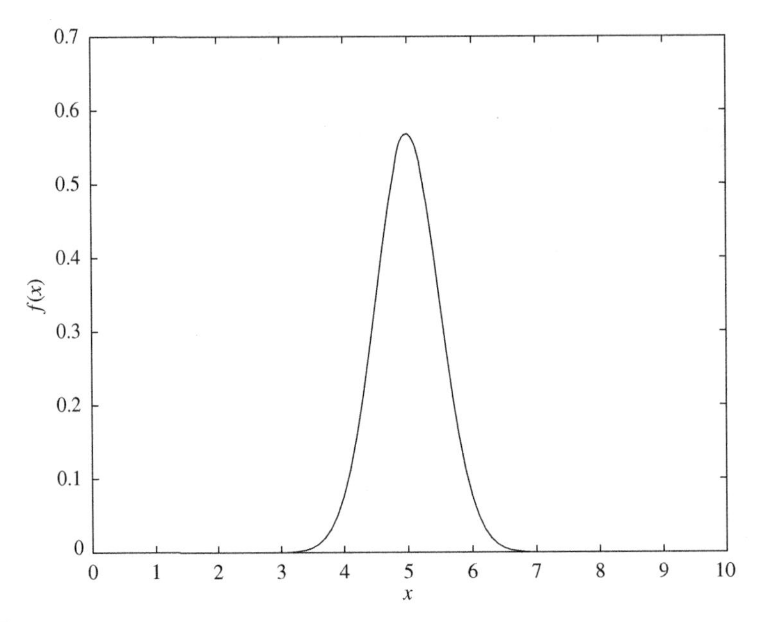

FIGURE A.1 One-dimensional Gaussian probability density function.

The *covariance* between two scalar random variables ξ and θ is defined as

$$\mathbf{COV}\{\xi, \theta\} = \mathbf{E}\{(\xi - \mathbf{E}\{\xi\})(\theta - \mathbf{E}\{\theta\})\}. \tag{A.5}$$

The variance of a random variable ξ is just a covariance of that variable with itself, i.e.

$$\sigma^2\{\xi\} = \mathbf{COV}\{\xi, \xi\} = \mathbf{E}\{(\xi - \mathbf{E}\{\xi\})^2\}. \tag{A.6}$$

Vector-valued Random Variables

Let us now assume that ξ is a vector-valued random variable, i.e.

$$\xi : \ \xi(\omega), \quad \omega \in \Omega, \ \xi(\omega) \in \mathcal{R}^\mu. \tag{A.7}$$

Then its mean is also a real-valued vector $m \in \mathcal{R}^\mu$ and its variance is a matrix called its *covariance matrix*,

$$\mathbf{COV}\{\xi\} = \mathbf{E}\{(\xi - \mathbf{E}\{\xi\})(\xi - \mathbf{E}\{\xi\})^{\mathrm{T}}\}. \tag{A.8}$$

It is important to note that *covariance matrices are positive definite symmetric matrices*, i.e.

$$z^{\mathrm{T}}\mathbf{COV}\{\xi\}z \geq 0, \quad \forall z \in \mathcal{R}^\mu.$$

The probability density function f_ξ of a vector-valued random variable is a scalar-valued multivariate function, i.e.

$$f_\xi : \mathcal{R}^\mu \rightarrow \mathcal{R}.$$

A *vector-valued random variable with multivariate Gaussian distribution*

$$\xi \sim \mathcal{N}(m, \Sigma) \tag{A.9}$$

with mean m and covariance matrix Σ is a random variable where each of its component is a scalar valued random variable is a $\xi_i, i = 1, \ldots, \mu$ as well as any set of its component random variables is normally distributed.

It is important to note that *the pair* (m, Σ) *is a sufficient statistic of a multivariate Gaussian distribution.* They are necessary and sufficient to describe a multivariate Gaussian distribution.

The shape of the 2D Gaussian probability density function is shown in Fig. A.2.

One of the most important properties of a Gaussian distribution is its *self-reproducibility*. It means that any linear combination of independent random variables with arbitrary but Gaussian distribution will be a random variable with Gaussian distribution.

Standard Normal (Gaussian) Distribution

A vector-valued ξ: $\xi(\omega), \quad \omega \in \Omega, \ \xi(\omega) \in \mathcal{R}^{\mu}$ random variable has a *standard normal (Gaussian) distribution* if its entries $\xi_i, \ i = 1, \ldots, r$ are independent normally distributed random variables with 0 mean and unit variance:

$$\xi_i \sim \mathcal{N}(0, 1), \qquad \xi \sim \mathcal{N}(0, \mathbf{I}) \tag{A.10}$$

where $\mathbf{I}$ is a unit matrix with appropriate dimension.

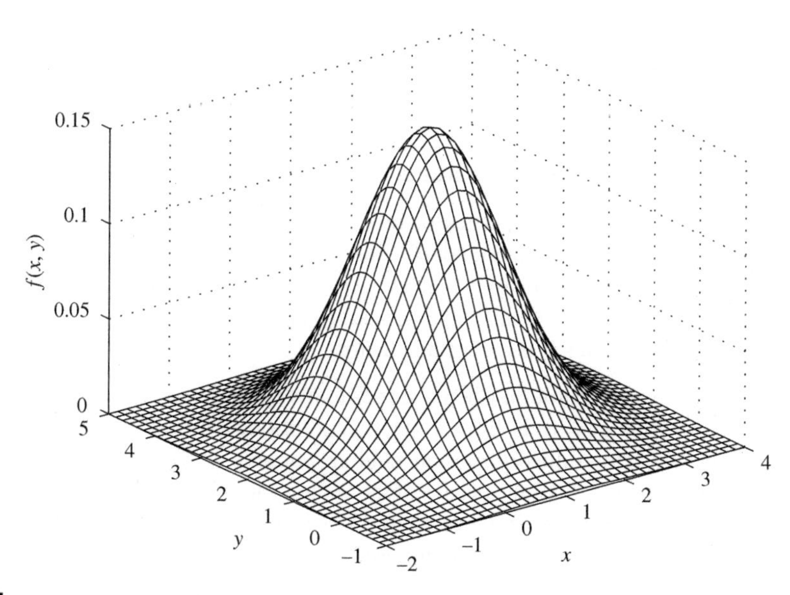

FIGURE A.2 Two-dimensional Gaussian probability distribution function.

A vector-valued ξ random variable has a standard normal (Gaussian) distribution [189] if and only if

1. the probability density function of ξ is in the form of

$$f_\xi(x) = (2\pi)^{-n/2} \exp(-\tfrac{1}{2}\|x\|_2^2), \qquad (A.11)$$

   where $\|x\|_2 = x^T x$ is the Euclidean or two norm of a vector x, or
2. with any given real vector $c \in \mathcal{R}^\nu$ the random variable $c^T\xi$ is a Gaussian random variable with zero mean and variance $\|c\|_2^2$, or
3. with any given orthonormal matrix $\mathbf{U} \in \mathcal{R}^{\nu\times\nu}$ the distribution of $\mathbf{U}\xi$ is the same as that of ξ and the random variable $\|\xi\|_2$ has a χ^2 distribution with the degree of freedom n.

Any of the properties above are necessary and sufficient for the random variable ξ having standard normal distribution.

A.1.2. Linear Transformation of Gaussian Random Variables

Let us assume that a vector-valued random variable $\xi(\omega) \in \mathcal{R}^n$ is given. If we transform this random variable to another one by applying a non-singular square transformation matrix $\mathbf{T} \in \mathcal{R}^{n\times n}$ to get

$$\eta = \mathbf{T}\xi, \qquad (A.12)$$

then the transformed random variable η has the following statistics:

$$\mathbf{E}\{\eta\} = \mathbf{TE}\{\xi\}, \quad \mathbf{COV}\{\eta\} = \mathbf{TCOV}\{\xi\}\mathbf{T}^{\mathbf{T}}. \qquad (A.13)$$

If, however, the random variable ξ has a Gaussian distribution $\mathcal{N}(m_\xi, \Delta_\xi)$ with the mean value m_ξ and covariance matrix Δ_ξ then the transformed random variable η is also normally distributed $\mathcal{N}(m_\eta, \mathbf{\Delta}_\eta)$ with

$$m_\eta = \mathbf{T}m_\xi, \qquad \Delta_\eta = \mathbf{T}\Delta_\xi\mathbf{T}^{\mathrm{T}}.$$

A.1.3. Standardization of Vector-valued Gaussian Random Variables

If one compares the properties of the vector-valued Gaussian random variables ξ and that of vector-valued standard Gaussian random variables $\xi(0, \mathbf{I})$ of the same dimension, then it can be seen that a general vector-valued Gaussian random variable can be generated by a suitable linear transformation [83] $\xi = \mathbf{A}\xi + m$ where $\mathbf{A} \in \mathcal{R}^{\nu\times\nu}$ constant matrix and m is a constant vector.

The inverse operation when one tries to obtain a vector-valued standard Gaussian random variable $\mathcal{N}(0, \mathbf{I})$ from a given general one ξ is called standardization. Let us have a vector-valued Gaussian random variable ξ with its mean vector m and covariance matrix Σ. If the entries of the vector-valued random variable ξ are linearly independent then the covariance matrix Σ is an invertible positive definite symmetric matrix. Therefore, it can be decomposed in the form of

$$\Sigma = \Sigma^{1/2}\Sigma^{1/2}, \quad \Sigma^{-1} = \Sigma^{-1/2}\Sigma^{-1/2}$$

with $\Sigma^{1/2}$ being a positive definite symmetric matrix, too. Then the vector-valued random variable

$$\xi_* = \Sigma^{-1/2}\xi - m \tag{A.14}$$

is a standard Gaussian vector-valued random variable.

A.2. HYPOTHESIS TESTING

Hypothesis testing is a standard procedure in mathematical statistics to get an evaluation of whether or not a particular relation holds for a random variable. The way to perform a hypothesis test is described below as a standard algorithmic problem.

HYPOTHESIS TESTING

Given:

- a *sample*

$$\mathbf{S}(\xi) = \{\xi_1, \xi_2, \ldots, \xi_n\} \tag{A.15}$$

  consisting of independent measurements as elements $\xi_i,\ i = 1, \ldots, n$ taken from a joint underlying random variable ξ;
- a *hypothesis* in the form of a relation on the statistics of the random variable in the sample, e.g.

$$H_0 : m = m_0 \tag{A.16}$$

  with m_0 is a given constant, the hypothetical value for the mean value m;
- an *assumption on the underlying distribution of the sample elements*, e.g. Gaussian with known variance Δ;
- a *significance level* $0 \leq \varepsilon \leq 1$ which is the level of confidence to which you test your hypothesis.

Question:
Is the hypothesis H_0 true on the significance level ε based on measurements in the sample $\mathbf{S}(\xi)$?

Method:
The hypothesis testing is usually done in three subsequent steps:

1. *Compute an estimate of the statistics present in the hypothesis relation*, e.g.

$$m = \mathbf{E}\{\xi\} \simeq \mathbf{M}\{\mathbf{S}(\xi)\} = \frac{\sum_{i=1}^{n} \xi_i}{n}, \tag{A.17}$$

   where $\mathbf{M}$ stands for the sample arithmetic mean.
2. *Construct a statistic of a normalized random variable with a known distribution*, e.g.

$$u = \frac{\mathbf{M}\{\mathbf{S}(\xi)\} - m_0}{\sqrt{\Delta}} \sim \mathcal{N}(0, 1). \tag{A.18}$$

3. *Compare the computed normalized statistics to its "critical" value (e.g. u_{crit}) corresponding to the given significance level ε*. The critical values are given usually in the form of tables. *If $u \leq u_{\text{crit}}$ then accept the hypothesis otherwise reject it.*

A.3. VECTOR AND SIGNAL NORMS

If we have objects of a non-scalar nature, for example vectors, matrices, functions or signals, we measure their magnitude by using *norms*. Norms are the extensions of a length of a vector applied to other objects of non-scalar nature forming a so-called *vector space*. The set of elements X is called a vector space if

- if $x_1,\ x_2 \in X$ then $x^* = (x_1 + x_2) \in X$,
- if $x_1 \in X$ then $x^* = ax_1 \in X$ for any $a \in \mathcal{R}$,
- there is a zero element $x_0 \in X$ for which $x_1 + x_0 = x_1$ holds for every $x_1 \in X$.

A scalar-valued function $\rho(\cdot) : X \to \mathcal{R}$ is a norm on a vector space X if

- $\rho(x) \geq 0, \quad \forall x \in X$,
- $\rho(x) = 0 \ \Leftrightarrow \ x = 0 \in X$, ($x = 0$ is the zero element),
- $\rho(x + y) \leq \rho(x) + \rho(y)$ (triangular inequality).

A.3.1. Vector Norms

The n-dimensional vectors with real-valued entries $x = [x_1, x_2, \ldots, x_n]^T$ form a vector space called $\mathcal{R}^n$. There are different known and useful norms defined over the vector space $\mathcal{R}^n$ which are as follows:

1. *Euclidean norm or 2-norm*

$$\|x\|_2 = \sqrt{\sum_{i=1}^{n} x_i^2} \tag{A.19}$$

2. *Maximum norm or ∞-norm*

$$\|x\|_\infty = \max_i(|x_i|) \tag{A.20}$$

A.3.2. Signal Norms

Discrete time linear systems are usually described by *linear difference equations* which contain *discrete time signals* $f(k)$, $k = 0, 1, \ldots$. Discrete time signals can be *sampled data* of continuous time signals. Discrete time signals form a vector space. A discrete time signal may be *scalar-valued or vector-valued*, i.e.

$$f(k) \in \mathcal{R}, \qquad F(k) \in \mathcal{R}^{\mu}.$$

We shall use two types of *norms of a discrete time signal*: the *infinity norm* and the *2-norm*

$$\|f\|_\infty = \sup_k |f(k)|, \qquad \|F\|_\infty = \sup_k \|F(k)\|_\infty, \tag{A.21}$$

$$\|f\|_2^2 = \sum_{k=-\infty}^{\infty} f^2(k), \qquad \|F\|_2^2 = \sum_{k=-\infty}^{\infty} \|F(k)\|_2^2. \tag{A.22}$$

The induced norm of an operator $\mathbf{Q}$ on the vector space X induced by a norm $\|\cdot\|$ on the same space is defined as

$$\|\mathbf{Q}\| = \sup_{\|x\|=1} \frac{\|\mathbf{Q}(x)\|}{\|x\|}. \tag{A.23}$$

A.4. MATRIX AND OPERATOR NORMS

Matrices can be seen as linear operators transforming vectors from a vector space like $\mathcal{R}^n$ to another vector space which can be the same or a different one like $\mathcal{R}^m$. For rectangular matrices when $m \neq n$ and $\mathbf{T} \in \mathcal{R}^{m\times n}$ we have different vector spaces and for square matrices when $n = m$ they are the same vector space.

The "magnitude" of a matrix is also characterized by its norm. Similarly to the case of vectors and matrices there are various norms applicable for matrices. The most important ones are the so-called *induced norms* where the matrix norm is derived using an already defined vector norm. The *induced N-norm* of a square matrix $\mathbf{T} \in \mathcal{R}^{n\times n}$ is defined as

$$\|\mathbf{T}\|_N = \sup_{x\in\mathcal{R}^n} \frac{\|\mathbf{T}x\|_N}{\|x\|_N}, \tag{A.24}$$

where $\|\cdot\|_N$ is a vector norm.

The norm of an operator $\mathbf{T}$ acting on a signal space is defined similarly but now we have a signal norm instead of a vector norm in $\|\cdot\|_N$.

A.4.1. Quadratic Forms and Definite Matrices

Quadratic forms are important expressions in loss functions of both parameter estimation and control problem formulations. A quadratic form is an expression containing a vector $x \in \mathcal{R}^n$ and a square matrix $\mathbf{Q} \in \mathcal{R}^{n\times n}$:

$$q_Q = x^{\mathrm{T}}\mathbf{Q}x. \tag{A.25}$$

Observe that the value of the quadratic form itself is a real number, that is

$$x^{\mathrm{T}}\mathbf{Q}x \in \mathcal{R}.$$

A quadratic form is *definite* if it has the same sign for every possible vector x. The matrix $\mathbf{Q}$ present in the quadratic form is called definite if the corresponding quadratic form is definite. In particular, a *quadratic matrix* $\mathbf{Q}$ *is called*

- *positive definite, if* $x^{\mathrm{T}}\mathbf{Q}x > 0$,
- *negative definite, if* $x^{\mathrm{T}}\mathbf{Q}x < 0$,
- *positive semidefinite, if* $x^{\mathrm{T}}\mathbf{Q}x \geq 0$,
- *negative semidefinite, if* $x^{\mathrm{T}}\mathbf{Q}x \leq 0$,
- *if neither of the above conditions hold it is indefinite*

for every possible $x \in \mathcal{R}^n$.

A.5. GRAPHS

Graphs in general are combinatorial objects used for describing the structure of sets and set systems. A *graph* G in the mathematical sense is specified by a pair of related sets

$$G = (V, E),$$

where $V = \{v_1, v_2, \ldots, v_n\}$ is the *set of vertices* and E is the *set of edges* which contains vertex-pairs, i.e.

$$E = \{(v_i, v_j) \mid v_i, v_j \in V\}.$$

A *directed graph* $\bar{G}$ is a graph whose edge set $\bar{E}$ contains *ordered vertex-pairs*. It means that $(v_i, v_j) \in E$ follows from $(v_j, v_i) \in E$ in the undirected case but for directed graphs $(v_j, v_i) \in \bar{E}$ does not imply $(v_i, v_j) \in \bar{E}$.

Graphs and directed graphs can be conveniently visualized by figures where we depict vertices as points or small circles and edges (v_i, v_j) as intervals joining the points labelled v_i and v_j or directed arrows pointing to v_j and starting at v_i in case of directed graphs.

A sequence $P = (v_1, \ldots, v_n)$, $v_i \in V$ of vertices in a graph $G = (V, E)$, forms a path in G, if $(v_i, v_{i+1}) \in E$, i.e. the subsequent elements in the sequence are joined by an edge. A *directed path* is a vertex sequence in a directed graph $\bar{G} = (V, \bar{E})$ where the subsequent elements are joined by a directed edge pointing to the direction of increasing sequence numbers, i.e. $(v_i, v_{i+1}) \in \bar{E}$. The *length of a (directed) path* is the number of edges one traverses while traversing the path.

A *(directed) circle* is a (directed) path where the starting and ending vertices coincide, i.e. $v_1 = v_n$. A *(directed) loop* is an edge whose starting and ending vertex is the same, i.e. (v_i, v_i). A (directed) loop is a circle of length 1.

A subgraph $G' = (V', E')$ of a (directed) graph $G = (V, E)$ with $V' \subseteq V$, $E' \subseteq E$ is *connected (strongly connected)* if there is at least one (directed path) connecting any vertex pair (v_1, v_2), $v_1, v_2 \in V'$. A strongly connected subgraph is called a *strong component* of the original graph.

Weighted graphs are graphs where we put a weight, usually a real or integer number to each of the edges. The weight function w is then defined as

$$w(v_i, v_j) = w_{ij}, \quad \forall (v_i, v_j) \in E.$$

The weighted graph G_w is denoted by $G_w = (V, E; w)$. *Weighted directed graphs* $\bar{G}_w = (V, \bar{E}; w)$ are defined analogously.

The *value of a (directed) path in a weighted (directed) graph* is the product of the weights on the edges one traverses while traversing the path, i.e.

$$w((v_1, \ldots, v_n)) = \prod_{i=1}^{n-1} w(v_i, v_{i+1}).$$

The basic notions of graphs are illustrated on a simple example shown in Fig. A.3.

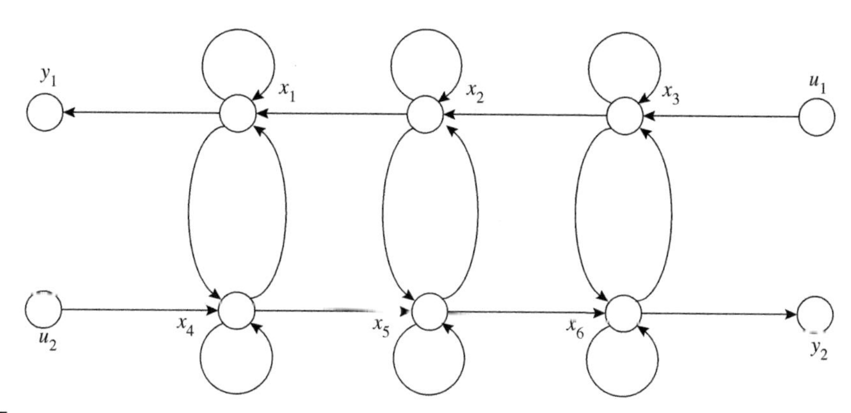

FIGURE A.3 A simple directed graph example.

EXAMPLE A.5.1 (A model structure graph as an example of a directed graph). Consider a directed graph

$$\bar{G} = (V, \bar{E}) \tag{A.26}$$

$$V = \{x_1, x_2, x_3, x_4, x_5, x_6;\ u_1, u_2;\ y_1, y_2\} \tag{A.27}$$

$$\bar{E} = \{(x_1, x_1),\ (x_2, x_2),\ (x_3, x_3),\ (x_4, x_4),\ (x_5, x_5),\ (x_6, x_6), \tag{A.28}$$

$$(x_1, x_4),\ (x_4, x_1),\ (x_2, x_5),\ (x_5, x_2),\ (x_3, x_6),\ (x_6, x_3), \tag{A.29}$$

$$(u_1, x_3),\ (x_3, x_2),\ (x_2, x_1),\ (x_1, y_1), \tag{A.30}$$

$$(u_2, x_4),\ (x_4, x_5),\ (x_5, x_6),\ (x_6, y_2)\} \tag{A.31}$$

$\bar{P} = (u_1, x_3, x_2, x_1, y_1)$ forms a directed path of length 4 in the graph. The edges belonging to the directed path are in Eq. (A.30).

There are several loops, e.g. (x_1, x_1), and several directed circles of length 2, e.g. (x_1, x_4, x_1) in this graph. The loops are collected in Eq. (A.28) and the circles of length 2 are in Eq. (A.29).

The subgraph G_x spanned by the state vertices

$$V_x = \{x_1, x_2, x_3, x_4, x_5, x_6\}$$

is strongly connected.

Note that the directed graph in Fig. A.3 is the structure graph of a countercurrent heat exchanger lumped with three lump pairs.

BIBLIOGRAPHY

[1] D.M. Himmelblau and K.B. Bishoff, *Process Analysis and Simulation*, Wiley, New York, 1968.
[2] M.E. Davis, *Numerical Methods and Modelling for Chemical Engineers*, Wiley, New York, Chichester, Brisbane, 1984.
[3] J.B. Riggs, *An Introduction to Numerical Methods for Chemical Engineers (2nd edn)*, Texas University Press, Texas, 1994.
[4] R.D. Rice and D.D. Do, *Applied Mathematics and Modelling for Chemical Engineers*, Wiley, New York, 1995.
[5] R. Aris, *Mathematical Modelling Techniques*, Pitman, London, 1978.
[6] M.M. Denn, *Process Modelling*, Longman, New York and London, 1986.
[7] M. Minsky, "Models, minds, machines", *Proceedings IFIP Congress*, pp. 45–49, 1965.
[8] Sdu Uitgevers, *Methods for the Calculation of Physical Effects, CPR14E (Pt2)*, Committee for Prevention of Disasters, The Netherlands, 1997.
[9] G.E. Davis, *A Handbook of Chemical Engineering 2nd edn*, Davis Bros., Manchester, England, 1904.
[10] W.H. Walker, W.K. Lewis, and W.H. McAdams, *Principles of Chemical Engineering*, McGraw-Hill Book Co. Inc., New York, London, 1923.
[11] J.C. Olsen, *Unit Processes and Principles of Chemical Engineering*, Macmillan and Co., London, 1932.
[12] T.J. Williams, *Systems Engineering for the Process Industries*, McGraw-Hill, New York, 1961.
[13] R.B. Bird, W.E. Stewart, and E.N. Lightfoot, *Transport Phenomena*, Wiley, New York, 1960.
[14] L. Ljung, *System Identification: Theory for the User*, Prentice Hall Inc., New Jersey, USA, 1987.
[15] R. Penrose, *The Emperor's New Mind*, Vintage, 1990.
[16] J.M. Honig, *Thermodynamics: Principles Characterizing Physical and Chemical Processes*, Elsevier, Amsterdam, Oxford, New York, 1982.
[17] G. Astarita, *Thermodynamics: An Advanced Textbook for Chemical Engineers*, Plenum Press, New York–London, 1989.

[18] A. Adetayo, J.D. Litster, and I.T. Cameron, "Steady-state modelling and simulation of a fertilizer granulation circuit", *Computers and Chemical Engineering*, **19**, pp. 383–393, 1995.

[19] P.C. Kapur, "Kinetics of granulation by non-random coalescence mechanism", *Chem. Eng. Science*, **27**, pp. 1863, 1972.

[20] A.A. Adetayo, *Modelling and Simulation of a Fertilizer Granulation Circuit*, PhD thesis, Department of Chemical Engineering, The University of Queensland, 1993.

[21] R.H. Perry and D. Green, *Perry's Chemical Engineer's Handbook (6th edn)*, McGraw-Hill, New York, 1985.

[22] R. Taylor and R. Krishna, *Multicomponent Mass Transfer*, John Wiley, New York, 1993.

[23] J.P. Holman, *Heat Transfer (5th edn)*, McGraw-Hill International, Singapore, 1981.

[24] J.R. Welty, *Engineering Heat Transfer*, Wiley, New York, London, 1974.

[25] F.P. Incropera and D.P. DeWitt, *Introduction to Heat Transfer (2nd edn)*, John Wiley International, New York, Singapore, 1990.

[26] J.M. Coulson, J.F. Richardson, J.R. Backhurst, and J.H. Harker, *Chemical Engineering Vol. 1 (5th edn)*, Butterworth Heinemann, Oxford, England, 1995.

[27] W.L. McCabe, J.C. Smith, and P. Harriott, *Unit Operations of Chemical Engineering (4th edn)*, McGraw-Hill Publishing Company, New York, 1985.

[28] H.C. Hottel and A.F. Sarofim, *Radiative Transfer*, McGraw-Hill, New York, 1967.

[29] J.M. Coulson, J.F. Richardson, and R.K. Sinnott, *Chemical Engineering Vol. 6 Design*, Pergamon Press, Oxford, 1983.

[30] Aspen Technology, *ASPEN PLUS User's Guide*, Aspen Technology, Cambridge, MA, 1997.

[31] Hyprotech, *HYSIM/HYSIS User's Guide*, Hyprotech Ltd., Calgary, Canada, 1997.

[32] IChemE, *PPDS Property Data System*, Institution of Chemical Engineers, UK, Rugby, UK, 1990.

[33] AIChE, *DIPPR Property Data System*, American Institute of Chemical Engineers, New York, USA, 1997.

[34] R.C. Reid, J.M. Prausnitz, and T.K. Sherwood, *The Properties of Gases and Liquids (3rd edn)*, McGraw-Hill, New York, 1977.

[35] S.M. Walas, *Phase Equilibria in Chemical Engineering*, Butterworths, Stoneham, MA 02180, 1985.

[36] S.I. Sandler, *Chemical and Engineering Thermodynamics*, Wiley and Sons, New York, 1989.

[37] J. Ingham, *Chemical Engineering Dynamics: An Introduction to Modelling and Computer Simulation*, Wiley-VCH, Cambridge, 2000.

[38] R. Gani and I.T. Cameron, "Modelling for dynamic simulation of chemical processes—the index problem", *Chemical Engineering Science*, **47**, pp. 1311–1315, 1992.

[39] K.M. Hangos and I.T. Cameron, "The formal representation of process system modelling assumptions and their implications", *Computers and Chemical Engineering*, **21**, pp. S823–S828, 1997.

[40] C.C. Pantelides, "The consistent initialization of differential-algebraic systems", *SIAM J. Sci. Stat. Comp.*, **9**(2), pp. 213–231, 1988.

[41] L.R. Petzold, B. Leimkuhler, and C.W. Gear, "Approximation methods for the consistent initialization of differential-algebraic equations", *SIAM J. Numer. Anal.*, **28**(1), pp. 205–226, 1991.

[42] W. Marquardt, A. Kroner, and E.D. Gilles, "Getting around consistent initialization of DAEs?", *Computers and Chemical Engineering*, **21**(2), pp. 145–158, 1997.

[43] E.C. Biscaia and R.C. Vieira, "Heuristic optimization for consistent initialization of DAEs", *Computers and Chemical Engineering*, **24**, pp. 2813–2191, 2000.

[44] R.B. Newell and P.L. Lee, *Applied Process Control: A Case Study*, Prentice-Hall, NJ, 1989.

[45] H.H. Robertson, *Solution of a Set of Reaction Rate Equations in Numerical Analysis*, Thompson Book Co., 1967.

[46] L.R. Petzold, "A description of dassl: A differential/algebraic system solver", *IMACS World Congress*, pp. xx–xx, 1982.

[47] P. Barton, "DAEPACK—a suite of numerical routines for continuous and hybrid differential-algebraic equations", 2001, http://yoric.mit.edu/daepack/daepack.html.

[48] I.T. Cameron, "Solution of differential-algebraic equations using diagonally implicit Runge-Kutta methods", *Inst. Maths Applics J. Numerical Analysis*, **3**, pp. 273–289, 1983.

[49] E. Hairer, C. Lubich, and M. Roche, *The Numerical Solution of Differential-Algebraic Systems by Runge–Kutta Methods*, Springer-Verlag, Berlin, Heidelberg, 1989.

[50] R. Williams, K. Burrage, I.T. Cameron, and M. Kerr, "A four-stage index-2 diagonally implicit Runge-Kutta method", *Applied Numerical Methods*, 2000 (in press).

[51] I. Duff, *Analysis of Sparse Systems*, D.Phil. Oxford University, 1972.

[52] R. Tarjan, "Depth first search and linear graph algorithms", *SIAM Journal Computing*, **1**, pp. 146–160, 1972.

[53] R.W.H. Sargent and A.W. Westerberg, "Speedup in chemical engineering design", *Trans. Inst. Chem. Eng.*, **42**, pp. 190–197, 1964.

[54] B. Roffel and J. Rijnsdorp, *Process Dynamics, Control and Protection*, Ann Arbor Science, UK, 1982.

[55] B.A. Finlayson, *The Method of Weighted Residuals and Variational Principles*, Academic Press, New York, 1972.

[56] G.D. Smith, *Numerical Solution of Partial Differential Equations; Finite Difference Methods*, Clarendon Press, Oxford, 1985.

[57] B.A. Finlayson, *Nonlinear Analysis in Chemical Engineering*, McGraw-Hill, New York, 1980.

[58] John Villadsen and M.L. Michelsen, *Solution of Differential Equation Models by Polynomial Approximation*, Prentice-Hall, NJ, 1978.

[59] C. Beckerman, L.A. Bertram, S.J. Pien, and R.E. Smelser (Eds.), "Micro/macro scale phenomena in solidification", in *Winter Annual Meeting of the American Society of Mechanical Engineers (Anaheim, CA, USA), Published by ASME*, pp. 1–151. New York (USA), 1992.

[60] A. Kaempf-Bernd, "Stresses and strains in two-phase materials", *Computational Materials Science*, **5**, pp. 151–156, 1996.

[61] T.D. Papathanasiou, "On the effective permeability of square arrays of permeable fiber tows", *International Journal of Multiphase Flow*, **23**, pp. 81–92, 1997.

[62] R.E. Showalter, "Diffusion models with microstructure", *Transport in Porous Media*, **6**, pp. 567–580, 1991.

[63] A. Claussen-Martin, "Area averaging of surface fluxes in a neutrally stratified, horizontally inhomogeneous atmospheric boundary layer", *Transport in Porous Media*, **6**, pp. 567–580, 1991.

[64] G.A. Robertson and I.T. Cameron, "Analysis of dynamic process models for structural insight and model reduction i: Structural identification measures", *Computers and Chemical Engineering*, **21**(5), pp. 455–473, 1997.

[65] G.A. Robertson and I.T. Cameron, "Analysis of dynamic process models for structural insight and model reduction II: A multistage compressor case study", *Computers and Chemical Engineering*, **21**(5), pp. 475–488, 1997.

[66] D. Maroudas, "Multiscale modelling of hard materials: Challenges and opportunities for chemical engineering", *AIChE Journal*, **46**(5), pp. 878–882, 2000.

[67] G. Martin, "Modelling materials far from equilibrium", *Current Opinion in Solid State and Materials Science*, **3**(6), pp. 552–557, 1998.

[68] G. Martin, "Simulation of solidification", *Current Opinion in Solid State and Materials Science*, **3**(3), pp. 217–218, 1998.

[69] van J. Leeuven (Ed.), *Handbook of Theoretical Computer Science, Vol. A., Algorithms and Complexity*, Elsevier- MIT Press, Amsterdam, 1990.

[70] T. Kailath, *Linear systems*, Prentice-Hall, NJ, 1980.

[71] K.J Reinschke, *Multivariable Control. A Graph-theoretic Approach*, Springer-Verlag, 1988.

[72] R. Johnson and D. Wichern, *Applied Multivariate Statistical Analysis*, Prentice-Hall, 1992.

[73] R.O. Gilbert, *Statistical Methods for Environmental Pollution Monitoring*, Van Nostrand Reinhold Co., New York, 1987.

[74] K.J. Astrom and B. Wittenmark, *Computer Controlled Systems*, Prentice Hall Inc., NJ, 1990.

[75] M. Basseville and I.V. Nikiforov, *Detection of Abrupt Changes. Theory and Application*, Prentice-Hall Inc., NJ, 1993.

[76] J.A. Romagnoli and M.C. Sanchez, *Data Processing and Reconciliation for Chemical Process Operations*, Academic Press, 1999.

[77] P. Eykhoff, *Trends and Progress in System Identification*, Pergamon Press, Oxford, 1981.

[78] G.K. Raju and C.L. Cooney, "Active learning from process data", *AIChE Journal*, **44**(10), pp. 2199–2211, 1998.

[79] J.-E. Haugen, O. Tomic, and K. Kvaal, "A calibration method for handling temporal drift of solid state gas-sensors", *Analytica Chimica Acta*, **407**(1–2), pp. 23–39, 2000.

[80] K.M. Thompson, "Developing univariate distributions from data for risk analysis", *Human and Ecological Risk Assessment*, **5**(4), pp. 755–783, 1999.

[81] T. Bohlin, *Interactive System Identification: Prospects and Pitfalls*, Springer-Verlag, Berlin, Heidelberg, New York, London, 1991.

[82] E. Walter and L. Pronzato, *Identification of Parametric Models*, Springer–Masson, Paris, Milan, Barcelona, 1997.

[83] S. S. Wilks, *Mathematical Statistics*, Wiley, New York, 1962.

[84] M. Schatzoff, "Exact distribution of wilks likelihood ratio test", *Biometrika*, **53**, pp. 347–358, 1966.

[85] L. Ljung, "Model validation and model error modeling", 1999, Report no: LiTH-ISY-R-2125 of the Automatic Control Group in Linköping Department of Electrical Engineering Linköping University.

[86] K.-R. Koch, *Parameter Estimation and Hypothesis Testing in Linear Models*, Springer-Verlag, 1999.

[87] P. Englezos, *Applied Parameter Estimation for Chemical Engineers (Chemical Industries, Vol. 81)*, Marcel Dekker, 2000.

[88] C.C. Heyde, *Quasi-Likelihood and Its Application: A General Approach to Optimal Parameter Estimation (Springer Series in Statistics)*, Springer, 1997.

[89] G.W. Barton, W.K. Chan, and J.D. Perkins, "Interaction between process design and control: the role of open-loop indicators", *J. Process Control*, **1**, pp. 161, 1991.

[90] A. Walsh, *Analysis and Design of Process Dynamics using Spectral Methods*, Ph.D. thesis, University of Queensland, Australia, 2000.

[91] B.M. Russel, J.P. Henriksen, S.B. Jorgensen, and R. Gani, "Integration of design and control through model analysis", *Computers and Chemical Engineering*. **24**(2–7), pp. 967–973, 2000.

[92] M.J. Mohideen, J.D. Perkins, and E.N. Pistikopoulos, "Optimal design of dynamic systems under uncertainty", *AIChE Journal*, **42**(8), pp. 2281, 1996.

[93] A.J. Groenendijk, A.C. Dimian, and P.D. Iedema, "Systems approach for evaluating dynamics and plantwide control of complex plants", *AIChE Journal*, **46**(1), pp. 133–145, 2000.

[94] N. Mahadevan and K.A. Hoo, "Wavelet-based model reduction of distributed parameter systems", *Chemical Engineering Science*, **55**(9), pp. 4271–4290, 2000.

[95] B. Faltings and P. Struss, *Recent Advances in Qualitative Physics*, The MIT Press, Cambridge, MA, 1992.

[96] C.J. Puccia and R. Levins, *Qualitative Modelling of Complex Systems: An Introduction to Loop Analysis and Time Averaging*, Harvard University Press, Cambridge, London, 1985.

[97] P. Rose and M.A. Kramer, "Qualitative analysis of causal feedback", in *Proc. 10th National Conf. an Artificial Intelligence (AAAI-91)*. Anahein, CA, 1991.

[98] D.S. Weld and J. deKleer, "A qualitative physics based on confluences", *Artificial Intelligence*, **24**, pp. 7–84, 1984.

[99] G. Stephanopoulos, *Chemical Process Control: An Introduction to Theory and Practice*, Prentice-Hall, 1984.

[100] D.E. Seborg, T.F. Edgar, and D.A. Mellichamp, *Process Dynamics and Control (Wiley Series in Chemical Engineering)*, John Wiley and Sons, 1989.

[101] S. Skogestad and I. Postlethwaite, *Multivariable Feedback Control*, Wiley, New York, 1996.

[102] E.D. Sontag, *Mathematical Control Theory: Deterministic Finite Dimensional Systems (Texts in Applied Mathematics 6.)*, Springer-Verlag, New York, 1990.

[103] J.L. Peterson, *Petri Net Theory and Modelling of Systems*, Prentice-Hall, Englewood Cliffs, NJ, 1981.

[104] R. David and H. Alla, *Petri Nets and Grafcet*, Prentice-Hall International (UK) Ltd., 1992.

[105] M. Gerzson and K. M. Hangos, "Analysis of controlled technological systems using high level petri nets", *Computers and Chemical Engineering*, **19**, pp. S531–S536, 1995.

[106] H. Alla, "Modelling and simulation of event-driven systems by petri nets", in *Workshop on Discrete Event Systems in Process Systems Engineering*. London (UK), 1996.

[107] K. Jensen and G. Rosenberg (Eds.), *High-level Petri Nets: Theory and Applications*, Springer Verlag, Berlin, Heidelberg, 1991.

[108] K. Jensen, *Coloured Petri Nets: Basic Concepts, Analysis Methods and Practical Use*, Springer-Verlag, Berlin, Heidelberg, 1992.

[109] C.G. Cassandras, S. Lafortune, and G. J. Oldser, "Introduction to the modelling, control and optimization of discrete event systems, in: A. Isidori (ed.)", *Trends in Control, A European Perspective*, pp. 217–292, 1995.

[110] R. David and H. Alla, "Petri nets for modelling of dynamic systems—a survey", *Automatica*, **30**, pp. 175–202, 1994.

[111] P. Philips, K.B. Ramkumar, K.W. Lim, H.A. Preisig, and M. Weiss, "Automaton-based fault detection and isolation", *Computers and Chemical Engineering*, **23**, pp. S215–S218, 1999.

[112] A. Sanchez, G. Rotstein, N. Alsop, and S. Machietto, "Synthesis and implementation of procedural controllers for event-driven operations", *AIChE Journal*, **45**, pp. 1753–1775, 1999.

[113] T. Nishi, A. Sakata, S. Hasebe, and I. Hashimoto, "Autonomous decentralized scheduling system for just-in-time production", *Computers and Chemical Engineering*, **24**, pp. 345–351, 2000.

[114] C. Azzaro-Pantel, L. Bernal-Haro, P. Bauder, S. Domenech, and L. Pibouleau, "A two-stage methodology for short-term batch plant scheduling: Discrete event simulation and genetic algorithm", *Computers and Chemical Engineering*, **22**, pp. 1461–1481, 1998.

[115] S. Viswanathan, C. Johnsson, R. Srinivasan, V. Venkatasubramanian, and K.E. Arzen, "Automating operating procedure synthesis for batch processes: Part I. knowledge representation and planning framework", *Computers and Chemical Engineering*, **22**, pp. 1673–1685, 1998.

[116] S. Viswanathan, C. Johnsson, R. Srinivasan, V. Venkatasubramanian, and K.E. Arzen, "Automating operating procedure synthesis for batch processes: Part II. implementation and application", *Computers and Chemical Engineering*, **22**, pp. 1687–1698, 1998.

[117] R. Srinivasan and V. Venkatasubramanian, "Automating hazop analysis for batch chemical plants: Part i. the knowledge representation framework", *Computers and Chemical Engineering*, **22**, pp. 1345–1355, 1998.

[118] R. Srinivasan and V. Venkatasubramanian, "Automating hazop analysis for batch chemical plants: Part II. algorithms and application", *Computers and Chemical Engineering*, **22**, pp. 1357–1370, 1998.

[119] A. Szucs, M. Gerzson, and K.M. Hangos, "An intelligent diagnostic system based on Petri nets", *Computers and Chemical Engineering*, **22**, pp. 1335–1344, 1998.

[120] S. Gonnet and O. Chiotti, "Modelling of the supervisory control system of a multi-purpose batch plant", *Computers and Chemical Engineering*, **23**, pp. 611–622, 1999.

[121] B. Lennartson, B. Egardt, and M. Tittus, "Hybrid systems in process control", *Proc. 33rd CDC*, **4**, pp. 3587–3592, 1994.

[122] S. Petterson and B. Lennartson, "Hybrid modelling focussed on hybrid Petri nets", *2nd European Workshop on Real-time and Hybrid Systems*, 1995.

[123] T.A. Henzinger, P.-H. Ho, and H. Wong-Toi, "Algorithmic analysis of nonlinear hybrid systems", *IEEE Transactions on Automatic Control*, **43**(4), pp. 540–554, 1998.

[124] V.D. Dimitriadis, N. Shah, and C.C. Pantelides, "Modelling and safety verification of discrete/continuous processing systems", *AIChE Journal*, **34**(4), pp. 1041–1059, 1997.

[125] Process Systems Enterprise Ltd., "gPROMS", 2000, http://www.psenterprise.com/products/products.html.

[126] Daesim Technologies, "Daesim Studio", 2001, http://www.daesim.com/.

[127] E. Hairer, S.P. Nørsett, and G. Wanner, *Solving Ordinary Differential Equations*, Springer-Verlag, Berlin, 1991.

[128] V.R. Marshall M.J. Hounslow, R.L. Ryall, "A discretized population balance for nucleation, growth and aggregation", *AIChEJ*, **34**(11), pp. 1821–1832, 1988.

[129] E. Hairer and G. Wanner, *Solving Ordinary Differential Equations II: Stiff and Differential-Algebraic Problems*, Springer-Verlag, 1996.

[130] H.A. Preisig, "A mathematical approach to discrete-event dynamic modelling of hybrid systems", *Computers and Chemical Engineering*, **20**, pp. S1301– S1306, 1996.

[131] G. Schilling and C.C. Pantelides, "Optimal periodic scheduling of multipurpose plants", *Computers and Chemical Engineering*, **23**, pp. 635–655, 1999.

[132] K. Wang, T. Lohl, M. Stobbe, and S. Engell, "Optimal periodic scheduling of multipurpose plants", *Computers and Chemical Engineering*, **24**, pp. 393–400, 2000.

[133] O. Abel and W. Marquardt, "Scenario-integrated modelling and optimization of dynamic systems", *AIChE Journal*, **46**(4), pp. 803–823, 2000.

[134] AusIMM, *A Comprehensive Dynamic Model of the Pierce-Smith Copper Converter: A First Step Towards Better Converter-Aisle Operation*, Mount Isa, 19–23 April 1998.

[135] S. Goto, "Equilibrium calculations between matte, slag and gaseous phases in copper making, copper metallurgy practice and theory", *IMM*, pp. 23–34, 1974.

[136] R. Williams and S. Pickering, "Modelling the UV light reactor rig", 1998, E1870 Report, Department of Chemical Engineering, The University of Queensland.

[137] E. Von Munch, *Mathematical Modelling of Prefermenters*, Ph.D. thesis, Department of Chemical Engineering, The University of Queensland, 1997.

[138] A. Adetayo, J.D. Litster, and M. Desai, "The effect of process parameters on drum granulation of fertilizers with broad size distributions", *Chemical Engineering Science*, **48**(23), pp. 3951–3961, 1993.

[139] J.D. Litster, F.Y. Wang, I.T. Cameron, and P.L. Douglas, "A distributed parameter approach to the dynamics of rotary drying processes", *Drying Techn.*, **11**(7), pp. 1641–1656, 1993.

[140] G.C. Brasil and M.M. Seckler, "A model for the rotary drying of granulator fertilisers", *Proc. 6th Int. Drying Symp.*, pp. 247–256, 1988.

[141] S.J. Friedman and W.P. Mashell, "Studies in rotary drying, parts 1 and 2", *Chem. Eng. Prog.*, **45**, **889**, pp. 482–493 and 573–588, 1949.

[142] W.J. Whiten, *Simulation and Model Building for Mineral Processing*, PhD thesis, University of Queensland, Australia, 1972.

[143] A.J. Lynch, *Mineral Crushing and Grinding Circuits*, Elsevier Scientific, New York, 1977.

[144] I.T. Cameron and F.Y. Wang, "Dynamics of fractionators with structured packing", *Chem.Eng.Commun.*, **119**, pp. 231–259, 1993.

[145] M. Sadeghi, *Dynamic Modelling and Simulation of Structured Packing Fractionators*, PhD thesis, University of Queensland, Brisbane, Australia, 1996.

[146] H. Takenchi, K. Onda and Y. Okumoto, "Mass transfer coefficients between gas and liquid in packed beds", *J. Chem. Eng. Japan*, **1**(1), pp. 56–62, 1968.

[147] J.A. Rocha, J.L. Bravo, and J.R. Fair, "Mass transfer in gauze packings", *Hydroc. Proc.*, **64**(1), pp. 91–95, 1985.

[148] L. Spiegel and W. Meier, "Correlations of the performance characteristics of various mellapak types", *Inst. Chem. Engrs. Symp. Ser.*, **104**, pp. A203–A215, 1987.

[149] R. Billet, "Modelling of fluid dynamics in packed columns", *IChemE Symp. Ser.*, **104**, pp. A171, 1987.

[150] J.R. Fair, J.L. Bravo, and J.A. Rocha, "A comprehensive model for the performance of columns containing structured packing", *AIChemE Symp. Ser.*, **128**, pp. A439, 1992.

[151] Aspen Technology Inc., "ASPEN PLUS", 2000, http://www.aspentech.com/.

[152] Hyprotech Ltd, "HYSYS integrated simulation for the continuous processing industries", 2000, http://www.hyprotech.com/products/.

[153] Mathworks Inc, "MATLAB software system", 2001, http://www.mathworks.com/.

[154] W. Marquardt, "Trends in computer-aided modeling", *Computers and Chemical Engineering*, **20**, pp. 591–609, 1996.

[155] Anne Krogh Jensen, *Generation of Problem Specific Simulation Models Within an Integrated Computer Aided System*, PhD thesis, Danish Technical University, 1998.

[156] B. Lohmann, B.A. Foss, and W. Marquardt, "A field study of the industrial modeling process", *Journal of Process Control*, **5**, pp. 325–338, 1998.

[157] JKTech Ltd, "SimMet—simulator for minerals processing circuits", 2001, http://www.jktech.com.au/.

[158] CAPE-OPEN, "Global CAPE-OPEN (GCO) standards in computer-aided process engineering", 2000, http://www.global-cape-open.org/.

[159] H.A. Preisig, "Modeller—an object oriented computer-aided modelling tool", in *4th International Conference on the Foundations of Computer-Aided Process Design*. 1995, pp. 328–331, CACHE Corporation, USA.

[160] ModKit, "Computer aided process modeling (ModKit)", 2000, http://www.lfpt.rwth-aachen.de/Research/Modeling/modkit.html.

[161] G. Stephanopoulos, G. Henning, and H. Leone, "MODEL.LA a modeling language for process engineering—1 the formal framework", *Computers and Chemical Engineering*, **8**, pp. 813–846, 1990.

[162] K.M. Hangos, R. Williams, and I.T. Cameron, "Assumption structures for model building", Tech. Rep. 091299, CAPE Centre, The University of Queensland, 1999.

[163] H. Rittel and W. Kunz, "Issues as elements of information systems", Tech. Rep. Working Paper 131, Institute of Urban and Regional Development, University of California, Berkeley, CA, 1970.

[164] R. Bañares-Alcantara and H.M.S. Lababidi, "Design support systems for process engineering: 1 and", *Computers and Chemical Engineering*, **19**, pp. 267–301, 1995.

[165] R. Bañares-Alcantara and J.M.P. King, "Design support systems for process engineering:", *Computers and Chemical Engineering*, **21**, pp. 263–276, 1997.

[166] Simulation Sciences Inc., "PRO 2 steady-state process simulator", 2000, http://www.simsci.com/.

[167] Chempute Software, "CHEMCAD process simulator", 2000, http://www.chempute.com.

[168] Intelligen Inc., "SuperPro designer simulation suite", 2000, http://www.intelligen.com.

[169] K.M. Westerberg, P.C Piela, T.G. Epperly, and A.W. Westerberg, "ASCEND: an object oriented computer environment for modeling and analysis: The modeling language.", *Computers and Chemical Engineering*, **15**, pp. 53–72, 1991.

[170] A.W. Westerberg, "How to ASCEND", 1998, http://www.ndim.edrc.cmu.edu/ascend/pdfhelp/howto-ascend.pdf.

[171] Modelica Design Group, "Modelica 1.3 design documentation", 2000, http://www.modelica.org/documents.html.

[172] Peter E.V. Fritson and Johan Gunnarsson, "An integrated modelica environment for modeling, documentation and simulation", in *Proceedings 1998 Summer Computer Simulation Conference SCSC98*. 1998, SCSC, Reno, Nevada, July 19–22.

[173] P. Barton, "ABACUSS II—a process modelling and simulation system for process engineering", 2001, http://yoric.mit.edu.au/abacuss2/abacuss2.html.

[174] G. Fischer and A.C. Lemke, "Construction kits and design environments: Steps towards human problem-domain communication", *Human-Computer Interaction*, **3**, pp. 179–222, 1988.

[175] G. Stephanopoulos, G. Henning, and H. Leone, "MODEL.LA a modeling language for process engineering—2 multifaceted modeling of processing systems", *Computers and Chemical Engineering*, **8**, pp. 847–869, 1990.

[176] Gensym Corporation, "G2 expert system", 2001, http://www.gensym.com/.

[177] K.M. Hangos and I.T. Cameron, "A formal representation of assumptions in process modelling", *Computers and Chemical Engineering*, p. (in print), 2001.

[178] S.I. Amari, "Mathematical foundations of neurocomputing", *Proceeding of the IEEE*, **78**, pp. 1443–1463, 1990.

[179] G. Cybenko, "Approximation by superpositions of sigmoidal function", *Math. Control, Signals and Systems*, **2**, pp. 303–314, 1989.

[180] L.B. Almedia F.M. Silva, "Acceleration techniques for the backpropagation algorithm", *Lecture Notes in Computer Science*, **412**, pp. 110–119, 1990.

[181] R. Haber and L. Keviczky, *Nonlinear System Identification—Input–Output Modeling Approach*, Kluwer Academic Publishers, Dordrecht, Boston, London, 1999.

[182] E.B. Baum and D. Haussler, "What size gives valid generalization?", *Neural Computations*, **1**, pp. 151–160, 1989.

[183] M.A. Kramer and J.A. Leonard, "Diagnosis using backpropagation neural networks—analysis and criticism", *Computers and Chemical Engineering*, **14**, pp. 1323–1338, 1990.

[184] D.R. Baughman and Y.A. Liu, *Neural Networks in Bioprocessing and Chemical Engineering*, Academic Press, 1996.

[185] M. Fullana, F. Trabelsi, and F. Recasens, "Use of neural net computing for statistical and kinetic modelling and simulation of supercritical fluid extractors", *Chemical Engineering Science*, **55**, pp. 79–95, 2000.

[186] G. Li Puma and P. Lock Yue, "A laminar falling film slurry photocatalitic reactor. part 1—model development", *Chemical Engineering Science*, **53**, pp. 2993–3006, 1998.

[187] H.Y. Wang and P. Joulain, "Numerical study of the turbulent burning between vertical parallel walls with a fire-induced flow", *Combustion Science and Technology*, **154**(1), pp. 119–161, 2000.

[188] S.P. Asprey and S. Macchietto, "Statistical tolls for optimal dynamic model building", *Computers and Chemical Engineering*, **24**, pp. 1261–1267, 2000.

[189] W. Feller, *An Introduction of Probability Theory and Applications*, Wiley, New York, 1967.

INDEX

Printed and bound by CPI Group (UK) Ltd, Croydon, CR0 4YY
08/05/2025
01864788-0002